AF794891

Mesoscopic Thermodynamics for Scientists and Engineers

Mesoscopic Thermodynamics for Scientists and Engineers

Mikhail A. Anisimov
University of Maryland
College Park, Maryland, US

Thomas J. Longo
University of Maryland
College Park, Maryland, US

WILEY

Copyright © 2024 by John Wiley & Sons Inc.. All rights reserved, including rights for text and data mining and training of artificial technologies or similar technologies.

Published by John Wiley & Sons, Inc., Hoboken, New Jersey.
Published simultaneously in Canada.

No part of this publication may be reproduced, stored in a retrieval system, or transmitted in any form or by any means, electronic, mechanical, photocopying, recording, scanning, or otherwise, except as permitted under Section 107 or 108 of the 1976 United States Copyright Act, without either the prior written permission of the Publisher, or authorization through payment of the appropriate per-copy fee to the Copyright Clearance Center, Inc., 222 Rosewood Drive, Danvers, MA 01923, (978) 750-8400, fax (978) 750-4470, or on the web at www.copyright.com. Requests to the Publisher for permission should be addressed to the Permissions Department, John Wiley & Sons, Inc., 111 River Street, Hoboken, NJ 07030, (201) 748-6011, fax (201) 748-6008, or online at http://www.wiley.com/go/permission.

Trademarks: Wiley and the Wiley logo are trademarks or registered trademarks of John Wiley & Sons, Inc. and/or its affiliates in the United States and other countries and may not be used without written permission. All other trademarks are the property of their respective owners. John Wiley & Sons, Inc. is not associated with any product or vendor mentioned in this book.

Limit of Liability/Disclaimer of Warranty: While the publisher and author have used their best efforts in preparing this book, they make no representations or warranties with respect to the accuracy or completeness of the contents of this book and specifically disclaim any implied warranties of merchantability or fitness for a particular purpose. No warranty may be created or extended by sales representatives or written sales materials. The advice and strategies contained herein may not be suitable for your situation. You should consult with a professional where appropriate. Further, readers should be aware that websites listed in this work may have changed or disappeared between when this work was written and when it is read. Neither the publisher nor authors shall be liable for any loss of profit or any other commercial damages, including but not limited to special, incidental, consequential, or other damages.

For general information on our other products and services or for technical support, please contact our Customer Care Department within the United States at (800) 762-2974, outside the United States at (317) 572-3993 or fax (317) 572-4002.

Wiley also publishes its books in a variety of electronic formats. Some content that appears in print may not be available in electronic formats. For more information about Wiley products, visit our web site at www.wiley.com.

Library of Congress Cataloging-in-Publication Data

Names: Anisimov, M. A. (Mikhail Alekseevich), author. | Longo, Thomas J., author.
Title: Mesoscopic thermodynamics for scientists and engineers / Mikhail A. Anisimov, Thomas J. Longo.
Description: Hoboken, New Jersey : Wiley, [2024] | Includes index.
Identifiers: LCCN 2024011117 (print) | LCCN 2024011118 (ebook) | ISBN 9781394241958 (hardback) | ISBN 9781394241972 (adobe pdf) | ISBN 9781394241965 (epub)
Subjects: LCSH: Thermodynamics. | Mesoscopic phenomena (Physics)
Classification: LCC QC311 .A55 2024 (print) | LCC QC311 (ebook) | DDC 536/.7–dc23/eng/20240412
LC record available at https://lccn.loc.gov/2024011117
LC ebook record available at https://lccn.loc.gov/2024011118

Cover Design: Wiley
Cover Image: Courtesy of Dr. M.A. Anisimov

Set in 9.5/12.5pt STIXTwoText by Straive, Chennai, India

To Our Families

Contents

Preface

This book originates from an elective course that one of the authors (M.A.A.) has developed and offered at the University of Maryland, College Park, to senior undergraduate and entry-level graduate students in engineering and science programs. The motivation to develop a new course was a result of a certain dissatisfaction with the contents of the traditional thermodynamics course that he had been teaching in the Department of Chemical and Biomolecular Engineering for almost 30 years.

Thermodynamics is a keystone in science and engineering, bridging the gap between fundamentals and applications. However, when one compares the content of conventional courses in thermodynamics with what scientists and engineers often do in practice, one may notice a gap-one that is widening every year. New emerging technologies and products deal with subjects such as biomolecular and soft-matter engineering, micro-reactor chemistry and microcapsule drug delivery, micro-fluidics and porous media, nanoparticles and nanostructures, supercritical-fluid extraction, and systems at the edge of stability. Engineers often must design processes or products under conditions where traditional thermodynamics may become insufficient, as in the case of strongly fluctuating and nano-size systems, or dissipative structures away from equilibrium. The scientific basis for such issues is not commonly addressed in traditional engineering thermodynamics and transport courses. The aim of *Mesoscopic Thermodynamics for Scientists and Engineers* is to fill this gap, at least partially.

Mesoscopic thermodynamics can be defined as a semi-phenomenological approach to phenomena in systems where a mesoscopic length-scale, intermediate between the atomistic scale and the macroscopic scale, emerges and where such a length explicitly affects the thermodynamic properties and phase behavior. There are numerous topics to be considered in mesoscale thermodynamics (equilibrium and nonequilibrium): finite-size and confined systems, fluctuations, wetting and adsorption, interfacial phenomena, microphase separation and nucleation, molecular self-assembly, self-organized criticality, and mesoscale dissipative structures. It would be unrealistic to expect from an introductory-level book to address all or even most of them. Each of these topics deserves a special monograph. Instead, this book emphasizes two major universal features (described in Part I) that must be introduced into the traditional thermodynamic framework in order to address mesoscopic phenomena. These features are fluctuations and local (spatially dependent) properties. To make the text self-sufficient for the readers, Part I also includes a summary of classical equilibrium and nonequilibrium thermodynamics (Chapter 2). Chapter 2 serves as an important background and guide to the core material of the book, but cannot substitute a comprehensive textbook on thermodynamics. In Part II, a universal thermodynamic approach to meso-heterogeneous systems is specified through examples, which are relevant to modern research and engineering practice. Most examples deal with meso-heterogeneous fluid systems. Each chapter in both parts is supplemented by exercises.

The readers are assumed to have an undergraduate background in thermodynamics, calculus, physics, and chemistry. Acquaintance with statistical mechanics would be useful, but is not required for comprehension of the material. All chapters include references to textbooks, monographs, and journal research articles for further, more advanced reading. Although the description of the topics is self-sufficient within the book, a supplementary conventional undergraduate text on thermodynamics would be useful, especially in solving the problems provided at the end of each chapter.

We are grateful to our colleague and collaborator Jan V. Sengers, Distinguished University Professor Emeritus at the University of Maryland, for fruitful discussions and his valuable comments on this text prior to publication. M.A.A. appreciates important contributions from his colleagues, collaborators, and former students, especially Christopher E. Bertrand (PhD, 2010), who helped in clarifying a number of issues addressed in this book.

College Park, MD
December 2023

Mikhail A. Anisimov and Thomas J. Longo

Notations, Acronyms, and Units

General Notations

Symbols	Designates (units)
$\mathcal{A}$	surface area (m^2)
A_t	total Helmholtz energy (J)
$A = A_t/n$ or A_t/N	Helmholtz energy per mole or per molecule (J/mol)
$A_0, B_0, \Gamma_0, D_0, \xi_0, \sigma_0$	asymptotic critical amplitudes of weak susceptibility, spontaneous order parameter, strong susceptibility, ordering field, correlation length, and surface tension, respectively
a and b	van der Waals constants ($J \cdot m^3/mol^2$) and (m^3/mol)
a_0, b_0, u_0, e_0, g_0	coefficients of the Landau expansion in the meanfield theory of phase transitions
$\boldsymbol{a}_i$	monomer random step (vector)
B	second virial coefficient (m^3/mol)
$\tilde{b}$	mobility of a Brownian particle ($m^2 \cdot K/s \cdot J$)
c	molar concentration (mol/L); also, speed of light (m/s)
c_0	gradient-term coefficient in the Landau-Ginzburg functional
C	third virial coefficient (m^3/mol^2)
C_P	isobaric molar or molecular heat capacity (J/mol · K)
C_V	isochoric molar or molecular heat capacity (J/mol · K)
$C(r)$	spatial correlation function
$C(t)$	temporal correlation function
D	mutual diffusion coefficient in a binary solution (m^2/s)
D_T	thermal diffusion coefficient (m^2/s)
d	number of dimensions (dimensionality) of space
E_t	total energy (J)
E_k	kinetic energy (J)
E_p	potential energy (J)
$\boldsymbol{F}$	force (vector) (N)
G_t	total Gibbs energy (J)
$G = G_t/n$ or G_t/N	Gibbs energy per mole or per molecule (J/mol)

$g = \lvert\mathbf{g}\rvert$	free-fall acceleration (m/s^2)
H_t	total enthalpy (J)
$H = H_t/n$ or H_t/N	enthalpy per mole or per molecule (J/K)
$\boldsymbol{h}$	magnetic field (vector) (A/m)
h	generalized field variable
$\hbar$	reduced (by 2π) Planck's constant (1.0546×10^{-34} J · s)
$\boldsymbol{j}_D$	mutual diffusion flux ($mol/m^2 \cdot s$)
$\boldsymbol{j}_Q$	heat flux ($J/m^2 \cdot s$)
$\mathcal{K}_0$	Krichevskiĭ parameter
$\mathcal{K}$	generalized Krichevskiĭ parameter
K	reaction equilibrium constant
$k_B = R/N_A$	Boltzmann's constant per molecule (1.38×10^{-23} J/K)
k_P	baro-diffusion ratio
k_T	thermo-diffusion ratio
ℓ	length (m)
$\boldsymbol{M}$	magnetization (vector) (A/m)
M_w	molecular weight (g/mol)
m	mass (kg)
n	number of moles
n_i	number of moles for species i
$N = nN_A$	number of molecules
N_A	Avogadro's number (6.02×10^{23} molecules per mole)
$N_i = n_i N_A$	number of molecules for species i
$\mathcal{N}$	degree of polymerization
n_λ	refractive index
N_G	Ginzburg number
P	pressure (Pa; J/m^3; N/m^2)
p_r	probability
Q	heat (J)
Q_{ij}	components of a tensor order parameter
$\boldsymbol{q}$	wave number (vector) (m^{-1})
R	gas constant (8.314 J/mol · K)
R_g	radius of gyration (m)
R_h	hydrodynamic radius of a Brownian particle
r	distance
$\bar{r}$	average intermolecular distance
r_0	molecular radius
S_t	total entropy (J/K)

$S = S_t/n$ or S_t/N	entropy per mole or per molecule (J/K)
T	temperature (K)
t	time
U_t	total Internal energy (J)
$U = U_t/n$ or U_t/N	internal energy per mole or per molecule (J/K)
$u = \|\boldsymbol{u}\|$	velocity (speed) (m/s)
u_s	thermodynamic speed of sound (m/s)
V_t	total volume (m^3)
$V = V_t/n$ (or $V = V_t/N$)	volume per mole or per molecule (m^3)
W	work (J)
w	probability density
$x \equiv x_2$	mole or molecular fraction of solute in a binary mixture
x_i	mole or molecular fraction of species i
y	effective "activity" of species
z	vertical coordinate
z_n	molecular coordination number
z_τ	dynamic critical exponent

Greek Notations

α_P	isobaric or volumetric expansivity (K^{-1})
α_H	Joule–Thompson coefficient (K/Pa)
α_s	sound attenuation (m^{-1})
$\alpha, \beta, \gamma, \delta, \nu, \vartheta, \eta$	critical exponents of weak susceptibility, spontaneous order parameter, strong susceptibility, ordering field, correlation length, surface tension, and correlation function, respectively
$\check{\alpha}$	Onsager's molar transport coefficient (mol/[m · s · J])
$\check{\beta}$	Onsager's cross transport coefficient (mol/[m · s · K])
$\check{\gamma}$	Onsager's heat transport coefficient (J/[m · s · K])
γ_i	activity coefficient of species i
$\check{\delta}$	Reaction kinetic coefficient (mol/[m^3 · s · J])
δ'	Tolman's length (m)
ε	dielectric constant
ϵ	energy of intermolecular interactions (J)
$\tilde{\zeta}$	bulk viscosity (Pa · s)
ζ	interfacial thickness (m); also, extent of reaction
$\tilde{\eta}$	dynamic shear viscosity (Pa · s)
θ	scattering angle
Θ	Theta (Flory) temperature
κ	thermal conductivity (J/m · s · K)
κ_T	isothermal compressibility (Pa^{-1})

κ_Π	osmotic compressibility (Pa^{-1})
λ	wavelength (m); also, a coupling constant
λ_s	sound wavelength (m)
μ	chemical potential of a pure substance (J/mol), equal to the Gibbs energy per molecule G in a single-component system
μ_{cxc}	chemical potential along phase coexistence (J/mol)
μ_0	chemical potential in zero-field ($\mu_0 = \mu_{cxc}$ for fluids for $T < T_c$)
μ_i	chemical potential of species i (J/mol)
$\mu_{21} = \mu_2 - \mu_1$	exchange chemical potential (J/mol)
ξ	correlation length of order-parameter fluctuations (m)
ξ_ψ	de Gennes correlation length of polymer-chain fluctuations (m)
Π	osmotic pressure (Pa)
$\rho = n/V_t$ or N/V_t	molecular or molar density (mol/m^3)
$\rho_i = \rho x_i$	partial molar density (molar concentration) of species i (mol/m^3)
$\rho A = A_t/V_t$	density of Helmholtz energy (J/m^3)
$\rho S = S_t/V_t$	density of entropy ($J/m^3 \cdot K$)
σ	surface tension (J/m^2;N/m)
τ	relaxation time (s)
v_0	volume of a molecule (m^3)
ϕ_i	volume fraction of species i
$\phi \equiv \phi_2$	volume fraction of polymers in a polymer solution
φ	order parameter; generalized density
φ_0	spontaneous order parameter (in zero ordering field)
$\hat{\chi}$	generalized dimensionless susceptibility
$\hat{\chi}_1$	strong scaling susceptibility
$\hat{\chi}_2$	weak scaling susceptibility
$\hat{\chi}_{12}$	cross scaling susceptibility
$\boldsymbol{\psi}$	de Gennes polymer-chain order parameter (vector)
ψ	critical exponent of shear viscosity
ϖ	Flory interaction parameter
$\Omega_t = -PV_t$	grand thermodynamic potential (J)

Special Notations

- In subscripts and superscripts: roman fonts indicate words, while italicized fonts indicate variables or functions.
- Circumflex (as in $\hat{\mu}$) designates reduced (dimensionless) thermodynamic properties.
- The inverted circumflex (as in $\check{\alpha}$) is used for Onsager transport coefficients.
- Over bar (as in $\overline{N}$) designates average equilibrium properties
- Subscript c (as in T_c) designates critical parameters.
- Superscript ∘ (as in μ_1°) designates a reference state.
- Bold (as in $\boldsymbol{h}$ or $\boldsymbol{r}$) designate vectors.

- Variation δ (as in $\delta\rho$) designates fluctuation or variation.
- Total differential d and partial differential ∂ indicate derivatives.
- Differential with a bar đ (as in đW) designates an inexact differential quantity of work or heat.
- Symbol Δ (as in ΔU) designates a property change or excess (as in ΔC_V).
- Symbol Å designates angstrom (1Å = 0.1 nm = 10^{-10} m).
- Brackets <> (as in $\langle\delta\rho\rangle$) designate statistical average of fluctuations or variations.
- The summation symbol, $\sum$, assumes a summation over the index i, such as in $\sum \mu_i N_i$.
- The symbol $\sim$ indicates "scales as" or "order of," the symbol $\approx$ indicates an approximate value, while the symbol $\cong$ indicates a rounded quantity.

Acronyms

$\overline{n}O\overline{m}$	4-n-alkyloxyphenyl-4′-n-alkyloxybenzoates ($\overline{n}$ and $\overline{m}$ are the number of carbon atoms in the alkyl groups)
$\overline{6}O\overline{6}NO_2$	4-n-hexyloxyphenyl-4′-n-hexyloxynitrobenzoate
BMOAB	4-n-butyl-4′-methoxyazoxybenzene
CHX	cyclohexane
CMC	critical micelle concentration
CP	critical point
DAG	dysprosium aluminum garnet
DCP	double critical point
DDAB	didodecyldimethylammonium bromide
DLS	dynamic light scattering
EOS	equation of state
Freon-113	trichlorotrifluoroethane
HLM	Halperin–Lubensky–Ma
LGCL	liquid–gas critical line (locus)
LLCL	liquid–liquid critical line (locus)
LGCP	liquid–gas critical point
LLCP	liquid–liquid critical point
LCEP	lower critical end point
LP	Lifshitz point
MBBA	4-methoxybenzylidene-4′-n-butylaniline
MC	Monte Carlo
MD	molecular dynamics
MFCP	meanfield critical point
NAC	nematic-smectic A-smectic C multicritical point
PBMA	poly(butyl methacrylate)
PEO	polyethylene oxide
PMMA	polymethylmethacrylate
PS	polystyrene
RG	renormalization group
RMS	root mean squared

SANS	small-angle neutron scattering
SAXS	small-angle X-ray scattering
SLS	static light scattering
SmA, SmC	smectic A, smectic C
S,S-MBBPC	*S,S*-(+)-4″-(methylbutyl) phenyl-4′-(methylbutyl) biphenyl carboxylate
TBA	tertiary butanol
TCP	tricritical point
TP	triple point
UCEP	upper critical end point
VDW	van der Waals

Greek Alphabet

Α	α	alpha	Ν	ν	nu
Β	β	beta	Ξ	ξ	xi
Γ	γ	gamma	Ο	ο	omicron
Δ	δ	delta	Π	π	pi
Ε	ϵ	epsilon	Ρ	ρ	rho
Ζ	ζ	zeta	Σ	σ	sigma
Η	η	eta	Τ	τ	tau
Θ	θ	theta	Υ	υ	upsilon
Ι	ι	iota	Φ	ϕ	phi
Κ	κ	kappa	Χ	χ	chi
Λ	λ	lambda	Ψ	ψ	psi
Μ	μ	mu	Ω	ω	omega

Expansions of Basic Functions

$$
\begin{aligned}
\frac{1}{1 \pm x} &= 1 \mp x + x^2 \pm x^3 + \cdots \\
e^{\pm x} &= 1 \pm x + \frac{1}{2!}x^2 \pm \frac{1}{3!}x^3 + \cdots \\
\ln(1 \pm x) &= \pm x - \frac{1}{2}x^2 \pm \frac{1}{3}x^3 - \frac{1}{4}x^4 \ldots \\
\sin x &= x - \frac{1}{3!}x^3 + \frac{1}{5!}x^5 \ldots \\
\cos x &= 1 - \frac{1}{2!}x^2 + \frac{1}{4!}x^4 \ldots \\
\tan x &= x + \frac{1}{3}x^3 + \frac{1}{15}x^5 + \cdots \\
\tanh x &= x - \frac{1}{3}x^3 + \frac{2}{15}x^5 + \cdots
\end{aligned}
$$

Out of Intense Complexities, Intense Simplicities Emerge
– Winston Churchill

Part I

Thermodynamic Approach to Meso-Heterogeneous Systems

1

Macro, Meso, Micro

This chapter includes:

- Definition of "mesoscale"
- Definition of "meso-thermodynamics"
- Typical objects of meso-thermodynamics
- Probing mesoscales by scattering techniques

1.1 Thinner than a Hair

One can roughly consider a length scale less than 10 μm (10,000 nm) and larger than 1 nm as mesoscopic. This is a huge range. The upper boundary is about 10 times smaller than the thickness of a human hair (Figure 1.1), and, usually, one needs a microscope, optical or electronic, to see objects smaller than that.

Below one micron, one enters a sub-micron world. A size between 1 and 100 nm can also be defined as nanoscopic. It should be noted that any classification is conditional and circumstantial: sometimes "thinner-than-a-hair" objects can be treated macroscopically, while some macroscopic objects, like polymer solutions or complex fluids, may require a mesoscopic approach.

A natural atomistic length scale is 1 Å = 0.1 nm. This is (approximately) the size of a hydrogen atom and the length of a carbon–carbon covalent bond. Whatever happens at the atomistic scale will be dominated by the material's intramolecular structure and by the details of the intermolecular interactions. Phenomena at the atomistic scale cannot be treated by thermodynamics. Yet, how could one describe a single polymer molecule which consists of millions of carbon–carbon bonds? Is this a thermodynamic system?

The principal question arises: is traditional (macro-scale) thermodynamics sufficient enough to describe mesoscale systems? Can one apply, without restriction, the laws and equations of classical thermodynamics to calculate the chemical reaction equilibria and design a micro-reactor, which is only 100 nm thick, or to describe a transition between the alternative structures of a protein molecule? What would be the definition of a "phase" in such systems? All biological cells are natural micro-reactors. Can we still use the concepts of chemical potential, osmotic pressure, surface tension, etc. for such objects? A microchip can contain only hundreds of atoms. What is the limiting size for which thermodynamics is still applicable? In a micro-porous medium, not only does the surface energy dominate over the bulk energy, but also the surface itself may be poorly defined. Is there still room for thermodynamics, beyond the trivial mass and energy balances?

Mesoscopic Thermodynamics for Scientists and Engineers, First Edition. Mikhail A. Anisimov and Thomas J. Longo.
© 2024 John Wiley & Sons, Inc. Published 2024 by John Wiley & Sons, Inc.

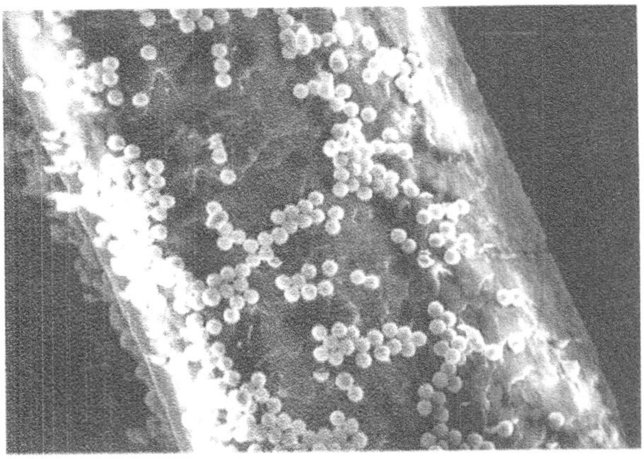

Figure 1.1 Electron microscopy image of polystyrene particles (3.4 μm diameter) on a human hair (Courtesy of Spherotech, Inc.). The hair (with a thickness of about 100 μm) can still be considered as a macroscopic object in most applications, while the polystyrene particles are mesoscopic.

1.2 Where Does the Size Matter?

Classical thermodynamics is essentially macroscopic. To define thermodynamic states and thermodynamic properties, one needs systems with a huge number of individual particles, such as atoms or molecules [1, 2]. Statistical mechanics defines thermodynamic states and properties only in the so-called "thermodynamic limit", where the total volume (V_t) and the number of particles (N) must tend to infinity, while the molecular number density of the substance, $\rho = N/V_t$, remains finite. Even a cubic millimeter of liquid water (1 mg) contains 3×10^{19} molecules. It is sufficient to regard such a small droplet of water as a macroscopic body and neglect the surface effects in considering its bulk properties. If the characteristic size of the object is ℓ, the surface energy is proportional to ℓ^2, while the bulk energy is proportional to ℓ^3. This is why one can measure the temperature of steam condensation as well as the vapor pressure in a small tube, and use these data in the design of an upscale distillation column, taller than 100 m.

Macroscopic thermodynamics eliminates the size of the system by reducing the extensive thermodynamic properties (which depend on size) by the number of particles, the number of kilograms, or the volume. Equations of thermodynamics usually operate with intensive properties, such as density, concentration, molar or specific volume, etc., which do not depend on the size. The actual length scale is restored only in the final stage of engineering design. The equations of state implicitly depend on atomistic length scales, the molecular size, and the range of intermolecular interactions. However, since these microscopic characteristics are integrated over a huge number of molecules, the parameters of common equations of state (EOS) are independent of the length scale.

There are at least three categories of popular objects in which macro-scale size-independent thermodynamics may breakdown:

1. Soft condensed-matter materials, such as complex fluids (polymer solutions and melts, micro-emulsions, gels, and liquid crystals) [3–15]. Most food and cosmetics are soft matter [16]. Humans (like other living organisms) are "soft-matter machines." All soft-matter materials, even being themselves macroscopic, have some kind of mesoscopic structure which is characterized by the existence of one or more mesoscales.
2. Systems that are essentially finite (practically speaking, sub-micron or nano-size) in one or more dimensions [17–19]. Examples include nanoparticles (if the particle is considered as a separate system), pores, films and layers, fibers, and threads. Meso-heterogeneous interfaces with

a thickness from 1 nm to microns also belong to this category. Sometimes such systems can be truly microscopic in a certain dimension, and they should thus be treated as two- or even one-dimensional objects. Here, the size really matters!

3. Systems in a near-critical state (i.e. in the vicinity of a critical point or a second-order phase-transition) [20–28]. Such systems are mesoscopically inhomogeneous as a result of strong thermal fluctuations. The mesoscopic length scale associated with these fluctuations is known as the "correlation length."

1.3 What Is Meso-Thermodynamics?

Each of these categories has been a subject of extensive experimental and theoretical investigations for the last 50 or more years. Thermodynamics of adsorption [29], wetting transitions [30], micro-phase separation in polymers [31], gelation [32, 33], and phase equilibria in confined fluids [34, 35] are just a few examples. Computer simulations of mesoscale systems become increasingly informative (see, for example, Refs. [36–41]). The existence and dominant role of mesoscale inhomogeneities is a common feature of all the systems mentioned above.

Mesoscopic thermodynamics (or in short, "*meso-thermodynamics*") can be defined as a semi-phenomenological approach to systems and phenomena in which a length, intermediate between the atomic scale and the macroscopic scale, emerges and where such a length explicitly affects the thermodynamic properties [42]. This science is still thermodynamics, as its objects are still governed by statistical mechanics but, in contrast to "macro-thermodynamics," it explicitly contains a mesoscopic length scale associated either with the structure of materials or with the actual finite size. Many ideas and concepts of mesoscopic thermodynamics were developed and explored in colloid and interface science long ago [42–44]. However, only more recently, a high degree of *universality* (and thus "simplicity"), originating from the very existence of the mesoscopic length scale, has been fully recognized.

Meso-thermodynamics also includes thermal fluctuations – spontaneous deviations from average (equilibrium) properties, assigned by thermodynamics. Fluctuations emerge in the apparent "violation" of the second law of thermodynamics. A famous astrophysicist Sir Arthur Eddington wrote: "... if your theory is found to be against the second law of thermodynamics I can offer you no hope..." [45]. However, the existence of fluctuations just demonstrates the statistical meaning of the second law. In fact, the statement of Sir Eddington is unconditionally correct only in the thermodynamic limit (for an infinite number of molecules).

Representation of a micro-heterogeneous system as an "ensemble" of small open systems was introduced and elaborated by Hill [46, 47]. This approach is important for small systems, where the surface energy plays a significant role and for the analysis of computer simulations that usually involve a relatively small number of molecules. In this book, we provide a different approach to unify the description of mesoscopic phenomena. This approach is based on the concept of a "local" coordinate-dependent free energy, known as the "Landau–Ginzburg functional" [1, 2], allowing for the unified treatment of "apparently" physically different mesoscopic heterogeneous systems, whose properties depend on an appropriate mesoscopic length scale.

The approach to introduce a mesoscale in thermodynamic properties goes back to van der Waals [48] and is sometimes referred to as "quasi-thermodynamics," "local thermodynamics," or "point-thermodynamics" [49]. This approach was generalized by Landau and Lifshitz [1, 2, 50] and was later elaborated for fluid interfaces by Cahn and Hilliard [51] and by Rowlinson and

(a) (b)

Figure 1.2 Two founders of mesoscopic thermodynamics: Johannes Diderik van der Waals (a) and Lev Davydovich Landau (b), see more information in the footnote. Source: Unknown author/Wikimedia Commons/Public Domain.

Widom [49]. Arguably, van der Waals and Landau (see Figure 1.2) may be viewed as the fathers of this branch of thermodynamics.[1]

A theory elaborating toward the microscopic description of condensed-matter systems is referred to as "density functional theory" (DFT) [52]. Unlike the microscopic description provided by DFT, the approach adopted in this book is semi-phenomenological, addressing the microscopic complexity through a few mesoscopic parameters (e.g. the correlation length). In short, the motto of this book is "finding simplicity in complexity." Typical models of meso-thermodynamics are coarse-grained models that do not have to account for every detail on the atomistic scale. Instead, such models emphasize universality. For example, the behavior of polymer solutions is basically determined not by a particular structure of the monomer units but by the fact that the polymer molecule is often a long, flexible curve in three-dimensional space. Another example is the near-critical state. The origin of critical-point universality lies in long-range correlations that involve a large number of molecules, so that the details of direct intermolecular interactions become unimportant.

In many cases, more than one mesoscale determines the properties of meso-heterogeneous systems. One of the mesoscales could be an instrumental scale, such as the wavelength of radiation, or the actual finite-size scale, such as the radius of a droplet. Another one could be inherent to the meso-heterogeneous structure, such as the size of a polymer molecule or the correlation length of density fluctuations. A competition of these mesoscales, which may produce new phenomena and modify the thermodynamic properties, can also be treated within the universal framework of meso-thermodynamics [42].

1 Van der Waals (1827–1923) was a Dutch physicist who won a Nobel Prize in 1910 for his work on equations of state for fluids. Landau (1908–1968) was a Soviet physicist who won a Nobel Prize in 1962 for his pioneering theoretical works on condensed matter, in particular, for the classical theory of phase transitions.

1.4 Probing Mesoscales

In order to experimentally probe the mesoscale properties of matter, one needs to generate an external perturbation in the corresponding range of the length scales. The response to this perturbation reveals the inherent mesoscale structure of the material. Thus, the most popular tools to study materials at mesoscales (in addition to electron and atomic-force microscopy images) are scattering techniques [53], such as static and dynamic light scattering (SLS and DLS, respectively) [54–56], small-angle X-ray scattering (SAXS), and small-angle neutron scattering (SANS) [57, 58]. Light scattering is caused by the interactions of photons with inhomogeneities of matter, such as colloid particles, polymer molecules in solution, or local inhomogeneities caused by thermal fluctuations. In neutron scattering, neutrons are scattered upon collisions with atomic nuclei, while in X-ray scattering, photons interact with the electronic clouds of atoms.

In a scattering experiment, the probing length scale, ℓ, is determined by the wavenumber $q = 2\pi/\ell$ of scattering, which is the magnitude of the scattering wave vector, $\boldsymbol{q}$. This vector is the difference between the wave vectors of incident and scattered radiation, $\boldsymbol{q}_0$ and $\boldsymbol{q}_s$, respectively, as $\boldsymbol{q} = \boldsymbol{q}_0 - \boldsymbol{q}_s$, see Figure 1.3. The scattering vector is controlled by the conservation of momentum of photons or neutrons in the process of scattering. The intensity of the scattering (counted as the number of photons or neutrons scattered per second) depends on the scattering wave number, and thus, characterizes the mesoscale structure of the material.

The scattering wave number and, consequently, the probing mesoscopic length scale can be tuned by the wavelength of radiation, λ, and by the angle θ, at which the scattered photons or neutrons are collected, as

$$q = \frac{4\pi}{\lambda}\sin\left(\frac{\theta}{2}\right) \tag{1.1}$$

Specifically, in light-scattering experiments, the wavelength of incident light in scattering media is $\lambda = \lambda_0 n_\lambda$, where λ_0 is the wavelength in vacuum and n_λ is the refractive index. Since the scattering-angle factor, $\sin(\theta/2)$, varies from 0 to 1, the length scale probing inhomogeneities of matter is limited by the wavelength of radiation. For $\lambda_0 \cong 438$ nm (blue light emitted by an Ar laser) and $n_\lambda \cong 1.33$ (water), the minimum the probing length scale (at $\theta = 180°$) is determined by the criterion $\ell^2 q^2 \sim 1$ in the measurements of the static light-scattering intensity (in this case, $q^2 \sim 0.001\ \text{nm}^2$). Consequently, SLS cannot detect inhomogeneities in materials at a scale much smaller than that. However, this task can be achieved in SAXS and SANS experiments, and, at certain conditions, in DLS experiments.

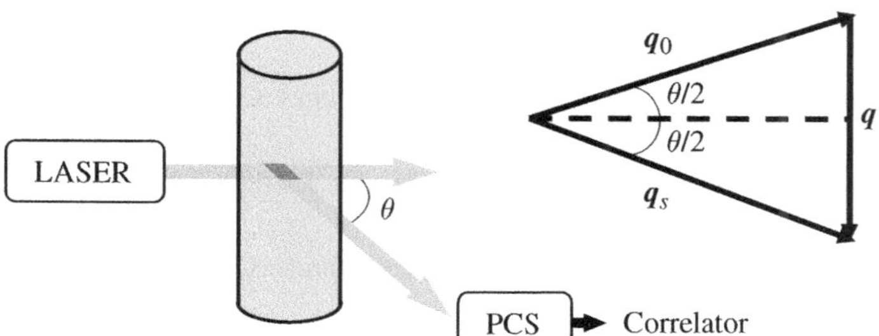

Figure 1.3 Schematics of dynamic light scattering and definition of the scattering wave vector, $\boldsymbol{q}$. Photons are collected in the photon counting system (PCS). The temporal correlations of the scattered photons are processed by the correlator. The scattering volume is shown by the darker area in the center of the flask.

In SAXS experiments, the wavelength of X-ray photons is about 0.1–0.2 nm; thus, the expected nanoscale inhomogeneities are studied at very low angles (typically 0.1°–10°). A SANS experiment (typically $\lambda_0 \approx 1$ nm) is the most convenient tool to investigate the structure of various substances at a mesoscopic scale of about 1–100 nm.

In a DLS experiment, unlike static light scattering, the correlation between the numbers of photons (LS intensity) scattered at two different times is measured. This correlation depends on the relaxation rate of the fluctuations of the inhomogeneities and, thus, on the dynamics of the mesoscale structure of these inhomogeneities. In particular, DLS can detect the random ("Brownian") motion of nanoparticles, as well as the relaxation rate of density and concentration fluctuations, that may have a characteristic length-scale even less than 1 nm.

The resolution of scattering techniques depends on the contrast. In light scattering, this is an optical contrast, the difference between the refractive index of the inhomogeneities and the average refractive index of the medium. In neutron scattering, the interaction with the atomic nuclei is characterized by the "scattering length density" (SLD), which strongly depends on the element. Importantly, hydrogen has a negative SLD, and thus in SANS experiments, hydrogen is replaced fully or partially by deuterium. In X-ray scattering experiments, the interactions of X-ray photons with electronic clouds is stronger for heavier atoms. For example, the X-ray contrast of a protein solution can be increased by replacing the light ions in the protein complex with heavy ions, such as substituting Ca^{2+} with Pb^{2+} [59].

References

1 L. D. Landau and E. M. Lifshitz, *Statistical Physics Part 1*, vol. 5 of Course of Theoretical Physics. Oxford: Pergamon, 3rd ed., 1980.

2 E. M. Lifshitz and L. P. Pitaevskil, *Statistical Physics: Part 2*, vol. 9 of Landau and Lifshitz Course of Theoretical Physics. Oxford: Pergamon, 2nd ed., 1981.

3 P.-G. de Gennes, *Scaling Concepts in Polymer Physics*. New York: Cornell University Press, 1st ed., 1979.

4 P.-G. de Gennes and J. Prost, *The Physics of Liquid Crystals*. Oxford: Oxford University Press, 2nd ed., 1995.

5 R. A. L. Jones, *Polymers at Surfaces and Interfaces*. Cambridge; New York: Cambridge University Press, 1999.

6 M. Daoud and C. E. Williams, eds., *Soft Matter Physics*. Berlin Heidelberg: Springer-Verlag, 1999.

7 I. Hamley, *Introduction to Soft Matter - Revised Edition*. Chichester: Wiley, 2nd ed., 2000.

8 R. A. L. Jones, *Soft Condensed Matter*. Oxford University Press, 2002.

9 S. Safran, *Statistical Thermodynamics of Surfaces, Interfaces, and Membranes*. Colorado: CRC Press, 2003.

10 M. Kleman and O. D. Laverntovich, *Soft Matter Physics: An Introduction*. Partially Ordered Systems. New York: Springer-Verlag, 2003.

11 J.-L. Barrat and J.-P. Hansen, *Basic Concepts for Simple and Complex Liquids*. Cambridge University Press, 2003.

12 A. V. Zvelindovsky, ed., *Nanostructured Soft Matter - Experiment, Theory, Simulation and Perspectives*. Dordrecht: Springer, 2007.

13 A. Fernandez-Nieves and A. M. Puertas, eds., *Fluids, Colloids and Soft Materials: An Introduction to Soft Matter Physics*. Wiley, 2016.

14 T. McLeish, *Soft Matter: A Very Short Introduction*. Oxford: Oxford University Press, 2020.

15 W. van Saarloos, V. Vitelli, and Z. Zeravcic, *Soft Matter: Concepts, Phenomena, and Applications*. Princeton, NJ: Princeton University Press, 2024.

16 R. Piazza, *Soft Matter: The Stuff that Dreams are Made of*. Springer, 2011.

17 V. Privman, ed., *Finite Size Scaling and Numerical Simulation of Statistical Systems*. Singapore: World Scientific, 1990.

18 J. Brankov, D. Danchev, and N. Tonchev, *Theory of Critical Phenomena in Finite-Size Systems*, vol. 9 of Modern Condensed Matter Physics. Singapore: World Scientific, 2000.

19 J. H. Fendler and I. Dékány, eds., *Nanoparticles in Solids and Solutions*. Nato Science Partnership Subseries: 3. Netherlands: Springer, 1996.

20 K. G. Wilson, "The renormalization group - Introduction," in *Phase Transitions and Critical Phenomena* (C. Domb and M. S. Green, eds.), vol. 6. New York: Academic Press, 1976.

21 H. E. Stanley, *Introduction to Phase Transitions and Critical Phenomena*. Oxford University Press, 1987.

22 M. E. Fisher, "Scaling, universality and renormalization group theory," in *Critical Phenomena* (F. J. W. Hahne, ed.), Lecture Notes in Physics, pp. 1–139. Berlin Heidelberg: Springer, 1983.

23 K. G. Wilson, "The renormalization group and critical phenomena," *Rev. Mod. Phys.*, vol. 55, pp. 583–600, 1983.

24 M. A. Anisimov, *Critical Phenomena in Liquids and Liquid Crystals*. New York: Gordon and Breach, 1991.

25 I. Herbut and S. Fraser, *A Modern Approach to Critical Phenomena*. Cambridge: Cambridge University Press, 2007.

26 M. Gitterman, *Phase Transitions: Modern Applications*. World Scientific, 2nd ed., 2013.

27 N. Hidetoshi and G. Ortiz, *Elements of Phase Transitions and Critical Phenomena*. Oxford: Oxford University Press, 2015.

28 M. E. Fisher, *Excursions in the Land of Statistical Physics*. World Scientific Publishing Co., 2017.

29 A. I. Rusanov, "Surface thermodynamics revisited," *Surf. Sci. Rep.*, vol. 58, no. 5, pp. 111–239, 2005.

30 P.-G. de Gennes, "Wetting: Statics and dynamics," *Rev. Mod. Phys.*, vol. 57, no. 3, pp. 827–863, 1985.

31 F. Bates and G. Fredrickson, "Block copolymer thermodynamics: Theory and experiment," *Annu. Rev. Phys. Chem.*, vol. 41, pp. 525–57, 1990.

32 J. E. Martin, J. Wilcoxon, and D. Adolf, "Critical exponents for the sol-gel transition," *Phys. Rev. A*, vol. 36, no. 4, pp. 1803–1810, 1987.

33 Y. Huang, X. Jin, H. Liu, et al., "A molecular thermodynamic model for the swelling of thermo-sensitive hydrogels," *Fluid Phase Equilib.*, vol. 263, no. 1, pp. 96–101, 2008.

34 A. V. Neimark and A. Vishnyakov, "Phase transitions and criticality in small systems: Vapor–liquid transition in nanoscale spherical cavities," *J. Phys. Chem. B*, vol. 110, no. 19, pp. 9403–9412, 2006.

35 I. Brovchenko and A. Oleinikova, *Interfacial and Confined Water*. Elsevier, 2008.

36 G. Ciccotti, K. Binder, and M. Ferrario, eds., *Computer Simulations in Condensed Matter: From Materials to Chemical Biology*, vol. 1. Physica-Verlag, 2007.

37 J. R. Davis and A. Z. Panagiotopoulos, "Monte Carlo simulations of amphiphilic nanoparticle self-assembly," *J. Chem. Phys.*, vol. 129, no. 19, p. 194706, 2008.

38 J. Olafsen, ed., *Experimental and Computational Techniques in Soft Condensed Matter Physics*. Cambridge: Cambridge University Press, 2010.

39 K. Binder, ed., *The Monte Carlo Method in Condensed Matter Physics*. Berlin Heidelberg: Springer, 2012.

40 F. H. Stillinger, *Energy Landscapes, Inherent Structures, and Condensed-Matter Phenomena*. Princeton, NJ: Princeton University Press, 2015.

41 M. P. Allen and D. J. Tildesley, *Computer Simulation of Liquids*. Oxford: Oxford University Press, 2017.

42 M. A. Anisimov and C. E. Bertrand, "Thermodynamics of fluids at meso and nano scales," in *Applied Thermodynamics of Fluids* (A. R. Goodwin, C. J. Peters, and J. V. Sengers, eds.), ch. 7, pp. 172–214. Cambridge: Royal Society of Chemistry, 2010.

43 D. H. Everett, *Basic Principles of Colloid Science*. Cambridge: The Royal Society of Chemistry, 1994.

44 J. N. Israelachvili, *Intermolecular and Surface Forces*. Amsterdam: Academic Press, 3rd ed., 2011.

45 A. S. Eddington, *The Nature of the Physical World*. Kessinger Publishing, LLC, 2010.

46 T. L. Hill, "Thermodynamics of small systems," *J. Chem. Phys.*, vol. 36, no. 12, pp. 3182–3197, 1962.

47 T. L. Hill, *Thermodynamics of Small Systems*. Courier Corporation, 1994.

48 J. D. Van der Waals, "The thermodynamic theory of capillarity under the hypothesis of a continuous variation of density," *Verhandel. Konink. Akad. Weten. Amsterdam*, vol. 20, no. 1, p. 197, 1893. (in Dutch, English translation by J. S. Rowlinson, *J. Stat. Phys.* (1979) 20, 197).

49 J. S. Rowlinson and B. Widom, *Molecular Theory of Capillarity*. New York: Dover Publications, 2003.

50 L. D. Landau and E. M. Lifshitz, *Mekhanika Sploshnykh Sred*. Mechanics of Continuous Media. Moscow: Ogiz, 1944. (in Russian). In the subsequent editions, the book's material was published as two separate volumes, *Fluid Mechanics* and *Theory of Elasticity*, but the section on the van der Waals theory of interfacial phenomena was omitted and not translated into English.

51 J. W. Cahn and J. E. Hilliard, "Free energy of a nonuniform system. I. Interfacial free energy," *J. Chem. Phys.*, vol. 28, p. 258, 1958.

52 J. Wu, *Square-Gradient Model for Inhomogeneous Systems: From Simple Fluids to Microemulsions, Polymer Blends and Electronic Structure*. Molecular Modeling and Simulation: Applications and Perspectives. Singapore: Springer, 2017.

53 B. Chu, *Laser Light Scattering: Basic Principles and Practice*. Academic Press, 2nd ed., 1991.

54 R. Pecora, ed., *Dynamic Light Scattering: Applications of Photon Correlation Spectroscopy*. New York: Springer, 1985.

55 J. Jeevanandam and M. K. Danquah, eds., *Research Advances in Dynamic Light Scattering*. Physics Research and Technology. Nova Science Publishers, 2020.

56 J. N. Weiss, *Dynamic Light Scattering Spectroscopy of the Human Eye*. Cham: Springer, 2022.

57 L. A. Feigin and D. I. Svergun, *Structure Analysis by Small-Angle X-Ray and Neutron Scattering*. Springer, 1987.

58 I. W. Hamley, *Small-Angle Scattering: Theory, Instrumentation, Data, and Applications*. Wiley, 2021.

59 A. Grishaev, N. J. Anthis, and G. M. Clore, "Contrast-matched small-angle X-ray scattering from a heavy-atom-labeled protein in structure determination: Application to a lead-substituted calmodulin-peptide complex," *J. Am. Chem. Soc.*, vol. 134, no. 36, p. 14686, 2012.

Exercises

1.1 Calculate the volume per molecule of liquid water at 300 K and 1 bar and compare this value with (a) the volume per molecule in water vapor at the same conditions and with (b) the volume of a single water molecule, approximated as a sphere of 0.29 nm diameter.

1.2 Give a reasonable estimate of the size of human DNA (both stretched and coiled).

1.3 How many molecules are contained at ambient conditions in a spherical droplet of water of 1 nm radius? What fraction of these molecules is on the surface of the droplet?

1.4 Explain why the surface energy of a particle becomes increasingly important (with respect to the total energy) upon decreasing the particle size.

1.5 Calculate the mass (in kg) of a single CO_2 molecule. Compare it with the mass of a red (633 nm wavelength) photon and with the mass of neutron.

1.6 Explain why nanoparticles cannot be seen through an optical microscope. What technique could you recommend to see such particles?

1.7 Under which conditions can a human hair be treated as a one-dimensional object (a "line")?

1.8 A polymer solution near the critical point of demixing is confined in a micron-sized pore. Identify the characteristic length scales of the system.

1.9 What scattering experiment is the most appropriate to probe the nanoscale structure of microemulsions?

1.10 Calculate the light-scattering wave number for the following experiment: A dilute aqueous solution of nanoparticles is investigated with a He-Ne laser light source at a scattering angle of 30° ($\lambda_0 = 633$ nm).

2

Basics of Molecular Thermodynamics

This chapter includes:

- Brief review of thermodynamic laws and state functions
- Phases of matter
- Thermodynamic equilibrium conditions and stability criteria
- Van der Waals fluid, lattice-gas, and regular-solution models
- Molecular interactions
- Elements of nonequilibrium thermodynamics

This chapter serves as a guide for the rest of the book; however, it cannot substitute for a systematic study of thermodynamics. There are many excellent texts on classical thermodynamics recommended for engineering and science majors at introductory, intermediate, and advanced levels, such as [1–8].

2.1 Laws of Thermodynamics

The transformation of energy (of any kind) and the relations between properties of substances are controlled and restricted by the laws of thermodynamics, namely the first law and the second law. Less conventional, but commonly referred to in the literature are the "zeroth law of thermodynamics," which claims the existence of thermodynamic equilibrium, and the "third law of thermodynamics," which states that the absolute entropy reaches a minimum value (ideally zero) at absolute zero temperature. The third law, based on statistical mechanics, fundamentally defines the absolute value of entropy.

The first law is the principle of the conservation of energy, formulated for systems containing a large number of particles (statistical systems). The first law is associated with the fundamental concept of the homogeneity of time, and as such, it has no restrictions – it is exact. According to the first law, the total energy change, ΔE_t, for a system undergoing heat and work interactions with the surroundings is

$$\Delta E_\mathrm{t} = \Delta E_\mathrm{k} + \Delta E_\mathrm{p} + Q + W \tag{2.1}$$

where ΔE_k and ΔE_p are the changes in the kinetic and potential energy of the system, while Q and W are the heat and work interactions, respectively. The change of the internal energy, ΔU_t, is given by

$$\Delta U_\mathrm{t} = \Delta E_\mathrm{t} - (\Delta E_\mathrm{k} + \Delta E_\mathrm{p}) = Q + W \tag{2.2}$$

Mesoscopic Thermodynamics for Scientists and Engineers, First Edition. Mikhail A. Anisimov and Thomas J. Longo.
© 2024 John Wiley & Sons, Inc. Published 2024 by John Wiley & Sons, Inc.

Equation (2.2) in differential form reads

$$dU_t = đQ + đW \tag{2.3}$$

The internal energy, being a state function, is independent of the thermodynamic path, such that its total differential is exact. In contrast, the differentials, $đQ$ and $đW$, are generally not exact since the heat and work are not state functions, depending on the path of the process. The work can be divided into two parts: "mechanical" work, W_m, and "chemical" work, W_{ch}. The differential mechanical work is defined as a scalar product of a force, $\boldsymbol{F}$, and an elementary displacement, $d\boldsymbol{r}$:

$$đW_m = \boldsymbol{F} \cdot d\boldsymbol{r} \tag{2.4}$$

For expansion or compression of an isotropic body

$$đW_m = -P\,dV_t \tag{2.5}$$

where P is the pressure and V_t is the total volume. The sign in Eq. (2.5) is conventional: it is adopted that work is negative for expansion and positive for compression. In the absence of work, since the differential of energy is exact, the differential heat exchange, $đQ$, also becomes an exact differential, and thus, the heat exchange will be the same for both reversible and irreversible processes.

The chemical work is associated with the change of the number of molecules. This concept does not exist in mechanics. The exact differential of the chemical work for a reversible exchange of species i is defined as,

$$dW_{ch} = \sum_{i=1}^{k} \mu_i\, dN_i \tag{2.6}$$

where μ_i and N_i are the chemical potential and number of particles of species i, respectively. If the system only exhibits heat transfer and work interactions with its surroundings, it is referred to as a "closed system," while if, in addition, the system may exchange molecules with its surrounding, it is referred to as an "open system." Figure 2.1 depicts a biological cell, an example of an open

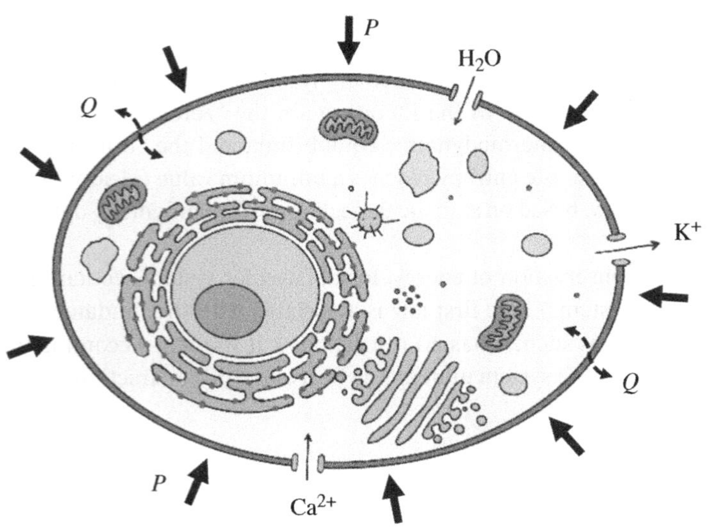

Figure 2.1 A biological cell is an example of an open system. This figure depicts a schematic representation of ion and molecular transport, heat transfer (Q) through the membrane, and pressure (P) acting on the surface. The supramolecular structures depicted inside the cell are: the nucleus, nucleolus, nuclear membrane, vacuoles, mitochondria, Golgi apparatus, centrosome, and ribosomes.

system. Molecules and ions may enter and leave the cell through transport channels, heat may be exchanged through the cell membrane, while the volume of the cell is subject to compression or expansion. In an open system, if there is no heat exchange and the total volume and number of molecules of all other species (except species i) are kept constant, the chemical potential, μ_i, is defined as the internal energy change when one molecule of species i is reversibly exchanged between the system and the surroundings.

In contrast to the first law, the second law is exact only in the thermodynamic limit, when the number of particles is infinite. The second law is associated with the asymmetry of time, the fundamental difference between the past and the future; this difference appears only in statistical (many-body) systems. One cannot recognize the future and the past in any event involving only two or three particles. This indicates that the second law has a statistical meaning, becoming less accurate with fewer molecules. The second law states that the total entropy change, ΔS_t, of an isolated system must be equal to zero (for reversible processes) or positive (for irreversible processes). The entropy of an isolated system reaches its maximum when the system approaches equilibrium. For a closed system $\mathrm{d}S_\mathrm{t} = \text{đ}Q/T$ (reversible processes) or $\mathrm{d}S_\mathrm{t} > \text{đ}Q/T$ (irreversible processes), where T is the temperature. Like the energy, the entropy is a state function; it does not depend on the way by which the state is approached. However, unlike the energy, the entropy is not a conserved property; an irreversibility inevitably causes generation of entropy (also known as "entropy production") in the system.

2.2 Thermodynamic Potentials and Legendre Transformations

For reversible processes in an open isotropic system, the first law, in differential form, becomes

$$\mathrm{d}U_\mathrm{t} = T\,\mathrm{d}S_\mathrm{t} - P\,\mathrm{d}V_\mathrm{t} + \sum_{i=1}^{k} \mu_i\,\mathrm{d}N_i \tag{2.7}$$

where μ_i, the chemical potential of species i, is defined as

$$\mu_i = \left.\frac{\partial U_\mathrm{t}}{\partial N_i}\right|_{S_\mathrm{t}, V_\mathrm{t}, N_{j\neq i}} \tag{2.8}$$

Since Eq. (2.7) interrelates differentials of state functions only, it can be applied to any process, reversible or irreversible (with the same initial and final states). The internal energy, whose total differential is given by Eq. (2.7), is an example of a characteristic state function, also known as a "thermodynamic potential" that generates other thermodynamic properties. The internal energy is a particular thermodynamic potential having the canonical ("natural") variables ("thermodynamic coordinates") S_t, V_t, and N_i, such that the total number of all particles is $N = \sum N_i$. The fundamental thermodynamic properties of the system: T, P, and μ_i ("thermodynamic forces"), are found by taking partial derivatives of this potential with respect to its canonical variables. In a single-component system, Eq. (2.7) is reduced to

$$\mathrm{d}U_\mathrm{t} = \mathrm{d}(NU) = N\,\mathrm{d}U + U\,\mathrm{d}N = T\,\mathrm{d}S_\mathrm{t} - P\,\mathrm{d}V_\mathrm{t} + \mu\,\mathrm{d}N \tag{2.9}$$

Since $T\,\mathrm{d}S_\mathrm{t} = T\,\mathrm{d}(NS) = TS\,\mathrm{d}N + TN\,\mathrm{d}S$ and $P\,\mathrm{d}V_\mathrm{t} = P\,\mathrm{d}(NV) = PV\,\mathrm{d}N + PN\,\mathrm{d}V$, it follows from Eq. (2.8) that the chemical potential of a one-component system can also be defined as

$$\mu = U - TS + PV \tag{2.10}$$

where $U = U_t/N$, $S = S_t/N$, and $V = V_t/N$ are the internal energy, entropy, and volume respectively, per molecule.

Common Thermodynamic Potentials

- Internal energy with canonical variables: S_t, V_t, and N_i

$$U_t = TS_t - PV_t + \sum \mu_i N_i \tag{2.11a}$$

$$dU_t = T\,dS_t - P\,dV_t + \sum \mu_i\,dN_i \tag{2.11b}$$

- Enthalpy with canonical variables: S_t, P, and N_i

$$H_t = U_t + PV_t + \sum \mu_i N_i \tag{2.12a}$$

$$dH_t = T\,dS_t + V_t\,dP + \sum \mu_i\,dN_i \tag{2.12b}$$

- Helmholtz energy with canonical variables: T, V_t, and N_i

$$A_t = U_t - TS_t \tag{2.13a}$$

$$dA_t = -S_t\,dT - P\,dV_t + \sum \mu_i\,dN_i \tag{2.13b}$$

- Gibbs energy with canonical variables: T, P, and N_i

$$G_t = A_t + PV_t = U_t - TS_t + PV_t = \sum \mu_i N_i \tag{2.14a}$$

$$dG_t = -S_t\,dT + V_t\,dP + \sum \mu_i\,dN_i \tag{2.14b}$$

- Grand thermodynamic potential with canonical variables: T, V_t, and μ_i

$$\Omega_t = A_t - \sum \mu_i N_i = -PV_t \tag{2.15a}$$

$$d\Omega_t = -S_t\,dT - P\,dV_t - \sum N_i\,d\mu_i \tag{2.15b}$$

In engineering practice, the number of moles of a species (defined by $n_i = N_i/N_A$, where $N_A \cong 6.02 \times 10^{23}$ is Avogadro's number), rather than the number of molecules, is commonly used as a variable. Consequently, the extensive thermodynamic functions may be considered per mole rather than per molecule, such that the chemical potential defined as $\mu_i = (\partial U_t/\partial n_i)|_{S_t,V_t,n_{j\neq i}}$ or $V = V_t/n$ will be N_A times larger than the same values calculated per molecule. Avogadro's number also converts Boltzmann's constant $k_B \cong 1.38 \times 10^{-23}$ J/K to the universal gas constant $R = k_B N_A \cong 8.314\,\text{J/mol} \cdot \text{K}$. The decision to use properties in units per mole or per molecule is just a question of convenience and should not cause confusion. In this book, occasionally, the same notations are used to represent the properties per molecule (preferable for nanoscale systems) or per mole; however, the difference is always stated.

One can obtain other thermodynamic potentials, each having its own canonical variables, by using *Legendre transformations*. Suppose a thermodynamic force, $\Phi(z, x)$, is a function of two canonical variables, z and x. There are two ways to describe this function at constant x as illustrated in Figure 2.2. The function is represented as a curve composed of an infinite number of coordinate pairs (z, Φ). Alternatively, it is also a curve composed of an infinite numbers of slopes (derivatives), $y = (\partial\Phi/\partial z)|_x$, and intercepts, $\Psi(y, x)$, in which $\Psi(y, x)$ is a new thermodynamic function of canonical variables y and x. Evaluating the instantaneous derivative,

$$y(x, z) = \frac{\Phi(z, x) - \Psi(z, x)}{z - 0} \tag{2.16}$$

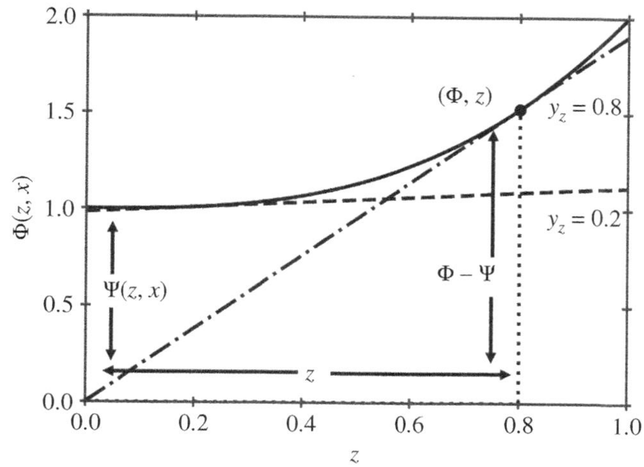

Figure 2.2 Illustration of a Legendre transform. At constant x, the function $\Phi(x, z)$ is a curve described by an infinite number of coordinate pairs (z, Φ), while also being described by an infinite number of slopes y and intercepts $\Psi(x, y)$ evaluated at each point along the curve, z. The Legendre transform between the functions $\Phi(x, z)$ and $\Psi(x, y)$ is $\Psi(x, y) = \Phi(x, z) - zy$.

and upon rearranging, one finds that

$$\Psi(y, x) = \Phi(z, x) - yz \tag{2.17}$$

which is composed of an infinite number of slopes, $z = (\partial\Psi/\partial y)|_x$, and intercepts $\Phi(z, x)$. Based on these transformations, the derivatives of the functions $\Phi(z, x)$ and $\Psi(y, x)$, y and z, are "conjugate" variables to one another. To produce a new thermodynamic potential by a Legendre transform, a product of the corresponding canonical variables (a thermodynamic "force" and a conjugate thermodynamic "coordinate," such as TS_t, PV_t, or $\mu_i N_i$), is added or subtracted. The most common thermodynamic potentials are defined by Eqs. (2.11–2.15).

Note that, in addition to Eq. (2.8), the chemical potential of species i can be equally defined as

$$\mu_i = \left.\frac{\partial H_t}{\partial N_i}\right|_{S_t,P,N_{j\neq i}} = \left.\frac{\partial A_t}{\partial N_i}\right|_{T,V_t,N_{j\neq i}} = \left.\frac{\partial G_t}{\partial N_i}\right|_{T,P,N_{j\neq i}} \tag{2.18}$$

Since

$$\mathrm{d}G_t = \sum \mu_i\,\mathrm{d}N_i + \sum N_i\,\mathrm{d}\mu_i \tag{2.19}$$

$$\Delta G_t = -S_t\,\mathrm{d}T + V_t\,\mathrm{d}P + \sum \mu_i\,\mathrm{d}N_i \tag{2.20}$$

the chemical potentials of species in a mixture are interrelated by

$$\boxed{\sum N_i\,\mathrm{d}\mu_i + S_t\,\mathrm{d}T - V_t\,\mathrm{d}P = 0} \tag{2.21}$$

Relation (2.21) is known as the Gibbs–Duhem equation. For a single-component system, the chemical potential $\mu_i = \mu$ becomes the Gibbs energy per molecule (or, if preferable, per mole), so in this case, $G_t = \mu N$, and

$$\mu \equiv G = G_t/N \tag{2.22}$$

Thermodynamic Potentials for $N = \sum N_i$ = constant

- Helmholtz Energy per molecule

$$A = \frac{A_t}{N} = \sum \mu_i x_i - PV_t \tag{2.23a}$$

$$dA = -S\,dT - P\,dV + \sum \mu_i\,dx_i \tag{2.23b}$$

- Gibbs energy per molecule

$$G = \frac{G_t}{N} = \sum \mu_i x_i \tag{2.24a}$$

$$dG = -S\,dT + V\,dP + \sum \mu_i\,dx_i \tag{2.24b}$$

- The grand potential per molecule is

$$\Omega = \frac{\Omega_t}{N} = -P\frac{V_t}{N} = -PV \tag{2.25a}$$

$$d\Omega = -P\,dV - S\,dT - \sum x_i\,d\mu_i \tag{2.25b}$$

There are two practically important cases that further specify the thermodynamic potentials, namely: if the total number of particles N (or the number of moles, n) is fixed or if the total volume V_t is fixed. In Eqs. (2.23)–(2.25), we provide examples of thermodynamic potentials in the case when N = constant. In these equations, where $x_i = N_i/N = n_i/n$ is the species molecular concentration (or mole fraction), such that $\sum x_i = 1$ and $S = S_t/N$ is the entropy per molecule. If the total volume is kept constant, the Helmholtz energy per unit volume (the "density of the Helmholtz energy") is

$$\frac{A_t}{V_t} = \rho\frac{A_t}{N} = \rho A \tag{2.26a}$$

$$d(\rho A) = -\rho S\,dT + \sum \mu_i\,d\rho_i \tag{2.26b}$$

where $\rho = N/V_t$ is the average molecular density and $\rho_i = \rho x_i = N_i/V_t$ is the partial molecular density (proportional to the "molarity," $n_i/V_t = N_i/V_t N_A$), such that $\rho = \sum \rho_i$. Thus, the chemical potential may also be defined as

$$\boxed{\mu_i = \left.\frac{\partial(\rho A_t)}{\partial \rho_i}\right|_{T,\rho_{j\neq i}}} \tag{2.27}$$

Thus, the chemical potential of species i can be defined through any thermodynamic potential that depends on the number of molecules or on the molecular density. The density of the grand potential may also be expressed through the pressure as

$$\frac{\Omega_t}{V_t} = -P \tag{2.28a}$$

$$dP = \rho S\,dT + \sum \rho_i\,d\mu_i \tag{2.28b}$$

$$\rho_i = \left.\frac{\partial P}{\partial \mu_i}\right|_{T,\mu_{j\neq i}} \tag{2.28c}$$

Thermodynamic Potentials with a Variable $1/T$

- Internal energy and enthalpy (Gibbs–Helmholtz Relations)

$$\frac{\partial(G_t/T)}{\partial(1/T)} = H_t \qquad \frac{\partial(A_t/T)}{\partial(1/T)} = U_t \tag{2.29}$$

$$\frac{\partial(A_t/T)}{\partial T} = -\frac{U_t}{T^2} \qquad \frac{\partial(G_t/T)}{\partial T} = -\frac{H_t}{T^2} \tag{2.30}$$

- Helmholtz energy and Gibbs energy

$$\mathrm{d}(A_t/T) = U_t\,\mathrm{d}(1/T) - (P/T)\mathrm{d}V_t + \sum(\mu_i/T)\mathrm{d}N_i \tag{2.31}$$

$$\mathrm{d}(G_t/T) = H_t\,\mathrm{d}(1/T) + V_t\,\mathrm{d}(P/T) + \sum \mu_i\,\mathrm{d}(N_i/T) \tag{2.32}$$

- Grand potential

$$\Omega_t/T = -(P/T)V_t \tag{2.33}$$

$$\mathrm{d}(\Omega_t/T) = U_t\,\mathrm{d}(1/T) - (P/T)\mathrm{d}V_t + \sum N_i\,\mathrm{d}(\mu_i/T) \tag{2.34}$$

Note that the expression for $\mathrm{d}P$, given by Eq. (2.28b), also follows from the Gibbs–Duhem relation Eq. (2.21). Also note that for the density of the Helmholtz energy, the canonical variables are the temperature and partial densities. In the particular case of a single-component system, the canonical variables are the density ρ (conjugate to the chemical potential μ) and the temperature T, while for the Helmholtz energy per mole or per molecule, the canonical variables are the volume per mole or per molecule $V = 1/\rho$ (conjugate to P) and T (conjugate to S).

Thermodynamic Potentials for $V_t = \sum N_i/V_i =$ constant

- Density of the Helmholtz Energy

$$\rho A = \frac{A_t}{V_t} = \sum \mu_i \rho_i - P \tag{2.35a}$$

$$\rho\,\mathrm{d}A = -\rho S\,\mathrm{d}T - \mathrm{d}P + \sum \mu_i\,\mathrm{d}\rho_i \tag{2.35b}$$

- Density of the Gibbs energy

$$\rho G = \rho G_t = \sum \mu_i \rho_i \tag{2.36a}$$

$$\rho\,\mathrm{d}G = -\rho S\,\mathrm{d}T + \mathrm{d}P + \sum \mu_i\,\mathrm{d}\rho_i \tag{2.36b}$$

- Density of the grand potential

$$\rho\Omega = \frac{\Omega_t}{V_t} = \rho G - \rho A = -P \tag{2.37a}$$

$$\rho\,\mathrm{d}\Omega = -\mathrm{d}P - \rho S\,\mathrm{d}T - \sum \rho_i\,\mathrm{d}\mu_i \tag{2.37b}$$

By using combinations of traditional variables and Legendre transformations, one may produce even more thermodynamic potentials. Examples of thermodynamic potentials with a constant $1/T$ are defined by Eqs. (2.29)–(2.34). A choice of thermodynamic variables depends on experimental or computer-simulation conditions as well as on the formulation of theoretical models. For example, popular "lattice" models that describe phase transitions in fluids and solids usually implement molecular density rather than molecular volume as a variable. Thermodynamic potentials used for such lattice models at constant total volume, V_t are defined by Eqs. (2.36a) and (2.37b). The proper choice of thermodynamic variables and potentials is important for testing the adequacy and accuracy of theoretical models with respect to the properties of real physical systems.

2.3 Maxwell Relations

Common Maxwell Relations

- In single-component systems:

$$\frac{\partial^2 A_t}{\partial T \partial V_t} = \left.\frac{\partial S_t}{\partial V_t}\right|_T = \left.\frac{\partial P}{\partial T}\right|_{V_t} \tag{2.38}$$

$$\frac{\partial^2 G_t}{\partial T \partial P} = -\left.\frac{\partial S_t}{\partial P}\right|_T = \left.\frac{\partial V_t}{\partial T}\right|_P \tag{2.39}$$

$$\frac{\partial^2 P}{\partial T \partial \mu} = -\left.\frac{\partial (\rho S)}{\partial \mu}\right|_T = \left.\frac{\partial \rho}{\partial T}\right|_\mu \tag{2.40}$$

- In binary mixtures:

$$\frac{\partial^2 A_t}{\partial V \partial x} = -\left.\frac{\partial P}{\partial x}\right|_{V_t,T} = \left.\frac{\partial \mu_{21}}{\partial V_t}\right|_{x,T} \tag{2.41}$$

$$\frac{\partial^2 G_t}{\partial P \partial x} = \left.\frac{\partial V_t}{\partial x}\right|_{T,P} = \left.\frac{\partial \mu_{21}}{\partial P}\right|_{x,P} \tag{2.42}$$

There are several tools used for manipulating thermodynamic functions and their derivatives. As mentioned above, all thermodynamic potentials are related by Legendre transformations. There are also famous relations between mixed (or "cross") derivatives of thermodynamic potentials, known as "Maxwell relations." These relations originate from the mathematical finding that the differential order of a state function is immaterial. One can immediately verify that in a single-component system, the mixed partial derivatives satisfy the Maxwell relations given by Eqs. (2.38)–(2.40). Meanwhile, the most common Maxwell relations for binary mixtures are given by Eqs. (2.41) and (2.42), obtained with use of the Gibbs–Duhem relation, Eq. (2.21).

Useful Calculus Rules in Thermodynamics

Inverse derivative

$$\left.\frac{\partial x}{\partial y}\right|_z = \left(\frac{\partial y}{\partial x}\right)_z^{-1}$$

Triple-product rule $F = F(x, y)$

$$\left.\frac{\partial F}{\partial x}\right|_y \left.\frac{\partial x}{\partial y}\right|_F \left.\frac{\partial y}{\partial F}\right|_x = -1$$

Chain rule $F = F[x, y(z)]$

$$\left.\frac{\partial F}{\partial y}\right|_x = \frac{\left.\frac{\partial F}{\partial z}\right|_x}{\left.\frac{\partial y}{\partial z}\right|_x}$$

Maxwell relations can also be applied to third and higher-order derivatives of thermodynamic potentials, such as

$$-T\frac{\partial^3 A_t}{\partial T^2 \partial V_t} = N\left.\frac{\partial C_V}{\partial V_t}\right|_T = T\left.\frac{\partial^2 P}{\partial T^2}\right|_{V_t} \tag{2.43}$$

In this example, the isochoric molar heat capacity, C_V, is given by

$$C_V = \left.\frac{\partial U}{\partial T}\right|_V = T\left.\frac{\partial S}{\partial T}\right|_V = -T\left.\frac{\partial^2 A}{\partial T^2}\right|_V \tag{2.44}$$

and is defined either per molecule or per mole. Similarly,

$$-T\frac{\partial^3 G_t}{\partial T^2 \partial P} = N\left.\frac{\partial C_P}{\partial P}\right|_T = -T\left.\frac{\partial^2 V_t}{\partial T^2}\right|_P \tag{2.45}$$

where the isobaric molar heat capacity, C_P, is given by

$$C_P = \left.\frac{\partial H}{\partial T}\right|_P = T\left.\frac{\partial S}{\partial T}\right|_P = -T\left.\frac{\partial^2 G}{\partial T^2}\right|_P \tag{2.46}$$

Another important second derivative (a "response function") is the isothermal compressibility, κ_T, given by

$$\kappa_T = -\frac{1}{V_t}\left.\frac{\partial V_t}{\partial P}\right|_T = \frac{1}{\rho}\left.\frac{\partial \rho}{\partial P}\right|_T \tag{2.47}$$

This is a measure of the system's response to compression or expansion.

It is convenient that, with use of the Maxwell relations, one can express thermodynamic properties, which are sometimes difficult to measure directly (such as derivatives of entropy), through easily measurable properties (like derivatives of volume or density). For example, the thermal volumetric expansivity, α_P, is given through

$$\alpha_P = \frac{1}{V_t}\left.\frac{\partial V_t}{\partial T}\right|_P = -\frac{1}{V_t}\left.\frac{\partial S_t}{\partial P}\right|_T \tag{2.48}$$

α_P and κ_T are related through the triple product rule,

$$\left.\frac{\partial P}{\partial T}\right|_V \left.\frac{\partial T}{\partial V}\right|_P \left.\frac{\partial V}{\partial P}\right|_T = \left.\frac{\partial P}{\partial T}\right|_V \frac{\kappa_T}{\alpha_P} = -1 \tag{2.49}$$

Another property that is important for engineering applications is the Joule–Thomson coefficient, $\alpha_H = \partial T/\partial P|_H$, which can be obtained from *Throttling Processes*, irreversible adiabatic expansion at constant enthalpy. This process is used in petrochemical industry to liquefy gases and in cryogenic applications (e.g. the production of liquid oxygen, nitrogen, and argon). Only when the Joule–Thomson coefficient, for a given gas at a specified temperature, is positive, the temperature will drop (provided that the pressure drops) and the gas will be liquefied at that temperature. In other words, a gas must be below its inversion temperature to be liquefied by a throttling effect. For this reason, this process cannot easily be used to liquefy helium, hydrogen, or neon. Using Legendre transformations and Maxwell relations, one can relate the Joule–Thomson coefficient to the isobaric heat capacity and the volumetric expansivity as

$$\alpha_H = \left.\frac{\partial T}{\partial P}\right|_H = -\frac{V}{C_P}(1 - \alpha_P T) \tag{2.50}$$

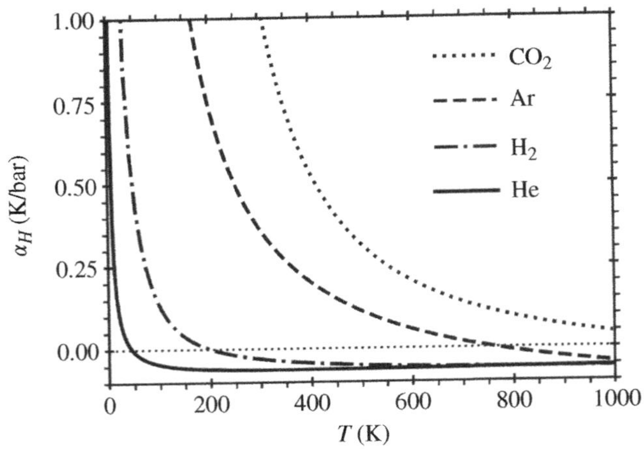

Figure 2.3 Temperature dependence of Joule–Thomson coefficients for selected fluids along the isobar, $P = 1.02$ bar.

Figure 2.3 shows the Joule–Thomson coefficient as a function of temperature for a selection of basic fluids.

2.4 "Fields" and "Coordinates," Equilibrium Conditions, Gibbs Phase Rule

All thermodynamic properties can be divided into two classes, thermodynamic "forces" (or "fields") and thermodynamic "coordinates." Thermodynamic coordinates are extensive properties, but they can be made intensive by reducing them by the number of moles or molecules. If the thermodynamic coordinates are considered per unit volume, the reduced thermodynamic coordinates are referred to as "generalized densities." Examples of generalized densities are the densities of all extensive properties, such as the molecular density, the energy and entropy per unit volume, the magnetization per unit volume, and the partial molecular density ("molarity"). Generalized densities are obtained from first-order partial derivatives of the thermodynamic potentials, specified per unit volume with respect to the appropriate thermodynamic field. Thermodynamic fields are defined through first-order partial derivatives of appropriate thermodynamic potentials with respect to the coordinates conjugate to these fields.

At equilibrium, a thermodynamic potential exhibits a minimum with respect to the change in its canonical variables. To move the system away from equilibrium requires external work. For example, consider a homogeneous system at constant temperature and pressure, if a part of this system is reversibly driven away from equilibrium, the external reversible work, W_{rev}, is given by the change of the Gibbs energy, $W_{rev} = \Delta G_t$.

Examples of Thermodynamic Fields

$$T = \left.\frac{\partial U_t}{\partial S_t}\right|_{V_t,N_i} = \left.\frac{\partial H_t}{\partial S_t}\right|_{P,N_i}$$
$$P = -\left.\frac{\partial U_t}{\partial V_t}\right|_{S_t,N_i} = -\left.\frac{\partial A_t}{\partial V_t}\right|_{T,N_i} \tag{2.51}$$

In single-component systems:

$$\mu \equiv G = \frac{G_t}{N} = \left.\frac{\partial U_t}{\partial N}\right|_{S_t,V_t} = \left.\frac{\partial H_t}{\partial N}\right|_{S_t,P} = \left.\frac{\partial A_t}{\partial N}\right|_{T,V_t} \tag{2.52}$$

In mixtures:

$$\mu_i = \left.\frac{\partial G_t}{\partial N_i}\right|_{T,P,\mu_{j\neq i}} \tag{2.53}$$

Consequently, at equilibrium, all fields must be uniform across the system, while the coordinates can be different in different coexisting phases. For example, if vapor and liquid phases coexist, the temperature and pressure are the same in the coexisting phases, while the entropy, energy, and molecular densities are (in general) different. For a system with any number of components, i, coexistence is achieved under the following conditions,

$$\boxed{T^{\tilde{\alpha}} = T^{\tilde{\beta}}, \qquad P^{\tilde{\alpha}} = P^{\tilde{\beta}}, \qquad \mu_i^{\tilde{\alpha}} = \mu_i^{\tilde{\beta}}} \tag{2.54}$$

where the superscripts indicate the property of any two coexisting phases, $\tilde{\alpha}$ and $\tilde{\beta}$.

Examples of Thermodynamic Coordinates

$$V_t = \left.\frac{\partial G_t}{\partial P}\right|_{T,N_i} = \left.\frac{\partial H_t}{\partial P}\right|_{V_t,N_i}$$
$$S_t = -\left.\frac{\partial G_t}{\partial T}\right|_{P,N_i} = -\left.\frac{\partial A_t}{\partial T}\right|_{V_t,N_i} = -\left.\frac{\partial \Omega_t}{\partial T}\right|_{V_t,N_i} \tag{2.55}$$

In single-component systems:

$$N = -\left.\frac{\partial \Omega_t}{\partial \mu}\right|_{V_t,T} \tag{2.56}$$

In mixtures:

$$N_i = -\left.\frac{\partial \Omega_t}{\partial \mu_i}\right|_{V_t,T,N_{j\neq i}} \tag{2.57}$$

Thermodynamic equilibrium can, in general, be considered in the presence of other physical fields, such as the electric and the magnetic fields, as well as under mechanical stresses (e.g. anisotropy in solids). The surface energy ($\sigma\mathcal{A}_t$) can also be included in the overall energy of the system, such that

$$dU_t = T\,dS_t - P\,dV_t + \sigma\,d\mathcal{A}_t + \sum \mu_i\,dN_i \tag{2.58}$$

where the surface tension (σ), being a thermodynamic field, is defined through partial derivatives of the thermodynamic potentials with respect to the surface area, $\mathcal{A}_t$, as

$$\sigma = \left.\frac{\partial U_t}{\partial \mathcal{A}_t}\right|_{S_t,V_t,N_i} = \left.\frac{\partial H_t}{\partial \mathcal{A}_t}\right|_{S_t,P,N_i} = \left.\frac{\partial A_t}{\partial \mathcal{A}_t}\right|_{T,V_t,N_i} = \left.\frac{\partial G_t}{\partial \mathcal{A}_t}\right|_{T,P,N_i} \tag{2.59}$$

By incorporating the interfacial tension, the external electric field ($\boldsymbol{E}$), and the magnetic field ($\boldsymbol{h}$), a more general Gibbs–Duhem relation for an isotropic system reads

$$\sum N_i\,d\mu_i + S_t\,dT - V_t\,dP + \mathcal{A}_t\,d\sigma + \varepsilon\boldsymbol{E}\cdot d\boldsymbol{E} + \boldsymbol{M}\cdot d\boldsymbol{h} = 0 \tag{2.60}$$

where ε is the dielectric constant, and $\boldsymbol{M}$ is the magnetization.

The number of phases, π, in a multicomponent system of chemically nonreactive species which may coexist in equilibrium is described by the Gibbs phase rule,

$$\boxed{\pi = i + 2 - f} \tag{2.61}$$

where i is the number of species and f is the number of thermodynamic degrees of freedom (the number of independent fields, such as T and P, describing this state). For example, in the phase diagram of carbon dioxide (representing the typical behavior of simple fluids), shown in Figure 2.4, the triple point (where the three phases of gas, liquid, and solid coexist) is invariant ($f = 0$), while any two-phase coexistence (solid–gas, solid–liquid, or liquid–gas) with one degree of freedom, P or T. The liquid–gas critical point is also invariant since two-phase coexistence is terminated at this point.

In Figure 2.5, the phase diagram of carbon dioxide is represented through the pressure and the density. When depicting the pressure–density phase diagram, the liquid–gas equilibrium states are represented by the two branches of the coexistence curve, as the difference between the densities of liquid and gas increase upon departure from the critical point. Also, at the triple point, three different density values indicate the coexistence of the three phases, solid, liquid, and gas, at the

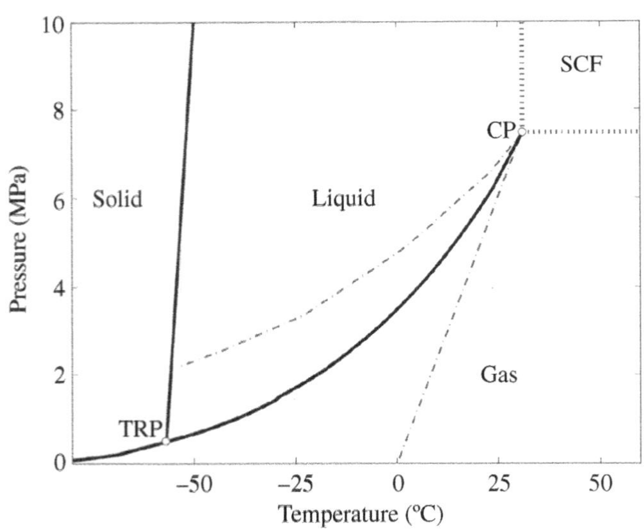

Figure 2.4 Pressure–temperature phase diagram of carbon dioxide at moderate pressures. TRP is the triple point, CP is the liquid–gas critical point, and SCF designates the super-critical fluid (loosely defined as the fluid state above the critical pressure and temperature). The dashed curves show the absolute stability limits of the gaseous state (upper branch) and the liquid state (lower branch).

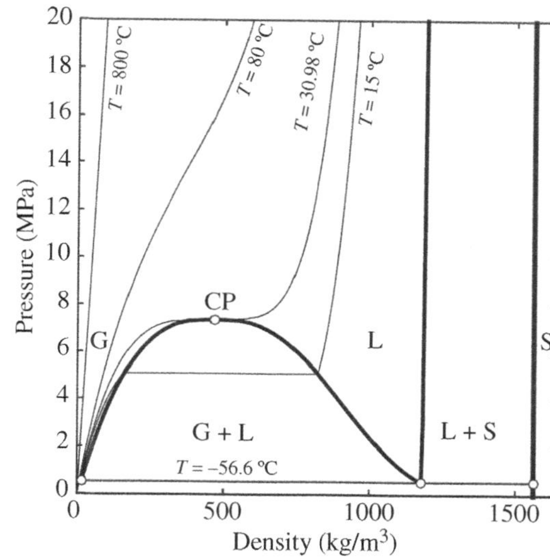

Figure 2.5 Pressure–density phase diagram of carbon dioxide. G, L, and S designate gaseous, liquid and solid states, respectively. At the triple point, three phases, with different densities, coexist at the same pressure and temperature.

same temperature and pressure. A gaseous state in coexistence with a liquid or a solid is often referred to as "vapor."

One may also note that at solid–liquid equilibrium, the density of carbon dioxide, like most other substances, is almost independent of pressure, corresponding to the relatively small difference in molar volumes between the two phases. For any two-phase ($\tilde{\alpha}$–$\tilde{\beta}$) equilibrium, as follows from the equality of the Gibbs energies of the two phases (Eq. (2.24b) for a single-component system), the slope of the transition line is determined by the ratio of the entropy change and the volume change as

$$\left.\frac{dP}{dT}\right|_{\tilde{\alpha}\tilde{\beta}} = \frac{S_{\tilde{\alpha}} - S_{\tilde{\beta}}}{V_{\tilde{\alpha}} - V_{\tilde{\beta}}} = \frac{\Delta H_{\tilde{\alpha}\tilde{\beta}}}{T(V_{\tilde{\alpha}} - V_{\tilde{\beta}})} \tag{2.62}$$

This is known as the "Clapeyron equation."

One should note that the Gibbs phase rule, given by Eq. (2.61), is formulated with the assumption that there is no chemical reaction in the system. In the presence of a chemical reaction, the concentration of species will not be an independent degree of freedom, as it is determined by thermodynamic equilibrium. Furthermore, the presence of an external field, such as the electric field, will effect the number of thermodynamic degrees of freedom, whereas the presence of a surface will not effect the number of degrees of freedom, since surface effects disappear in the thermodynamic limit ($V_t/A_t \to \infty$).

2.5 States of Matter, Crystal Polymorphism, and Fluid Polyamorphism

If a liquid is quenched below the temperature of solidification, its viscosity may increase so rapidly that the substance, instead of being crystallized, enters a glassy state. Glass is a nonequilibrium, noncrystalline state of matter that appears rigid like a solid on a short time scale, but can flow like a liquid on a longer time scale, eventually crystallizing. A common window glass pane demonstrates this generic property of the glass state. Vitrification, the transition from liquid to glass,

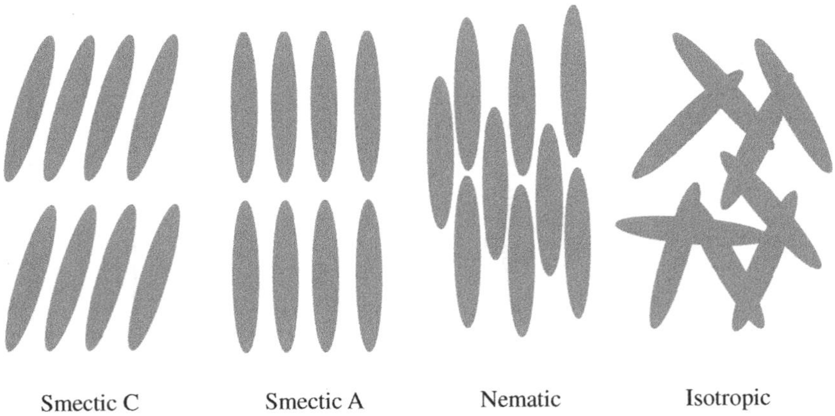

Figure 2.6 Four major phases of liquid crystals depending on orientational and partial-translational order: isotropic liquid (no orientational or translational order), nematic liquid crystal (orientational order without translational order), smectic A liquid crystal (long-ranged orientational order with one-dimensional translational order), and smectic C liquid crystal (long-ranged tilted orientational order with one-dimensional translational order). There are more known liquid-crystal phases with more complex structures.

is not a thermodynamic phase transition, although the properties sharply (though continuously) change at the temperature of vitrification [9, 10]. Isotropic liquids (generally, fluids) and glasses are amorphous, having irregular spatial arrangements of molecules, and thus, they lack long-ranged molecular translational and orientational order. In contrast, solid crystals lack the symmetry of fluids and glasses. They have edges and faces, diffract X-rays, and melt to liquid or sublimate to gas through a well-defined transition temperature. Some liquids, containing highly anisotropic molecules (shaped like "rods" or like "discs") may exhibit orientational or partially translational (in one or two dimensions) molecular order. Such states, usually observed between the isotropic-liquid and solid-crystal states, are known as "liquid crystals" (see Figure 2.6) [11]. The transitions between liquid crystalline phases and other phases of matter are thermodynamically well defined [12].

Typically, pure substances may be found with only one gaseous or liquid state, while their solid state may exist in various crystalline forms. In solids, this ability to adopt more than one crystal structure is called crystal polymorphism. Polymorphism can exist as a result of the difference in crystal packing or from different conformational forms of the same molecule. For instance, graphite and diamond are both made of carbon atoms, but are constructed with entirely different crystalline structures. Also, silica is known to form many polymorphs, the most famous of which is quartz, and water can adopt several forms of ice (see Figure 2.7). There is also the phenomenon of liquid-crystal polymorphism, in which anisotropic molecules in a liquid phase could exhibit orientational order, but, fully or partially, lack translational order [13].

Finding two distinct isotropic liquid forms from a single component substance is more unusual since liquids lack the long-range order common in crystals. Yet, this phenomenon, known as "liquid polyamorphism" [14, 15], is continuously being discovered in various substances. It has already been observed in sulfur, phosphorous, liquid carbon, cerium, helium, and high-pressure hydrogen, and it is predicted to exist in selenium, tellurium, silicon, silica, and, most remarkably, in supercooled water (see Figure 2.7) [16, 17]. Supercooled liquid water is assumed to exist in two forms, a low-density and a high-density form. The existence of these two alternative structures could result in metastable liquid–liquid separation in supercooled water. The hypothesized

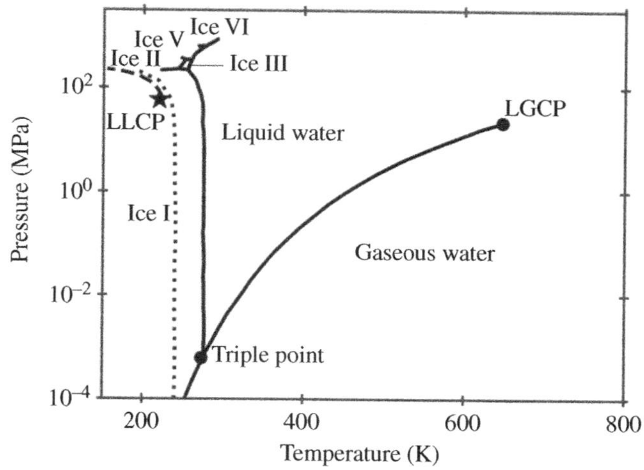

Figure 2.7 Ice polymorphism and the hypothesized polyamorphism in supercooled liquid water. The solid curves indicate phase transitions between solid, liquid, and gaseous states of water. LGCP is the liquid–gas critical point, while LLCP is the hypothesized liquid–liquid critical point (predicted below 100 MPa). The dashed curve shows the predicted liquid–liquid phase transition below the temperatures of spontaneous homogeneous ice nucleation (dotted curve), thus being inaccessible to direct experiments.

liquid–liquid metastable coexistence is not directly accessible to bulk-water experiments because it is presumably located a few degrees below the line of homogeneous ice nucleation. Thus, the experimental verification of the existence of liquid polyamorphism in water still remains as an extremely challenging problem [17].

Liquid polyamorphism is found in simulations of several microscopic models of water. In real supercooled water, the existence of the liquid–liquid critical point is supported by the pronounced increase of the isothermal compressibility, isobaric expansivity, and heat capacity upon supercooling [17]. Most recently, X-ray ultrafast (50–100 fs) pulses were able to detect two alternative liquid phases of liquid water below the temperature of homogeneous crystallization (in what is commonly referred to as "no-man's land") [18]. The existence of liquid polyamorphism in water is also indirectly supported by the discovery of two glassy states of water, a low-density glass and a high-density glass [19], see Figure 2.8. Indeed, one can assume that if the crystallization of these glasses upon melting would not occur, then the coexistence of the glassy states could become the liquid–liquid coexistence upon further heating [20].

Fluid polyamorphism ultimately originates from the complex interactions between molecules or supramolecular structures. However, it may be phenomenologically modeled through the reversible interconversion of two alternative molecular or supramolecular states [21–23]. The application of this "two-state" approach to the variety of polyamorphic substances could be just as a useful phenomenology or it may reflect the true microscopic origin of fluid polyamorphism. Indeed, there are a few substances, such as hydrogen, sulfur, phosphorous, and carbon, where the existence of alternative liquid or dense-fluid states is explicitly induced by a reversible chemical reaction: dimerization in hydrogen [24] or polymerization in sulfur, phosphorus, and carbon [15, 25]. Fluid polyamorphism in hydrogen is illustrated in Figure 2.9. One can see that the fluid–fluid critical point is located in the enormously compressed (with respect to the liquid–gas critical point) supercritical-fluid region.

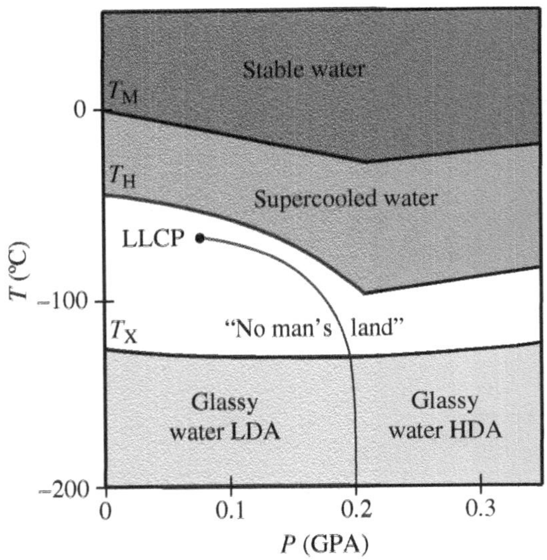

Figure 2.8 Phase diagram of noncrystalline water. Source: Adapted from Ref. [17]. "No man's land" indicates the region in which only crystalline ices have been observed so far. It is enclosed by the homogeneous liquid-crystallization line, T_H, from the top and the glass-crystallization line, T_X, from the bottom. Below T_X, the high-density (HDA) and low-density (LDA) amorphous (glassy) states coexist. The line of this coexistence smoothly continues into "no man's land." LLCP indicates the hypothesized liquid–liquid critical point.

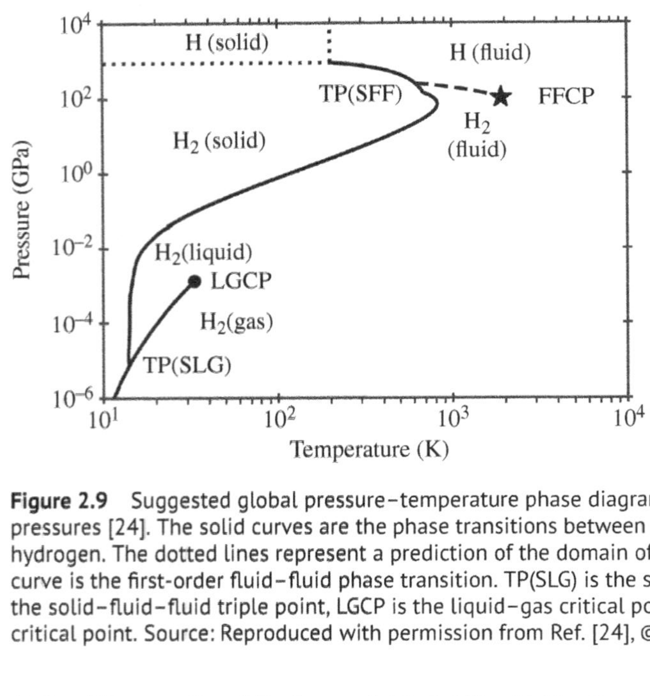

Figure 2.9 Suggested global pressure–temperature phase diagram of hydrogen from low to extreme pressures [24]. The solid curves are the phase transitions between solid, liquid, and gaseous states of hydrogen. The dotted lines represent a prediction of the domain of solid metallic hydrogen. The dashed curve is the first-order fluid–fluid phase transition. TP(SLG) is the solid–liquid–gas triple point, TP(SFF) is the solid–fluid–fluid triple point, LGCP is the liquid–gas critical point, and FFCP is the predicted fluid–fluid critical point. Source: Reproduced with permission from Ref. [24], © AIP Publishing.

2.6 Phase Equilibria and Stability Criteria in Single-Component Substances

While equilibrium conditions are controlled by thermodynamic forces (i.e. the first-order derivatives of thermodynamic potentials), the stability of a thermodynamic system is determined by the second-order derivatives, convexity vs. concavity of appropriate thermodynamic potentials. If the temperature, volume, and number of molecules are selected as independent variables, the stability criteria can be formulated through the second-order derivatives of the Helmholtz energy at constant number of molecules (or moles). The stability criteria of single-component substances are expressed in Eqs. (2.63) and (2.64), in which the isochoric heat capacity per molecule is given by Eq. (2.44).

Stability Criteria for Single-Component Substances

- Mechanical Stability

$$\left.\frac{\partial^2 A_\mathrm{t}}{\partial V_\mathrm{t}^2}\right|_{T,N} = -\left.\frac{\partial P}{\partial V_\mathrm{t}}\right|_{T,N} \geq 0 \tag{2.63}$$

- Thermal Stability

$$-\left.\frac{\partial^2 A_\mathrm{t}}{\partial T^2}\right|_{V,N} = \frac{NC_V}{T} \geq 0 \tag{2.64}$$

Equation (2.63) is known as the "mechanical stability" criterion. The simple meaning of mechanical stability is that compression or expansion at constant temperature and number of molecules should never cause a decrease or increase of density, respectively, such that the isothermal compressibility $\kappa_T = -(1/V_\mathrm{t})(\partial V_\mathrm{t}/\partial P)|_T$ is always positive. The second stability criterion, given by Eq. (2.64), is known as the "thermal stability" criterion, meaning that heating or cooling at constant volume and number of molecules will never cause a decrease or increase of temperature, respectively. It follows from the mechanical and thermal stability conditions that the isobaric heat capacity, defined by Eq. (2.46), cannot be smaller than the isochoric heat capacity because of the thermodynamic relation

$$C_P - C_V = -T\left(\left.\frac{\partial P}{\partial T}\right|_V\right)^2 \left.\frac{\partial V}{\partial P}\right|_T \tag{2.65}$$

The surface in $P - V - T$ space, for which the mechanical stability criterion, Eq. (2.63), becomes an equality ($\partial P/\partial V|_T = 0$), that bounds the mechanical stability limit of a phase is referred to as the "spinodal." Inside the spinodal, the substance is unstable. The two branches of the liquid–gas spinodal (lower, unstable liquid, and upper, unstable gaseous) and the liquid–gas coexistence for a single-component substance is shown in Figure 2.4 by the dashed and solid curves, respectively. The coexistence curve and the two branches of the spinodal merge at the liquid–gas critical point, where the two coexisting phases (liquid and gas) become identical. The critical point is defined by two independent conditions,

$$\left.\frac{\partial P}{\partial V}\right|_T = 0 \quad \text{and} \quad \left.\frac{\partial^2 P}{\partial V^2}\right|_T = 0 \tag{2.66}$$

where the former condition is the same as for the spinodal.

2.6.1 The Principle of Corresponding States

The phase behavior of simple fluids demonstrates a significant degree of universality if it is represented in terms of variables reduced by their critical parameters. This apparent universally, known as the "principle of corresponding states" was originally formulated by van der Waals on the basis of his equation of state (EOS) [27]. Later, Guggenheim[1] analyzed the experimental coexistence

---

1 Edward Armand Guggenheim (1901–1970) was a British physical chemist, distinguished for his important contributions to thermodynamics and statistical mechanics.

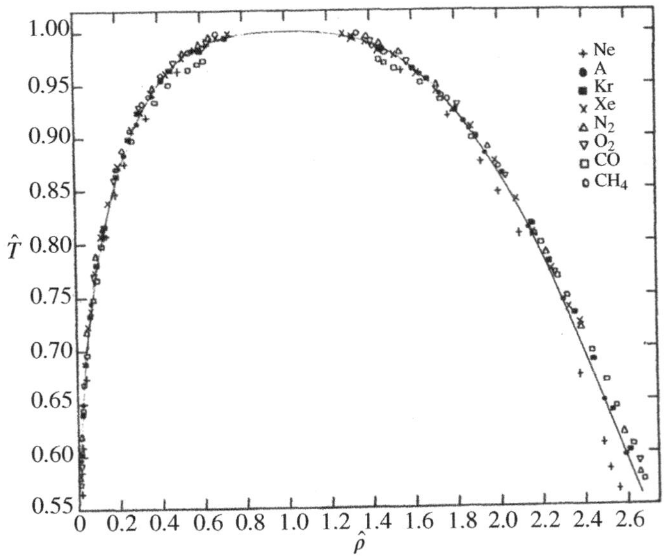

Figure 2.10 Phase behavior of several simple fluids in reduced variables, $\hat{T} = T/T_c$ and $\hat{\rho} = \rho/\rho_c$, where T_c and ρ_c are the critical temperature and critical density, respectively. In the critical region (between $\hat{T} = 0.95$ and $\hat{T} = 1$), the universal coexistence is described by a cubic parabola, $\hat{T} - 1 \sim (\hat{\rho} - 1)^3$. Source: Reproduced with permission from Ref. [26]. © American Institute of Physics.

curves of many fluids and found that the liquid–gas coexistence followed a cubic parabola, in contradiction to the quadratic parabola predicted by the van der Waals EOS.[2] [26]. The difference between these alternative shapes – a cubic parabola being much "flatter" than a quadratic parabola – is of fundamental importance. Much later this shape was explained as being caused by critical fluctuations (discussed in Chapter 8). Figure 2.10 demonstrates the near-universal temperature–density phase behavior of many simple fluids. While the principle of corresponding states is applicable to several groups of similar fluids, such as the noble gases, it is not exact. If, however, the fluids could be described by simple equations of state, with two interaction parameters (such as the van der Waals EOS), then the principle of corresponding states would become exact.

2.6.2 Maxwell's Equal Area Rule

Determining the locations of the domains bounded by the coexistence curve (also referred to as the "binodal") and the spinodal is crucial to understanding the phase behavior of a system. The spinodal can be calculated from the violation of the stability criterion, Eq. (2.63), but the binodal may not necessarily have an analytical expression. It may be calculated via a variety of methods, but in this section, we will focus on the Maxwell construction, referred to as "Maxwell's equal area" rule.

Along two-phase coexistence, the chemical potentials for both phases must be equal, as indicated by the equilibrium conditions expressed in Eq. (2.54). In accordance with the Gibbs–Duhem equation Eq. (2.21), along a selected isotherm, the derivative of the chemical potential is $d\mu = V\,dP$. Consequently, the integral from an equilibrium point in phase $\tilde{\alpha}$ to the point in the coexisting

2 In fact, Guggenheim rediscovered this important behavior of fluids half a century after it was original found by Dutch scientists from the van der Waals' school [27].

phase $\tilde{\beta}$, at the same temperature and pressure, must be zero, as given by

$$\boxed{\int_{\tilde{\alpha}}^{\tilde{\beta}} \mathrm{d}\mu = \int_{\tilde{\alpha}}^{\tilde{\beta}} V\,\mathrm{d}P = 0} \tag{2.67}$$

Consider, for instance, a P–V diagram shown in Figure 2.11. In accordance with the principle of corresponding states, the phase diagram is shown in variables reduced by the critical parameters. The isotherms presented in Figure 2.11 are calculated from the van der Waals EOS, discussed in detail in Section 2.7.1. For a selected isotherm below the critical temperature, $T < T_\mathrm{c}$, Maxwell's equal area rule, Eq. (2.67), assigns the location of the coexisting liquid and gaseous states as indicated by the dashed lines.

Maxwell's equal area rule may also be formulated for the equilibrium condition of equal pressures in the two coexisting phases, at constant temperature and chemical potential, as given by

$$\int_{\tilde{\alpha}}^{\tilde{\beta}} \mathrm{d}P = \int_{\tilde{\alpha}}^{\tilde{\beta}} \rho\,\mathrm{d}\mu = 0 \tag{2.68}$$

The condition, given by Eq. (2.68), is illustrated in Figure 2.12.

The concept of the stable, metastable, and unstable phases may be further illustrated by introducing the excess density of the Helmholtz energy defined as,

$$\Delta(\rho A) = \rho A - \mu_\mathrm{cxc}\rho \tag{2.69}$$

where μ_cxc is the chemical potential along coexistence. Only the behavior of the excess Helmholtz energy is relevant for obtaining fluid-phase equilibrium. As indicated in Figure 2.13, above the critical point, only one state is stable, while below the critical point, there are two stable states, located at the liquid and gas coexistence, with the same value of $\Delta(\rho A)$. This representation will be

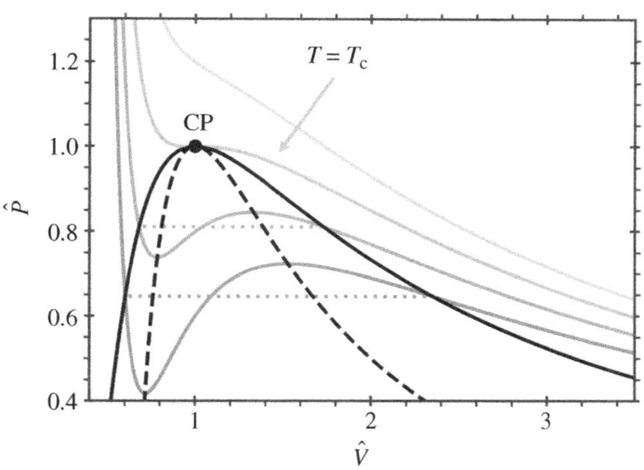

Figure 2.11 Supercritical ($T > T_\mathrm{c}$), critical ($T = T_\mathrm{c}$), and two subcritical ($T < T_\mathrm{c}$) liquid–gas isotherms for the van der Waals equation of state, expressed through reduced variables: $\hat{P} = P/P_\mathrm{c}$, $\hat{T} = T/T_\mathrm{c}$, and $\hat{V} = V/V_\mathrm{c}$, where P_c, T_c, and V_c are the critical pressure, temperature, and molecular volume, respectively. Unstable states are within the two branches of the spinodal (dashed black), while metastable states are between the two branches of both the spinodal and the binodal (solid black). To the left and right of the binodal are stable states where the system is in the liquid or gaseous phase, respectively. Maxwell's equal-area rule for obtaining the volumes of coexisting phases is illustrated by the dashed gray lines, where the temperature increases for lighter colored curves.

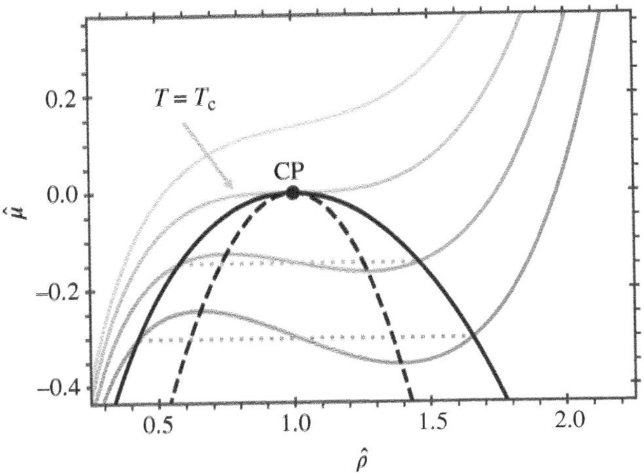

Figure 2.12 Supercritical, critical, and two subcritical liquid–gas isotherms for the van der Waals equation of state (gray curves), expressed through reduced variables: $\hat{\mu} = \mu/k_BT_c$, $\hat{T} = T/T_c$, and $\hat{\rho} = \rho/\rho_c$, where ρ_c is critical molecular density. The spinodal and binodal are indicated by the dashed and solid black curves, respectively. Maxwell's equal-area rule for obtaining the densities of coexisting phases determines the location of the dashed gray lines (increasing in temperature for lighter colored curves). The absolute value of the chemical potential at the critical point is adopted as zero.

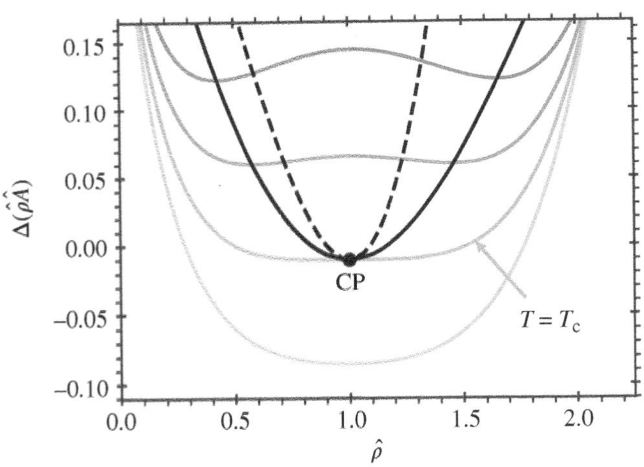

Figure 2.13 The behavior of the excess density of the Helmholtz energy as a function of the density, calculated for the van der Waals EOS and expressed through the reduced variables: $\hat{A} = A/k_BT_c$, $\hat{\rho} = \rho/c$, and $\hat{T} = T/T_c$. The binodal and spinodal are indicated by the solid and dashed black curves, respectively.

further used to connect with fluid-phase equilibria in the general thermodynamic description of phase transitions.

2.6.3 Metastability

Thermodynamic states are stable if the stability criteria are satisfied, but they may not necessarily be in equilibrium if any of the thermodynamic fields are not equal to their equilibrium values. Such states could temporarily exist between the binodal and spinodal, as illustrated in Figures 2.4

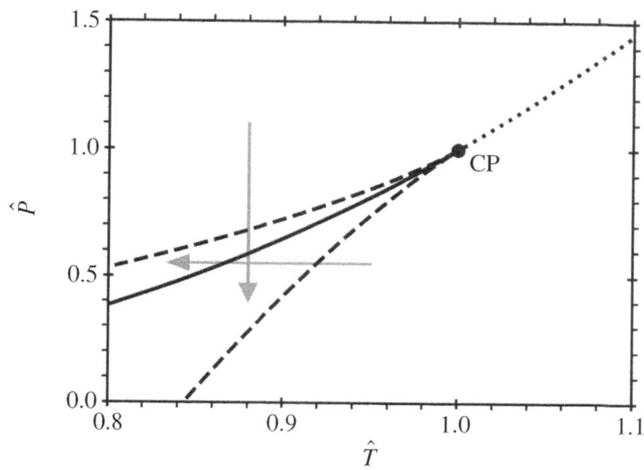

Figure 2.14 Thermodynamic paths (gray arrows: horizontal for supercooling along an isobar and vertical for over-compression along an isotherm) to access metastable states in a van der Waals fluid. The solid curve is the liquid–gas coexistence, the dashed curves are the two branches of the spinodal, and the dotted curve is the critical isochore. The solid black circle represents the location of the critical point (CP).

and 2.11–2.13. These states are designated as "metastable" [9, 28]. Examples of metastable states are liquids supercooled below their equilibrium freezing temperatures, liquids overheated above their equilibrium boiling temperatures, solutions supersaturated above their equilibrium solubility limits, etc. The area of metastability in fluids shrinks as the system approaches its critical state.

The size of the metastable region at any given pressure is determined by the difference between the temperature of the phase coexistence, T_{cxc}, and the temperature of the stability limit (spinodal), T_{s}. There are multiple ways to reach the metastable region. For example, a gas may be supercooled below the condensation point at constant pressure or a liquid may be superheated above the boiling point at constant pressure. Alternatively, a gas may be over-compressed along an isotherm or a liquid may be stretched below the saturation pressure at constant temperature. Along an isochore, the liquid could be supercooled and stretched, while the gas could be supercooled and over-compressed to reach the metastable region. Some of these alternative ways for a fluid to reach a metastable state are illustrated in Figure 2.14.

2.7 Equations of State

The quantitative description of fluid-phase equilibria requires an explicit EOS, $P = P(V, T)$ or $P = P(\rho, T)$. The two most theoretically studied examples are the van der Waals fluid and the lattice-gas model. The reader should be aware that these are only two simple examples of a variety of equations of state that have been used to describe the properties of matter in engineering practice.

2.7.1 Van der Waals Equation and Virial Expansion

The van der Waals EOS accounts for the molecular interactions and qualitatively describes the transition between liquid and gaseous states, as given by

$$P = \frac{Nk_{\text{B}}T}{V_{\text{t}} - Nb} - \frac{N^2a}{V_{\text{t}}^2} \tag{2.70}$$

In this equation, a and b are referred to as the van der Waals constants, which are substance-dependent, representing the attraction and the repulsion between molecules, respectively. In terms of molecular density, the van der Waals equation reads

$$P = \frac{\rho k_B T}{1 - b\rho} - a\rho^2 \tag{2.71}$$

In the limit of an infinitely dilute gas, $\rho \to 0$, the van der Waals equation reduces to the ideal-gas EOS, $P = \rho k_B T$. Expanding the pressure in powers of the parameter $b\rho$ (for a nonideal dilute gas where $b\rho \ll 1$), one obtains

$$P = \rho k_B T(1 + B\rho + C\rho^2 + \cdots) \tag{2.72}$$

In this expansion, known as the "virial expansion," the coefficients B and C are referred to as the "virial coefficients," the second and the third, respectively. Specifically, for the van der Waals EOS,

$$B = b - \frac{a}{k_B T} \tag{2.73}$$

and

$$C = b^2 \tag{2.74}$$

Note that the coefficient b is proportional to the volume of a molecule. One may verify that all of the higher-order virial coefficients starting with C, derived from the van der Waals equation, depend only on the constant b. This feature is associated with the quadratic dependence of the attractive part of the EOS on the density, and, in turn, it corresponds to the approximation that multibody interactions may be represented as a superposition of pair interactions.

The van der Waals constants may be expressed through the critical parameters. From the critical-point conditions, Eq. (2.66), one obtains

$$a = \frac{27}{64}\frac{k_B T_c^2}{P_c} \tag{2.75}$$

and

$$b = \frac{1}{8}\frac{k_B T_c}{P_c} = \frac{1}{3\rho_c} \tag{2.76}$$

In accordance with the principle of corresponding states (Section 2.6.1), reducing the pressure, temperature, and density, by their critical values, one obtains the universal (parameter-free) form of the van der Waals EOS as

$$\hat{P} = \frac{8\hat{T}\hat{\rho}}{3 - \hat{\rho}} - 3\hat{\rho}^2 \tag{2.77}$$

where $\hat{P} = P/P_c$, $\hat{T} = T/T_c$, and $\hat{\rho} = \rho/\rho_c$. Note that among the three critical parameters, for the van der Waals EOS, only two are independent. In fact, any two-parameter EOS may be represented in a universal form by reducing P, T, and ρ by their critical values.

The chemical potential may be obtained via the Legendre transform, $\mu = A + PV$, where $P = -\partial A/\partial V|_T$, such that

$$\mu = PV - \int P\,\mathrm{d}V \tag{2.78}$$

Therefore, for the van der Waals fluid

$$\mu = \frac{k_B T}{1 - b\rho} + k_B T \ln\left(\frac{\rho}{1 - b\rho}\right) - 2a\rho + f(T) \tag{2.79}$$

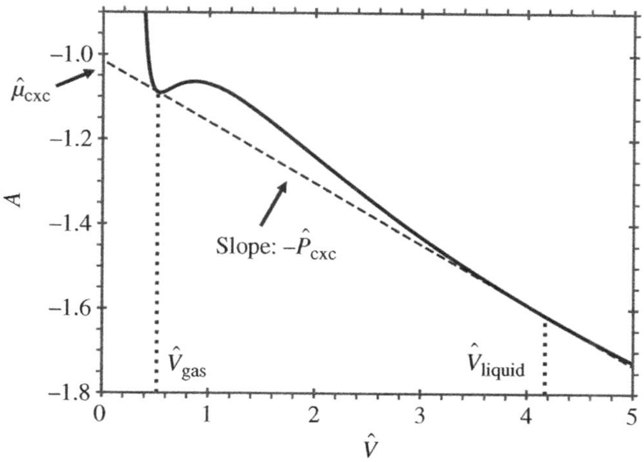

Figure 2.15 Helmholtz energy (per mole) as a function of volume for the van der Waals equation of state at $\hat{T} = 0.8$. The properties are reduced by the critical parameters ($\hat{A} = A/k_BT_c$, $\hat{V} = V/V_c$, $\hat{\mu} = \mu/k_BT_c$, $\hat{P} = P/P_c$, and $\hat{T} = T/T_c$). The intercept and slope of the dashed line connecting the two coexisting states (indicated by dotted lines) are the chemical potential, $\hat{\mu}_{cxc}$, and pressure, $\hat{P}_{cxc}$, at coexistence respectively.

where $f(T)$ is a function of temperature, which represents the contribution from the ideal-gas heat capacity and depends on the selection of the absolute values of the energy and the entropy. Note that this function, which effects the temperature dependence of the chemical potential along a selected isochore, does not effect calculations of fluid-phase equilibria. One may also note that the binodal, obtained from Maxwell's equal area rule, is asymmetric with respect to the critical isochore, see (for instance) Figure 2.11. This asymmetry is reflected in the density dependence of the chemical potential and the asymmetric shape of the phase coexistence.

The behavior of the derivatives of the Helmholtz energy and the density of the Helmholtz energy, the pressure and the chemical potential, for the van der Waals fluid are illustrated in Figures 2.11–2.13. The connection between the chemical potential and the Helmholtz energy at a selected temperature is shown in Figure 2.15.

2.7.2 Lattice Gas Model

Another important model that describes the transition between liquids and gases is the lattice-gas model, which was first introduced by Frenkel[3] in 1931 [29]. This model plays a special role in the physics of condensed matter as it can be applied to very different systems and phenomena, thus bridging the gap between the physics of simple and complex fluids as well as solid-state physics. In the lattice-gas model, the volume of the system is divided into cells of molecular size, ℓ_0. These cells are arranged in a lattice with coordination number z_n, which defines the number of the nearest interacting neighbors ($z_n = 6$ for a simple cubic lattice). In its simplest version, each cell in the lattice gas is either occupied by one molecule or empty (Figure 2.16).

The molecular density in the lattice-gas model is dimensionless, being defined as $\rho = \ell_0^3(N/V_t)$, where N is the number of occupied cells. The nearest-neighboring molecules interact by short ranged attractive forces. Empty cells do not interact with their neighbors.

3 Yakov (Jacov) Il'ich Frenkel (1894–1952) was a Soviet physicist who introduced the lattice-gas model and the concept of excitations in crystals ("holes", identified as "quasiparticles").

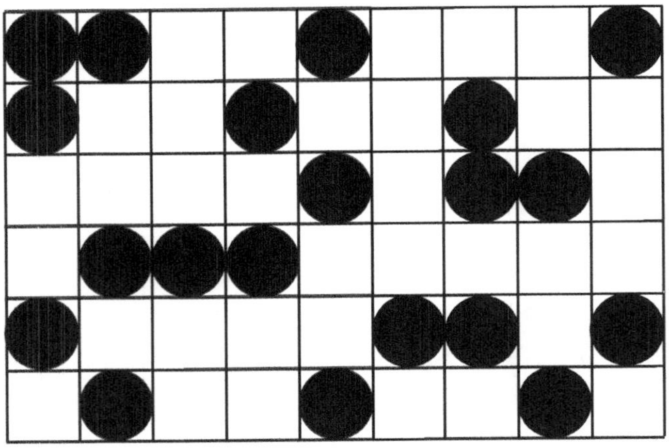

Figure 2.16 A two-dimensional representation of the lattice gas model.

The analytical solution for all of the possible intermolecular configurations of the three-dimensional lattice gas has not been obtained so far, although accurate numerical calculations are available. The simplest approximate analytical solution can be obtained in the so-called "meanfield approximation," in which the energy associated with multibody attractions is represented by a single interaction parameter, ω. This approximation is equivalent to the way by which the van der Waals fluid accounts for the attraction between species through the constant a. Note that the van der Waals constant b, which accounts for the molecular volume and, correspondingly, intermolecular repulsion, is represented in the lattice gas by ℓ_0^3.

The EOS for the lattice gas in the meanfield approximation reads

$$P = -k_B T \ln(1-\rho) - \omega\rho^2 \tag{2.80}$$

The chemical potential is obtained from Eq. (2.78), such that, for the lattice-gas model, it is given as

$$\mu = k_B T \ln\left(\frac{\rho}{1-\rho}\right) + \omega(1-2\rho) + f(T) \tag{2.81}$$

where $f(T)$ is a function of temperature only. The density of the Helmholtz energy is obtained from the Legendre transform, $\rho A = \rho\mu - P$, as

$$\rho A = k_B T\left[\rho\ln\rho + (1-\rho)\ln(1-\rho)\right] + \omega\rho(1-\rho) + \rho f(T) \tag{2.82}$$

One may notice that the excess density of the Helmholtz energy, Eq. (2.69), is symmetric with respect to ρ, such that the condition of the fluid-phase equilibrium (the binodal) in the lattice gas, unlike the van der Waals fluid, can be found from

$$\left.\frac{\partial[\Delta(\rho A)]}{\partial\rho}\right|_T = 0 \tag{2.83}$$

giving an analytic expression for the binodal as

$$T_{cxc} = \frac{\omega(2\rho-1)}{k_B\ln\left[\rho/(1-\rho)\right]} \tag{2.84}$$

Since $-(\partial P/\partial V)|_{N,T} = \rho^2(\partial\mu/\partial\rho)|_{V,T}$ via a Maxwell relation, the spinodal is obtained from the mechanical stability criterion, Eq. (2.63), as $(\partial\mu/\partial\rho)|_T = 0$, such that

$$T_{sp} = 2(\omega/k_B)\rho(1-\rho) \tag{2.85}$$

while the liquid–gas critical parameters are obtained from the critical-point condition, given in Eq. (2.66), as

$$\rho_c = 1/2, \qquad T_c = \omega/2k_B, \qquad P_c = -k_B T_c \left[\ln(1/2) + 1/2\right] \tag{2.86}$$

where $P_c \cong 0.193 k_B T_c$.

The phase behavior of the lattice-gas model is qualitatively similar to that given by the van der Waals model, presented in Figure 2.13. However, unlike the van der Waals fluid, the liquid and gas coexistence of the lattice-gas model is completely symmetric with respect to the critical density. Real fluids do not possess the symmetry of the lattice-gas model. Nonetheless, lattice models are extremely useful computationally and theoretically to reproduce the essential physics of phase transitions in condensed matter.

One may connect the discrete lattice-gas model with the continuous behavior of real fluids described by the van der Waals EOS. A crossover between the symmetric lattice-gas model and the asymmetric van der Waals fluid is given by a "fine-lattice discretization" [31]. This procedure introduces a discretization parameter, $\zeta = \ell/\ell_0 \geq 1$, where ℓ is the diameter of a spherical molecule and ℓ_0 is the lattice spacing. Rescaling the van der Waals EOS by the transformations $\rho' = b\rho$ and $a/b = \omega$, the EOS for this crossover model reads

$$P = \zeta k_B T \ln\left[1 + \frac{\rho'}{\zeta(1-\rho')}\right] - \omega(\rho')^2 \tag{2.87}$$

The lattice-gas limit corresponds to $\zeta = 1$, while at $\zeta \to \infty$, the van der Waals EOS is recovered.

As stated earlier, the lattice gas and the van der Waals equations of state are the two simplest examples of more realistic, but commonly more empirical, equations of state, which in addition to liquid and gaseous phases may describe a solid state. The common phase behavior of a single-component substance is shown in Figure 2.17 in temperature–density coordinates. The fluid state may be distinguished between two substates by the "Zeno line," which for the van der

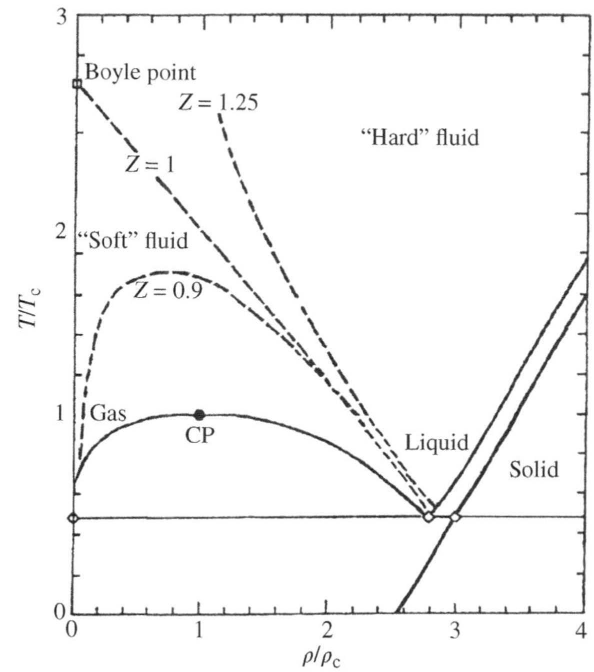

Figure 2.17 Temperature–density phase diagram for methane, presented in reduced units [3]. Coexistence between gas, liquid, and solid phases is shown as solid lines. Compressibility factors, Z, are shown by the dashed curves. For $Z < 1$, the fluid is "soft," while for $Z > 1$, the fluid is "hard." The curve for $Z = 1$ is the Zeno line. The Zeno line goes from the Boyle Point (the extrapolation of the Zeno line to zero density) to the density of liquid at the triple point. At the triple point, three phases, solid, liquid, and gas, coexist with different densities (shown by open circles) at the same temperature and pressure. Source: Adapted from Ref. [33].

Waals EOS, is given by,

$$Z \equiv \frac{P}{\rho k_B T} = \frac{1}{1-b\rho} - \frac{a\rho}{k_B T} = 1 \tag{2.88}$$

where Z is known as the "compressibility factor." The Zeno line ($Z = 1$) separates a "hard" fluid ($Z > 1$), where the intermolecular repulsion between molecules dominates, from a "soft" fluid ($Z < 1$), where the intermolecular attractions dominate. Only hard fluids may exhibit crystallization (phase transitions to a solid state). Remarkably, the Zeno line for simple fluids could be realistically represented by a linear function of density predicted by the van der Waals EOS. At the critical point, the compressibility factor of a van der Waals fluid is $Z_c = 0.375$, while for methane $Z_c = 0.286$.

2.8 Thermodynamics of Mixtures

Until now, we have considered only single-component systems. In this section, the basics of classical thermodynamics will be extended to binary mixtures and, in brief, ternary mixtures.

2.8.1 Chemical Potentials and Osmotic Pressure

Typically, in a binary solution, component 1 represents the solvent, while component 2 represents the solute. However, this definition is often conventional, depending on convenience. Throughout this book, we define the molecular (mole) fraction, x, through component 2 as $x \equiv x_2$, while $1 - x \equiv x_1$ (due to the conservation of the number of molecules).

The total Gibbs energy, at constant temperature and pressure, for a binary solution is

$$G_t = \mu_1 N_1 + \mu_2 N_2 \tag{2.89}$$

where μ_1 and μ_2 are the chemical potentials of the solvent and solute, respectively. If the total number of molecules is constant, the Gibbs energy per molecule and its differential are given by

$$G = \frac{G_t}{N} = \mu_1(1-x) + \mu_2 x = \mu_1 + \mu_{21} x \tag{2.90}$$

$$dG = S\,dT + V\,dP + \mu_{21}\,dx \tag{2.91}$$

where $\mu_{21} = \mu_2 - \mu_1$ is referred to as the "exchange chemical potential." This field variable plays an important role in the thermodynamics of binary mixtures as it is conjugate to the molecular fraction, x. The chemical potentials μ_1 and μ_2 are given through derivatives of the Gibbs energy as

$$\mu_1 = G - x\left.\frac{\partial G}{\partial x}\right|_{T,P} \tag{2.92}$$

$$\mu_2 = G + (1-x)\left.\frac{\partial G}{\partial x}\right|_{T,P} \tag{2.93}$$

The Gibbs energy of a binary mixture can be expressed as

$$G = \mu_1^\circ(1-x) + \mu_2^\circ x + \Delta G_{mix} = \mu_1^\circ + \mu_{21}^\circ x + \Delta G_{mix} \tag{2.94}$$

where $\mu_1^\circ = \mu_1(T, P, x = 0)$ and $\mu_2^\circ = \mu_2(T, P, x = 1)$ are the chemical potentials of pure solvent and solute, respectively, while ΔG_{mix} is the Gibbs energy of mixing, and may be represented as, $\Delta G_{mix} = \Delta H_{mix} - T\Delta S_{mix}$, with contributions from the enthalpy and entropy of mixing, respectfully.

Molecular Fraction and Partial Density

- Molecular fraction, x

$$\begin{aligned} x \equiv x_2 &= \frac{N_2}{N_1 + N_2} = \frac{N_2}{N} = \frac{n_2}{n} \\ 1 - x &= x_1 = \frac{N_1}{N} = \frac{n_1}{n} \end{aligned} \tag{2.95}$$

where $x_1 + x_2 = 1$.

- Partial density ("molar concentration")

$$\begin{aligned} c \equiv \rho_2 &= \rho x = \tfrac{N_2}{V_t} \\ \rho - c &\equiv \rho_1 = \tfrac{N_1}{V_t} = \rho(1 - x) \end{aligned} \tag{2.96}$$

where $\rho_1 + \rho_2 = \rho$.

For a binary system with a constant total number of molecules, the Gibbs–Duhem relation, Eq. (2.21), becomes

$$\rho x \,\mathrm{d}\mu_{21} + \rho \,\mathrm{d}\mu_1 + \rho S \,\mathrm{d}T - \mathrm{d}P = 0 \tag{2.97}$$

Hence, the mole fraction and the molar concentration ($c \equiv \rho x$) are obtained as

$$x = -\left.\frac{\partial \mu_1}{\partial \mu_{21}}\right|_{T,P} \quad \text{and} \quad c = \left.\frac{\partial P}{\partial \mu_{21}}\right|_{T,\mu_1} \tag{2.98}$$

If the solute is separated from the solvent by a semipermeable membrane, the phenomenon of osmosis arises when the solvent molecules can diffuse through the membrane, while the solute cannot. This process is of vital importance in biology as the membranes of biological cells are semipermeable. As a result of osmosis, an osmotic pressure is established, which is defined as

$$\Pi = \rho[\mu_1^\circ(T, P) - \mu_1(T, P, x)] \tag{2.99}$$

Note that, due to the Gibbs–Duhem relation,

$$\left.\frac{\partial \mu_{21}}{\partial c}\right|_{T,P} = \left.\frac{1}{c}\frac{\partial \Pi}{\partial c}\right|_{T,P} = \frac{1}{c^2 \kappa_\Pi} \tag{2.100}$$

The property $\kappa_\Pi = (1/c)(\partial c/\partial \Pi)|_{P,T}$ is known as the "osmotic compressibility."

The chemical potentials of the components in a nonideal binary mixture are represented as

$$\mu_1 = \mu_1^\circ + k_B T \left[\ln(1 - x) + \ln \gamma_1\right] \tag{2.101}$$

$$\mu_2 = \mu_2^\circ + k_B T \left[\ln x + \ln \gamma_2\right] \tag{2.102}$$

The activity coefficients of the solvent and solute, $\gamma_1(T, P, x)$ and $\gamma_2(T, P, x)$ respectively, represent the nonideality of the solution. In an ideal solution ($\gamma_1 = \gamma_2 = 1$), there is no contribution to the Gibbs energy of mixing from the heat of mixing ($\Delta H_{\text{mix}} = 0$), and the entropy of mixing is represented by

$$\Delta S_{\text{mix}} = k_B T \left[x \ln x + (1 - x) \ln(1 - x)\right] \tag{2.103}$$

In a nonideal solution, the connection between the activity coefficients, at constant P and T, is constrained by a version of the Gibbs–Duhem relation:

$$x \,\mathrm{d} \ln \gamma_2 + (1 - x) \mathrm{d} \ln \gamma_1 = 0 \tag{2.104}$$

This relation is commonly used to check on the thermodynamic consistency of the experiments performed on binary liquid solutions.

2.8.2 Thermodynamic Stability of Mixtures

For the general case of a multicomponent system, additional stability criteria (known as the "mixture stability" or "material stability"), are expressed in a matrix form, which contains the second-order derivatives of the Helmholtz energy with respect to the number of molecules of each species [3, 32]. Consider, for example, a binary mixture of two species. Typically, in these solutions, the mixture may become unstable before the mechanical stability condition, $(\partial P/\partial V_t)|_{T,x} \leq 0$, is violated; meaning that, the "mechanical spinodal" is generally located inside the mixture stability limit.

Krichevskiĭ Parameter

The thermodynamic parameter $\mathcal{K}_0$ that controls the behavior of dilute supercritical-fluid solutions is the "Krichevskiĭ parameter," defined as

$$\mathcal{K}_0 \equiv \lim_{x\to 0} \left.\frac{\partial P}{\partial x}\right|_{T,V} = \frac{\mathrm{d}T_c}{\mathrm{d}x_c}\left[\frac{\mathrm{d}P_c}{\mathrm{d}T_c} - \left.\frac{\partial P}{\partial T}\right|_{V,x_c\to 0}\right] \tag{2.105}$$

where the derivatives $\mathrm{d}T_c/\mathrm{d}x_c$ and $\mathrm{d}P_c/\mathrm{d}T_c$ are evaluated along the critical locus. The high selectivity of supercritical fluid solvents due to small variations of temperature and pressure is largely determined by the Krichevskiĭ parameter. Depending on the magnitude and the sign of the Krichevskiĭ parameter, the critical lines emanate from the liquid–gas critical loci of pure solvent into different directions.

For the case of a binary mixture with a fixed total number of molecules ($N = N_1 + N_2 =$ constant), the mixture stability condition is given by

$$\boxed{\left.\frac{\partial^2 A}{\partial x^2}\right|_{V,T} \left.\frac{\partial^2 A}{\partial V^2}\right|_{T,x} - \left(\frac{\partial^2 A}{\partial x \partial V}\right)_T^2 \geq 0} \tag{2.106}$$

Using a Maxwell relation,

$$\left.\frac{\partial^2 A}{\partial x \partial V}\right|_T = \left.\frac{\partial \mu_{21}}{\partial V}\right|_{T,x} = -\left.\frac{\partial P}{\partial x}\right|_{T,V} \tag{2.107}$$

one can also represent the mixture stability condition in terms of first derivatives of the exchange chemical potential, $\mu_{21} = \mu_2 - \mu_1 = (\partial G/\partial x)|_{P,T}$ and pressure, as

$$\boxed{-\left.\frac{\partial \mu_{21}}{\partial x}\right|_{V,T} \left.\frac{\partial P}{\partial V}\right|_{T,x} - \left(\frac{\partial P}{\partial x}\right)_{T,V}^2 \geq 0} \tag{2.108}$$

Furthermore, this condition can be represented either as

$$\left.\frac{\partial^2 G}{\partial x^2}\right|_{P,T} = \left.\frac{\partial \mu_{21}}{\partial x}\right|_{P,T} = \left.\frac{\partial \mu_{21}}{\partial x}\right|_{V,T} + \left(\left.\frac{\partial P}{\partial x}\right|_{T,V}\right)^2 \left.\frac{\partial V}{\partial P}\right|_{T,x} \geq 0 \tag{2.109}$$

or

$$\left.\frac{\partial^2 (A - \mu_{21}x)}{\partial V^2}\right|_{T,\mu_{21}} = -\left.\frac{\partial P}{\partial V}\right|_{T,\mu_{21}} = -\left.\frac{\partial P}{\partial V}\right|_{T,x} - \left(\left.\frac{\partial P}{\partial x}\right|_{T,V}\right)^2 \left.\frac{\partial x}{\partial \mu_{21}}\right|_{V,T} \geq 0 \tag{2.110}$$

where $A - \mu_{21}x$ is the Legendre transform of the Helmholtz energy from $A = A(T, V, x)$ to being a function of T, V, and μ_{21}.

Both the phase behavior for fluid–gas phase equilibria and fluid–solid phase equilibria are strongly affected by nonideality and, consequently, by the mutual solubilities of the components in the fluid and solid phases. A comprehensive discussion on fluid phase equilibria in binary and ternary mixtures with many examples can be found in a monograph of Rowlinson and Swinton [33]. Below, only a few characteristic phase diagrams for liquid–gas and solid–liquid equilibria in binary mixtures are presented.

While both stability criteria, Eqs. (2.106) and (2.108), are thermodynamically equivalent, the mixture stability criterion, given by Eq. (2.108), is especially convenient to represent thermodynamic stability of dilute solutions, since it smoothly transforms into the material stability criterion, given by Eq. (2.63), in the limit $x \to 0$ as in this limit $(\partial x/\partial \mu_{21})|_{V,T} \to x/k_BT$. In Eq. (2.109) for the mixture stability criterion, the term $(\partial P/\partial x|_{T,V})^2(\partial V/\partial P)_{T,x}$ diverges in the limit $x \to 0$ compensating the divergence of $(\partial \mu_{21}/\partial x)|_{T,V}$. Therefore, the property $(\partial P/\partial x)|_{T,V}$ plays an important role in the relation between the mixture and mechanical stability criteria. In an infinitely dilute solution, this property, $\mathcal{K}_0 \equiv \lim_{x\to 0}(\partial P/\partial x)|_{T,V}$ was coined as the "Krichevskiĭ parameter"[4] by Johanna Levelt Sengers[5] [34, 35] (see more information in the inset above). Note that in the special case of an azeotropic mixture, $(\partial P/\partial x)|_{T,V} = 0$, such that the criteria given by Eqs. (2.63) and (2.108) become equivalent. In these two specific cases, the mechanical and the mixture stability criteria coincide.

Critical Dilute-Solution Dilemma

In a dilute solution, the thermodynamic expression for the derivative of the exchange chemical potential with respect to concentration reads

$$\left.\frac{\partial \mu_{21}}{\partial x}\right|_{P,T} = \frac{k_BT}{x} + \frac{(\partial P/\partial x|_{V,T})^2}{(\partial P/\partial V)|_{T,x}} = 0 \tag{2.111}$$

In a dilute solution, the chemical potential of the solute is dominated by the logarithm of concentration, as given by

$$\mu_{21} \approx k_BT\ln x \to \lim_{x\to 0}\left.\frac{\partial \mu_{21}}{\partial x}\right|_{P,T} = \frac{k_BT}{x} \to \infty \tag{2.112}$$

However, at the mixture critical point $(\partial \mu_{21}/\partial x)|_{P,T} = 0$, which for an infinitely dilute-solution appears to contradict the above condition. This is the so-called "critical dilute-solution dilemma." This dilemma was resolved by Krichevskiĭ, who showed that the critical-mixture condition and the dilute-solution limit can be reconciled by considering the divergence of the isothermal compressibility at the critical point of the solvent. The inverse isothermal compressibility vanishes at the critical point of the pure solvent as

$$\left.\frac{\partial P}{\partial V}\right|_{T,x} \to -\frac{\mathcal{K}_0^2}{\rho_c k_BT}x \tag{2.113}$$

where $\mathcal{K}_0 \equiv \lim_{x\to 0}(\partial P/\partial x)_{V,T}$ is the Krichevskiĭ Parameter. Thus, in an infinitely-dilute solution, the two stability conditions: mechanical, given by Eq. (2.109), and mixture, given by Eq. (2.110), become equivalent [34, 35].

The inequalities given by Eqs. (2.106) and (2.108) become equalities at the mixture spinodal, which separates the stable states of the mixture from the unstable states. Along the binary-mixture

4 Issak Ruvimovich Krichevskiĭ (1901–1993) was a Soviet and Russian chemist. He made fundamental contributions to thermodynamics of fluid mixtures, in particular, fluid–fluid separation and critical phenomena at high pressures [27].

5 Johanna Maria Henrica (Anneke) Levelt Sengers (1929–2024) was a Dutch and American physicist who made outstanding contributions to studies of the critical states of fluids and fluid mixtures.

spinodal

$$\left.\frac{\partial \mu_{21}}{\partial x}\right|_{P,T} = 0 \tag{2.114}$$

or, equivalently,

$$\left.\frac{\partial P}{\partial V}\right|_{T,\mu_{21}} = 0 \tag{2.115}$$

If, in addition,

$$\left.\frac{\partial^2 \mu_{21}}{\partial x^2}\right|_{P,T} = 0 \tag{2.116}$$

or, equivalently,

$$\left.\frac{\partial^2 P}{\partial V^2}\right|_{T,\mu_{21}} = 0 \tag{2.117}$$

the mixture stability criteria, given by Eqs. (2.109) and (2.114) define the critical points of phase equilibria (liquid–gas, liquid–liquid, or solid–solid) in binary mixtures.

It may be seen from the definitions of the chemical potentials of each species, Eqs. (2.101) and (2.102), that in the liquid phase of an ideal solution, $\gamma_1 = \gamma_2 = 1$, $\mu_1^\circ = \mu_1^\circ(P,T)$, and $\mu_2^\circ = \mu_2^\circ(P,T)$, the mixture stability criterion, given by Eq. (2.109), is always satisfied, such that no liquid–liquid separation can occur. Similarly, the mixture stability criterion given by Eq. (2.110) is always satisfied in an ideal-gas mixture (a particular case of an ideal solution), such that any fluid–fluid separation cannot occur.

The binary-mixture spinodal merges with the fluid–fluid binodal (fluid–fluid coexistence) at the mixture critical point. Phase coexistence in binary fluid mixtures could be either liquid–gas, liquid–liquid, or gas–gas terminating at the liquid–gas, liquid–liquid, or supercritical fluid–fluid ("gas-gas") critical points, respectively. These critical points, according to the Gibbs phase rule, given by Eq. (2.61), have one degree of freedom (unlike the invariant critical points of each individual component). This means that the critical points in a binary fluid form a critical locus in P–T–x (or P–V–x) space. We note that, the critical point conditions, given by Eqs. (2.114) and (2.116), are more suitable to represent liquid–liquid stability, while the conditions, given by Eqs. (2.115) and (2.117), are more natural to represent liquid–gas stability in binary fluids. Since the locus of the liquid–gas critical points emanates from the liquid–gas critical point of each pure component, the conditions of the latter case smoothly transform into those given by Eq. (2.66) for the single-component limit.

Liquid–gas equilibria in binary mixtures are terminated at the liquid–gas critical locus (LGCL), which terminates at the liquid–gas critical points of the pure components. If the nonideality of the mixture is weak and the critical parameters of the components are not significantly different, the critical locus is often a continuous line that connects the critical points of the two pure components. If the nonideality is strong, the critical locus can be interrupted at a so-called "critical end point," while another branch can continue to a higher pressure. A classification of the possible shapes of critical loci in binary fluids was first suggested by Van Konynenburg and Scott [36, 40] (presented in Figure 2.18) and is discussed in detail by Rowlinson and Swinton [33].

Binary fluids belonging to Type I critical behavior do not exhibit liquid–liquid phase separation, while their LGCL smoothly connects the component's critical points. The simplest example of a Type I system is a mixture of two isotopes forming an ideal solution. The difference in the liquid–gas critical points of isotopes is a purely quantum effect, proportional to $\hbar^2(1/m_2 - 1/m_1)$, where $\hbar$

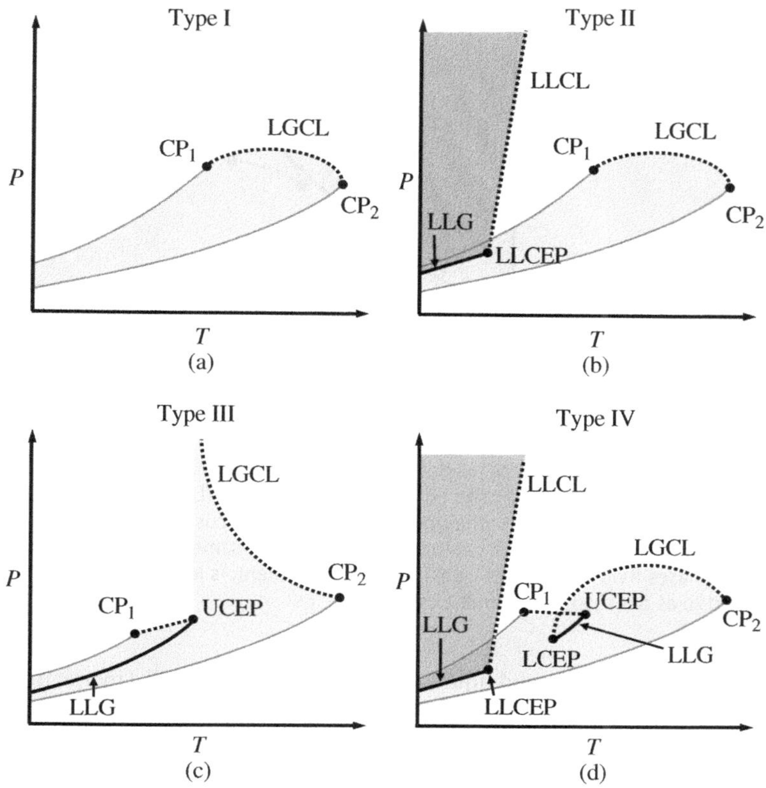

Figure 2.18 Four types of critical loci and phase behavior of binary fluid mixtures according to the classification of Van Konynenburg and Scott [36]. CP_1 and CP_2 are the liquid–gas critical points of individual species, LGCL is the liquid–gas critical locus, LLCL is the liquid–liquid critical locus, LLG is the liquid–liquid–gas triple line, LLCEP is the liquid–liquid critical end point, LCEP is the lower liquid–gas critical end point, and UCEP is the upper liquid–gas critical end point. Liquid vapor coexistence is indicated by the light gray shaded area, while liquid–liquid coexistence is indicated by the dark gray shaded area. Note that in Type III behavior, at the UCEP temperature, the two fluid states do not mix for any pressure, and the corresponding LGCL is sometimes referred to as the "gas-gas critical locus" at very high pressures. In Type IV behavior, the LLG line between the LCEP and UCEP indicates three-fluid phase coexistence.

is Plank's constant and m_1 and m_2 are the molecular masses of the components [41, 42]. Consequently, along the entire LGCL, $\mathcal{K}_0^2 = (\partial P/\partial x)^2_{V,T} \propto \hbar^4(1/m_2 - 1/m_1)^2$, while the inverse isothermal compressibility along the critical locus behaves linearly with respect to the concentration as

$$\kappa_T^{-1} \propto x(1-x)\hbar^4\left(\frac{1}{m_2} - \frac{1}{m_1}\right)^2 \tag{2.118}$$

Other examples of Type I systems include mixtures of noble gases, isomers of hydrocarbons, and mixtures of hydrocarbons with similar molecular sizes. As the molecules of each species become more dissimilar, the liquid–gas critical locus becomes increasingly concave, potentially becoming discontinuous.

The particular case of Type II critical behavior is often observed in nonideal binary mixtures. Two special cases of Type II critical behavior, depending on the interplay between the intermolecular interactions, are illustrated in Figure 2.19, with the scenario where the LGCL contains a critical

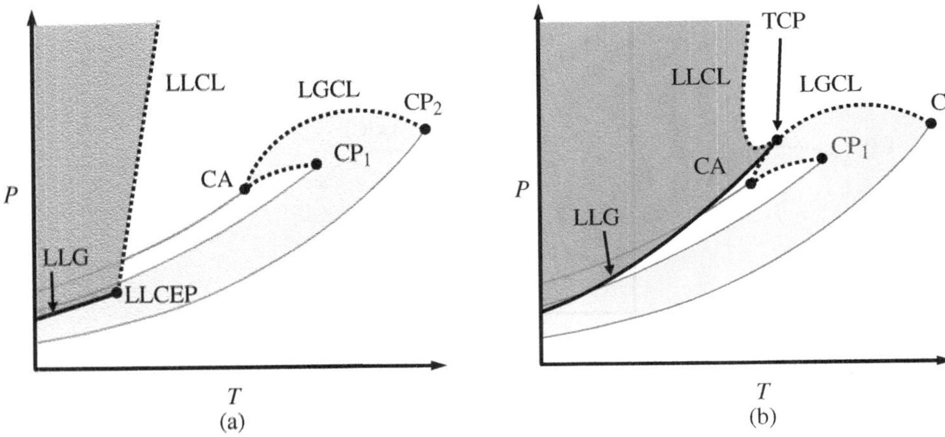

Figure 2.19 Two special cases of Type II critical behavior (as defined in Figure 2.18b) that may emerge depending on the intermolecular interactions [22, 37]. (a) The liquid–gas critical locus (LGCL) contains a critical azeotrope (CA), where $dP_c/dx_c = 0$. (b) Three lines: the liquid–liquid critical locus (LLCL), liquid–liquid–gas (LLG) triple line, and the liquid–gas critical locus merge with the same slope at a single point. This point, when three fluid phases (two liquids and a gas) become equivalent, is known as a "fluid tricritical point" (TCP), also referred to as an "asymmetric tricritical point".

azeotrope (a) and the scenario where the LGCL and liquid–liquid critical locus (LLCL) merge in a tricritical point (three fluid phases become identical) (b), are shown.

The temperature-concentration phase diagram of a nonideal mixture of two miscible liquids for two isobars of a system exhibiting Type II critical behavior is illustrated in Figure 2.20. Figure 2.20a

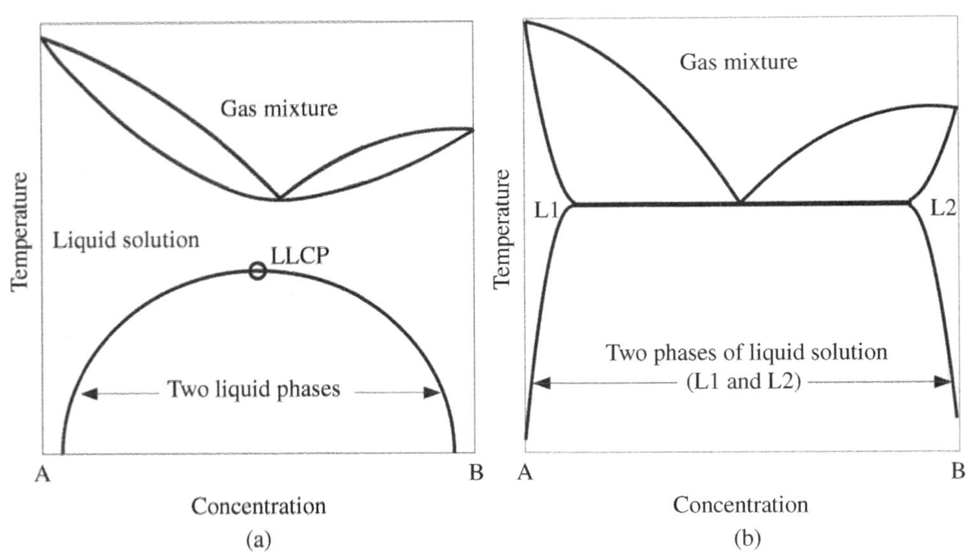

Figure 2.20 Liquid–gas and liquid–liquid equilibria in a binary solution exhibiting Type II behavior with partial solubility of the components along two isobars: above and below the LLCEP (see Figure 2.19a). (a) Complete mutual solubility above the LLCP and with a liquid–gas azeotrope (minimum boiling temperature). (b) Partial (incomplete) mutual solubility of species in the liquid state. If the liquid species are completely immiscible, the points L1 and L2 would move to the vertical axis.

depicts the case when the isobar is above the liquid–liquid critical end point (LLCEP), while Figure 2.20b illustrates the case when the isobar is below the LLCEP. In the latter case, two liquids and a gas coexist at the triple point.

A notable example of a Type III system with immiscible components in a certain range of concentrations is the mixture of carbon dioxide and water. The high-pressure fluid–fluid phase behavior of such systems is sometimes referred to as "gas-gas equilibrium" [27]. Type IV and a variant of Type IV, without independent liquid–liquid separation, designated as Type V, are remarkable because of the possibility to tune the gap between the LCEP (where liquid–liquid critical state coexists with gaseous noncritical state) and UCEP (where liquid–gas critical state coexists with liquid noncritical state) by adding a smoothing impurity. When the LCEP and the UCEP merge, the system reaches a tricritical point where three fluid phases become identical [33].

Solid solutions may also exhibit phase separation, as illustrated by the phase diagram shown in Figure 2.21. If the species are completely immiscible in the solid state, while being miscible in the liquid state, then the three phases, two solids (consisting of pure species) and liquid solution, coexist at the so-called "eutectic point," see Figure 2.21a. If the species are only partially immiscible in the solid state, the coexistence between the two solid phases may be terminated at a solid–solid critical demixing point (Figure 2.21b).

Phase separation in solid solutions could also originate from the solubility of gases in solids. An example, being an important application to hydrogen storage, is a solution of atomistic hydrogen in palladium [43]. Hydrogen molecules dissociate on the surface of palladium and the atoms easily penetrate into the palladium solid matrix. The solid phase separation of this solid solution occurring below room temperature is shown in Figure 2.22. The equilibrium and stability conditions for the solid mixtures are the same as for fluid mixtures. The algorithm to calculate the phase equilibria in fluid and solid mixtures is presented in Appendix 2.A.

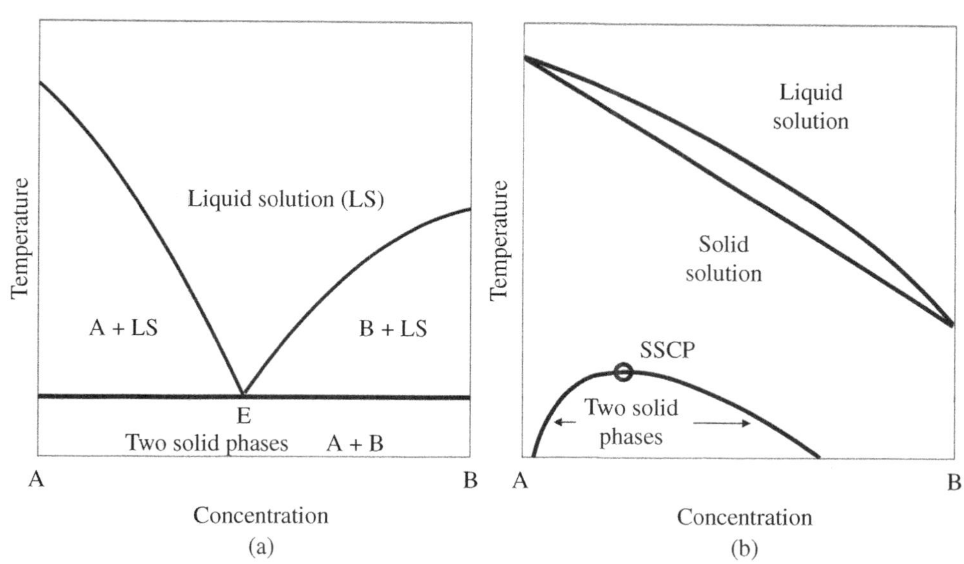

Figure 2.21 Solid–liquid and solid–solid equilibria in binary solutions. (a) Species are completely immiscible as solids but are completely miscible in the liquid state. The eutectic point (E) is where the three phases, liquid solution and two solids, coexist. (b) Species partially immiscible in the solid state, while being completely miscible in the liquid state.

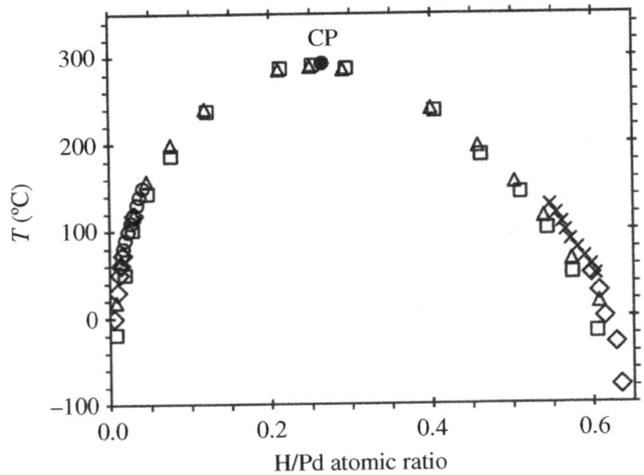

Figure 2.22 Solid–solid immiscibility in solutions of atomistic hydrogen in palladium [43]. CP designates the location of the critical point of solid–solid phase equilibria. Various symbols are experimental data from different sources. Source: Adapted from Ref. [43].

2.8.3 Regular Solution Model

The lattice-gas model could also be reformulated to describe phase separation in a symmetric, nearly incompressible, liquid solution (referred to as a "regular solution"). The regular-solution may be modeled through a weakly-compressible lattice, and in this case, the lattice cells are occupied by species 1 or species 2, such that the role of the density in the lattice-gas model is played by the molecular fraction in the regular-solution model. The weak compressibility of the lattice is accounted for by the pressure dependence of the lattice size, $\ell = \ell(P)$. Therefore, the Gibbs energy per molecule, as given for a binary mixture by Eq. (2.94), in the regular-solution model is expressed as

$$G = \mu_1^\circ + \mu_{21}^\circ x + k_B T\,[x \ln x + (1-x)\ln(1-x)] + \omega x(1-x) \tag{2.119}$$

where the term in square brackets represents the entropy of mixing, as given by Eq. (2.103), and the last term represents the enthalpy of mixing.

In this model, nonideality originates from the enthalpy of mixing, assumed to be independent of temperature; thus, the nonideality parameter, ω, is a function of pressure only. The weak compressibility in the regular-solution model is accounted for by the pressure dependence of $\mu_1^\circ = \mu_1^\circ(T,P)$, $\mu_{21}^\circ = \mu_{21}^\circ(T,P)$, and $\omega = \omega(P)$ in the Gibbs energy. The nonideality parameter, ω, has a simple physical meaning. It depends on the intermolecular interactions between species: ϵ_{11}, ϵ_{22}, and $\epsilon_{12} = \epsilon_{21}$, as $\omega = (z_n/2)(2\epsilon_{12} - \epsilon_{11} - \epsilon_{22})$, where z_n is the coordination number. Phase separation occurs only if $2\epsilon_{12} - \epsilon_{11} - \epsilon_{22} > 0$; meaning that, unlike species effectively repel each other.

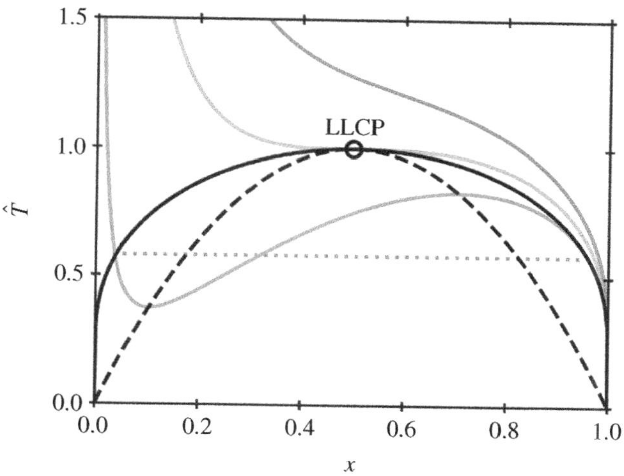

Figure 2.23 Temperature-molecular fraction phase diagram for the regular-solution model, in units of reduced temperature ($\hat{T} = T/T_c$). The binodal and spinodal are indicated by the solid and dashed black curves, respectively. The gray curves are lines of constant reduced exchange chemical potential, $\hat{\mu}_{21} = \mu_{21}/k_B T_c$. In physical-chemistry practice, the curves of constant chemical potential are obtained from the partial pressures of the components in the gaseous phase (assumed to be ideal gases).

The exchange chemical-potential is given by the first derivative of the Gibbs energy, $\mu_{21} = (\partial G/\partial x)|_{T,P}$, as

$$\mu_{21} = \mu_{21}^{\circ} + k_B T \ln\left(\frac{x}{1-x}\right) + (1 - 2\omega)x \tag{2.120}$$

In this symmetric model, the liquid–liquid coexistence (binodal) is obtained from the condition, $\mu_{21} - \mu_{21}^{\circ} = 0$, and is described by Eq. (2.84) with ρ replaced by x. Liquid–liquid equilibria in the regular-solution model is illustrated in Figure 2.23. The limit of the mixture stability (spinodal) is obtained from Eq. (2.114). Applying the critical point conditions, given by Eqs. (2.114) and (2.116), the critical temperature of liquid–liquid demixing is $T_c = \omega/2k_B$ and the critical molecular fraction is $x_c = 1/2$.

Note that Eq. (2.119) for the Gibbs energy in the regular-solution model is mathematically equivalent to Eq. (2.82) for the density of the Helmholtz energy in the lattice-gas model of a single-component substance, replacing the density, ρ, by the molecular fraction, x. Consequently, the equations for the binodal and spinodal in the regular-solution model are mathematically equivalent to those of the lattice-gas model, Eqs. (2.84) and (2.85). In fact, the lattice-gas model becomes mathematically equivalent to the incompressible regular-solution model if the lattice cells are fully occupied by two different species. This similarity in the description of physically different systems, if their thermodynamic behavior is represented by appropriate thermodynamic coordinates (such as the behavior of single-component substances and binary-fluid mixtures), will further be referred to as "isomorphism."

Beyond the Regular-Solution Model

Applications of the regular-solution model are limited to a relatively narrow case of binary mixtures where the molecules of each species have simple van der Waals interactions and similar molecular sizes. Such systems could exhibit only an upper LLCP, as shown in Figure 2.20a, e.g. mixtures of nitroethane and low-molecular-weight hydrocarbons. However, most mixtures demonstrate significant deviations from the predictions of the regular-solution model, such as fluid-phase asymmetry and the possibility of liquid–liquid separation upon heating (with a lower LLCP) or even closed-loop liquid–liquid coexistence (with a lower and an upper LLCP).

Phenomenologically, these deviations could be described by introducing temperature and molecular fraction dependence into the nonideality interaction parameter, $\omega = \omega(P, T, x)$, in Eq. (2.119), such that mixing may be asymmetric and the entropy of mixing would be nonideal. If the nonideal part of the entropy of mixing is negative and sufficient enough to overcome the positive contribution to the enthalpy of mixing, binary liquids may separate upon heating and exhibit a lower LLCP or closed loop demixing with both upper and lower critical points.

A particular limiting case, when $\omega = \omega_0 T$ (with ω_0 being a constant), such that the enthalpy of mixing (given by Eq. 2.29) is zero, and the Gibbs energy of mixing contains only the entropy of mixing, is known as an "athermal solution." In this case, phase separation may occur at any temperature (when $\omega_0 \geq 2k_B$) due to the nonideal entropy of mixing.

Additionally, in many binary fluids, their exists both liquid–gas and liquid–liquid phase coexistence, exhibiting Type II–Type IV behavior (see Figure 2.18). Qualitatively, this behavior could be modeled either by a mixture of two van der Waals fluids or by a compressible binary lattice, in which the cells are occupied by two species or unoccupied, see Figure 2.24.

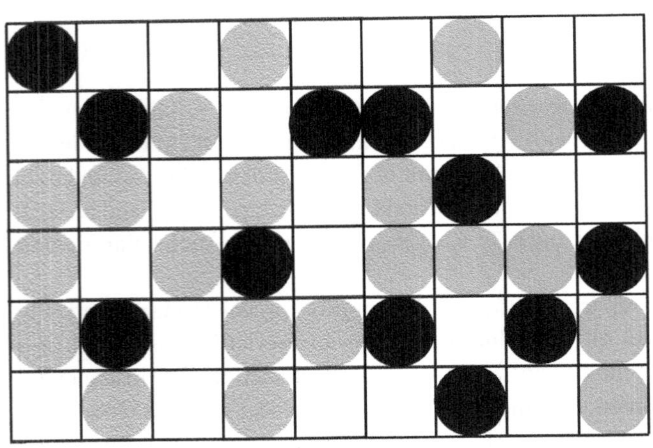

Figure 2.24 A two-dimensional representation of the compressible binary lattice model, where gray and black circles indicate the two species.

It follows from Eq. (2.92) that μ_1 is given by

$$\mu_1 = \mu_1^\circ + k_B T\left[\ln(1-x) + \omega x^2\right] \tag{2.121}$$

$$\mu_2 = \mu_2^\circ + k_B T\left[\ln x + \omega(1-x)^2\right] \tag{2.122}$$

A comparison with the general expressions for μ_1 and μ_2, as given by Eqs. (2.101) and (2.102), shows that the activity coefficients of species 1 and 2 are $\ln \gamma_1 = \omega x^2$ and $\ln \gamma_2 = \omega(1-x)^2$. Note that, in general, the phase behavior of real binary mixtures is asymmetric, such that $\mu_{21} \neq \mu_{21}^{\circ}$ along phase coexistence, and thus, the equilibrium conditions will be obtained from the equality of chemical potentials in each phase, Eq. (2.54).

The osmotic pressure in the regular-solution model is

$$\Pi = \ell_0^{-3}(\mu_1^{\circ} - \mu_1) = -\ell_0^{-3} k_B T[\ln(1 - x) + \omega x^2] \tag{2.123}$$

Assuming a dilute solution ($x \ll 1$) and expanding $\ln(1 - x)$ in powers of x, as $\ln(1 - x) \cong -x - (1/2)x^2 - (1/3)x^3$, one obtains the virial expansion for the osmotic pressure in a dilute regular solution,

$$\Pi = \ell_0^{-3} k_B T(1 + Bx + Cx^2 + \cdots) \tag{2.124}$$

where the virial coefficients are

$$B = \frac{1}{2} - \frac{\omega}{k_B T} \quad \text{and} \quad C = \frac{1}{3} \tag{2.125}$$

For an infinitely dilute-solution, the expansion given by (2.124) reduces to

$$\Pi = \ell_0^{-3} k_B T \tag{2.126}$$

This expression is known as the "van't Hoff formula" and is an analog of $P = \rho k_B T$ (the ideal-gas EOS).

2.8.4 Phase Equilibria in Ternary Systems

A ternary liquid system, with three components, often exhibits similar phase behavior to the binary mixture. Due to the additional independent concentration variable, constrained by $x_1 + x_2 + x_3 = 1$, the phase behavior is typically displayed on a triangular phase diagram. For example, consider the phase diagram of the ternary system of water, cyclohexane (CHX), and *t*-butanol (TBA), shown in Figure 2.25. Each vertex of the triangle represents the pure components (water, hydrocarbon, or alcohol). The edges connecting each corner represent the binary mixtures of any composition of two species. Inside the triangle depicts a mixture of all three components [44]. Above the binodal (at constant temperature and pressure) the system is in a homogeneous state, while below the binodal, the system is separated into two liquid phases. For any other temperature (at constant pressure), the different triangles will form a prism, and the critical points will form a line.

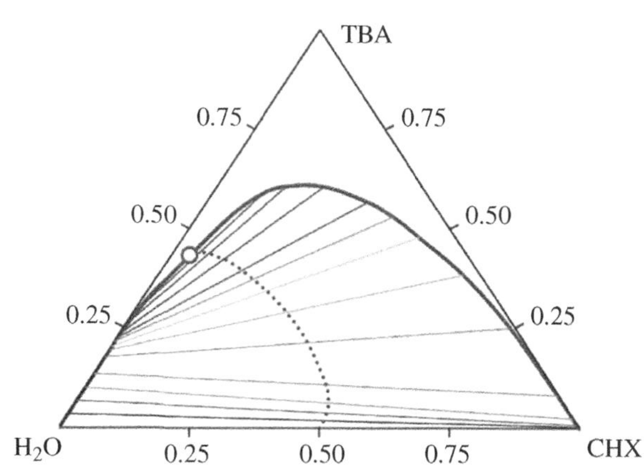

Figure 2.25 Ternary phase diagram of water, cyclohexane (CHX) and *tert*-butanol (TBA) at constant temperature (20 °C) and pressure in terms of mass fractions. The thick curve shows the phase separation boundary (binodal), the circle indicates the critical point of liquid–liquid demixing, the thin lines are referred to as "tie-lines" (which connect the compositions of coexisting phases along the binodal), and the dotted line connects the midpoints of the tie-lines. Source: Reproduced with permission from Ref. [45], © American Chemical Society.

Stability Criteria: Jacobian Matrix

The stability criteria for a compressible binary mixture, Eq. (2.106), and an incompressible ternary mixture, Eq. (2.130), are obtained through the determinant of a Jacobian matrix.

- The Jacobian is generally a matrix of derivatives of two functions, $\Psi(x, y)$ and $\Phi(x, y)$, with respect to two independent variables x and y, as

$$J = \begin{pmatrix} \frac{\partial \Psi}{\partial x} & \frac{\partial \Psi}{\partial y} \\ \frac{\partial \Phi}{\partial x} & \frac{\partial \Phi}{\partial y} \end{pmatrix} \tag{2.127}$$

- For a binary mixture ($\Psi = \Phi = A$) with two independent variables x and V, the stability criterion is given by the determinant of the Jacobian matrix of second derivatives of A as

$$\begin{vmatrix} \frac{\partial^2 A}{\partial x^2} & \frac{\partial^2 A}{\partial x \partial V} \\ \frac{\partial^2 A}{\partial V \partial x} & \frac{\partial^2 A}{\partial V^2} \end{vmatrix} \geq 0 \tag{2.128}$$

- For an incompressible ternary mixture ($\Psi = \Phi = G$) with two independent variables, x_1 and x_2, the stability criterion is given by the determinant of the Jacobian of second derivatives of G as

$$\begin{vmatrix} \frac{\partial^2 G}{\partial x_1^2} & \frac{\partial^2 G}{\partial x_1 \partial x_2} \\ \frac{\partial^2 G}{\partial x_2 \partial x_1} & \frac{\partial^2 G}{\partial x_2^2} \end{vmatrix} \geq 0 \tag{2.129}$$

For a compressible ternary mixture, the Jacobian would have nine components. Jacobians are also useful for transforming between the partial thermodynamic derivatives taken at different constant fields or coordinates.

The phase behavior of an incompressible ternary mixture, may be modeled by an incompressible ternary lattice (Figure 2.26). The incompressible ternary lattice is isomorphic to the compressible binary lattice model, provided that the role of the empty spots (see Figure 2.24) is played by the concentration of the additional component and the pressure is played by the osmotic pressure of the third component. The mixture stability criteria for an incompressible ternary mixture is obtained from the determinant of the "Jacobian," a matrix of the second derivatives of the Gibbs energy with respect to molecular fractions of each species. Adopting Euler's notation that $\partial_x \Psi = \partial \Psi / \partial x$, then the stability condition reads

$$\partial_{x_1 x_1} G \partial_{x_2 x_2} G - \left(\partial_{x_1 x_2} G\right)^2 \geq 0 \tag{2.130}$$

where $\partial_{x_1 x_2} G = \partial_{x_2 x_1} G$. The critical point condition is obtained when the criterium given by Eq. (2.130) is equal to zero and from the zero of the determinant of the Jacobian of third-order derivatives (3×3 matrix), given as

$$\partial_{x_1 x_1 x_1} G \left(\frac{\partial_{x_1 x_2} G}{\partial_{x_1 x_1} G}\right)^3 - 3\partial_{x_1 x_1 x_2} G \left(\frac{\partial_{x_1 x_2} G}{\partial_{x_1 x_1} G}\right)^2 + 3\partial_{x_1 x_2 x_2} G \left(\frac{\partial_{x_1 x_2} G}{\partial_{x_1 x_1} G}\right)^3 - \partial_{x_2 x_2 x_2} G = 0 \tag{2.131}$$

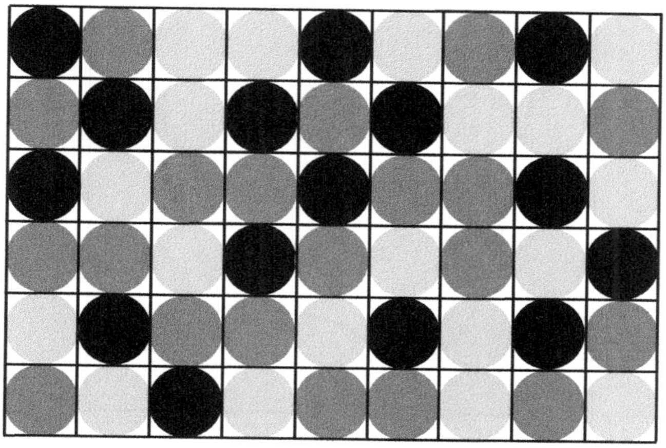

Figure 2.26 A two-dimensional representation of an incompressible ternary lattice mixture, where the light gray, dark gray, and black circles indicate the three species.

2.9 Chemical-Reaction Equilibria

If a chemical reaction occurs between the components of a binary mixture, the concentrations cannot be treated as independent variables as they are constrained by the chemical-reaction equilibrium condition.

An equilibrium (reversible) chemical reaction between two states can be represented in the form

$$\sum \nu_i a_i = 0 \tag{2.132}$$

where a_i are chemical symbols of the reacting substances and ν_i are stoichiometric coefficients. By convention, the reactants are assigned to be negative, while the products are assigned to be positive. At chemical-reaction equilibrium (at constant P and T), the total Gibbs energy attains a minimum, such that

$$\left.\frac{\partial G_t}{\partial \zeta}\right|_{P,T} = 0 \tag{2.133}$$

where ζ is the extent of the reaction (also referred to as the "reaction coordinate") defined through the relation between the actual number of molecules of species i, N_i, at any stage of the reaction and the initial number of molecules, N_i°, at $\zeta = 0$:

$$N_i = N_i^\circ + \nu_i \zeta \tag{2.134}$$

According to the definition of the Gibbs energy, presented in Eq. (2.14b), at constant pressure and temperature, $dG_t = \sum \mu_i \, dN_i$. Then, with use of Eq. (2.134), it follows from Eq. (2.133) that the chemical-reaction equilibrium condition restricts the chemical potentials of each species by

$$\sum \mu_i \nu_i = 0 \tag{2.135}$$

Since the chemical potentials may be expressed through the activity coefficients as

$$\mu_i = \mu_i^\circ(T, P) + k_B T \ln(\gamma_i x_i) \tag{2.136}$$

then Eq. (2.135) can be rewritten in the form

$$\boxed{\frac{\sum \nu_i \mu_i^\circ}{k_B T} = -\ln K(T, P)} \tag{2.137}$$

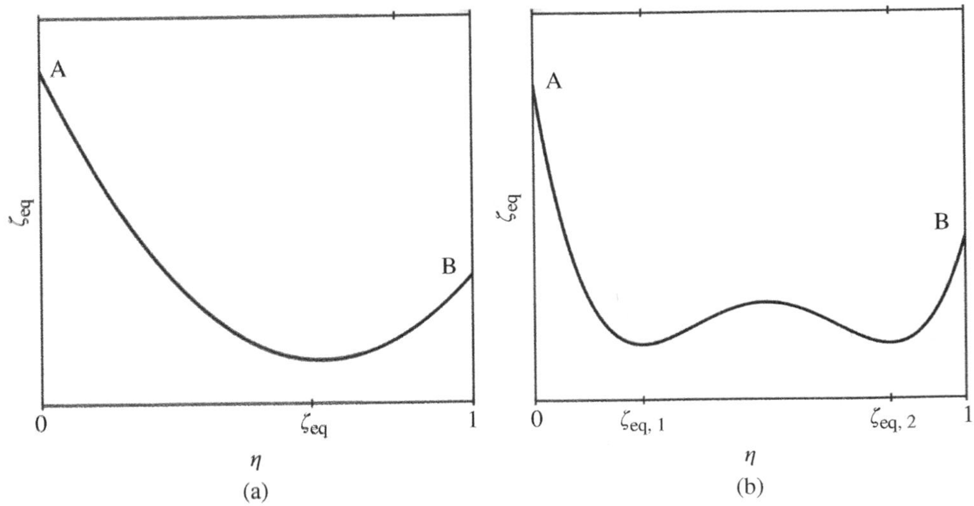

Figure 2.27 Gibbs energy as a function of the reaction coordinate, ζ, for two different scenarios of a fluid mixture with an A $\rightleftharpoons$ B reaction. (a) A homogeneous solution of species. (b) A phase-separated solution of species. Chemical-reaction equilibrium, ζ_{eq}, is at the minimum of the Gibbs energy with respect-to-reaction coordinate.

where $K(T, P) = \Pi(\gamma_i x_i)^{\nu_i}$ is the reaction-equilibrium constant. Equation (2.137) indicates that the algebraic sum of all Gibbs energies of products and reactants control the equilibrium concentrations of species. If the equilibrium constant is known, all thermodynamic characteristics of the reaction, such as the heat of reaction (enthalpy change and volume change), can be obtained through standard thermodynamic relations between the Gibbs energy and its derivatives.

Figure 2.27a,b illustrates two possible scenarios for chemical-reaction equilibrium between two species in a binary solution. Figure 2.27a depicts the behavior of an homogeneous solution of products and reactants, in which the chemical-reaction equilibrium condition defines a single minimum of the Gibbs energy. Figure 2.27b illustrates the effect of demixing in the product-reactant solution, which (at certain conditions) may produce two minima in the Gibbs energy, corresponding to the coexistence of two phases. Note that the two coexisting phases are separated by a thermodynamic barrier (shown by the maximum in the Gibbs energy in Figure 2.27b). However, the rate of the reaction toward equilibrium is determined by an additional kinetic barrier (not shown), referred to as the "activation energy," which does not affect the chemical-reaction equilibrium condition.

As mentioned in Section 2.5, the possibility of liquid-state polyamorphism in a single-component fluid can phenomenologically be explained by chemical-reaction equilibrium between two interconvertible molecular species or supramolecular structures. In this case, the chemical reaction between two species, A and B, is a simple interconversion reaction of the form, A $\rightleftharpoons$ B, and the chemical-reaction constant may be expressed in the linear approximation as

$$\ln K = \varepsilon + sT + \upsilon P \tag{2.138}$$

where ε, s, and υ represent the energy, entropy, and volume change of the reaction, respectively.

In the simple case of the regular-solution model, the reaction coordinate is the molecular fraction, and from the chemical-reaction equilibrium condition,

$$\left.\frac{\partial G}{\partial x}\right|_{T,P} = -\ln K + \ln\left(\frac{x}{1-x}\right) + \frac{\omega}{k_B T}(1-2x) = 0 \tag{2.139}$$

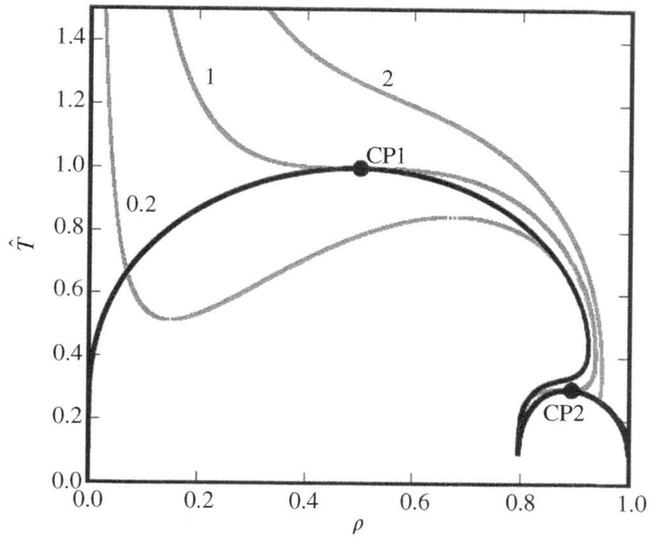

Figure 2.28 Temperature–density phase diagram for a polyamorphic substance in which liquid–liquid demixing as modeled by the regular-solution. Thick black curves depict the liquid–gas and liquid–liquid coexistence terminated by the critical points (CP1 and CP2), respectively. Gray curves are isobars at $\hat{P} = 0.2$, $\hat{P} = 1$, and $\hat{P} = 2$. The selection of the energy, enthalpy, and entropy (ε, υ, and s, respectively) of the reaction were chosen to imitate the hypothesized polyamorphic behavior of supercooled water. Source: Adapted from Ref. [21].

the molecular fraction is no longer an independent variable, being a function of temperature and pressure, $x = x(T, P)$. Meaning that the phase behavior of the interconverting binary mixture is thermodynamically equivalent to that of a single-component fluid. Correspondingly, the density, $\rho = \rho(T, P)$, may be calculated from the derivative of the Gibbs energy with respect to pressure as

$$\rho = \left(\frac{\partial G}{\partial P}\right)_T^{-1} = \left[\frac{1}{\rho_1^\circ} - k_B T x \left.\frac{\partial(\ln K)}{\partial P}\right|_T\right]^{-1} \tag{2.140}$$

where $\rho_1^\circ = (\partial\mu_1^\circ/\partial P)|_T$ is the density of pure species 1. Figure 2.28 illustrates the polyamorphic behavior of a system described by Eq. (2.140) where the molecular fraction, x, is obtained from the chemical-reaction equilibrium condition, Eq (2.139) and the parameters of the interconversion reaction, Eq. (2.138), ε, υ, and s, were chosen to mimic the hypothesized phase behavior of supercooled water (see Section 2.5).

2.10 Connection to Partition Functions in Statistical Mechanics

In the late 19th century, Maxwell, Boltzmann, and Gibbs founded statistical mechanics, a branch of physics that describes the observed thermodynamic properties in terms of the statistics consisting of large molecular ensembles. Statistical mechanics bridges the gap between the purely phenomenological description of matter by axiomatic thermodynamics and a molecular theory, based on the knowledge of intermolecular interactions [46–48].

Unlike Newtonian mechanics, which generally considers the behavior of a single body (or a small number of bodies), statistical mechanics considers many interacting bodies via a probability distribution over all microscopic states. In classical statistical mechanics, the molecular motion is described by Newtonian laws, whereas in quantum theory, the molecular motion is described by the Schrödinger equation [49, 50]. For most engineering applications to meso-heterogeneous systems, classical statistics is sufficient.

All thermodynamic potentials are connected to the corresponding statistical ensembles, as originally introduced by Gibbs, depending on the choice of canonical variables. For example, the so-called "microcanonical ensemble" describes an isolated system at constant energy, volume, and number of particles, which thermodynamically defines the entropy of the system. Alternatively, a closed system at constant energy, volume (or pressure), and number of particles is referred to as the "canonical ensemble," defining the Helmholtz (or Gibbs) energy. Finally, an open system at constant energy, volume, and chemical potential is referred to as the "grand canonical ensemble," defining the grand thermodynamic potential.

Example: Application to Ideal Gas

Calculations of the thermodynamic potentials and their derivatives through partition functions require substantial information about the intramolecular interactions involved in the system and can give exact results only for simple models, such as the ideal-gas model. In an ideal gas, molecules do not interact and the partition function, Z_A, depends only on the translational and the intramolecular contributions to the energy of the system. Moreover, at low and moderate temperatures, the molecular structure is usually stable and the intermolecular contribution to the partition function can be approximated as a constant. In this approximation, for a classical monoatomic ideal gas [3]

$$\boxed{A_t = -k_B T \ln Z_A = -Nk_B T \ln V_t - \frac{3}{2}Nk_B T \ln T - NS_0 T + NU_0} \tag{2.141}$$

where S_0 and U_0 are the absolute values of the entropy and energy, respectively. Correspondingly, other thermodynamic functions of the ideal gas in this approximation read

$$U_t = k_B T^2 \left.\frac{\partial \ln Z_A}{\partial T}\right|_V = \frac{3}{2}Nk_B T + NU_0 \tag{2.142}$$

$$H_t = U_t + PV_t = U_t + Nk_B T = \frac{5}{2}Nk_B T + NU_0 \tag{2.143}$$

$$\begin{aligned} G_t &= A_t + PV_t = A_t + Nk_B T \\ &= Nk_B T \ln V_t + \frac{3}{2}Nk_B T \ln T + Nk_B T + NS_0 T + NU_0 \end{aligned} \tag{2.144}$$

$$\begin{aligned} S_t &= k_B \ln Z_A + k_B T \left.\frac{\partial \ln Z_A}{\partial T}\right|_{V_t} \\ &= k_B N \ln V_t + \frac{3}{2}k_B N \ln T + \frac{3}{2}k_B N + NS_0 \end{aligned} \tag{2.145}$$

The heat capacity per molecule of a monoatomic ideal gas in this approximation is a constant, given by

$$C_V = T \left.\frac{\partial S}{\partial T}\right|_V = \frac{3}{2}k_B \tag{2.146}$$

$$C_P = C_V + T\left(\left.\frac{\partial P}{\partial T}\right|_V\right)^2 \left.\frac{\partial V}{\partial P}\right|_T = C_V + k_B = \frac{5}{2}k_B \tag{2.147}$$

Summing the probabilities of all possible microscopic states, gives a quantity referred to as the "partition function." In the microcanonical ensemble, the partition function is the number of possible microstates, w, related to the entropy by the third law of thermodynamics,

$$S_\mathrm{t} = k_\mathrm{B} \ln w \tag{2.148}$$

In the canonical ensemble, the partition function is

$$\boxed{Z_A = \sum_k e^{-E_{k,i}/k_\mathrm{B}T}} \tag{2.149}$$

where $E_{k,i}$ is the energy of a microstate, k, for each species, i. The Helmholtz energy is obtained as

$$A_\mathrm{t} = U_\mathrm{t} - TS_\mathrm{t} = -k_\mathrm{B}T \ln Z_A \tag{2.150}$$

where the internal energy and entropy are found as

$$U_\mathrm{t} = k_\mathrm{B}T^2 \left.\frac{\partial \ln Z_A}{\partial T}\right|_{V_\mathrm{t}} \tag{2.151}$$

$$S_\mathrm{t} = k_\mathrm{B} \ln Z_A + k_\mathrm{B}T \left.\frac{\partial \ln Z_A}{\partial T}\right|_{V_\mathrm{t}} \tag{2.152}$$

In the grand canonical ensemble, the partition function reads

$$\boxed{Z_\Omega = \sum_k e^{-(E_{k,i} - \mu_i N_{k,i})/k_\mathrm{B}T}} \tag{2.153}$$

Thus, the grand thermodynamic potential is given by

$$\Omega_\mathrm{t} = -k_\mathrm{B}T \ln Z_\Omega \tag{2.154}$$

For a single-component system, the statistical ensembles and corresponding thermodynamic functions are summarized in Table 2.1.

In some applications, like in computer simulations of a closed system with fixed N_i, T, and P, the partition function reads

$$\boxed{Z_G = \sum_{k,i} e^{-(E_{k,i} - PV_\mathrm{t})/k_\mathrm{B}T}} \tag{2.155}$$

The corresponding thermodynamic potential is the Gibbs energy

$$G_\mathrm{t} = \sum \mu_i N_i = A_\mathrm{t} + PV_\mathrm{t} = -k_\mathrm{B}T \ln Z_G \tag{2.156}$$

Relations between the partition functions belonging to different statistical ensembles are called "Laplace transformations." Each Laplace transformation relates to a corresponding Legendre

Table 2.1 Summary of statistical ensembles in single-component systems.

	Microcanonical	**Canonical**	**Grand canonical**
Variables	N, U_t, V_t	N, T, V_t	T, V_t, μ
Statistical functions	Number of microstates, w	Canonical partition function Z_A (Eq. 2.149)	Grand partition function Z_Ω (Eq. 2.153)
Thermodynamic functions	Entropy $S_\mathrm{t} = k_\mathrm{B} \ln w$	Helmholtz energy $A_\mathrm{t} = -k_\mathrm{B}T \ln Z_A$	Grand potential $\Omega_\mathrm{t} = -k_\mathrm{B}T \ln Z_\Omega$

transformation as a thermodynamic potential relates to its corresponding partition function. Since the probability distributions are multiplicative, while the thermodynamic potentials are additive, mathematically, Laplace transformations relate to Legendre transformations as exponentials relate to logarithms.

It follows from Eq. (2.142) that the internal energy in an ideal gas is equally distributed between the three translational molecular degrees of freedom, such that each degree of freedom possesses an internal energy, $k_B T/2$. If the ideal gas is not monoatomic, additional contributions from microscopic states, associated with rotational and vibrational degrees of freedom should also be included in the partition function.

2.11 Intermolecular Interactions

With an increase of density at constant temperature, the pressure of a gaseous fluid increasingly deviates from the ideal-gas pressure, $P = \rho k_B T$. This deviation is caused by attractive and repulsive interactions between molecules. A competition between the kinetic energy of molecular motion and the potential energy of molecular interaction is responsible for the existence of different phases of matter.

In principle, all intermolecular interactions have electromagnetic origin, stemming from the fact that a molecule is comprised of positively charged nuclei surrounded by a cloud of negatively charged electrons. Depending on the molecular structure and, mainly, on the intermolecular separation of the charges, the intermolecular forces can be roughly divided into three groups, ionic forces (Columbic electrostatic attraction between opposite ions), dipole–dipole forces between two polar molecules that possess electric dipoles, and "dispersion" forces between nonpolar molecules that possess instantaneous ("virtual") dipoles.

Dipole–dipole forces can exist between two permanent dipoles of polar molecules or even when a polar molecule induces a dipole in its nonpolar neighbor though the molecular electrostatic field. Both kinds of molecular forces, dipole–dipole and dispersion, often make comparable contributions to the interaction potential energy. Ionic forces are long-ranged as the Columbic energy decays with the distance r between the molecules as r^{-1}. The force of intermolecular interaction, $\boldsymbol{F}$, and the pair intermolecular potential energy, U_p, referred to as the "intermolecular potential," are related by

$$\boldsymbol{F} = -\frac{d}{d\boldsymbol{r}} U_p(r) \tag{2.157}$$

Therefore, the electrostatic force decays as r^{-2}. The electrostatic interactions are exerted over a relatively long range.

However, because of statistical averaging over all possible dipole–dipole orientations in an isotropic fluid, the attractive part of the intermolecular potentials for dipole–dipole and dispersion interactions decays with the distance between molecules much faster than a Columbic potential, as r^{-6}. Thus, such interactions are short ranged. On the other hand, the repulsions between molecules are always extremely short ranged, since they are associated with the inability of molecules to physically overlap. These features of intermolecular attraction and repulsion are adopted in two pair interaction potentials, commonly used in analytical theory and computer simulations, the Lennard-Jones (LJ) potential and the square-well (SW) potential. Both potentials are especially suitable to represent the properties of substances consisting of simple nonpolar molecules. The LJ potential, presented in Figure 2.29a, contains two parameters: the collision diameter σ, which defines the distance between two molecules for which $U_p(r) = 0$ and the minimum value $-\epsilon$, which defines the equilibrium separation between molecules. The LJ intermolecular potential is given by

$$U_p^{LJ} = 4\epsilon \left[\left(\frac{\sigma}{r} \right)^{12} - \left(\frac{\sigma}{r} \right)^{6} \right] \tag{2.158}$$

In certain situations, where the repulsive interactions need to be modeled more sharply, this part of the potential can be modified by replacing $(\sigma/r)^{12}$ by $e^{-r/\sigma}$. The short-ranged nature of the intermolecular attraction and the extremely sharp repulsion enables one to represent the pair potential in a rectilinear form, a collection of straight-line segments. The simplest rectilinear pair potential is the square-well potential, shown in Figure 2.29b, given by

$$U_p^{SW} = \begin{cases} \infty & 0 \le r < \sigma \\ -\epsilon & \sigma \le r < \sigma + d \\ 0 & r \ge \sigma + d \end{cases} \tag{2.159}$$

where d is the width of the square well.

The square-well potential is especially popular in computer simulations, particularly for Monte Carlo (MC) and Molecular Dynamics (MD). Despite its simplicity, the square-well potential provides excellent results for simulating thermodynamic properties of simple fluids, such as argon.

The number of neighboring molecules that interact through short-ranged intermolecular forces depends on the coordination number. These multibody interactions are often approximated as a superposition of two-body interactions represented by a pair potential for each pair of molecules in the first coordination sphere.

Statistical mechanics provides an expression for the second virial coefficient, B, through a spatial integral of the intermolecular pair potential for a monoatomic fluid [41],

$$B(T) = \int_0^{\infty} \left(1 - e^{-U_p/k_B T}\right) \mathrm{d}V \tag{2.160}$$

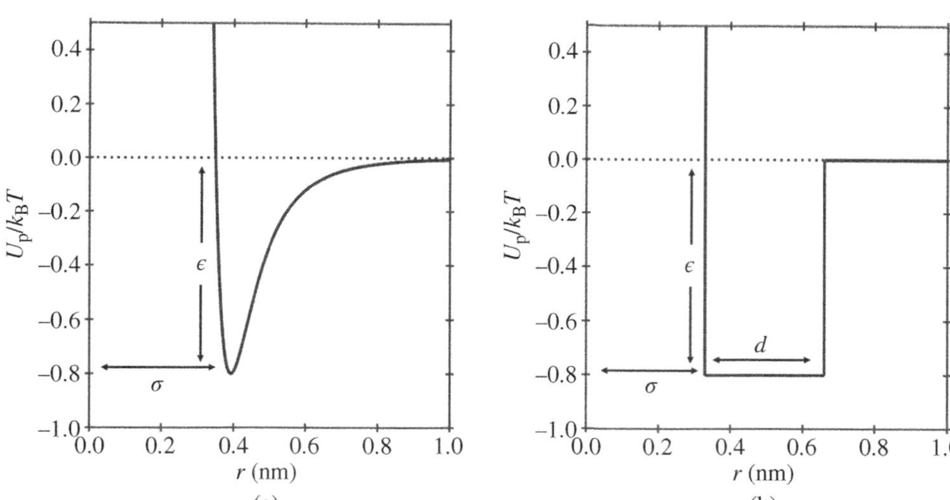

Figure 2.29 Two forms of pair intermolecular potentials, U_p. (a) Lennard–Jones potential and (b) square-well potential.

Dividing the range of integration into two parts, one obtains

$$B(T) = 2\pi \int_0^{\sigma} \left(1 - e^{-U_p/k_B T}\right) r^2\,dr + 2\pi \int_{\sigma}^{\infty} \left(1 - e^{-U_p/k_B T}\right) r^2\,dr \tag{2.161}$$

where σ is the collision diameter of a molecule. Since the potential will be large ($U_p \gg k_B T$) within the collision diameter ($r < \sigma$), the first integral can be estimated for the LJ potential as $(2/3)\pi\sigma^3$. Since, for the van der Waals fluid, the second virial coefficient is related to the van der Waals constants, a and b, by Eq. (2.73), the repulsive part of B is

$$b = \frac{2}{3}\pi\sigma^3 \tag{2.162}$$

Outside the collision diameter ($r \geq \sigma$), the potential is small and negative, $e^{-U_p/k_B T} \cong 1 - U_p/k_B T$, such that the second integral can be estimated for the LJ potential as $-(16/9)\pi\epsilon\sigma^3/(k_B T)$, where ϵ is the depth of the pair potential. For a van der Waals fluid, this describes the attractive interactions and is given by

$$a = \frac{16}{9}\pi\sigma^3\epsilon \tag{2.163}$$

With use of Eqs. (2.75) and (2.76) for the relations between the van der Waals constants and the critical parameters, T_c, V_c, and P_c, one finds that the critical parameters may also be related to the interaction parameters of the LJ potential as

$$V_c = 2\pi\sigma^3, \quad T_c = \frac{64\epsilon}{81k_B}, \quad \text{and} \quad P_c = \frac{4\epsilon}{27\pi\sigma^3} \tag{2.164}$$

The same procedure could be used to obtain the van der Waals constants a and b from the SW potential.

2.12 Ideal Gas and Condensed Matter; Quasiparticles

The ideal-gas model is the first approximation for any statistical ensemble containing many particles. The most important mathematical feature of the ideal-gas EOS is its linearity as a function of density. Interestingly, one can recognize the ideal-gas model in the description of physical systems other than fluids. In particular, radiation can be viewed as an "ideal gas" of photons, which follows classical Boltzmann statistics at low frequencies (radio waves) and follows quantum statistics at high frequencies (visible and ultraviolet light, X-rays and gamma-rays).

In addition, a solid state can be viewed as containing two kinds of "ideal gases," namely free electrons in metals and collective thermal oscillations (known as "phonons") in all solids. The "gas" of electrons follows quantum statistics at low temperatures and classical Maxwell–Boltzmann statistics at very high temperatures [13]. Conductive solids may not only be represented as a "gas" of individual electrons. For example, in semiconductors, electrons may be excited from one layer to another, leaving behind empty spots, called "holes." The motion of these holes within the semiconductor can be modeled as an ideal gas of positively charged "quasiparticles."

Quasiparticles are collective excitations, which microscopically may contain real interacting particles. Many quasiparticles may be described as "bosons" or "fermions" depending on the quantization of the particles "spin." Quantum spin is an intrinsic angular momentum inherent to each particle. Bosons have an integer spin of 0, 1, 2, etc., while fermions have a half-integer spin of 1/2, 3/2, etc. For example, photons and He^4 atoms are bosons, while electrons and He^3 atoms, having an uncompensated spin, are fermions. In superconductors, a pair of electrons form another type of quasiparticle, known as "Cooper pairs." Cooper pairs are boson particles or quasiparticles with zero quanta of spin.

A key difference between fermions and bosons emerges from the *Pauli exclusion principle*, that two fermions cannot occupy the same quantum state. Since bosons are not excluded from occupying the same quantum state, then depending on the conditions (typically close to absolute zero temperature), bosons may "condensate" into the lowest quantum state, forming a "Bose–Einstein Condensate." The condensation of Cooper pairs indicates the emergence of superconductivity in metals and the emergence of superfluidity in helium.

An alternative form of quasiparticles present in condensed matter are phonons. Phonons are collective thermal oscillations of strongly-interacting atoms, whose collective motion generates acoustic waves in the material. Therefore, in accordance with the quantum mechanical principle of wave-particle duality, phonons may be viewed as quasiparticles of sound, similar to photons being quasiparticles of light. The elasticity of solids (linear deformation as a response to an external force, known as "Hook's law") is directly associated with the absence of interactions in the "ideal gas" of phonons.

If a quasiparticle (boson or fermion) has energy $\epsilon \ll k_B T$, the "gas" of quasiparticles may be described by classical Boltzmann statistics, like an ordinary molecular gas. However, if $\epsilon \gg k_B T$, the "gas" of quasiparticles is described by Fermi–Dirac statistics (for fermions) or Bose–Einstein statistics (for bosons). In accordance with quantum statistics for a canonical ensemble, the number of phonons, N_p, in a given quantum state is determined by the Planck function

$$N_p = \frac{1}{e^{\epsilon/k_B T} \pm 1} \tag{2.165}$$

where the "+" state corresponds to Fermi–Dirac statistics and the "−" state corresponds to Bose–Einstein statistics. For a given photon or phonon, each quasiparticle has an energy $\epsilon = \hbar\omega$, where $\hbar$ is Planck's constant and ω is the frequency of oscillations. At high temperatures, $\epsilon \ll k_B T$, such that

$$N_p = \frac{k_B T}{\epsilon} \tag{2.166}$$

This corresponds to the classical equipartition of energy among the three translational and three vibrational degrees of freedom (see Section 2.10). Thus, the internal energy of a phonon in a solid is $U = U_0 + 3k_B T$, and the total heat capacity of a nonconducting solid is

$$NC_V = 3Nk_B \tag{2.167}$$

This expression is known as the "law of Dulong and Petit." At low temperatures, $\epsilon \gg k_B T$, the phonon gas obeys Bose–Einstein statistics with internal energy, $U \propto T^4$ and heat capacity, $C_V \propto T^3$ (known as "Debye's law"[6]). A crossover between the low-temperature (quantum) behavior and the high-temperature (classical) behavior is given by an interpolation between these two limits (originally derived by Debye), as presented in Figure 2.30. For details on the statistical mechanics of quasiparticles, see Refs. [41, 47, 49].

There is no thermodynamic phase transition between isotropic liquid and gaseous states above the critical point as liquids and gases have the same symmetry with respect to translations and rotations in space. However, thermodynamics itself cannot explain the obvious similarity between the

---

6 Peter Joseph William Debye (1884–1966) was a Dutch-American physical chemist and physicist. He received a Nobel Prize in Chemistry in 1936 for his contributions to the knowledge of molecular structure through his investigations of dipole moments and electron scattering.

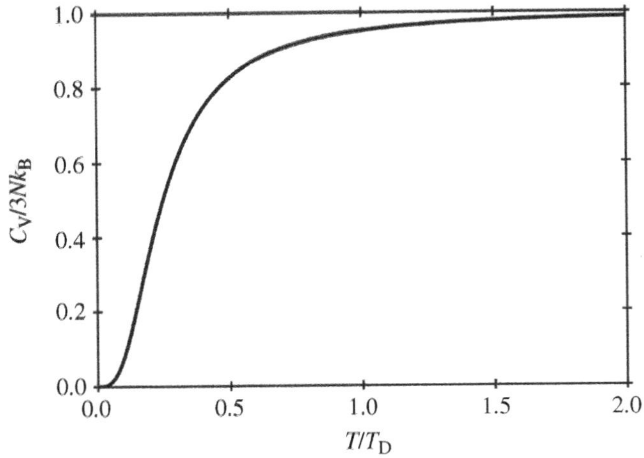

Figure 2.30 Debye's interpolation for the heat capacity of solids, where N is the number of phonon modes in the lattice and T_D is the Debye temperature, $T_D \approx \hbar\omega_D/k_B$, in which ω_D is a characteristic cut-off frequency that is proportional to the speed of sound, u_s, divided by the characteristic length scale of the phonons, ℓ_{ph} (of the order of the size of the lattice cell), as $\omega_D \propto u_s/\ell_{ph}$.

properties of liquids and solids, such as their comparably weak compressibility and similar magnitude of their heat capacities. Also, both solids and liquids are characterized by the existence of transverse phonons, which gives rise to the *viscoelasticity* of liquids at high frequencies. The line that qualitatively separates the regions in the fluid state into "gas-like" (no transverse phonons) and "liquid-like" (transverse phonons and viscoelasticity at hypersonic frequencies) is known as the "Frenkel line" [53]. This concept is based on the assumption that the particles in a liquid state,

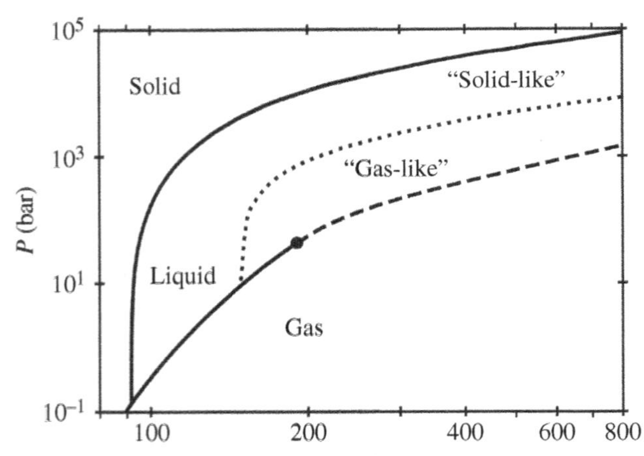

Figure 2.31 The Frenkel line (dotted) in methane illustrates the separation of the short-ranged molecular order in liquids, being "solid-like" above and "gas-like" below the line. The solid curves indicate phase coexistence. The dashed curve is the minimum speed of sound. Source: Adapted from Refs. [51, 52]. This curve, as well as other lines of extrema of the response functions emanating from the critical point (CP), such as the maxima of isothermal compressibility, isobaric heat capacity, and isobaric expansivity, asymptotically merge with the critical isochore with the same slope in P–T coordinates as the slope of the saturation-pressure curve.

due to the existence of short-ranged ordering in the positions of neighboring molecules, behave in a similar manner to a crystal, i.e. the positional order decays after the particles demonstrate oscillatory motion. However, while particles oscillate around the same location in a crystalline lattice, in liquids, the particles will change their location after several oscillations [52]. The estimated Frenkel line for methane is shown in Figure 2.31. It is located at a significantly higher pressure than the line of the minimum speed of sound which is close to the critical isochore.

2.13 Elements of Nonequilibrium Thermodynamics

In the absence of equilibrium, at least one of the thermodynamic driving forces (thermodynamic fields) is not uniform throughout the system. Gradients of thermodynamics forces, such as gradients of temperature and chemical potentials, will cause currents (fluxes) of energy and matter to restore the equilibrium. These processes are essentially irreversible; they generate dissipation of energy and, thus, cannot be treated by equilibrium (reversible) thermodynamics. Moreover, the relaxation to equilibrium occurs temporally and, hence, with a certain rate. The rates, described by kinetics of physical and chemical processes, cannot be obtained in equilibrium thermodynamics, as they depend not only on the thermodynamic driving forces but also on "resistances" caused by various kinds of "internal frictions," which are macroscopic consequences of a microscopic potential-energy landscape.

Calculation of the resistances requires microscopic kinetic theory based on nonequilibrium statistical mechanics. However, to consider macroscopic, or even mesoscopic, phenomena close to equilibrium, one may use a phenomenological approach, based on *nonequilibrium thermodynamics*, also known as the "thermodynamics of irreversible processes" [54–56]. Nonequilibrium thermodynamics is closely connected with fluid mechanics, which is based on the principle of the conservation of momentum, formulated by the Navier–Stokes equation and, in turn, is fundamentally related to the concept of the homogeneity of space. The approach to nonequilibrium thermodynamics of binary fluids (not far from equilibrium), adopted in this book, was developed by Landau and Lifshitz and can be found in more detail in Ref. [57].

Consider a binary-fluid mixture. The presence of concentration and temperature gradients in the fluid generates a diffusion flux, $\boldsymbol{j}_D$, of matter (mutual diffusion of solvent and solute molecules) and a heat flux, $\boldsymbol{j}_Q$, of energy, given in the linear approximation as

$$\boldsymbol{j}_D = -\check{\alpha}\nabla\mu_{21} - \check{\beta}\nabla T \tag{2.168}$$

$$\boldsymbol{j}_Q = -\check{\beta}\nabla\mu_{21} - \check{\gamma}\nabla T + \mu_{21}\boldsymbol{j}_D \tag{2.169}$$

Note that the fluxes, like any gradient of physical properties, are vectors. The diffusion flux is defined here as the number of molecules of solute transported by diffusion through a unit area per unit time. Also note that both fluxes depend on both gradients of chemical potential and temperature. In addition, the heat flux contains a direct contribution from the diffusion flux. The coefficients: $\check{\alpha}$, $\check{\beta}$, and $\check{\gamma}$ are known as "transport coefficients" (or "Onsager's kinetic coefficients" [58]). The symmetry in the fluxes (the same coefficient $\check{\beta}$ appears in both equations) is known as "Onsager's reciprocal relation".[7]

Eliminating the gradient of the exchange chemical potential from the heat flux, one can obtain

$$\boldsymbol{j}_D = -\check{\alpha}\nabla\mu_{21} - \check{\beta}\nabla T \tag{2.170}$$

7 Lars Onsager (1903–1976) was an American physical chemist and physicist born in Norway. He received a Nobel Prize in Chemistry in 1968 for his discovery of the reciprocal relations bearing his name.

$$\boldsymbol{j}_Q = -\kappa \boldsymbol{\nabla} T + \left(\mu_{21} + \frac{\breve{\beta} T}{\breve{\alpha}} \right) \boldsymbol{j}_D \tag{2.171}$$

where κ is the thermal conductivity, defined as

$$\kappa = \breve{\gamma} - \breve{\beta}^2 T / \breve{\alpha} \tag{2.172}$$

Indicating that if the diffusion flux is absent, the heat is transported by pure thermal conduction.

The gradient of the exchange chemical potential, μ_{21}, can be expressed through gradients of pressure, temperature, and molecular fraction as

$$\boldsymbol{\nabla} \mu_{21} = \left. \frac{\partial \mu_{21}}{\partial x} \right|_{P,T} \boldsymbol{\nabla} x + \left. \frac{\partial \mu_{21}}{\partial P} \right|_{T,x} \boldsymbol{\nabla} P + \left. \frac{\partial \mu_{21}}{\partial T} \right|_{P,x} \boldsymbol{\nabla} T \tag{2.173}$$

Using a Maxwell relation $(\partial \mu_{21}/\partial P)|_{T,x} = (1/N)(\partial V_t/\partial x)|_{P,T} = (\partial V/\partial x)|_{P,T}$, one obtains

$$\boldsymbol{j}_D = -\rho D \left[\boldsymbol{\nabla} x + \frac{k_T}{T} \boldsymbol{\nabla} T + \frac{k_P}{P} \boldsymbol{\nabla} P \right] \tag{2.174}$$

$$\boldsymbol{j}_Q = -\kappa \boldsymbol{\nabla} T + \left[\left. \frac{\partial \mu_{21}}{\partial x} \right|_{P,x} - T \left. \frac{\partial \mu_{21}}{\partial T} \right|_{P,x} + \mu_{21} \right] \boldsymbol{j}_D \tag{2.175}$$

where the new transport coefficients, D, k_T, and k_P, are defined as follows:

$$D = \frac{\breve{\alpha}}{\rho} \left. \frac{\partial \mu_{21}}{\partial x} \right|_{P,T} \tag{2.176}$$

is the mutual-diffusion coefficient,

$$k_T = \frac{T}{\rho D} \left[\breve{\alpha} \left. \frac{\partial \mu_{21}}{\partial T} \right|_{P,x} + \breve{\beta} \right] \tag{2.177}$$

is the thermo-diffusion ratio, and

$$k_P = P \frac{(\partial V/\partial x)|_{P,T}}{(\partial \mu_{21}/\partial x)|_{P,T}} \tag{2.178}$$

is the baro-diffusion ratio. Note that the baro-diffusion ratio is a purely thermodynamic quantity.

In the absence of gradients of temperature and pressure, diffusion is caused by the gradient of concentration only. It is important to note that gradients of concentrations do not necessarily cause a diffusion flux. The driving force of diffusion is the gradient of the chemical potential. Coexisting phases in equilibrium usually have different concentrations and densities, but since the chemical potentials (as well as their temperatures and pressures) are equal, there are no macroscopic fluxes crossing the interface. In the absence of a diffusion flux ($\boldsymbol{j}_D = 0$) and in the presence of a stationary heat flux, an inhomogeneous nonequilibrium state could persist in the binary mixture with a concentration gradient caused by the gradient of the temperature and pressure in accordance with Eq. (2.174).

The time evolution of the concentration and the temperature (at constant pressure) is obtained from the divergence of the diffusion flux as

$$\frac{\partial x}{\partial t} = -\boldsymbol{\nabla} \cdot \boldsymbol{j}_D = D \left[\nabla^2 x + \frac{k_T}{T} \nabla^2 T \right] \tag{2.179}$$

and

$$\frac{\partial T}{\partial t} - \frac{k_T}{C_P} \left. \frac{\partial \mu_{21}}{\partial x} \right|_{P,T} \frac{\partial x}{\partial t} = D_T \nabla^2 T \tag{2.180}$$

The ratio $D_T = \kappa/\rho C_P$ is called the "thermal diffusivity" or "temperature conductivity." The thermal diffusivity is measured in the same units as the mutual diffusion coefficient, m^2/s, and controls the relaxation of temperature in the same manner that the diffusion coefficient controls the relaxation of concentration. In a single-component system, $\partial x/\partial t = 0$ and $\partial T/\partial t = D_T \nabla^2 T$.

In a dilute solution of solute molecules, the exchange chemical potential is

$$\mu_{21} \cong \mu_{21}^{\circ}(P, T) + k_B T \ln x + k_B T \ln \gamma_2^{\infty} \tag{2.181}$$

where γ_2^{∞} is the solute activity coefficient, approaching a constant in the limit $x \to 0$. Therefore, it follows from Eq. (2.171) that the diffusion coefficient of independent particles, reduced (by $k_B T$), referred to as the "mobility," is

$$\tilde{b} = \frac{D}{k_B T} = \frac{\breve{\alpha}}{\rho x} = \frac{\breve{\alpha}_0}{\rho} \tag{2.182}$$

The mobility remains finite in an infinitely dilute solution, since Onsager's coefficient, $\breve{\alpha} = \breve{\alpha}_0 x(1 - x)$, vanishes symmetrically in each dilute solution limit as x and $1 - x$. For a spherical particle with radius, R_h, often called the "hydrodynamic radius," the mobility and the diffusion coefficient are

$$\tilde{b} = \frac{1}{6\pi\tilde{\eta}R_{\text{h}}} \tag{2.183}$$

and

$$\boxed{D = k_B T \tilde{b} = \frac{k_B T}{6\pi\tilde{\eta}R_{\text{h}}}} \tag{2.184}$$

where $\tilde{\eta}$ is the shear viscosity. Equation (2.184) is known as the Stokes–Einstein relation.

2.14 Absolute Values of Energy and Entropy

There are two important consequences of the absolute values of the internal energy and entropy, which depend on the reference values U_0 and S_0 (often referred to as the "zero-points"), see Eqs. (2.142) and (2.145). The reference value of the internal energy, U_0, is always arbitrary and does not affect any physical properties; thus, the absolute values of all thermodynamic potentials are always arbitrary, except for the grand thermodynamic potential, $\Omega_t = -PV_t$. In particular, the absolute values of the chemical potentials for individual species may be shifted by arbitrary constants without any physical consequence.

However, the consequence of the reference value of the entropy, S_0, is more subtle. The third law of thermodynamics (originating from quantum theory), enables one to define the absolute value of the entropy through integration of experimentally obtained heat capacity data from 0 K to a given temperature; thus, the reference value of the entropy is generally not arbitrary. This means that such commonly used thermodynamic derivatives as $(\partial\mu/\partial T)|_P = -S$, in a single-component system, or $(\partial\mu_{21}/\partial T)|_{P,x} = -(\partial S/\partial x)|_{P,T}$, in a binary mixture, cannot always be shifted by an arbitrary constant without physical consequences.

Remarkably, however, in most applications of classical thermodynamics, the choice of the reference value of the entropy is conventional, depending on convenience. Since the chemical potential always contains a function that depends on temperature only (e.g. the function $f(T)$ in Eqs. (2.79) and (2.81) for the chemical potentials of the van der Waals and lattice-gas models, respectively), which does not contribute to the EOS, $P = P(\rho, T)$. Therefore, in calculations of phase equilibria, the absolute value of the entropy is arbitrary. For example, in engineering practice, the entropy of liquid water is assigned to be zero at 0 °C (as given in most "Steam Tables"), while in applications of the theory of critical phenomena in fluids, the entropy at the critical point is commonly assigned to be equal to the slope of the liquid–gas saturation curve, $S_c = (1/\rho_c)(dP/dT)_{cxc}$, so that

$$\left.\frac{d\mu}{dT}\right|_{cxc} = -S_c + \frac{1}{\rho_c}\left.\frac{dP}{dT}\right|_{cxc} = 0 \tag{2.185}$$

at this point.

Also, in nonequilibrium thermodynamics, among the Onsager kinetic coefficients, only the mobility, $\check{\alpha}$, and particular combinations of $\check{\beta}$ and $\check{\gamma}$ – the thermal conductivity, given by Eq. (2.172), and the thermal-diffusion ratio, given by Eq. (2.177), – are uniquely defined, being independent of the zero-point of the entropy. To illustrate this concept, consider Eq. (2.177) in the form,

$$k_T\frac{\rho D}{T} = -\check{\alpha}\left.\frac{\partial S}{\partial x}\right|_{P,T} + \check{\beta} \tag{2.186}$$

in which the Maxwell relation, $(\partial\mu_{21}/\partial T)|_{P,x} = -(\partial S/\partial x)|_{P,T}$, was used. One finds that since the left-hand side of this equation is uniquely defined, being independent of the difference in zero-points of entropy of each species, $S_{0,2} - S_{0,1}$, then the right-hand side should also be independent of this difference. Therefore, if $\check{\alpha}(\partial S/\partial x)|_{P,T}$ is shifted by an arbitrary constant, $\check{\alpha}(S_{0,2} - S_{0,1})$, then the Onsager coefficient, $\check{\beta}$, must be shifted by the same arbitrary constant [57].

Appendix 2.A Calculating Phase Equilibria in Binary Mixtures

In this appendix, we present examples of a general algorithm for calculating two-phase equilibrium of binary mixtures as shown in Figures 2.20a and 2.21a,b. The conditions for liquid–gas equilibrium at constant temperature and pressure are given by

$$\mu_1(\text{liquid}) = \mu_1(\text{gas}) \qquad \text{and} \qquad \mu_2(\text{liquid}) = \mu_2(\text{gas}) \tag{2.A.1}$$

We first consider the case of complete mutual solubility in the liquid state below the liquid–gas phase transition (Figure 2.20a). We assume that the liquid state is a nonideal solution, while the gaseous state is an ideal-gas mixture. In this case, the equilibrium condition may be expressed at constant temperature and pressure for each species as

$$\mu_1^\circ(\text{liquid}) + RT\ln\gamma_1x_1 = \mu_1^\circ(\text{gas}) + RT\ln y_1 \tag{2.A.2}$$

$$\mu_2^\circ(\text{liquid}) + RT\ln\gamma_2x_2 = \mu_2^\circ(\text{gas}) + RT\ln y_2 \tag{2.A.3}$$

where x_1, x_2, y_1, and y_2 are the molecular fractions of species 1 and 2 in the coexisting liquid and gaseous phases and γ_1 and γ_2 are the corresponding activity coefficients. At constant pressure and temperature, the difference in chemical potentials of each pure species $\Delta\mu_1^\circ = \mu_1^\circ(\text{liquid}) - \mu_1^\circ(\text{gas}) = 0$ and $\Delta\mu_2^\circ = \mu_2^\circ(\text{liquid}) - \mu_2^\circ(\text{gas}) = 0$. Approximating the heats of the phase transition between the liquid and gaseous phases for pure species, ΔH_1° and ΔH_2°, and the activity coefficients, γ_1 and γ_2, to be independent of temperature, Eqs. (2.A.2) and (2.A.3) give

$$\ln\left(\frac{\gamma_1x_1}{y_1}\right) = \frac{\Delta H_1^\circ}{RT_1}\left(\frac{T_1 - T}{T}\right) \tag{2.A.4}$$

$$\ln\left(\frac{\gamma_2x_2}{y_2}\right) = \frac{\Delta H_2^\circ}{RT_2}\left(\frac{T_2 - T}{T}\right) \tag{2.A.5}$$

where T_1 and T_2 are the boiling points of each pure species. Equations (2.A.4) and (2.A.5) define two branches of the liquid–gas coexistence on the phase diagram presented in Figure 2.20a.

The conditions for liquid–gas equilibrium in the case of near-complete immiscibility of the two liquid species ($\gamma_1 x_1 = \gamma_2 x_2 \approx 1$), such as a mixture of water and oil, gives the simplification of Eqs. (2.A.4) and (2.A.5), as

$$\ln y_1 = \frac{\Delta H_1^\circ}{RT_1}\left(\frac{T - T_1}{T}\right) \tag{2.A.6}$$

$$\ln y_2 = \frac{\Delta H_2^\circ}{RT_2}\left(\frac{T - T_2}{T}\right) \tag{2.A.7}$$

The same procedure could be used to describe the solid–liquid phase equilibrium. In the case when both the solid and liquid state are nonideal solutions, the two-phase equilibrium conditions give

$$\ln\left(\frac{\gamma_1^\ell x_1}{\gamma_1^s z_1}\right) = \frac{\Delta H_1^\circ}{RT_1}\left(\frac{T_1 - T}{T}\right) \tag{2.A.8}$$

$$\ln\left(\frac{\gamma_2^\ell x_2}{\gamma_2^s z_2}\right) = \frac{\Delta H_2^\circ}{RT_2}\left(\frac{T_2 - T}{T}\right) \tag{2.A.9}$$

where z_1 and z_2 are the fraction of species in the solid state, while γ_1^s, γ_2^s, γ_1^ℓ, and γ_2^ℓ are the activity coefficients of each species in the solid or liquid state, respectively. In this case, the heats of the phase transition of the pure species, ΔH_1° and ΔH_2° are the heats of melting. These equations describe the solid–liquid transition presented by the upper curves in Figure 2.21b.

Lastly, for the case of complete immiscibility in the solid state ($\gamma_1^s z_1 = \gamma_2^s z_2 = 1$) and complete miscibility in the liquid phases, Eqs. (2.A.8) and (2.A.9) simplify to

$$\ln \gamma_1^\ell x_1 = \frac{\Delta H_1^\circ}{RT_1}\left(\frac{T_1 - T}{T}\right) \tag{2.A.10}$$

$$\ln \gamma_2^\ell x_2 = \frac{\Delta H_2^\circ}{RT_2}\left(\frac{T_2 - T}{T}\right) \tag{2.A.11}$$

being analogous to Eqs. (2.A.6) and (2.A.7). Note that, in this case, the two branches of the solid–liquid coexistence, calculated from Eqs. (2.A.10) and (2.A.11), cross at the eutectic point, where three phases (two pure solid species and their liquid solution) coexist (Figure 2.21a).

References

1 I. Prigogine and R. Defay, *Chemical Thermodynamics*. Longmans, 1965.

2 R. Kubo, *Thermodynamics: An Advanced Course with Problems and Solutions*. Amsterdam: North-Holland, 1968.

3 J. W. Tester and M. Modell, *Thermodynamics and its Applications*. New Jersey: Prentice-Hall, 3rd ed., 1996.

4 J. M. Prausnitz, R. N. Lichtenthaler, and E. G. de Azevedo, *Molecular Thermodynamics of Fluid-Phase Equilibria*. New Jersey: Prentice-Hall, 3rd ed., 1999.

5 S. I. Sandler, *Chemical, Biomolecular, and Engineering Thermodynamics*. Wiley, 4th ed., 2006.

6 F. Weinhold, *Classical and Geometrical Theory of Chemical and Phase Thermodynamics*. Wiley, 2009.

7 J. P. O'Connell and J. M. Haile, *Thermodynamics: Fundamentals for Applications.* Cambridge University Press, 2011.
8 M. Smith, H. V. Ness, and M. M. Abbott, *Introduction to Chemical Engineering Thermodynamics.* McGraw-Hill, 9th ed., 2022.
9 P. G. Debenedetti, *Metastable Liquids: Concepts and Principles.* Princeton University Press, 1996.
10 J.-L. Barrat and J.-P. Hansen, *Basic Concepts for Simple and Complex Liquids.* Cambridge University Press, 2003.
11 P. J. Collings, *Liquid Crystals: Nature's Delicate Phase of Matter*. Princeton University Press, 2002.
12 P.-G. de Gennes and J. Prost, *The Physics of Liquid Crystals.* Clarendon Press, 1993.
13 C. Kittel, *Introduction to Solid State Physics.* Wiley, 2005.
14 H. E. Stanley, "Liquid polyamorphism," in *Advances in Chemical Physics* (S. A. Rice and A. R. Dinner, eds.), vol. 152. Wiley, 2013.
15 H. Tanaka, "Liquid–liquid transition and polyamorphism," *J. Chem. Phys.*, vol. 153, p. 130901, 2020.
16 P. G. Debenedetti, "One substance, two liquids?," *Nature*, vol. 392, pp. 127–128, 1998.
17 P. Gallo, K. Amann-Winkel, C. A. Angell, et al., "Water: A tale of two liquids," *Chem. Rev.*, vol. 116, no. 13, pp. 7463–7500, 2016.
18 A. Nilsson, "Origin of the anomalous properties in supercooled water based on experimental probing inside "no-man's land"," *J. Non-Cryst. Solids X*, vol. 14, p. 100095, 2022.
19 K. Amann-Winkel, K. H. Kim, N. Giovambattista, et al., "Liquid–liquid phase separation in supercooled water from ultrafast heating of low-density amorphous ice," *Nat. Commun.*, vol. 14, p. 442, 2023.
20 P. Gallo, J. Bachler, L. E. Bove, et al., "Advances in the study of supercooled water," *Eur. Phys. J. E*, vol. 44, p. 143, 2021.
21 M. A. Anisimov, M. Duska, F. Caupin, et al., "Thermodynamics of fluid polyamorphism," *Phys. Rev. X*, vol. 8, p. 011004, 2018.
22 F. Caupin and M. A. Anisimov, "Minimal microscopic model for liquid polyamorphism and waterlike anomalies," *Phys. Rev. Lett.*, vol. 127, p. 185701, Oct. 2021.
23 T. J. Longo and M. A. Anisimov, "Phase transitions affected by natural and forceful molecular interconversion," *J. Chem. Phys.*, vol. 156, p. 084502, 2022.
24 N. R. Fried, T. J. Longo, and M. A. Anisimov, "Thermodynamic modeling of fluid polyamorphism in hydrogen at extreme conditions," *J. Chem. Phys.*, vol. 157, p. 101101, 2022.
25 N. A. Shumovskyi, T. J. Longo, S. V. Buldyrev, et al., "Modeling fluid polyamorphism through a maximum-valence approach," *Phys. Rev. E*, vol. 106, p. 015305, 2022.
26 E. A. Guggenheim, "The principle of corresponding states," *J. Chem. Phys.*, vol. 13, pp. 253–261, 1945.
27 J. M. Levelt Sengers, *How Fluids Unmix: Discoveries by the School of Van Der Waals and Kamerlingh Onnes.* Amsterdam: Royal Netherlands Academy of Arts and Sciences, 2002.
28 V. P. Skripov, *Metastable Liquids.* Wiley, 1974.
29 J. Frenkel, "On the transformation of light into heat in solids. I," *Phys. Rev.*, vol. 37, pp. 17–44, 1931.
30 D. Ben-Amotz and D. R. Herschbach, "Correlation of Zeno ($Z = 1$) line for supercritical fluids with vapor-liquid rectilinear diameters," *Isr. J. Chem.*, vol. 30, pp. 59–68, 1990.
31 S. Moghaddam, Y. C. Kim, and M. E. Fisher, "Convergence of fine-lattice discretization for near-critical fluids," *J. Phys. Chem. B*, vol. 109, pp. 6824–6837, 2005.
32 J. H. Hildebrand and R. L. Scott, *Regular Solutions.* Prentice-Hall, 1962.

33 J. S. Rowlinson and F. L. Swinton, *Liquids and Liquid Mixtures*. London: Butterworth Scientific, 3rd ed., 1982.

34 J. M. Levelt Sengers, "Solubility near the solvent's critical point," *J. Supercrit. Fluids*, vol. 4, no. 4, pp. 215–222, 1991.

35 J. M. H. Levelt Sengers, "Critical behavior of fluids: Concepts and applications," in *Supercritical Fluids: Fundamentals for Application* (E. Kiran and J. M. H. L. Sengers, eds.), pp. 3–38. Dordrecht: Springer Netherlands, 1994.

36 P. H. van Konynenburg and R. L. Scott, "Critical lines and phase equilibria in binary van der Waals mixtures," *Philos. Trans. R. Soc. A*, vol. 540, p. 298495, 1980.

37 J. A. Schouten, C. A. Ten Seldam, and N. J. Trappeniers, "The two-component lattice-gas model," *Physica*, vol. 73, no. 3, pp. 556–572, 1974.

38 I. Lawrie and S. Sarbach, "Theory of tricritical points," in *Phase Transitions and Critical Phenomena* (C. Domb and M. S. Green, eds.), vol. 9. New York: Academic Press, 1984.

39 C. M. Knobler and R. L. Scott, "Multicritical points in fluid mixtures: Experimental studies," in *Phase Transitions and Critical Phenomena* (C. Domb and M. S. Green, eds.), vol. 9. New York: Academic Press, 1984.

40 R. L. Scott, "Models for phase equilibria in fluid mixtures," *Acc. Chem. Res.*, vol. 20, no. 3, pp. 97–107, 1987.

41 L. D. Landau and E. M. Lifshitz, *Statistical Physics Part 1*, vol. 5 of Course of Theoretical Physics. Oxford: Pergamon, 3rd ed., 1980.

42 M. A. Anisimov, "Thermodynamic properties of binary solutions along the critical liquid-gas equilibrium line," *Russ. J. Phys. Chem.*, vol. 45, pp. 439–444, 1971.

43 F. Manchester, A. San-Martin, and J. Pitre, "The H-Pd (hydrogen-palladium) system," *J. Phase Equilib.*, vol. 15, pp. 62–83, 1994.

44 J. P. Novák, J. Matouš, and J. Pick, *Liquid–Liquid Equilibria*. Elsevier, 1987.

45 A. A. Novikov, A. P. Semenov, V. Monje-Galvan, et al., "Dual action of hydrotropes at the water/oil interface," *J. Phys. Chem. C*, vol. 121, no. 30, pp. 16423–16431, 2017.

46 K. Huang, *Statistical Mechanics*. Wiley, 2nd ed., 1987.

47 R. Kubo, H. Ichimura, T. Usui, et al., *Statistical Mechanics*. Elsevier Science, 1990.

48 L. E. Reichl, *A Modern Course in Statistical Physics*. Wiley, 2nd ed., 1998.

49 E. M. Lifshitz and L. P. Pitaevskiĭ, *Statistical Physics: Part 2*, vol. 9 of Landau and Lifshitz Course of Theoretical Physics. Oxford: Pergamon, 2nd ed., 1981.

50 D. J. Griffiths, *Introduction to Quantum Mechanics*. Cambridge University Press, 3rd ed., 2018.

51 Y. D. Fomin, V. N. Ryzhov, E. N. Tsiok, and et al., "Dynamics, thermodynamics and structure of liquids and supercritical fluids: Crossover at the Frenkel line," *J. Phys.: Condens. Matter*, vol. 30, no. 13, p. 134003, 2018.

52 C. Cockrell, V. Brazhkin, and K. Trachenko, "Transition in the supercritical state of matter: Review of experimental evidence," *Phys. Rep.*, vol. 941, pp. 1–27, 2021.

53 J. Frenkel, *Kinetic Theory of Liquids*. Oxford University Press, 1947.

54 S. R. de Groot and P. Mazur, *Non-Equilibrium Thermodynamics*. North-Holland, 1963.

55 S. Kjelstrup, D. Bedeaux, E. Johannessen, and et al., *Non-Equilibrium Thermodynamics for Engineers*. World Scientific, 2010.

56 D. Bedeaux, S. Kjelstrup, and J. V. Sengers, eds., *Nonequilibrium Thermodynamics with Applications*. Experimental Thermodynamics Volume X, Cambridge: Royal Society of Chemistry, 2015.

57 L. D. Landau and E. M. Lifshitz, *Fluid Mechanics*, vol. 6 of Course of Theoretical Physics. Oxford: Pergamon Press, 3rd ed., 1966.

58 L. Onsager, "Reciprocal relations in irreversible processes. I.," *Phys. Rev.*, vol. 37, pp. 405–426, 1931.

Exercises

2.1 Formulate the first law for an open single-component system in terms of total heat exchange, mechanical work, and chemical work.

2.2 Define the chemical potential of a single-component system through the fundamental thermodynamic equation combining the first and second laws of thermodynamics.

2.3 Derive the Gibbs–Helmholtz equations for a single-component system.

2.4 Obtain the thermodynamic relation between the molar heat capacity of an individual substance at constant pressure (C_P) and the isochoric molar heat capacity (C_V). Under which conditions does $C_P = C_V$?

2.5 Recall that $C_P = -T(\partial^2 G/\partial T^2)|_P = T(\partial S/\partial T)|_P$ is the isobaric molar heat capacity. Prove that the thermodynamic condition $C_P \geq C_V$ follows from the mechanical stability criteria, while $C_V \geq 0$ is independent of the thermal stability condition.

2.6 Under which condition(s) does the isobaric heat capacity diverge, $C_P \to \infty$? When is the isobaric heat capacity zero?

2.7 Derive the Clausius–Clapeyron equation, $\ln P_{\text{cxc}} = A + B/T$, where P_{cxc} is the gas saturation pressure. Why is the Clausius–Clapeyron equation approximate, while the Clapeyron equation Eq. (2.62) is exact? What is the physical meaning of the constant B? Formulate all the approximations and assumptions you used.

2.8 Derive the expression for the Joule–Thomson coefficient. Under which conditions is this coefficient zero?

2.9 Explain the Gibbs phase rule. How many phases can coexist in a ternary system?

2.10 Derive the relation between the mole fraction and the mass fraction. Plot (on one graph) this relation for the solutions of methanol and decanol in water. What is the difference?

2.11 Prove that the condition of material stability for a binary solution, Eq. (2.63), coincides with the condition of mechanical stability Eq. (2.106) in the limit $x \to 0$.

2.12 If the chemical potential of solvent, μ_1, is chosen to serve as a field-dependent thermodynamic potential for a binary mixture, what canonical variables belong to this thermodynamic function? *Hint:* use the Legendre transformation $G - \mu_{21}x = \mu_1$ and obtain the total differential.

2.13 Derive the Gibbs–Duhem equation for a binary solution.

2.14 If $(\partial\mu_{21}/\partial x_2)|_{P,T} \geq 0$, then why is $(\partial\mu_2/\partial x_2)|_{P,T} \geq 0$ and $(\partial\mu_1/\partial x_1)|_{P,T} \geq 0$ as well? Use the Gibbs–Duhem equation to prove this.

2.15 The standard molar Helmholtz energy for a binary mixture has the canonical variables: T, V, and x. Using a Legendre transformation, develop an analog of the molar Helmholtz energy through the total differential with the canonical variables: T, V, and μ_{21}.

2.16 Represent the osmotic pressure of a binary fluid mixture as a power series in terms of the solute concentration (solute partial density). Obtain the expressions for the second and third virial coefficients.

2.17 Explain the protocol to compute the phase diagram for the liquid–gas phase equilibria in nonideal binary solutions. Plot temperature as a function of concentration in liquid and gaseous states.

2.18 The regular-solution model for the molar Gibbs energy reads

$$G = \mu_1^\circ(1 - x) + \mu_2^\circ x + \Delta G_{\text{mix}} = \mu_1^\circ + \mu_{21}^\circ x + \Delta G_{\text{mix}} \tag{2.187}$$

where μ_1°, μ_2°, and $\mu_{21}^\circ = \mu_2^\circ - \mu_1^\circ$ are the bulk properties of species 1 and 2, while the Gibbs energy of mixing, ΔG_{mix}, for the two species is

$$\Delta G_{\text{mix}} = -T\Delta S_{\text{mix}} + \Delta H_{\text{mix}} = RT[(1 - x)\ln(1 - x) + x\ln x] + \omega x(1 - x) \tag{2.188}$$

where $\omega = 2RT_c$ is an interaction (nonideality) constant. Adopting the critical temperature, $T_c = 300$ K, obtain the equations for the liquid–liquid spinodal and binodal as well as the conditions for the liquid–liquid critical point (LLCP) of demixing. Plot the temperature as a function of the molecular fraction, x, for both the binodal and spinodal on the same plot. Plot ΔG_{mix} at $T = 310$ K and $T = 290$ K. Identify the stable, metastable, and unstable states on the diagram.

2.19 Using Maxwell's equal area rule, obtain the binodal for the van der Waals equation of state. Plot the binodal on a temperature–density and a pressure–temperature diagram.

2.20 Represent the van der Waals and lattice-gas equations of state in terms of reduced (by the critical parameters) variables.

2.21 (a) Obtain and plot the van der Waals spinodal for carbon dioxide. (b) Obtain the lattice-gas spinodal and binodal and plot them in terms of reduced variables. (c) Compare the van der Waals and lattice-gas spinodals. What is the difference between them?

2.22 Derive the chemical potential (Gibbs molar energy) of the van der Waals fluid.

2.23 Find the condition $PV/k_BT = 1$ (known as the "Zeno line") for the van der Waals fluid. Explain the linearity of the Zeno line as a function of density.

2.24 Calculate the second virial coefficient for the van der Waals fluid obeying the Lennard–Jones potential. Calculate the critical parameters expressed as functions of the well depth, ϵ, and the collision diameter, σ.

2.25 Plot the second virial coefficient of argon as a function of temperature by assuming the van der Waals equation of state for argon. Explain the physical consequences of the second virial coefficient changing from negative to positive upon increasing the temperature.

2.26 Consider a phonon, a quasiparticle that originates from thermal vibration of a crystal lattice. These vibrations generate acoustic waves with energy of a phonon $E = \hbar\omega$ where ω is the frequency of vibration and $\hbar$ is the reduced (by 2π) Planck's constant. Assuming a classical "ideal gas" of phonons in the solid state, what would be the temperature dependence of the frequency of a vibration?

2.27 Derive the thermo-diffusion and baro-diffusion ratios. What kinetic coefficients effect the thermo-diffusion ratio? Explain: Why does the baro-diffusion ratio not depend on kinetic coefficients?

2.28 Calculate the baro-diffusion coefficient, k_P, for a mixture of toluene and benzene ($x = 0.5$) at 1 bar and 300 K. Assume the mixture is an ideal solution. Explain why the coefficient does not depend on the absolute value of entropy.

2.29 Consider a closed vial containing a binary liquid mixture. The temperature in the vial is not uniform. Is this system in equilibrium? What can you say about the concentration of the mixture. Is it spatially uniform? Under which condition would you expect a steady-state pattern of concentration inhomogeneities?

2.30 Derive the relation between the thermal conductivity and the Onsager transport coefficients: $\kappa = \hat{\gamma} - (\hat{\beta}^2 T/\hat{\alpha})$. Explain why κ and $\hat{\alpha}$ do not depend on the absolute value of the entropy, while $\hat{\gamma}$ and $\hat{\beta}$ do.

2.31 Obtain a relation between the "zero point" of the chemical potential, μ_0, and the absolute values of the entropy and internal energy for an ideal gas. Is the zero point of the chemical potential uniquely defined?

3

Fluctuations and Brownian Motion

This chapter includes:

- Origin of thermal fluctuations
- Fluctuations of thermodynamic properties
- Fluctuations of a generalized density
- Brownian motion
- Probing fluctuations and Brownian motion

A key feature that determines the thermodynamic behavior of mesoscopic systems is the existence of thermal fluctuations, caused by random molecular motion.

3.1 Probability and Work of Fluctuations

Fluctuations are spontaneous random deviations in the physical properties from their average equilibrium values. Fluctuations of thermodynamic properties are caused by thermal molecular motion and are inherent to all statistical systems. Macroscopic thermodynamics ignores fluctuations, because in the thermodynamic limit ($N \to \infty$), they do not affect thermodynamic properties. Thus, in practice, in macroscopic (although technically finite) systems, the effects of fluctuations can usually be neglected.

However, the situation drastically changes when the system becomes very small or the system (even a macroscopic one) is near the limit of thermodynamic stability. In these two cases, fluctuations may become very large and may play a dominant role in the behavior of thermodynamic and transport properties[1]. A general approach to introduce fluctuations into thermodynamics of equilibrium systems is given by statistical mechanics. In this chapter, we conceptually follow the approach developed by Landau and Lifshitz [1].

Let us consider an arbitrarily small, but still mesoscopic (i.e. containing a large number of microscopic particles), domain of an isolated system. Macroscopically, this domain is in thermal equilibrium with its surroundings at temperature, $\overline{T}$, and pressure, $\overline{P}$. Thermal molecular motion

1 Note that fluctuations in fluids away from equilibrium always become large and long ranged. This phenomenon, one of the major discoveries of recent decades, has been predicted theoretically and confirmed experimentally [J. R. Dorfman, T. R. Kirkpatrick, and J. V. Sengers, "Generic long-range correlations in molecularfluids," *Annu. Rev. Phys. Chem.*, Vol. 45, pp. 213–239, 1994.] [J. M. Ortiz de Zrate and J. V. Sengers, *Hydrodynamic Fluctuations in Fluids and Fluid Mixtures*, Elsevier, Amsterdam, 2006]. In this book, we consider only fluctuations in systems that are in thermodynamic equilibrium.

Mesoscopic Thermodynamics for Scientists and Engineers, First Edition. Mikhail A. Anisimov and Thomas J. Longo.
© 2024 John Wiley & Sons, Inc. Published 2024 by John Wiley & Sons, Inc.

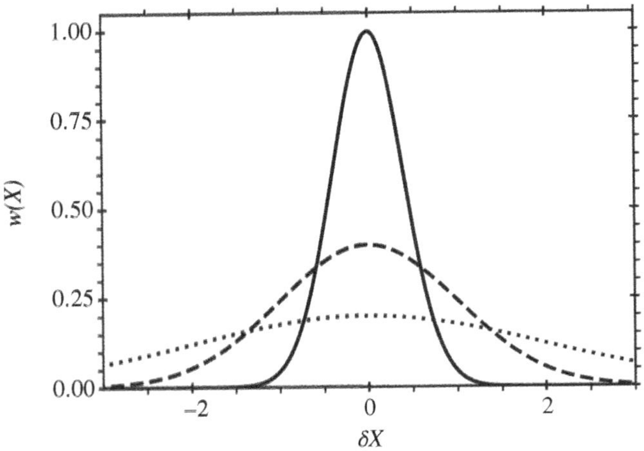

Figure 3.1 One-dimensional Gaussian probability-distribution function obtained from Eq. (3.2) for three different values of the standard deviation, $\sigma = \sqrt{\langle(\delta X)^2\rangle}$: $\sigma = 0.4$ (solid), $\sigma = 1$ (dashed), and $\sigma = 2$ (dotted).

causes fluctuations of various thermodynamic properties of the domain. If the domain has a fixed volume, the number of particles (and thus, the density) would fluctuate. Alternatively, if the domain has a fixed number of particles, then the volume would fluctuate.

Assume that fluctuations of the property X, $\delta X = X - \overline{X}$ (where $\overline{X}$ is the equilibrium value of X), are random, small, noninteractive, and thus, Gaussian. Thus, the mean square of a fluctuation, $\langle(\delta X)^2\rangle$, is the same as the square of the standard deviation, σ, in a Gaussian distribution (see Figure 3.1)

$$\sigma^2 = \langle(\delta X)^2\rangle = \int_{-\infty}^{\infty} (\delta X)^2 w(X)\, \mathrm{d}X \tag{3.1}$$

The Gaussian probability-distribution function (known as the "probability density"), $w(X)$, for a one-dimensional distribution is given by

$$\boxed{w(X) = \frac{1}{\sqrt{2\pi\langle(\delta X)^2\rangle}} \exp\left\{-\frac{(\delta X)^2}{2\langle(\delta X)^2\rangle}\right\}} \tag{3.2}$$

The probability distribution has a maximum when $\overline{X} = 0$ and decreases exponentially as the deviation increases (equally positive or negative). The probability, p_r, to find the value of X between any two values, X_1 and X_2, is given by

$$p_r = \int_{X_1}^{X_2} w(X)\, \mathrm{d}X \tag{3.3}$$

as illustrated in Figure 3.2. The probability to observe a value of δX within one standard deviation of the mean is approximately 68% and within two standard deviations is about 95%. The total area under the curve $w(X)$ is normalized such that the probability to observe any value of δX is 100%,

$$p_r = \int_{-\infty}^{\infty} w(X)\, \mathrm{d}X = 1 \tag{3.4}$$

One may see that from Eqs. (3.1) and (3.2), if fluctuations are Gaussian, then the only parameter that controls the probability distribution is the mean square value of the fluctuation. How could we relate this parameter to fluctuations of thermodynamic properties?

Strictly speaking, fluctuations exist in apparent "violation" of the second law of thermodynamics. Any fluctuation will decrease the total entropy, S_t, of the isolated system with respect to its

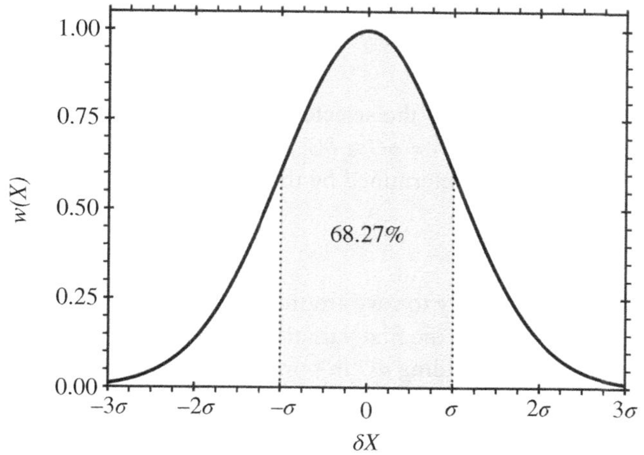

Figure 3.2 The probability to find the value of X deviating from its average within $\pm\sigma$ is about 68%. This is about 68% of the total area under the curve $w(X)$.

equilibrium (maximum) value. Hence, the probability density of a fluctuation is

$$\boxed{w \propto e^{\delta S_t/k_B}} \tag{3.5}$$

where δS_t is the negative change of the entropy caused by a fluctuation. The question could arise: what happens to the total entropy of the system, if fluctuations continuously emerge at every possible moment and at every possible point within the system? The answer is given by nonequilibrium thermodynamics: once a fluctuation inhomogeneity has emerged, it will generate a gradient of a thermodynamic field causing a flux of matter or heat toward equilibrium. Hence, the fluctuation will eventually dissipate, relaxing toward equilibrium. This dissipation is accompanied by a positive change of the entropy, δS_t, which restores the total entropy to its maximum (equilibrium) value. Therefore, the second law is not violated!

According to the first law, given by Eq. (2.2), for a closed system,

$$\Delta U_t = Q + W \tag{3.6}$$

where $\Delta U_t = 0$ for an isolated system, $Q = \overline{T}\delta S_t$ that denotes a reversible heat exchange of the selected domain with its surroundings in the system, and $W = -Q$ is the work of the surroundings required to create a fluctuation inhomogeneity. Thus, the work of a fluctuation, W_f, is expressed through the entropy change due to the fluctuation,

$$W_f \equiv W = -\overline{T}\delta S_t \tag{3.7}$$

Since fluctuations (by definition) are deviations from equilibrium for the entire system, then the heat Q originates from internal (fluctuation induced) processes and is compensated by the work of fluctuations. Since the fluctuations cause the total entropy of the system to decrease, then the reversible work of a fluctuation is always positive. Thus, the work of fluctuations can be regarded as the work of an "external force" (with respect to the fluctuating domain) originating from the thermal molecular motion of the surroundings.

Consider variations of the total thermodynamic properties, namely the internal energy (δU_t^*), the volume (δV_t^*), and the entropy (δS_t^*), within the selected mesoscopic domain (denoted with the superscript *), assuming that the temperature and pressure of the surroundings have equilibrium values, $\overline{T}$ and $\overline{P}$. When the mesoscopic domain is driven from equilibrium at constant temperature and pressure, for a single-component substance, the fluctuation-induced increase of the total Gibbs energy of the domain, $G_t^* = U_t^* + \overline{P}V_t^* - \overline{T}S_t^*$, may be expressed in variational form as

$$\delta G_t^* = \delta U_t^* - \overline{T}\delta S_t^* + \overline{P}\delta V_t^* \tag{3.8}$$

Consider the case when the number of molecules within the selected domain, $N = N^*$, is constant, then the work of a fluctuation per molecule is $W_f/N^* = \delta G \equiv \delta G_t^*/N^*$. Since the equilibrium conditions, at constant temperature and pressure, are determined by the minimum of the Gibbs energy (see Section 2.4), the work of a fluctuation is

$$W_f = N^*\delta G = N^*[\delta U - \overline{T}\delta S + \overline{P}\delta V] \tag{3.9}$$

Since thermal molecular motion causes the Gibbs energy to vary around its minimum value, the work of the fluctuation, being always positive, is zero in the first-variational order approximation and appears only in the second-variational order. Expanding δU in powers of δS and δV around the equilibrium state gives the second-order approximation

$$W_f = N^*\delta G = \frac{N^*}{2}\left[\frac{\partial^2 U}{\partial S^2}(\delta S)^2 + 2\frac{\partial^2 U}{\partial S \partial V}(\delta S\delta V) + \frac{\partial^2 U}{\partial V^2}(\delta V)^2\right] \tag{3.10}$$

By representing the variations of δT and δP through the partial derivatives of the internal energy, with use of a Maxwell relation, $(\partial T/\partial V)|_S = -(\partial P/\partial S)|_V$, then

$$\delta T = \delta\left(\frac{\partial U}{\partial S}\right)\bigg|_V = \frac{\partial^2 U}{\partial S^2}\bigg|_V \delta S + \frac{\partial T}{\partial V}\bigg|_S \delta V \tag{3.11}$$

and

$$\delta P = \delta\left(\frac{\partial U}{\partial V}\right)\bigg|_S = \frac{\partial^2 U}{\partial V^2}\bigg|_S \delta V + \frac{\partial P}{\partial S}\bigg|_V \delta S \tag{3.12}$$

and one obtains the work of a fluctuation in the form

$$W_f = \frac{N^*}{2}\left[\delta\left(\frac{\partial U}{\partial S}\right)_V \delta S + \delta\left(\frac{\partial U}{\partial V}\right)_S \delta V\right] = \frac{N^*}{2}(\delta S\delta T - \delta P\delta V) \tag{3.13}$$

Expressing variations of entropy and pressure through the variations of temperature and volume as

$$\delta S = \frac{C_V}{T}\delta T + \frac{\partial P}{\partial T}\bigg|_V \delta V \tag{3.14}$$

$$\delta P = \frac{\partial P}{\partial T}\bigg|_V \delta T + \frac{\partial P}{\partial V}\bigg|_T \delta V \tag{3.15}$$

one obtains the work of a fluctuation, reduced by $k_B T$, consisting of two parts

$$\boxed{\frac{W_f}{k_B T} = \frac{N^*}{2}\left[\frac{C_V}{2k_B T^2}(\delta T)^2 - \frac{(\partial P/\partial V)|_T}{2k_B T}(\delta V)^2\right]} \tag{3.16}$$

This equation represents the work needed to create a fluctuation of volume at constant temperature (illustrated by Figure 3.3) and the work needed to create a fluctuation of temperature at constant volume.

Hence, the probability density of a fluctuation that involves both temperature and volume variations, given by Eq. (3.5), is

$$\boxed{w \propto e^{-W_f/k_B T} = \exp\left\{-\frac{N^* C_V}{2k_B T^2}(\delta T)^2 + \frac{N^*(\partial P/\partial V)|_T}{2k_B T}(\delta V)^2\right\}} \tag{3.17}$$

Comparing this equation with the expression for a Gaussian distribution, Eq. (3.2), one obtains three important results. First, the mean-square fluctuations of volume per molecule is given by

$$\langle(\delta V)^2\rangle = -\frac{k_B T}{N^*}\frac{\partial V}{\partial P}\bigg|_T \tag{3.18}$$

while the fluctuations of volume of the selected domain are given by

$$\langle(\delta V_{\rm t}^*)^2\rangle = -N^* k_{\rm B} T \left.\frac{\partial V}{\partial P}\right|_T \tag{3.19}$$

which increases with an increase of the number of molecules in the domain. Second, the mean-square fluctuations of the temperature read

$$\langle(\delta T)^2\rangle = \frac{k_{\rm B} T^2}{N^* C_V} \tag{3.20}$$

and third, the cross correlation between the fluctuations of temperature and volume is

$$\langle \delta T \delta V \rangle = 0 \tag{3.21}$$

Thus, indicating that fluctuations of temperature and volume are statistically independent. Note that the smaller the selected domain, with a selected number of particles (N^*), the larger the fluctuations of volume per molecule and temperature.

3.2 Fluctuations of Density

Consider a mesoscopic domain with a fixed volume, $V_{\rm t}^*$, in an isolated system. The number of particles in the domain, N, (and, hence, the molecular density, $\rho = N/V_{\rm t}^*$) can fluctuate. In contrast to the previously selected domain (with a constant number of molecules), this mesoscopic domain may be considered as an open subsystem within the overall isolated system. An example of such a domain is the scattering volume defined by the geometry of a light scattering spectrometer (see Figure 3.3). The equilibrium conditions for this subsystem are determined from the minimum of the grand thermodynamic potential per unit volume, $\Omega^*/V_{\rm t}^* = -P$. For a single-component system, $-P = \rho U - \overline{T}\rho S - \overline{\mu}\rho$, and the fluctuation-induced change of the grand potential of the mesoscopic domain, when it is driven from equilibrium at constant temperature and chemical potential of the surroundings, is

$$-P = \delta(\rho U) - \overline{T}\delta(\rho S) - \overline{\mu}\delta\rho \tag{3.22}$$

Expanding the density of the internal energy, ρU, in terms of the temperature and the density, one obtains the work of fluctuations around the minimum of the grand potential in a form similar to Eq. (3.13) as

$$\delta W_{\rm f} = -PV_{\rm t}^* = \frac{1}{2} V_{\rm t}^* \left[\delta(\rho S)\delta T + \delta\mu\delta\rho\right] \tag{3.23}$$

Therefore, when the number of particles fluctuates at constant volume and temperature, one obtains the work of fluctuations as

$$\boxed{\frac{W_{\rm f}}{k_{\rm B}T} = \frac{V_{\rm t}^*}{2k_{\rm B}T}\left[\frac{\rho C_V}{T}(\delta T)^2 + \left.\frac{\partial \mu}{\partial \rho}\right|_T (\delta\rho)^2\right]} \tag{3.24}$$

Comparing this expression with the Gaussian distribution Eq. (3.2), one obtains the same expression for the fluctuations of temperature, as given by Eq. (3.20),

$$\langle(\delta T)^2\rangle = \frac{k_{\rm B} T^2}{\rho V_{\rm t}^* C_V} = \frac{k_{\rm B} T^2}{N^* C_V} \tag{3.25}$$

since $\rho V_t^* = N^*$. Also, note that the fluctuations of density and temperature are statistically independent, as $\langle \delta T \delta \rho \rangle = 0$. Also, one finds that the density fluctuations are given by

$$\langle (\delta \rho)^2 \rangle = \frac{k_B T}{V_t^*} \left. \frac{\partial \rho}{\partial \mu} \right|_T \tag{3.26}$$

Expressing the derivative $(\partial \rho / \partial \mu)|_T$ through the isothermal compressibility, κ_T, given by Eq. (2.47), as

$$\kappa_T = \frac{1}{\rho} \left. \frac{\partial \rho}{\partial P} \right|_T = \frac{1}{\rho^2} \left. \frac{\partial \rho}{\partial \mu} \right|_T \tag{3.27}$$

one finds that the mean-squared fluctuations of the density, given by Eq. (3.26), may also be expressed as

$$\boxed{\langle (\delta \rho)^2 \rangle = \frac{k_B T}{V_t^*} \rho^2 \kappa_T} \tag{3.28}$$

An important property to evaluate the magnitude of the density fluctuations is the product, $\langle (\delta \rho)^2 \rangle V_t^*$, which is uniquely determined by the thermodynamic state of the system, being independent of the selected domain. As the condition

$$\left. \frac{\partial P}{\partial \rho} \right|_T = \rho \left. \frac{\partial \mu}{\partial \rho} \right|_T = 0 \tag{3.29}$$

defines the mechanical stability limit (spinodal), the mean-square value of the density fluctuations depends on how close the substance is to instability. In particular, one should expect diverging fluctuations of the density at the liquid-gas critical point and along the liquid–gas spinodal.

Correspondingly, one could obtain the fluctuations of the number of particles in the selected volume as

$$\langle (\delta N)^2 \rangle = k_B T V_t^* \left. \frac{\partial \rho}{\partial \mu} \right|_T = N k_B T \left. \frac{\partial \rho}{\partial P} \right|_T \tag{3.30}$$

It is important to note that the fluctuations of the number of molecules and fluctuations of the total volume (like any fluctuation of extensive properties) diverge when the number of molecules or volume tends to infinity (in the thermodynamic limit). In particular, for an ideal gas ($P = \rho k_B T$), Eq. (3.30) is reduced to

$$\langle (\delta N)^2 \rangle = N \tag{3.31}$$

However, as this result depends on the selected volume, it does not have much practical use; instead, the intensive properties are more useful. For example, the relative root-mean-square (RMS) fluctuation of the number of molecules in an ideal gas is given by

$$\sqrt{\frac{\langle (\delta N)^2 \rangle}{N^2}} = \sqrt{\frac{1}{N}} \tag{3.32}$$

This equation is known as the "noise" formula for a RMS fluctuation of non-interacting particles, quasiparticles, or (more generally) independent degrees of freedom.

Similarly, the relative fluctuations of molecular volume and density, given by reducing Eqs. (3.18) and (3.28) by V^2 and ρ^2, respectively, yield

$$\frac{\langle (\delta V)^2 \rangle}{V^2} = \frac{\langle (\delta \rho)^2 \rangle}{\rho^2} = \frac{k_B T}{V_t^*} \kappa_T \tag{3.33}$$

Therefore, the relative fluctuations of the molecular volume and density are identical.

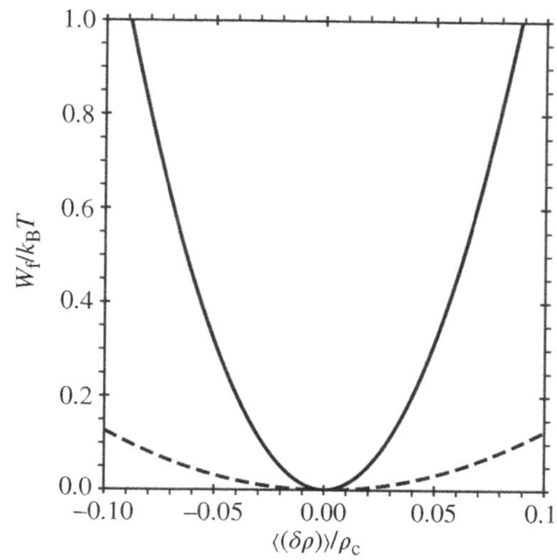

Figure 3.3 Work of the relative fluctuations of the molecular density (or volume) in a domain of a van der Waals fluid containing, on average, 100 molecules at distances to the liquid-gas critical temperature: $\Delta\hat{T} = (T - T_c)/T_c \cong 0.03$ (solid) and $\Delta\hat{T} \cong 0.003$ (dashed).

Figure 3.3 demonstrates the work of fluctuations with respect to the variations of the reduced molecular volume or density in the van der Waals fluid. Note that the work of a fluctuation decreases upon approaching the critical temperature.

3.3 Fluctuations of Entropy

Based on the approach presented in Sections 3.1 and 3.2, one can obtain the fluctuations of other thermodynamic quantities. For example, consider variations of pressure and entropy to be the independent variables in Eq. (3.13) for the work of fluctuations at constant N^*. Following the same procedure, one obtains the fluctuations of the entropy per molecule as

$$\boxed{\langle(\delta S)^2\rangle = \frac{k_B C_P}{N^*}} \tag{3.34}$$

These fluctuations are of special interest for meso-thermodynamics. Since $\langle(\delta S)^2\rangle$ is proportional to the easily observable thermodynamic property, C_P, an anomaly in the isobaric heat capacity may indicate strong order-disorder fluctuations. Additionally, the fluctuations of pressure are given by

$$\langle(\delta P)^2\rangle = -\frac{k_B T}{N^*}\left.\frac{\partial P}{\partial V}\right|_S \tag{3.35}$$

The mean-square fluctuations of pressure are proportional to the square of the thermodynamic speed of sound in the zero frequency limit, $u_S^2 = (1/M_w)(\partial P/\partial\rho)|_S$, where M_w is the molecular (or molar) weight. Also, note that in analogy with Eq. (3.21), fluctuations of entropy and pressure are independent as $\langle\delta S\delta P\rangle = 0$.

Unlike the fluctuations of volume and temperature, density and temperature, or entropy and pressure, the fluctuations of both thermodynamic "coordinates" of the entropy and volume as well as the entropy and density are correlated by

$$\langle\delta V\delta S\rangle = \frac{k_B T}{N^*}\left.\frac{\partial V}{\partial T}\right|_P \tag{3.36}$$

Similarly, if variations of the entropy per unit volume (the entropy density), $\delta(\rho S)$, and variations of the chemical potential, $\delta\mu$, are the independent variables in Eq. (3.23), for the work of fluctuations at constant volume, V_t^*, one obtains the fluctuations of the chemical potential as

$$\langle(\delta\mu)^2\rangle = \frac{k_B T}{V_t^*}\frac{\partial\mu}{\partial\rho}\bigg|_S \tag{3.37}$$

In addition, the fluctuations of the entropy density are given by

$$\boxed{\langle[\delta(\rho S)]^2\rangle = \frac{k_B \rho C_\mu}{V_t^*}} \tag{3.38}$$

where $C_\mu = (1/T)(\partial S/\partial T)|_\mu$ is the heat capacity at constant chemical potential. Note that C_μ is a thermodynamic property that depends on the zero point of the entropy (see Section 2.14) and is related to the isochoric heat capacity as

$$C_\mu = C_V - \left(\frac{\partial\mu}{\partial T}\right)_V^2 \frac{\partial V}{\partial\mu}\bigg|_T \tag{3.39}$$

Unlike the well-defined property $(\partial P/\partial T)|_V$ in the difference $C_P - C_V$, given by Eq. (2.65), the derivative $(\partial\mu/\partial T)|_V$ may be shifted by selecting the zero point of the entropy. If, $(\partial\mu/\partial T)|_V = 0$, then $C_\mu = C_V$.

3.4 Fluctuations of Concentration in a Binary Mixture

Let us consider a fluctuation of the number of molecules of a solute when the temperature, pressure, and total number of molecules in the domain, $N_1 + N_2 = N^*$, are constants. Then, $\delta N_2 = -\delta N_1$, and the molecular fraction is

$$\delta x = \frac{\delta N_2}{N^*} = -\frac{\delta N_1}{N^*} \tag{3.40}$$

The work of fluctuations is given by the variation of the Gibbs energy at constant temperature and pressure as,

$$W_f = N^*\delta G = N^*\mu_{21}\delta x \tag{3.41}$$

Expanding δG in powers of δx at constant pressure and temperature, one obtains in the lowest (second-variation order) approximation

$$W_f = \frac{1}{2}N^*\frac{\partial\mu_{21}}{\partial x}\bigg|_{P,T}(\delta x)^2 \tag{3.42}$$

Consequently, for the fluctuations of molecular fraction, one finds

$$\langle(\delta x)^2\rangle = \frac{k_B T}{N^*}\frac{\partial x}{\partial\mu_{21}}\bigg|_{P,T} \tag{3.43}$$

If one expresses the variation of the Gibbs energy through the partial density, $c = \rho x$, then the fluctuations of this property, with use of the Gibbs–Duhem equation given by Eq. (2.97), are obtained as

$$\boxed{\langle(\delta c)^2\rangle = \frac{k_B T}{V^*}\frac{\partial c}{\partial\mu_{21}}\bigg|_{P,T} = \frac{k_B T c}{V_t^*}\frac{\partial c}{\partial\Pi}\bigg|_{P,T}} \tag{3.44}$$

One may also express Eq. (3.44) in reduced form through the osmotic compressibility, κ_Π given by Eq. (2.100), as

$$\frac{\langle(\delta c)^2\rangle}{c^2} = \frac{k_B T}{V_t^*}\kappa_\Pi \tag{3.45}$$

Note that the product, $\langle(\delta c)^2\rangle V_t^*$, is uniquely determined by the thermodynamic state of the system, being independent of the selected domain.

Fluctuations of Thermodynamic Properties Part I: Thermodynamic Coordinates and Fields

- Volume and pressure for pure substances (constant N^*)

$$\langle(\delta V_t)^2\rangle = -N^* k_B T \left.\frac{\partial V}{\partial P}\right|_T; \qquad \langle(\delta P)^2\rangle = \frac{k_B T}{N^*}\left.\frac{\partial P}{\partial V}\right|_{S_t}$$

- Entropy and temperature for pure substances

$$\langle(\delta S_t)^2\rangle = N^* k_B C_P; \qquad \langle(\delta T)^2\rangle = \frac{k_B T^2}{N^* C_V}$$

- Number of particles and chemical potential for pure substances (constant V_t^*)

$$\langle(\delta N)^2\rangle = V_t^* k_B T \left.\frac{\partial \rho}{\partial P}\right|_T; \qquad \langle(\delta \mu)^2\rangle = \frac{k_B T}{V_t^*}\left.\frac{\partial \mu}{\partial \rho}\right|_S$$

- Density and entropy density for pure substances

$$\langle(\delta \rho)^2\rangle = \frac{k_B T}{V_t^*}\rho^2 \kappa_T; \qquad \langle[\delta(\rho S)]^2\rangle = \frac{k_B \rho C_\mu}{V_t^*}$$

- Molecular Fraction and concentration in a binary mixture

$$\langle(\delta x)^2\rangle = \frac{k_B T}{N^* \left(\partial \mu_{21}/\partial x\right)|_{P,T}}; \qquad \langle(\delta c)^2\rangle = \frac{k_B T}{V_t^*}\kappa_\Pi$$

3.5 Fluctuations of a Generalized Density

The results of the fluctuations of the thermodynamic densities (introduced in Section 2.4) may be generalized to any density-like variable, φ_i, which represents a density of the property i per unit volume. Let h_i be a generalized field conjugate to φ_i and let h_j be another independent generalized field of the property j, conjugate to another generalized density φ_j. The generalized density may not necessarily be a scalar, but it can also be a vector or a tensor. The number of independent fields and densities depends on the number of independent degrees of freedom in the system. Thus, the fluctuations of φ_i are given as

$$\boxed{\langle(\delta \varphi_i)^2\rangle = \frac{k_B T}{V_t^*}\chi_i} \tag{3.46}$$

where χ_i is a generalized *susceptibility*, defined as

$$\chi_i = \left.\frac{\partial \varphi_i}{\partial h_i}\right|_{h_j} \tag{3.47}$$

whose units depend on the units of the generalized density and the generalized field. For example, if the generalized density is $\varphi_i = \rho$ and the generalized field is $h_i = \mu$, then the susceptibility is $\chi_i = \rho^2\kappa_T$, see Eq. (3.28). Alternatively, if c is the generalized density, $\varphi_i = c$, then the role of the generalized susceptibility in a binary system is played by the osmotic compressibility, see Eq. (3.44).

For a ferromagnet, the generalized density is the magnetization per unit volume, while the generalized field is the magnetic field. For some other density-like variables, such as the energy density or the enthalpy density, the corresponding susceptibilities will be a combination of several thermodynamic derivatives. The fluctuations of the generalized density will always diverge at an appropriate stability limit (spinodal) defined as

$$\left.\frac{\partial h_i}{\partial \varphi_i}\right|_{h_j} = 0 \tag{3.48}$$

Note that depending on the number of independent degrees of freedom, there can be more than one secondary field, h_j.

Fluctuations of Thermodynamic Properties Part II: Energy and Enthalpy

- Internal energy for pure substances

$$\langle(\delta U_t)^2\rangle = -N^* k_B T\left[T\left.\frac{\partial P}{\partial T}\right|_V - P\right]^2 \left.\frac{\partial V}{\partial P}\right|_T + N^* k_B T^2 C_V$$

obtained from the expansion of the internal energy with V and T as independent variables.

- Enthalpy for pure substances

$$\langle(\delta H_t)^2\rangle = -N^* k_B T V^2 \left.\frac{\partial P}{\partial V}\right|_S + N^* k_B T^2 C_P$$

obtained from the expansion of the internal energy with P and S as independent variables.

The correlation of any density-variable fluctuations can be obtained from

$$\boxed{\langle \delta\varphi_i \delta\varphi_j\rangle = \frac{k_B T}{V_t^*}\chi_{ij}} \tag{3.49}$$

where the cross susceptibility is defined as $\chi_{ij} = \left.(\partial\phi_i/\partial h_j)\right|_{h_i}$. Also, as indicated in Sections 3.1–3.4, one may note that while the fluctuations of two generalized densities or fields are correlated, the fluctuations of a density and a field variable are uncorrelated, $\langle \delta\varphi_i \delta h_i\rangle = 0$. Since in a generalized ideal gas of particles or quasiparticles, the susceptibility is

$$\chi_i = \frac{\varphi_i}{k_B T} \tag{3.50}$$

Equation (3.46) yields the relative RMS fluctuation of the generalized density for the generalized ideal gas, in a form equivalent to Eq. (3.31), as

$$\sqrt{\frac{\langle(\delta\varphi_i)^2\rangle}{\varphi_i}} = \frac{1}{\sqrt{N_i^*}} \tag{3.51}$$

where N_i^* is equal to the number of particles or quasiparticles in the selected domain. This is a general result for the relative RMS fluctuations of any statistically independent degrees of freedom.

Fluctuations of Thermodynamic Properties Part III: Cross-Term Fluctuations

- Volume–temperature and volume–pressure cross terms for pure substances

$$\langle \delta V_t \delta T \rangle = 0; \qquad \langle \delta V_t \delta P \rangle = -k_B T$$

- Entropy–volume and entropy–temperature cross terms for pure substances

$$\langle \delta S_t \delta V_t \rangle = k_B T \left. \frac{\partial V_t}{\partial T} \right|_P; \qquad \langle \delta S_t \delta T \rangle = k_B T$$

- Temperature–pressure cross term for pure substances

$$\langle \delta T \delta P \rangle = \frac{k_B T^2}{N^* C_V} \left. \frac{\partial P}{\partial T} \right|_{V_t}$$

- Concentration–temperature and concentration–pressure cross terms in a binary mixture

$$\langle \delta c \delta T \rangle = 0; \qquad \langle \delta c \delta P \rangle = 0$$

3.6 Fluctuations of a Generalized Field

The fluctuations of temperature, given by Eq. (3.20), is a particular case of the fluctuations of a field variable. It may be rewritten in the form

$$\langle (\delta T)^2 \rangle = \frac{k_B T}{V_t^*} \left. \frac{\partial T}{\partial (\rho S)} \right|_\rho \tag{3.52}$$

and generalized to any field variable, $\hat{h}_i$, as

$$\langle (\delta \hat{h}_i)^2 \rangle = \frac{k_B T}{V_t^*} \chi_j \tag{3.53}$$

where $\chi_j = \partial T / \partial(\rho S)|_\rho = T/\rho C_V$ (for the case of temperature fluctuations), which is a "secondary" susceptibility defined, in general, as

$$\chi_j = \left. \frac{\partial \phi_i}{\partial h_i} \right|_{\phi_j} \tag{3.54}$$

Note that in contrast to the "primary" generalized susceptibility, given by Eq. (3.47), as a derivative at constant secondary field, h_j, the "secondary" susceptibility is the derivative taken at constant density, φ_j, conjugate to h_j. We note that at the stability limit, while the fluctuations of density-like variables always diverge, the fluctuations of field variables may either remain finite or vanish.

3.7 Relaxation of Fluctuations

When a small domain in the system exhibits a fluctuation, this domain is no longer in equilibrium with the rest of the system. This means that there appears to be an imbalance of thermodynamic forces, which drives this domain back to equilibrium. This process is known as the "relaxation" of fluctuations. How does it happen at the mesoscopic scale?

Consider, for instance, an isolated system. A fluctuation of a property in any mesoscopic part of the system will decrease the total entropy. This means that in this domain, at least one of the thermodynamic fields appears to be different from its equilibrium value. Thermal fluctuations cause the emergence of random gradients of the thermodynamic fields at a mesoscopic scale. These gradients will cause mesoscopic currents (or fluxes) of energy and matter to dissipate the fluctuation and restore equilibrium.

The random mesoscopic fluxes caused by fluctuations are not macroscopic currents of energy and matter, which occur in a macroscopic nonequilibrium system. However, there is an important assumption, known as "Onsager's principle," that fluxes caused by fluctuations obey the same laws of nonequilibrium thermodynamics as the corresponding macroscopic fluxes. For example, a fluctuation of concentration disappears (or "relaxes") by diffusion with the macroscopically defined diffusion coefficient, given by Eq. (2.176). A fluctuation of entropy will relax through the diffusion of heat, given by the macroscopically defined temperature diffusivity, see Eq. (2.180). However, thermodynamic and kinetic properties at the mesoscale may become dependent on the mesoscopic length scale. Therefore, Onsager's principle could be generalized to:

> Fluxes caused by mesoscopic fluctuations obey the same laws of nonequilibrium thermodynamics as the corresponding macroscopic fluxes where the Onsager coefficients depend on the length scale of the fluctuations.

One of the most important consequences of the generalization of Onsager's principle is that it enables one to obtain the length-scale dependence of the kinetic coefficients from the fluctuational fluxes obtained by scattering experiments in macroscopically equilibrium systems.

Application of Fick's Second Law

Consider one-dimensional oscillations of concentration in the z direction, caused by thermal fluctuations, given by $x(t,z) = \cos\left(\frac{2\pi}{\lambda}z\right)e^{-t/\tau}$, where the fluctuations amplitude is determined by thermodynamics and the wavelength is selected by a spectrometer. Fick's second law, expressed in Eq. (3.55),

$$\frac{\partial x}{\partial t} = D\nabla^2 x$$

can be used to determine the characteristic diffusion-relaxation time. Taking temporal and spatial derivatives, one finds

$$-\frac{1}{\tau}\cos\left(\frac{2\pi}{\lambda}z\right)e^{-t/\tau} = -D\left(\frac{2\pi}{\lambda}\right)^2\cos\left(\frac{2\pi}{\lambda}z\right)e^{-t/\tau}$$

Simplifying this equality, one obtains the diffusion relaxation time, $\tau = \lambda^2/(4\pi^2 D)$. Note the significant prefactor $(4\pi^2)$.

How could one obtain the characteristic relaxation time of a fluctuation at a certain length scale? Consider the relaxation of a single concentration fluctuation of the solute, δx, in one dimension, relaxing over a distance ℓ. According to Eq. (2.179), at constant temperature,

$$\boxed{\frac{\partial(\delta x)}{\partial t} = D\frac{\partial^2(\delta x)}{\partial \ell^2}} \tag{3.55}$$

This equation is often referred to as "Fick's Second Law of Diffusion." To solve this differential equation, one needs to specify the initial and boundary conditions. If at the initial instant ($t = 0$), the excess amount of a component, x_0, is concentrated at $\ell = \ell_0$, the subsequent spatial and temporal distribution of the concentration has the form

$$\boxed{\delta x(\ell, t) = \frac{x_0 \ell_0}{\sqrt{4\pi D t}} e^{-(\delta\ell)^2/4Dt}} \tag{3.56}$$

where $\delta\ell = \ell - \ell_0$. Figure 3.4 demonstrates the temporal and spatial distribution of concentration in a fluctuation as given by Eq. (3.56). At any given time t,

$$\delta x(\ell) = \frac{x_0 \ell_0}{\sqrt{2\pi\langle(\delta\ell)^2\rangle}} e^{-(\delta\ell)^2/2\langle(\delta\ell)^2\rangle} \tag{3.57}$$

One may notice that this distribution has the same form as the Gaussian probability-distribution function, given by Eq. (3.2), with the standard deviation

$$\langle(\delta\ell)^2\rangle = 2Dt \tag{3.58}$$

and, at any given length-scale ℓ, the exponential decay of the concentration is given by

$$\delta x(t) = \frac{x_0}{\sqrt{2\pi t/\tau}} e^{-\tau/2t} \tag{3.59}$$

where the characteristic time, $\tau = \ell^2/2D$, is the relaxation time. For diffusion relaxation, the relaxation time and the characteristic time scale are defined by the same equation, Eq. (3.56). If the length scale is selected at a fixed "instrumental" length scale, such as by the scattering wavenumber, the relaxation time will be defined by the diffusion coefficient. Similarly, if the time is the "instrumental" time scale, fixed by the frequency of a sound wave in an acoustic experiment, the characteristic length scale will also be defined by the diffusion coefficient.

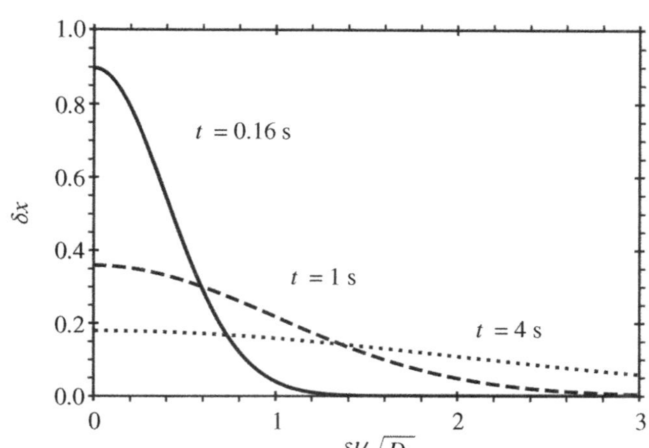

Figure 3.4 Temporal and spatial distribution of concentration as given by Eq. (3.56) for three selected times. At time, $t = 0$, the concentration (the probability to find the species at $\delta\ell = 0$) $\delta x = 1$ is the integral of a delta function centered on $\delta\ell = 0$.

3.8 Fluctuations of Anisotropy

There is another kind of fluctuation in a macroscopically isotropic fluid which is also associated with thermal molecular motion. Dipole–dipole interactions make mutual orientation of molecular dipoles energetically favorable. Even without permanent molecular dipoles, if molecules are anisotropic in shape, such as disc-like or rod-like molecules, the dispersion forces will be anisotropic, causing a preferable mutual orientation of neighboring molecules. Since mutual molecular orientations are usually entropically unfavorable, restricting the number of accessible microscopic states, many substances consisting of essentially anisotropic molecules (such as aromatic hydrocarbons), remain macroscopically isotropic in the liquid state.

If one considers the molecules to be simple rods, the average orientational order can be characterized by a scalar property

$$\hat{Q} = \frac{1}{2}\left\langle\left(3\cos^2\theta - 1\right)\right\rangle \tag{3.60}$$

where θ is the angle between the molecular axes (see Figure 3.5) and the direction of preferable orientation, caused by the interactions between molecules, known as the "director." Since, in isotropic fluids, the molecular orientations are chaotic, then after averaging $\langle\cos^2\theta\rangle = 1/3$, so that $\hat{Q} = 0$.

Upon cooling the system of such molecules, the entropic contribution to the free energy decreases. If the energy of the orientational interaction between molecules becomes sufficient to overcome the entropic disordering of orientations (thus making $\langle\cos^2\theta\rangle > 1/3$), then the parameter $\hat{Q} > 0$ may emerge, and the liquid phase will manifest a macroscopic anisotropy. Such a phase is called a "nematic phase" and the substances with the ability to form the nematic phase are known as nematic liquid crystals (see Figure 2.6). Despite the existence of macroscopically defined anisotropy, the translational molecular order (which is the signature of solids) in nematic liquid crystals is absent. In the nematic phase, the absence of translational order makes fluctuations of the director unconstrained, and thus, macroscopically significant. These fluctuations generate a well-known opacity of nematic liquid crystals, which disappears in the presence of an electric or a magnetic field [2]. The interactions between electric fields and the nematic director are used in the design of liquid crystal displays (LCDs).

In isotropic fluids, including, but not limited to, the isotropic phase of nematic liquid crystals (which usually exhibits a transition to an isotropic liquid upon heating), $\hat{Q}$ takes the equilibrium value $\hat{Q} = 0$. However, even in the absence of a macroscopic anisotropy, in any selected mesoscopic domain of the system, there exists random fluctuations of the local orientational order – fluctuations of anisotropy. In reality, molecules that form liquid crystals are not simple

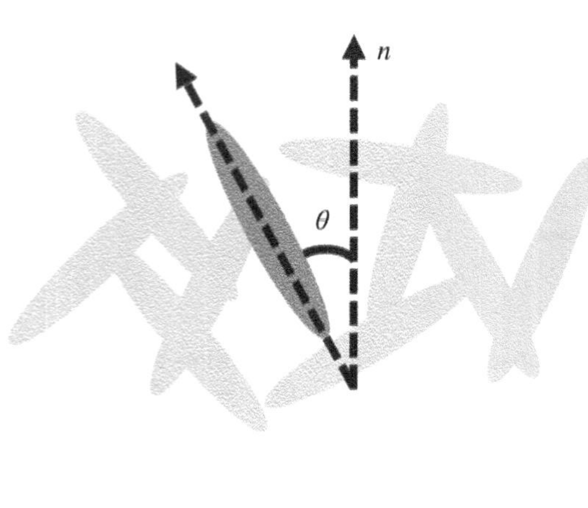

Figure 3.5 Orientation of an anisotropic molecule with respect to the "director," ***n***, the direction of preferable orientation in a liquid phase.

rods but complex three-dimensional formations. The molecular orientation, in this more general case, is characterized by a second-rank tensor, with a modulus given by Eq. (3.60).

The mean-square fluctuations of anisotropy per unit volume in an isotropic fluid follows the universal expression given by Eq. (3.46) for the generalized density as

$$\langle(\delta\hat{Q})^2\rangle = \frac{k_B T}{V^*}\hat{\chi}_{\hat{Q}} \tag{3.61}$$

where $\hat{\chi}_{\hat{Q}}$ is the susceptibility of the system to exhibit orientational order. Generally, $\hat{\chi}_{\hat{Q}}$ is proportional to the molecular anisotropy, usually defined through the anisotropic part of the tensor of molecular polarizability.

However, there is an important difference between the diffusive relaxation of density and concentration, on one hand, and the structural relaxation of anisotropy, on the other hand. Unlike the number of molecules or energy, the structure (e.g. mutual molecular orientations) is not a conserved property. The lack of structural order in a selected mesoscopic domain does not necessarily correspond to the excess of such order nearby. Therefore, unlike length scale-dependent diffusion of matter and heat, given by Eqs. (2.174) and (2.175), the relaxation equation for the property characterizing the orientational order, $\hat{Q}$, reads [3, 4]

$$\frac{\partial\hat{Q}}{\partial t} = -\frac{\delta\hat{Q}}{\tau} \tag{3.62}$$

where $\delta\hat{Q}$ is the deviation of $\hat{Q}$ from its equilibrium value $\hat{Q}_{eq}$. In accordance with Onsager's principle (Section 3.7), Eq. (3.62) may be expressed in the form, $\partial(\delta\hat{Q})/\partial t = -\delta\hat{Q}/\tau$, which can describe the relaxation of the fluctuations of anisotropy in the isotropic phase, where $\hat{Q}_{eq} = 0$. The relaxation time, τ, is proportional to the orientational susceptibility, $\hat{\chi}_{\hat{Q}}$, as

$$\tau \approx \tau_0 \hat{\chi}_{\hat{Q}} \tag{3.63}$$

The microscopic time of reorientation of individual molecules, τ_0 is proportional to the resistance, a friction coefficient, which could be reasonably assumed to be proportional to the shear viscosity, $\tilde{\eta}$. If the molecules are simple rods, the characteristic time of re-orientation via molecular rotation can be estimated as,

$$\tau_0 \sim \frac{\tilde{\eta}\ell_0^3}{k_B T} \sim \frac{\tilde{\eta}}{\langle u^2\rangle\rho} \tag{3.64}$$

where ℓ_0 is the length of a rod and u is the fluctuating thermal velocity of the molecule with a weight, M_w. The mean-square fluctuation of each Cartesian component of the thermal velocity, $u = \sqrt{u_x^2 + u_y^2 + u_z^2}$, is given by

$$\frac{1}{3}\langle(\delta u)^2\rangle = \frac{k_B T}{M_w} \tag{3.65}$$

Thus, one may estimate τ_0 as $\tau_0 \sim M_w\tilde{\eta}/3k_B T$.

Similar relaxation dynamics, indicated by the spatially independent relaxation time, are expected for the fluctuations of the reaction-coordinate near chemical-reaction equilibrium (Section 2.9), such as the equilibrium interconversion of two species, folding/unfolding of a protein molecule, or a molecular isomerization reaction.

3.9 Brownian Motion

How do the thermal fluctuations affect the random motion of a Brownian particle in a viscous medium? A Brownian particle is a mesoscopic body; it may contain millions of molecules and may exceed the size of a biological cell. Consequently, a Brownian particle has a radius R_0, that is much larger than the size of a molecule, r_0. The random motion of a Brownian particle is caused by random collisions with the molecules within the fluid medium. The motion of the Brownian particle is controlled by its diffusion in solution. In a dilute solution, Brownian particles do not mutually interact, and their concentration is represented by the probability of finding the particle in any particular part of the system.

For a three-dimensional Brownian random walk (Figure 3.6), the probability to find the particle at time, t, within a spherical layer $4\pi r^2\,dr$ is obtained from the Gaussian distribution as

$$w(r,t)\,dr = \frac{1}{8(\pi Dt)^{3/2}} e^{-r^2/4Dt} r^2\,dr \tag{3.66}$$

(note the similarities with the one-dimensional Gaussian distribution, Eq. (3.56)) where the distance $|r|$ is expressed through Cartesian coordinates, x, y, and z, as

$$r^2 = r_x^2 + r_y^2 + r_z^2 \tag{3.67}$$

Hence, one expects to find the particle at a time t, away from its initial position ($r = 0$), within one standard deviation, as

$$\langle r^2 \rangle = \int_0^\infty r^2 w(r,t)\,dr = 6Dt \tag{3.68}$$

where D is the diffusion coefficient of Brownian particles in a medium with shear viscosity, $\tilde{\eta}$. Note that for a one-dimensional random walk, $\langle r^2 \rangle = 2Dt$, and for a two-dimensional random walk, $\langle r^2 \rangle = 4Dt$.

The diffusion coefficient of noninteracting spherical Brownian particles is given by the Stokes–Einstein formula,[2] Eq. (2.184), in which the "hydrodynamic radius" R_h is equivalent to the

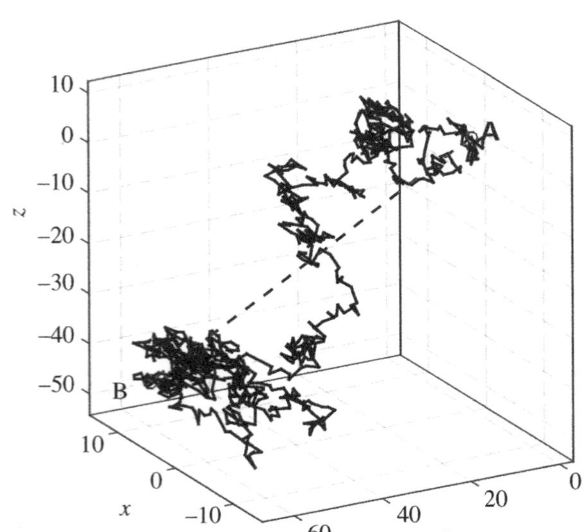

Figure 3.6 Simulation of three-dimensional Brownian motion of a particle with 1000 random steps. A and B are the initial and final positions of the particle. The distance between A and B (represented by the dashed line) is defined by Eq. (3.70).

2 Sir George Gabriel Stokes (1819–1903) was a British physicist and mathematician, born in Ireland. He made seminal contributions to fluid mechanics and optics.

geometric radius, R_0, such that

$$\boxed{D = \frac{k_B T}{6\pi\tilde{\eta}R_0}} \tag{3.69}$$

The physical meaning of Brownian motion can be illustrated by considering a random walk with n independent steps. Each step lasts a time τ_0, so that $t = \tau_0 n$. Therefore, the mean-square distance is proportional to the total number of steps as

$$\langle r^2 \rangle = 6Dn\tau_0 \tag{3.70}$$

where the time of each step can be estimated as

$$\tau_0 = \frac{R_0}{\sqrt{\langle u^2 \rangle}} = \frac{R_0}{\sqrt{3k_B T/m}} \tag{3.71}$$

where the mean-square fluctuations of the thermal velocity, $\langle u^2 \rangle$, is given for a particle of mass m by Eq. (3.65). Thus, the time of each microscopic step of a Brownian particle does not depend on the viscosity of the medium, but strongly depends on the particle size. If, for a spherical particle, $m \propto R_0^3$, then $\tau_0 \propto R_0^{5/2}$.

3.10 Brownian Motion Under Gravity

The role of the shear viscosity, $\tilde{\eta}$, in Brownian motion can be better understood if we consider some external force acting on the particle (for example, the force of gravity, $m\mathbf{g}$, where $\mathbf{g}$ is the free-fall acceleration). In a steady state, the drag force, exerted by the fluid on a moving particle, balances this force. When the velocity is small, the drag force is proportional to the velocity where the coefficient of proportionality is related to an internal friction (viscosity). In other words, the velocity of the Brownian particle driven by gravity is given by [4]

$$\boldsymbol{u} = b(\Delta m)\mathbf{g} \tag{3.72}$$

where the mobility, $\tilde{b} \propto 1/\tilde{\eta}$, and Δm is the difference between the mass of the particle and that of the fluid of the same volume. For a spherical particle, the mobility is given by Eq. (2.183). The two different motions of the particle, namely the random three-dimensional Brownian motion and the directed one-dimensional motion caused by gravity, are independent. For Brownian motion, the diffusion time scales as $t \propto r^2$, while for directed motion $t \propto r$. Therefore, for such conditions, the probability distribution for three-dimensional Brownian motion under gravity has the form of a Gaussian probability distribution,

$$w \propto \frac{1}{8(\pi D\tau)^{3/2}} \exp\left\{-\frac{r_x^2 + r_y^2 + \left(r_z'\right)^2}{4Dt}\right\} \tag{3.73}$$

where $r_z' = r_z + \tilde{b}(\Delta m)gt = r_z + ut$, and $D = \tilde{b}k_B T$.

The shape of the probability distribution of the location of a Brownian-particle at any fixed vertical coordinate remains unchanged under gravity, but the maximum of the probability density moves down with a constant speed, given by Eq. (3.72). Therefore, the mean-square displacement of a Brownian particle under gravity can be found as

$$\langle r^2 \rangle = \langle r_x^2 \rangle + \langle r_y^2 \rangle + \left\langle \left(r_z + \frac{D(\Delta m)g}{k_B T}t\right)^2 \right\rangle \tag{3.74}$$

such that

$$\boxed{\langle r^2 \rangle = 6D\tau + \left(\frac{D(\Delta m)g}{k_B T} t \right)^2} \tag{3.75}$$

Thus, there are three characteristic velocities for Brownian motion under gravity: the average thermal velocity, $\sqrt{\langle u^2 \rangle} = \sqrt{3k_B T/m}$, which is determined by the temperature and mass of the particles; the average velocity $6D/\sqrt{\langle r^2 \rangle}$ of the random walk, traveled by the particles over the distance $\sqrt{\langle r^2 \rangle}$; and the velocity of the directed particle flow under gravity, $\tilde{b}(\Delta m)g$. For a $1\,\mu$m polystyrene particle suspended in water, one can obtain the following order-of-magnitude estimates:

- Average thermal velocity ~1 mm/s
- Velocity of gravitational drift ~1 μm/s
- Speed of Brownian motion ~1 nm/s

in which the "speed of Brownian motion" refers to the distance of the average displacement per time of observation. As expected, the thermal motion is the fastest, while Brownian motion is the slowest.

Driven by gravity, the Brownian particles are gradually accumulated at the bottom of the vessel. However, this process eventually stops as this closed system achieves a final equilibrium distribution of the particle density as a function of height. As the particles, which have a concentration, $c = N_2/V$, are macroscopically in equilibrium with the surroundings, their chemical potential, μ_2, must be constant for all parts of the system. This means that at constant temperature, the osmotic pressure is balanced by the hydrostatic pressure. Applying the Gibbs–Duhem equation, Eq. (2.21), in a dilute solution, when $c \ll 1$, at constant temperature and pressure, one obtains

$$d\mu_2 = -\frac{\rho}{c}\, d\mu_1 + (\Delta m)g\, dh = 0 \tag{3.76}$$

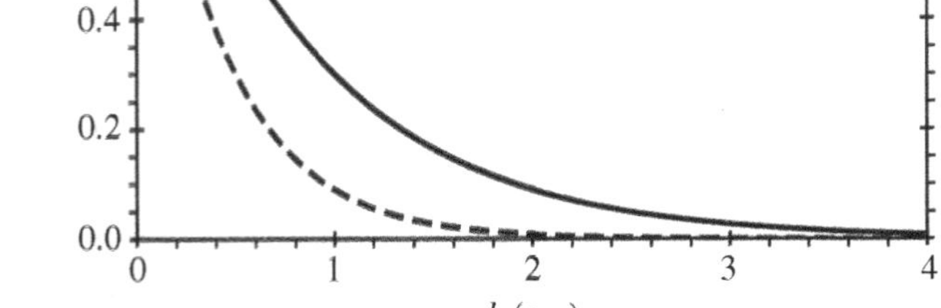

Figure 3.7 Decay of the relative concentration of one-micron (solid curve) and two-micron (dashed curve) polystyrene Brownian particles suspended in water upon increasing distance from the bottom of the container.

where h is the height of the container. Since, for noninteracting particles in a dilute solution $\rho\, d\mu_1 = -k_B T\, dc$, and upon integration of Eq. (3.76) from $h = 0$ (where $c = c_0$) to h, one obtains the concentration of the particles as a function of height in the form

$$\boxed{c = c_0 e^{-(\Delta m)gh/k_B T}} \tag{3.77}$$

where c_0 is the concentration at height $h = 0$. An example concentration distribution as a function of height for polystyrene Brownian particles in water is presented in Figure 3.7. This distribution is known as Boltzmann's distribution of concentration under gravity, which was initially obtained for the isothermal distribution of air molecules in the Earth's atmosphere.

3.11 Probing Fluctuations and Brownian Motion

In this section, we discuss two practical methods to probe inhomogeneities, in particular by light scattering and acoustic spectroscopy. The most popular experimental methods to observe and study mesoscopic inhomogeneities in liquid media are static and dynamic light scattering (SLS and DLS). Light scattering techniques are simple, relatively inexpensive, and commonly used in scientific, engineering, and medical practice [5–9]. Other popular and powerful methods to study complex fluids are small-angle neutron and X-ray scattering (SANS and SAXS), see Section 1.4; however, both SANS and SAXS are less available in laboratory practice as they require more sophisticated and expensive equipment [7, 10, 11].

3.11.1 Rayleigh Molecular Scattering

The density and concentration fluctuations in solution produce optical inhomogeneities of the medium causing fluctuations of the dielectric constant, ε. The dielectric constant at an optical-frequency range is directly related to the refractive index, n, as $\varepsilon = n^2$. Therefore, the mean-squared fluctuations of the dielectric constant in a single-component fluid caused by fluctuations of density at constant temperature are

$$\langle(\delta\varepsilon)^2\rangle = \left(\frac{\partial\varepsilon}{\partial\rho}\right)_T^2 \langle(\delta\rho)^2\rangle \tag{3.78}$$

In a binary-liquid solution, the fluctuations of concentration produce additional contributions to the fluctuations of the dielectric constant

$$\langle(\delta\varepsilon)^2\rangle = \left(\frac{\partial\varepsilon}{\partial c}\right)_{P,T}^2 \langle(\delta c)^2\rangle \tag{3.79}$$

Moreover, the fluctuations of the orientations of anisotropic molecules cause fluctuations of optical anisotropy in the isotropic medium (see Section 3.8).

If the isotropic inhomogeneities are much smaller than the wavelength of light, the scattering intensity of vertically polarized light does not depend on the scattering angle, such scattering is known as "Rayleigh scattering".[3] The scattering angle determines the wavenumber of scattering, defined by Eq. (1.1). Scattering of monochromatic vertically polarized light caused by fluctuations

3 Lord Rayleigh, John William Strutt (1842–1919) was a British mathematician and physicist who made extensive contributions in physics, optics, and acoustics. He received a Nobel Prize in Physics in 1904 for his discovery of argon, simultaneously with William Ramsay, a Scottish chemist (1852–1916) who received the Nobel Prize in Chemistry.

of density and concentration is also vertically polarized. In contrast to isotropic light scattering, the scattering caused by the fluctuations of anisotropy is partially depolarized. Measuring the depolarization ratio is the method used to distinguish between isotropic and anisotropic scattering. In particular, the horizontally polarized light scattering at $\theta = 90°$ emerges solely from the fluctuations of anisotropy, such that the ratio of the intensities of the horizontally polarized, I_h, and vertically polarized, I_v, scattering at $\theta = 90°$ is $I_h/I_v = 3/4$. In liquids consisting of molecules with significant anisotropy, such as benzene, even if they do not form liquid crystalline phases, the light scattering contribution from fluctuations of anisotropy could be significant (up to ~50%) [5].

In a static light-scattering experiment, the intensity of scattering (the number of scattered photons per second) and the depolarization ratio are measured at a selected scattering wave number. Since fluctuations of physical properties are proportional to the corresponding thermodynamic susceptibilities, the intensity is a direct measure of these susceptibilities. In a dynamic light-scattering experiment, the measurable quantity is the time-dependent intensity–intensity correlation function of scattered photons. This function provides information on the relaxation rate of fluctuations.

The idea that fluctuations are the cause of light scattering originates from the work of Smoluchowski[4] [12]. The thermodynamic theory of light scattering by spatially independent fluctuations of density and concentration was developed by Einstein just two years later [13]. According to this theory, for vertically polarized incident light, the light-scattering intensity caused by fluctuations of density in a single-component fluid reads as

$$I = I_0\langle(\delta\varepsilon)^2\rangle V_t^* = I_0\left(\frac{\partial\varepsilon}{\partial\rho}\right)_T^2\langle(\delta\rho)^2\rangle V_t^* \tag{3.80}$$

where V_t^* is the scattering volume. The prefactor, I_0, depends on: the intensity of incident light, the wavelength of incident light (as λ_0^{-4}), and on the particular geometry of the optical design. Since the absolute value of I_0 is usually unknown, the intensity of light scattering is shown in practice in arbitrary units or relative to that of a standard value for a well-studied substance, such as benzene, for which the "absolute" intensity value, I/I_0, has been established [5]. Since the mean-squared fluctuations of density are proportional to the isothermal compressibility, see Eq. (3.26), the Einstein formula,[5] Eq. (3.80), can be written in the form

$$\boxed{I = I_0\left(\frac{\partial\varepsilon}{\partial\rho}\right)_T^2 k_B T\rho^2\kappa_T} \tag{3.81}$$

Another light-scattering method to measure the intensity of density or concentration fluctuations is turbidity measurements. In these measurements, the intensity of light traveling through a medium along a path ℓ is attenuated as $I = I_0 e^{-h_t\ell}$, where h_t is the extinction coefficient or "turbidity." If scattering is caused by inhomogeneities that are much smaller than the wavelength of light, the extinction coefficient is directly expressed through the Rayleigh scattering coefficient at 90°, as $R_{\theta=90°} = I/I_0$, as $h_t = (16/3)\pi R_{\theta=90°}$ [5].

The condition that the size of the inhomogeneities must be much smaller than the scattering wavelength is a severe restriction on the Einstein formula, Eq. (3.81), which does not describe

4 Marian Smoluchowski (1872–1917), ethnically Polish, was a physicist who was born in the Austro-Hungarian Empire and worked in Krakow when southern Poland was part of the Empire. He demonstrated that the blue color of the sky is a consequence of light scattering in the atmosphere. He was also the first physicist to ascribe the phenomenon of critical opalescence to large density fluctuations.

5 Albert Einstein (1879–1955), born in the German Empire and worked in Europe and in the United States, was not only best known for developing the theory of relativity, but also for explaining Brownian motion and for developing the thermodynamic theory of molecular light scattering (following the work of Smoluchowski).

the behavior of the system when fluctuations of inhomogeneities are truly mesoscopic. In this equation, the light-scattering intensity is independent of the light-scattering wavenumber, tuned by the light-scattering angle. The modification of Einstein's formula, to describe larger inhomogeneities, will be addressed in Chapters 4 and 8.

3.11.2 Spectrum of Molecular Scattering

The Einstein formula has been confirmed by accurate static light-scattering experiments in molecular fluids with relatively small fluctuational inhomogeneities. However, this formula, being purely thermodynamic, cannot address the relaxation of fluctuations. It turns out that in condensed-fluid media, the fluctuating number of molecules per unit volume can relax via two independent dynamic processes: first, through diffusive motion, and, second, through the propagation of density waves. Density waves are the collective motion of molecules caused by random fluctuations of pressure. These waves are phonons propagating through condensed matter with the speed of sound (see Section 2.12). In solids, where the diffusion of molecules is usually negligible, local density fluctuations can be attributed almost solely to phonons. In liquids, the relative contributions of phonons into the relaxation of density fluctuations depends on the thermodynamic properties.

The separation of density fluctuations into two parts can be illustrated by the following thermodynamic arguments. Consider the density as a function of entropy and pressure. Then, the fluctuations of density can be represented as

$$\delta\rho(S,P) = \left.\frac{\partial\rho}{\partial S}\right|_P \delta S + \left.\frac{\partial\rho}{\partial P}\right|_S \delta P \tag{3.82}$$

Due to the statistical independence of δS and δP, $\langle\delta S\delta P\rangle = 0$ (see Section 3.3), the mean square fluctuations of the density can be written as

$$\langle(\delta\rho)^2\rangle = \left(\frac{\partial\rho}{\partial S}\right)_P^2 \langle(\delta S)^2\rangle + \left(\frac{\partial\rho}{\partial P}\right)_S^2 \langle(\delta P)^2\rangle \tag{3.83}$$

Therefore, density fluctuations can be represented as the sum of the isobaric fluctuations of entropy and the adiabatic fluctuations of pressure. Taking into account

$$\left.\frac{1}{\rho}\frac{\partial\rho}{\partial S}\right|_P = \left.\frac{T}{C_P}\frac{\partial\rho}{\partial P}\right|_T \tag{3.84}$$

and using Eqs. (3.34) and (3.35) for the mean-square fluctuations of entropy and pressure, one obtains from Eq. (3.80), the scattering intensity expressed in two parts

$$I = I_\mathrm{R} + 2I_\mathrm{B} = I_0\left(\frac{\partial\varepsilon}{\partial\rho}\right)_T^2 k_\mathrm{B}T\left[\frac{T\alpha_P^2}{\rho C_P} + \kappa_S\right] \tag{3.85}$$

where I_R is referred to as the "Rayleigh line" intensity, expressed through the isobaric expansivity $\alpha_P = \rho^{-1}(\partial\rho/\partial T)|_P$ and the isobaric heat capacity, while $2I_\mathrm{B}$ is referred to as the two "Brillouin line" intensities, expressed through the adiabatic compressibility, $\kappa_S = \rho^{-1}(\partial\rho/\partial P)|_S$, see Figure 3.8. Note that from the ratio of the Rayleigh scattering intensity to the Brillouin scattering intensity, one obtains

$$\boxed{\frac{I_\mathrm{R}}{2I_\mathrm{B}} = \frac{\kappa_T - \kappa_S}{\kappa_S} = \frac{C_P}{C_V} - 1} \tag{3.86}$$

This expression is known as the "Landau–Placzek ratio" [5].

The physically different nature of the relaxation dynamics of the adiabatic fluctuations of pressure and the isobaric fluctuations of entropy determines the spectrum of light scattered by the fluctuations of density. Adiabatic fluctuations of pressure are phonons propagating with a speed of sound, $u_s = \lambda_s\omega/2\pi$, where λ_s is the wavelength of sound and ω is the frequency. Photons are scattered in all directions due to the collisions with phonons. A light-scattering spectrometer, analyzing light scattering for a certain wave number, q (given by Eq. (1.1)), probes only those phonons that have the corresponding wavelength $\lambda_s = 2\pi/q$. Collisions of photons with moving phonons changes the frequency of scattered light due to the Doppler effect. The change in the frequency produces two satellites in the spectrum of the scattered light with frequencies $\omega_0 \pm \Delta\omega$, symmetrically located on the two sides of the initial light frequency, ω_0, as illustrated in Figure 3.8,

$$\Delta\omega = u_s q = \frac{2\pi}{\lambda_s}\sin\left(\frac{\theta}{2}\right) = 2\omega_0\frac{u_s}{c}\sin\left(\frac{\theta}{2}\right) \tag{3.87}$$

Equation (3.87) was obtained by Brillouin[6] [5] in 1922,[7] such that this effect is commonly called "Brillouin scattering" (also known as "Mandelstam-Brillouin[8] scattering"), and the spectrum satellites are referred to as "Brillouin components."

The frequency shift in the Brillouin scattering is very small relative to the frequency of incident light, since the ratio $\Delta\omega/\omega_0$ is proportional to the ratio of the speed of sound in liquid (on the

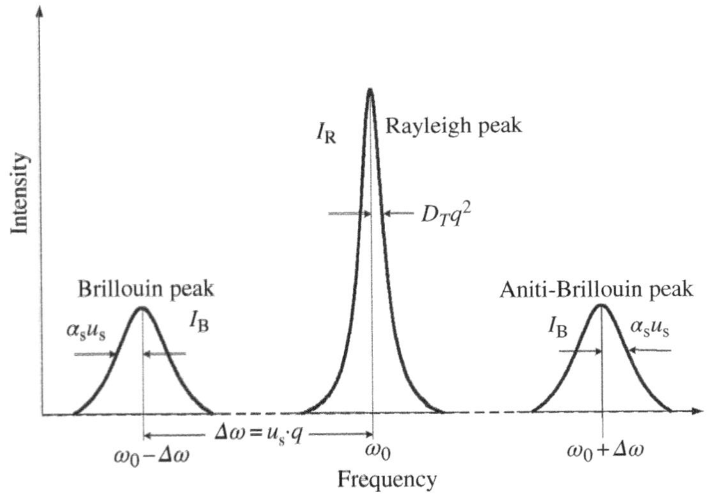

Figure 3.8 Schematic representation of the spectrum of isotropic molecular scattering in a single-component liquid.

6 Léon Nicolas Brillouin (1889–1969) was a French physicist who made contributions to quantum mechanics, radio-wave propagation in the atmosphere, solid-state physics, and information theory.

7 In fact, this equation was obtained even earlier, in 1918, by Mandelstam in Russia, but was published later because of the Russian civil war [5]. Consequently, in Russian literature this effect is called "Mandelstam-Brillouin scattering." Experimentally, Eq. (3.87) was confirmed in 1930 by E. F. Gross – also in Russia.

8 Leonid Isaakovich Mandelstam (1879–1944) was a Russian and Soviet physicist. He founded one of the most famous Russian scientific schools, in which he mentored three Nobel Prize laureates (Igor Tamm, Vitaly Ginzburg, and Andrei Sakharov). In addition to his discovery of Brillouin scattering, he was a co-discoverer of inelastic combinational scattering of light. This phenomenon is named after the Indian physicist, Chandrasekhara V. Raman (1888–1970, Nobel Prize in Physics in 1930), who first published observations of the effect in 1928. It was first observed just one week before Raman by Soviet physicists L. I. Mandelstam and G. S. Landsberg; however, because of a delay in mailing from the Soviet Union to the United Kingdom, their paper was published after Raman [14].

order of ~10^3 m/s) to the speed of light in liquid (~2×10^8 m/s). The frequency of red light is about 4×10^{15} Hz and, consequently, the frequency of elastic density waves, probed at the scattering angle of 60°, is about 3×10^{10} Hz. This is why the observation of such a small shift in the light frequency requires a high-resolution spectrometer, usually a Fabry–Pérot interferometer [5]. For sound propagation, the frequency ~ 10^{10} Hz is very high, such that it is referred to as the "hyper-sound frequency." The speed of hypersound may be significantly different from the speed of low-frequency sound, given by the thermodynamic equation, $u_s^2 = (M_w \rho \kappa_S)^{-1}$, because of the existence of relaxation processes.

The Brillouin components are not infinitely sharp lines; they have a certain linewidth, δ_B, associated with the attenuation of elastic density waves, given by

$$\delta_B/2 = \alpha_s u_s \tag{3.88}$$

where α_s is the absorption coefficient of sound in a viscous medium [5], given by

$$\boxed{\alpha_s = \frac{\omega^2}{2u_s^3 M_w \rho}\left[\frac{4}{3}\tilde{\eta} + \tilde{\zeta} + \kappa M_w\left(\frac{1}{C_V} - \frac{1}{C_P}\right)\right]} \tag{3.89}$$

which has units of inverse length. In Eq. (3.89), in addition to the previously introduced transport properties, the shear viscosity $\tilde{\eta}$ and the thermal conductivity κ, one finds a new transport coefficient $\tilde{\zeta}$, called the "bulk viscosity" or "second viscosity." The bulk viscosity is phenomenologically defined through the excess sound absorption that cannot be accounted for by the shear viscosity and thermal conductivity. Usually, the contribution of the bulk viscosity to the sound absorption is attributed to relaxation processes – in particular, to the relaxation of fluctuations.

The sound absorption coefficient in a binary mixture contains an additional contribution due to the temperature and pressure gradients occurring in the sound wave, resulting in a flux of matter represented by the thermo-diffusion and the baro-diffusion ratios, k_T and k_P, respectively, see Eqs. (2.177) and (2.178). Consequently, an extra term, α_D, should be added to the sound absorption coefficient, Eq. (3.89) [4]. In the lowest order approximation,

$$\alpha_D \approx \frac{\omega^2}{2u_s} M_w D \left.\frac{\partial \rho}{\partial x}\right|_{P,T} \left(\frac{k_P}{P} + \frac{2k_T}{\rho^2 C_{P,x}} \left.\frac{\partial \rho}{\partial T}\right|_{P,x}\right) \tag{3.90}$$

The contributions to the sound absorption from the baro- and thermo-diffusion are often insignificant since in most binary liquids the prefactor $\partial\rho/\partial x|_{P,T} \ll \rho$.

It is seen in Eqs. (3.87)–(3.90) that both the frequency shift and the linewidth of the Brillouin components vanish in the zero scattering angle limit, as $q = \omega/u_s$. However, the linewidth vanishes faster, causing the Brillouin lines to be seen as sharper at smaller scattering angles. At large scattering angles, the width of the Brillouin components is typically of the same order as the frequency shift.

The adiabatic pressure fluctuations represent only a part of the total fluctuations of density. Another contribution is provided by isobaric fluctuations of entropy. These kinds of fluctuations cause the unshifted ("central") component, known as the Rayleigh line, in the spectrum of molecular light scattering, as illustrated in Figure 3.8.

Relaxation of the isobaric entropy to its equilibrium value occurs by the diffusion of heat, given by Eq. (2.175), which reads for a single-component substance, as $\boldsymbol{j}_Q = -\kappa \nabla T$. Hence, the rate of the diffusive relaxation of entropy fluctuations, $\Gamma_T = 1/\tau_T$, is given by

$$\Gamma_T = D_T q^2 \tag{3.91}$$

This rate determines the half-width of the Rayleigh component, $\delta_R/2 = \Gamma_T$. Since the thermal diffusivity is given by $D_T = \kappa/\rho C_P$, then the linewidth will become very narrow near the liquid-gas critical point, at which the isobaric heat capacity diverges.

At the liquid–gas critical point, the isothermal compressibility diverges, as strongly as C_P (see Eq. (2.65)); thus, the Landau–Placzek ratio, given by Eq. (3.86), also diverges. Alternatively, in almost incompressible solids, $C_P \approx C_V$, such that the central line practically disappears. Remarkably, in liquid water (at the density maximum, 4° C), $C_P = C_V$ exactly, because the isobaric thermal expansivity, $\alpha_P = 0$, at this temperature and remains very small around it. This is why the central line is practically invisible in water at ambient conditions.

In addition to isotropic fluctuations of density and concentration, molecular orientations and vibrations produce a broad, continuous, and partially depolarized spectrum, known as the "wings of the Rayleigh line" with a maximum at ω_0. This spectrum is associated with the relaxation of local anisotropy in macroscopically isotropic media, caused by fluctuations of mutual orientation of anisotropic molecules (see Section 3.8). In contrast to the diffusive relaxation Eq. (3.91), the rate of the relaxation of the fluctuations of anisotropy, $\Gamma_{\text{an}} = 1/\tau$ (where τ is given by Eq. (3.63)), manifested by the width of this relatively broad spectrum does not depend on the wave number unless the orientational susceptibility, $\hat{\chi}_Q$, itself becomes q-dependent. This, more subtle, effect will be discussed in Chapter 4. The fluctuations of anisotropy in an isotropic liquid become especially pronounced near the transition to the nematic phase of liquid crystals, which possess macroscopic anisotropy. Since the relaxation rate of these fluctuations near the transition temperature is slow, the width of the corresponding wing may become as narrow as the central line of unpolarized isotropic scattering, such that it would be undetectable by an optical spectrometer. For measuring such narrow widths of the scattering spectra, methods exploring temporal correlations between scattered photons or neutrons, like DLS techniques, are commonly used.

3.11.3 Dynamic Light Scattering

Even away from the critical point, the rate of relaxation of entropy fluctuations is too slow and the linewidth of the unshifted Rayleigh component is too narrow to be measured with an optical spectrometer because of restrictions associated with the finite size of the aperture. For a typical value of $D_T \cong 10^{-5}\ \text{m}^2/\text{s}$ in liquids at ambient conditions, the linewidth of the Rayleigh component lies in the 100 MHz frequency range. This is two orders of magnitude narrower than the linewidth and the shift of the Brillouin components at the scattering angles of 90°.

The problem of detecting the linewidth of the unshifted Rayleigh component is even more challenging in binary fluids where fluctuations of the concentration cause additional, often dominant, scattering. The relaxations of concentration fluctuations is diffusive, similar to the relaxation of heat. The relaxation rate of concentration fluctuations depends on the coefficient of mutual diffusion D, defined by Eq. (3.69), as

$$\Gamma_{\text{D}} = \frac{1}{\tau_{\text{D}}(q)} = Dq^2 \tag{3.92}$$

The practical difference between Eqs. (3.91) and (3.92) is that the mutual diffusion coefficient, D, in binary-liquid solutions is usually two orders of magnitude smaller than the thermal diffusivity coefficient, D_T. Correspondingly, the linewidth of the Rayleigh component (caused by concentration fluctuations) lies in the MHz frequency range.

Moreover, there are two important examples of an extremely slow diffusion and, consequently, extremely narrow Rayleigh components. The first example is a sharp decrease in the diffusion of heat, the major contribution to the relaxation of density fluctuations, at the liquid–gas critical point in single-component fluids, because of the diverging isobaric heat capacity, and in the mutual diffusion at the fluid–fluid (liquid–gas or liquid–liquid) critical point in binary fluids, because of the divergence of the osmotic compressibility, Eq. (2.100). The effect of the sharp decrease of the relaxation rate of fluctuations near the critical point is called, "critical slowing down," and is discussed in detail in Chapter 8. The second example is light scattering caused by Brownian particles. In accordance with the Stokes–Einstein formula, given by Eq. (2.184), the diffusion coefficient of Brownian particles is inversely proportional to their size. For particles with a radius of 50 nm, the diffusion coefficient is two orders of magnitude smaller than that of a molecular solution (far away from the critical point) with the same viscosity. As a result, the linewidth of the Rayleigh component caused by critical fluctuations or by Brownian motion is reduced to the MHz, or even to the kHz, frequency range.

In order to measure the rate of relaxation of density or concentration fluctuations as well as the diffusion rate of Brownian particles, instead of using optical spectroscopy, another technique, known as photon-correlation spectroscopy or DLS is widely used in science and engineering. The physical characteristics of the scattering medium measured in a DLS experiment is the dynamic intensity-intensity autocorrelation function of scattered photons, $g_2(t) = \langle I(t)I(t+\delta t)\rangle$, where δt is known as the "delay time." The "brain" of a DLS spectrometer is the digital correlator that analyzes the correlation of the number of photons scattered between the time, t, and the time, $t+\delta t$. The time-dependent correlation function for a single-exponential relaxation process is given by

$$g_2(t) - 1 = A_\Gamma e^{-2\Gamma t} \tag{3.93}$$

where A_Γ is the amplitude of the correlation function, which depends on the intensity and coherence of scattering. The diffusive character of the relaxation is verified by measuring the rate, Γ, as a function of q^2. Figure 3.9 illustrates the diffusive character of Brownian motion [15]; the rate follows the linear q^2 dependence, given by Eq. (3.92). If there are two independent contributions

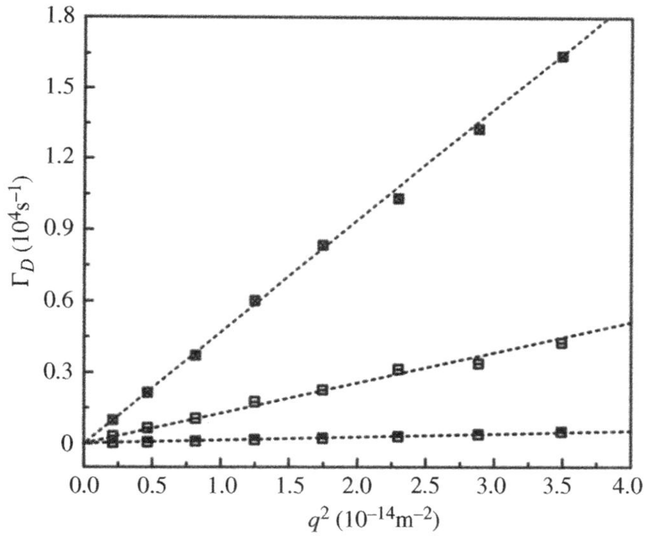

Figure 3.9 Decay rates $\Gamma_D(q^2)$, given by Eq. (3.92), observed in a semidilute solution of polyethylene oxide (PEO) at a concentration of 1% mass fraction at 25°C as a function of q^2. The three modes correspond to the relaxation of the PEO concentration fluctuations (upper), the Brownian diffusion of smaller PEO clusters (middle), and the Brownian diffusion of larger PEO aggregates (lower). Source: Reproduced with permission from Ref. [15], © American Chemical Society.

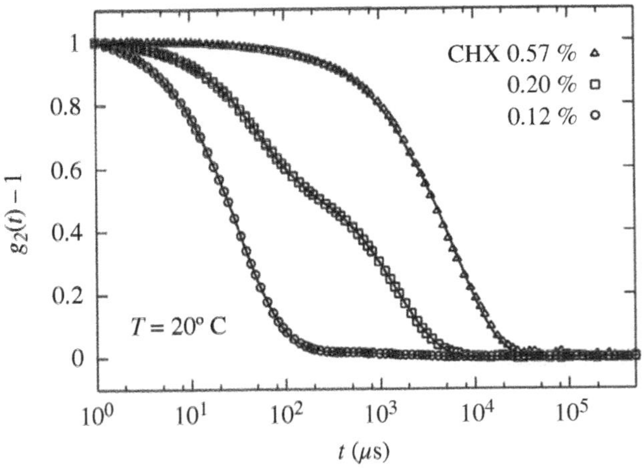

Figure 3.10 DLS correlation functions for the mesoheterogeneous quasi-binary system 2,6-lutidine(29%)-H_2O/D_2O with the addition of different amounts of cyclohexane (CHX), 0.12, 0.2, and 0.57 mass%. At the smallest concentration, the contribution of insoluble mesoscopic inhomogeneities associated with the addition of CHX to the scattering intensity is minor, and the correlation function is a single-exponential decay of the concentration fluctuations. At the intermediate concentration of CHX, the contributions of the concentration fluctuations and inhomogeneities are comparable, and the correlation function demonstrates a typical example of two-exponential diffusive dynamics. At the highest concentration of CHX, the contribution of inhomogeneities is much more significant, and again, the correlation function could be approximated as a single-exponential decay. Source: Reproduced with permission from Ref. [16], © Elsevier.

to the Rayleigh component, such as from Brownian motion and from molecular relaxation of concentration fluctuations, then the correlation function is approximated by a double exponential

$$g_2(t) - 1 = A_{\Gamma_1} e^{-2\Gamma_1 t} + A_{\Gamma_2} e^{-2\Gamma_2 t} \tag{3.94}$$

with two amplitudes, A_{Γ_1} and A_{Γ_2}, and two characteristic diffusion rates, Γ_1 and Γ_2. Figure 3.10 demonstrates a typical two-exponential dynamic correlation function obtained in the aqueous solution of lutidine in the presence of mesoscopic droplets of cyclohexane (CHX) [16]. The two relaxation processes are easily discernible only if the corresponding time scales are well separated and both amplitudes are comparable. DLS experiments require a statistically large number of optical inhomogeneities in the scattering volume to measure the decay of the concentration fluctuations. For example, in a light-scattering volume of a small fraction of a cubic millimeter, the number of Brownian particles in a dilute solution could still be on the order of thousands or tens-of-thousands.

There is another photon-correlation technique, known as fluorescence correlation spectroscopy (FCS), which analyzes the temporal fluctuations of fluorescence intensity. The advantage of FCS with respect to DLS is that the intensity-correlation function may be obtained for a very small number of molecules (nanomolar to picomolar concentrations). An important application of FCS is measuring the characteristic rates of folding/unfolding of a protein molecule [17].

In recent decades, various imaging techniques have been developed which show the fluctuations directly. A typical example could be found in ref. [18] and the references therein. Also, another method has been developed to track the diffusion of an individual Brownian particle, known as nanoparticle tracking analysis (NTA). Both DLS and NTA measure the Brownian motion of nanoparticles whose diffusion coefficient is related to the particle size through the Stokes–Einstein equation, Eq. (2.184). However, in contrast to DLS, which statistically analyzes fluctuations of the concentration of Brownian particles through the time-dependent correlation function, in an NTA experiment, the motion of an individual particle is recorded by video [19].

3.11.4 Acoustic Spectroscopy

An experimental method of probing the relaxation of fluctuations, alternative to DLS, is a study of the propagation of sound at different frequencies. This method is sometimes called "acoustic spectroscopy." The wave equation of sound propagation,

$$\frac{\partial^2 P}{\partial t^2} = \frac{1}{u_s^2}\frac{\partial^2 P}{\partial r^2} \tag{3.95}$$

contains both the distance and time to the same power, while the diffusion of heat or matter, given by Eqs. (2.174) and (2.175), scales r^2 as t. As a result, in a sound wave, the pressure and temperature change adiabatically, since the density and concentration may not have enough time to follow the change of pressure, resulting in the dissipation of energy in the form of sound attenuation.

The application of acoustic spectroscopy to the study of relaxation processes is based on the competition between the relaxation rate, $1/\tau$, and the frequency of sound, ω. In the low-frequency limit ($\omega\tau \ll 1$), the fluctuations of a structure have enough time to relax in the sound wave despite changes of pressure and temperature. In this limit, the speed of sound is described by the thermodynamic property, the adiabatic compressibility, κ_S, as $u_s(\omega \to 0) = (M_w \rho \kappa_S)^{-1/2}$. The speed of sound, measured in the kHz frequency range, can usually be regarded as thermodynamic unless the relaxation process is extraordinary slow, such as the case of "critical slowing-down" or if measured in a medium with an extremely high viscosity.

Alternatively, in the high-frequency limit ($\omega\tau \gg 1$), a molecular structure (or chemical-reaction equilibrium) is "frozen" in a state away from equilibrium where the speed of sound, $u_s(\omega \to \infty)$, is usually higher than its thermodynamic value. The dependence of the speed of sound on its frequency is referred to as "sound dispersion." For the relaxation of a nonconserved property that is also spatially independent, such as structural anisotropy or a chemical-reaction coordinate, interpolation between the low- and high-frequency limits is obtained in the classical relaxation theory[9] [4, 5]. The relaxation theory considers a complex frequency-dependent function, the square of the complex sound velocity, $\tilde{u}_s^2$, expressed as

$$[\tilde{u}_s(\omega)]^2 = [u_s(\omega \to 0)]^2 \left(1 + \frac{i\omega\tau}{1 + i\omega\tau}\frac{\Delta u_s^2}{[u_s(\omega \to 0)]^2}\right) \tag{3.96}$$

where $\Delta u_s^2 = u_s^2(\omega \to \infty) - u_s^2(\omega \to 0)$ is the maximum dispersion of the speed of sound (typically, $\Delta u_s^2/[u_s(\omega \to 0)]^2 \approx 0.1$–$0.2$). The real part of this complex function gives the frequency-dependent speed of sound, while the imaginary part gives the sound attenuation. Separating the real and imaginary parts one obtains,

$$\frac{u_s^2(\omega) - u_s^2(\omega \to 0)}{\Delta u_s^2} = \frac{\omega^2\tau^2}{1 + \omega^2\tau^2} \tag{3.97}$$

Measuring the speed of sound at different frequencies and assuming that the hypersound velocity measured through the shift of the Brillouin components in the spectrum of light scattering is equal to $u_s(\omega \to \infty)$, one can obtain the relaxation time, τ, of the process responsible for sound dispersion.

9 The classical theory of acoustic relaxation for a nonconserved spatially independent property was originally developed in 1937 by L. I. Mandelstam and M. A. Leontovich.

In addition to the speed of sound, the attenuation of sound waves, described by the sound absorption coefficient, given by Eq. (3.89), is also measured as a function of frequency. The existence of a relaxation process causes additional dissipation of sound, which is associated with the frequency-dependent bulk viscosity, $\tilde{\zeta} = \tilde{\zeta}(\omega, \tau)$, in Eq. (3.89). The classical theory of acoustic relaxation predicts for the excess sound absorption multiplied by the sound wavelength, $\Delta\alpha_s\lambda_s = \Delta\alpha_s(u_s/\omega)$, the following behavior [4, 5]

$$\Delta\alpha_s\lambda_s = \frac{\omega\tilde{\zeta}}{2M_w\rho u_s^2} \approx \frac{\Delta u_s^2}{2u_s^2}\left(\frac{\omega\tau}{1+\omega^2\tau^2}\right) \tag{3.98}$$

where u_s^2 is the interpolation between $u_s^2(\omega \to 0)$ and $u_s^2(\omega \to \infty)$, as given by Eq. (3.97). The product $\Delta\alpha_s\lambda_s$ is a function of $\omega\tau$, such that in the thermodynamic limit ($\omega\tau \ll 1$), $\Delta\alpha_s\lambda_s \propto \omega\tau$, while in the high-frequency limit ($\omega\tau \gg 1$), $\Delta\alpha_s\lambda_s \propto 1/\omega\tau$. Equation (3.98) has a maximum at $\omega\tau = \sqrt{u_s(\omega \to 0)/u_s(\omega \to \infty)} \sim 1$. The predictions of the classical relaxation theory are presented in Figure 3.11.

As follows from Eq. (3.98), the classical relaxation theory predicts the dependence of the bulk viscosity on the frequency and the relaxation time as

$$\tilde{\zeta} \approx M_w\rho\frac{\tau\Delta u_s^2}{1+\omega^2\tau^2} = M_w\rho\tau\Delta u_s^2 f_\zeta(\omega\tau) \tag{3.99}$$

where $f_\zeta(\omega\tau)$ is a universal function of the reduced frequency, $\omega\tau$. In the zero-frequency limit ($\omega\tau \to 0$), $\tilde{\zeta} = M_w\rho\tau\Delta u_s^2$, being independent of frequency, while in the high-frequency limit ($\omega\tau \to \infty$), the bulk viscosity scales as $\tilde{\zeta} \approx M_w\rho(\Delta u_s^2/\omega^2\tau)$, asymptotically vanishing as $\sim 1/\omega^2$.

Comparing Eq. (3.99) with the general estimate for the relaxation time of a nonconserved property, $\tau \approx \tau_0\hat{\chi}$ Eq. (3.63), one can further clarify the physical meaning of the bulk viscosity. Consequently, at low frequencies, the bulk viscosity can be expressed as

$$\tilde{\zeta} \approx \tilde{\eta}\hat{\chi}\frac{\Delta u_s^2}{\langle u^2 \rangle} \tag{3.100}$$

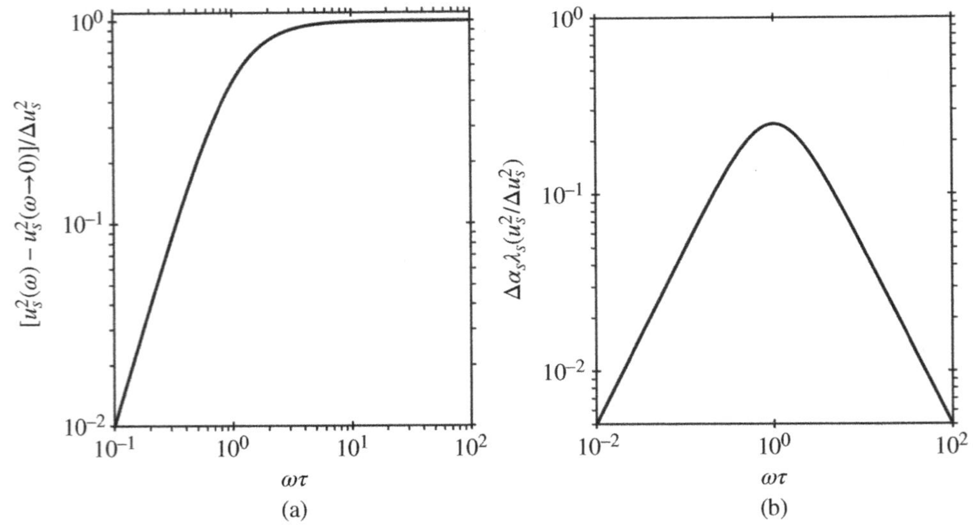

Figure 3.11 Universal functions for the normalized dispersion of the speed of sound (a) and the normalized excess sound absorption (b) as predicted by the classical relaxation theory, given by Eqs. (3.97) and (3.98), respectively.

where $\langle u^2 \rangle$ is the mean-square thermal velocity, given by Eq. (3.65). One can notice that the bulk viscosity is proportional to the shear viscosity and the generalized susceptibility. Also, the bulk viscosity only exists if the speed of sound exhibits a dispersion as a result of a relaxation process. The predictions of the classical relaxation theory were confirmed by measurements of the excess sound attenuation, associated with the bulk viscosity, as a function of frequency for various isotropic liquids containing anisotropic molecules, such as benzene, which exhibit significant fluctuations of local anisotropy [5].

References

1 L. D. Landau and E. M. Lifshitz, *Statistical Physics Part 1*, vol. 5 of Course of Theoretical Physics. Oxford: Pergamon Press, 3rd ed., 1980.

2 P.-G. de Gennes and J. Prost, *The Physics of Liquid Crystals*. Oxford University Press, 2nd ed., 1993.

3 E. M. Lifshitz and L. P. Pitaevskiĭ, *Physical Kinetics*, vol. 10 of Landau and Lifshitz's Course of Theoretical Physics. Oxford: Pergamon Press, 1981.

4 L. D. Landau and E. M. Lifshitz, *Fluid Mechanics*, vol. 6 of Course of Theoretical Physics. Oxford, London: Pergamon Press, 3rd ed., 1966.

5 I. L. Fabelinskiĭ, *Molecular Scattering of Light*. New York: Plenum Press, 1968.

6 B. Chu, *Laser Light Scattering: Basic Principles and Practice*. Academic Press, 2nd ed., 1991.

7 R. Pecora, ed., *Dynamic Light Scattering: Applications of Photon Correlation Spectroscopy*. New York: Springer, 1985.

8 J. Jeevanandam and M. K. Danquah, eds., *Research Advances in Dynamic Light Scattering*. Physics Research and Technology. Nova Science Publishers, 2020.

9 J. N. Weiss, *Dynamic Light Scattering Spectroscopy of the Human Eye*. Cham: Springer, 2022.

10 B. J. Berne and R. Pecora, *Dynamic Light Scattering: With Applications to Chemistry, Biology, and Physics*. Wiley, 1976.

11 I. W. Hamley, *Small-Angle Scattering: Theory, Instrumentation, Data, and Applications*. Wiley, 2021.

12 M. v. Smoluchowski, "Molekular-kinetische theorie der opaleszenz von gasen im kritischen zustande, sowie einiger verwandter erscheinungen," *Ann. Phys.*, vol. 25, pp. 205–226, 1908. Eng. Transl. "Molecular-kinetic theory of the opalescence of gases in the critical state, as well as some related phenomena".

13 A. Einstein, "Theorie der opaleszenz von homogenen flüssigkeiten und flüssigkeitsgemischen in der nähe des kritischen zustandes," *Ann. Phys.*, vol. 33, pp. 231–250, 1910. Eng. Transl. "On the theory of the opalescense of homogeneous fluids and liquid mixtures near the critical state"'.

14 I. L. Fabelinskiĭ, "Seventy years of combination (Raman) scattering," *Phys. Usp.*, vol. 41, no. 12, p. 1229, 1998.

15 X. Zheng, M. A. Anisimov, J. V. Sengers, et al., "Mesoscopic diffusion poly(ethylene oxide) in pure and mixed solvents," *J. Phys. Chem. B*, vol. 122, pp. 3454–3464, 2017.

16 A. E. Robertson, D. H. Phan, J. E. Macaluso, et al., "Mesoscale solubilization and critical phenomena in binary and quasi-binary solutions of hydrotropes," *Fluid Phase Equilib.*, vol. 407, pp. 243–254, 2016.

17 R. Rigler, *Fluorescence Correlation Spectroscopy: Theory and Applications*. Berlin Heidelberg: Springer-Verlag, 2012.

18 A. Oprisan, S. A. Oprisan, B. Rayleigh, J. J. Hegseth, Y. Garrabos, C. Lecoutre-Chabot, and D. Beysens, "Dynamic structure factor of density fluctuations from direct imaging very near (both above and below) the critical point of SF_6," *Phys. Rev.* E, vol. 86, 061501, 7pp., 2012.
19 V. Filipe, A. Hawe, and W. Jiskoot, "Critical evaluation of nanoparticle tracking analysis (NTA) by nanosight for the measurement of nanoparticles and protein aggregates," *Pharm. Res.*, vol. 27, pp. 796–810, 2010.

Exercises

3.1 Explain the origin of fluctuations of thermodynamic properties. Do the fluctuations violate the second law of thermodynamics?

3.2 Derive the expression for mean-square relative fluctuations of density. Apply this result for an ideal gas.

3.3 Obtain the mean-square fluctuations of selected volume with a constant number of molecules. How does the volume fluctuations depend on size? What is the difference between density fluctuations and volume fluctuations?

3.4 Plot the work (in units of $k_B T$) of a density fluctuation in a nano-domain of an ideal gas (10 nm size) as a function of the relative deviation of density from its equilibrium value.

3.5 Using the expression for the mean-square fluctuations of density, obtain the expression for the density fluctuations of a van der Waals fluid in a fixed volume.

3.6 Using the expression for the mean-square fluctuations of density of a van der Waals fluid in a selected volume $V_t^* = 100b$, obtain the mean-square fluctuations of density as a function of the distance to the critical temperature, $\Delta\hat{T} = (T - T_c)/T_c$, along the critical isochore, $\rho = \rho_c$. *Hint:* begin with the Van der Waals EOS in reduced variables.

3.7 Consider a nano-droplet of liquid water suspended in vapor at 300 K. Calculate and plot the relative RMS fluctuation of density (in %) in the droplet as a function of its radius in the range from 0.5 nm to 5 nm.

3.8 (a) Derive the expression for the mean-square fluctuations of the deviation from verticality for a pendulum under gravity. *Hint:* Assume small deviations of the pendulum from verticality, such that for a fluctuation of $\langle\theta\rangle$, $\cos\langle\theta\rangle \approx 1 - \langle\theta\rangle^2/2$. (b) A microchip engineer designs a precision device that requires a vertical suspension of a platinum spherical particle of 15-micron radius, provided that the standard deviation from verticality would not exceed 0.005 degree. The environment is air at ambient conditions. What minimal length of the suspension is required? Formulate the assumptions you made. What material could be good for the suspension thread?

3.9 (a) Derive the expression for the mean-square fluctuation of molecular concentration (molecular or mole fraction) in a binary mixture by calculating the work of the fluctuation. Express the result through the osmotic compressibility. (b) Using this expression, obtain the fluctuations of concentration in an ideal solution and in a dilute solution. Explain the difference. Show that the result obtained for a dilute solution is equivalent to the "noise" formula.

3.10 Using the expression for the mean-square fluctuations of a generalized density (an extensive property per unit volume), formulate the expression for a mean-square fluctuation of magnetization (taken per unit volume of the material).

3.11 Estimate the RMS thermal fluctuations of the speed of nitrogen molecules in an ideal-gas atmosphere at 300 K. Compare the answer with the average thermal speed of these molecules.

3.12 Estimate the RMS fluctuations of the speed of noninteracting polymer molecules (molecular weight = 6 million g/mol) executing Brownian motion in an ideal-gas atmosphere at 300 K. What will be the distribution of density (the density profile) of the polymer molecules caused by gravity?

3.13 What will be the molecular density of the polymer considered in Exercise 3.12 at a height of 3 cm, if its density on the ground level is 1 million particles per cubic centimeter? Plot the polymer concentration as a function of height.

3.14 What is the difference between the mobility of molecules and the mutual diffusion coefficient of species in a binary-liquid solution? When are these two properties proportional to each other?

3.15 Obtain the expressions for the mutual diffusion coefficient in an ideal solution and in the dilute-solution limit.

3.16 Calculate the diffusion coefficient of a carbon Brownian particle (with a radius of 1 micron) in water at 300 K using the Stokes–Einstein formula. Compare the result with typical values of the mutual diffusion coefficient in binary-liquid solutions (about 10^{-9} m^2/s) and explain the difference.

3.17 Determine an order of magnitude estimate of the time scale during which a Brownian particle (considered in Exercise 3.16) performs a half-rotation (180° around its axis) due to random collisions with surrounding molecules.

3.18 An experimentalist is looking at the particle considered in Exercise 3.16 through a high-resolution microscope. The particle is placed in a thin liquid layer between two glass plates and exhibits two-dimensional Brownian motion. Suppose you initially observed this particle in the center of the screen. Within which area and with what probability would you expect to find this particle around the initial position in one hour?

3.19 Compare and explain the difference between the average speed of thermal molecular motion, the average speed of Brownian motion, and the speed of sedimentation caused by gravity for a particle considered in Exercise 3.16.

3.20 Derive Fick's second law (the evolution of solute concentration upon time) from Fick's first law (diffusion flux) for one-dimensional diffusion. *Hint:*

$$J = \frac{dN_2}{Adt} \quad \text{and} \quad \frac{dN_2}{Adr} = \frac{dN_2}{dV} = c$$

3.21 Explain the differences between Rayleigh scattering and Brillouin scattering.

3.22 Using the mean-square fluctuations of entropy and pressure as well as the Einstein formula, derive the Landau–Placzek ratio. Calculate the relation between the isothermal compressibility, κ_T, and adiabatic compressibility, κ_S, of water at its density maximum (4 °C).

3.23 Estimate the relaxation rate of concentration fluctuations of a dilute solution of Brownian particles, with radius $R \approx 0.1\,\mu$m, suspended in water at ambient conditions measured from a dynamic light scattering experiment at a scattering angle of $\theta = 30°$ with a He–Ne laser light source ($\lambda_0 = 633$ nm).

3.24 Estimate a frequency shift of the Brillouin components in the light-scattering spectrum for benzene in ambient conditions at $\theta = 90°$. The source of light is a Ne–He laser (633 nm). Obtain the effective relaxation time of the molecular structure in benzene. Given that the thermodynamic speed of benzene at ambient condition is 1298 m/s, while in the high-frequency limit, the speed of sound is 1440 m/s. Additionally, assume that the sound velocity at this condition is in the hyper sound regime ($\omega\tau \gg 1$).

4

Landau Theory of Phase Transitions

This chapter includes:

- Classification of phase transitions
- Concept of the order parameter
- Landau expansion of the free energy
- Mesoscopic susceptibility and correlation length
- Landau–Ginzburg functional for the mesoscopic free energy
- Meanfield approximation and Ginzburg criterion
- Meanfield dynamics of phase transitions

4.1 First-Order and Second-Order Phase Transitions

Thermodynamic classification of phase transitions was originally proposed by Ehrenfest[1] [1] and generalized by Landau in 1935–1937 [2]. According to this classification, phase transitions may be separated into two categories: *first-order transitions* and *second-order transitions*. At a first-order transition, the thermodynamic potential (specified in accordance to the nature of the transition) is continuous at the transition point, but its first derivatives (e.g. molar volume, molar entropy, molar enthalpy, and concentration) are discontinuous. Examples of first-order transitions are: boiling-condensation in fluids, melting-crystallization, phase separation in mixtures, and most of the transitions between different solid polymorphs as well as between liquid crystalline phases. One important feature of first-order phase transitions is the possibility of forming metastable states.

As an example, the first-order transition between the anisotropic and isotropic phases in a nematic liquid crystal is illustrated in Figure 4.1. The anisotropy of the refractive index, Δn_λ, is a measure of the symmetry of the system, which exhibits a finite discontinuity at the transition point.

At a second-order phase transition, not only are the thermodynamic potentials continuous at the transition point, but their first derivatives are also continuous. However, the second derivatives of the potentials (like the heat capacity, volumetric isobaric expansion, isothermal compressibility, etc.) are discontinuous. Examples of second-order phase transitions are the critical points of fluid–fluid or solid–solid separation, the transition between the paramagnetic state and the ferromagnetic state at the Curie point (emergence of spontaneous magnetization), the

1 Paul Ehrenfest (1880–1933) was an Austrian theoretical physicist who made major contributions to the field of statistical mechanics and quantum mechanics.

Mesoscopic Thermodynamics for Scientists and Engineers, First Edition. Mikhail A. Anisimov and Thomas J. Longo.
© 2024 John Wiley & Sons, Inc. Published 2024 by John Wiley & Sons, Inc.

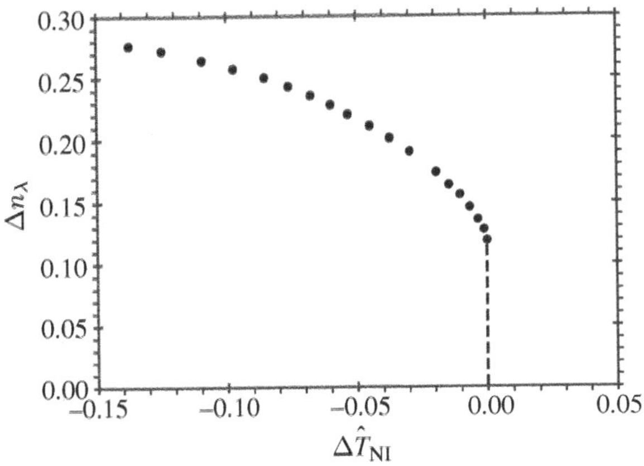

Figure 4.1 Discontinuity of the anisotropy of the refractive index, Δn_λ ($\lambda = 632.8$ nm) in the nematic phase of BMOAB where the distance to the nematic-isotropic phase transition temperature is $\Delta\hat{T}_{NI} = (T - T_{NI})/T_{NI}$. Source: Adapted from Ref. [4]/Gordon and Breach Science Publishers.

appearance of superconductivity[2] (in metals) and superfluidity[3] (in liquid helium), and certain phase transitions in solid and liquid crystals.

For both first-order and second-order phase transitions, the physical fields are uniform across the system as required by thermodynamic equilibrium. Sometimes, a line of first-order transitions in P–T space will stop and continue as a line of second-order transitions. The point that separates the second-order branch and the first-order branch on the transition line is referred to as a "tricritical point."

For first-order phase transitions, the relation between the discontinuities of the entropy and volume is given by the Clapeyron equation, Eq. (2.62), while for second-order transitions, the relation between the discontinuities of the second derivative of the Gibbs energy is given by Ehrenfest's equation,

$$\boxed{\left.\frac{\mathrm{d}P}{\mathrm{d}T}\right|_{\tilde{\alpha}\tilde{\beta}} = \frac{\Delta C_P}{TV\Delta\alpha_P} = \frac{\Delta\alpha_P}{\Delta\kappa_T}} \tag{4.1}$$

where $(\mathrm{d}P/\mathrm{d}T)|_{\tilde{\alpha}\tilde{\beta}}$ is the slope of the phase-transition line and the symbol Δ indicates the difference in the corresponding property (i.e., the isobaric heat capacity, volumetric expansivity, and isothermal compressibility) in each phase, $\tilde{\alpha}$ and $\tilde{\beta}$.

Despite being simple and elegant, the Ehrenfest classification does not answer the following fundamental questions: why is there no critical point on the crystallization-melting line? Why is there no phase transition between liquid and gas above the critical point? In what sense are the critical points of fluids equivalent or similar to second-order phase transitions? The answers to these questions are given by the Landau theory of phase transitions discussed throughout the remainder of this chapter.

4.2 The Concept of the Order Parameter

The concept of the order parameter, introduced by Landau, can be used to phenomenologically describe a broad range of phase transitions by considering the symmetry break of the system at the

2 The transition to superconductivity in pure metals at low temperatures was discovered in 1911 by Heike Kamerlingh Onnes (1853–1926), a Dutch experimental physicist, for which he was awarded a Nobel Prize in Physics in 1913.

3 The transition to superfluidity in helium-4 was discovered in 1937 by Pyotr Leonidovich Kapitsa (1894–1984) [3], a Soviet experimental physicist, for which he was awarded a Nobel Prize in Physics in 1978.

Figure 4.2 Two-dimensional representation of the spins in the Ising model. In the analogy with the lattice-gas model (see Figure 2.16), each spin state (up or down) could be considered as a particle or an empty spot.

phase transition point (such as the alignment of spins in a magnetic system or the spatial phase separation in a fluid). The order parameter reflects this symmetry and is defined to be zero in the disordered (e.g. isotropic and homogeneous) state and becomes nonzero in the ordered (e.g. anisotropic or inhomogeneous) state. For example, a simple model that describes the transition between a paramagnetic state (disordered random spin state) and a ferromagnetic state (ordered aligned spin state) is the Ising model, depicted in Figure 4.2. The Ising model serves as a reference system to describe physically different phase transitions in a universal way.

The Ising model is an incompressible lattice with identical magnetic or electric dipoles ("spins") at each site. The dipoles may be directed either upward or downward. Neighboring dipoles in equilibrium force each other to be parallel (ferromagnetic ordering) or antiparallel (anti-ferromagnetic ordering). At high temperatures, the entropic contribution to the overall free energy is large and the average magnetization (summation of all spin-vectors) is zero. At low temperatures, the contribution to the free energy from the internal energy is large and there is a spontaneous magnetization even in the absence of an applied magnetic field (the "zero field" condition). At the critical point, also referred to as the "Curie" point,[4] the magnetization emerges starting at zero value. For example, in iron, the Curie critical temperature is 1043 K. A sign of the interaction energy of neighboring dipoles determines whether a parallel or an antiparallel orientation is favored. In antiferromagnetic materials, the average magnetization is zero, but the sub-lattice magnetization behaves just like the full magnetization in ferromagnetic materials.

Therefore, in a simple magnetic system, the ferro-para transition across the Curie critical temperature (T_c) can be described based on the symmetry of aligned spins. In zero magnetic field, the spontaneous order parameter of this system is the magnetization density, $\varphi_0 = \langle \hat{M} \rangle$, where $\hat{M} = M/M_0$ and M_0 is the magnetization density at zero temperature. It is $\varphi_0 = 0$ in the paramagnetic (unaligned-spins) phase where $T > T_c$, and it is $\varphi_0 \neq 0$ in the ferromagnetic (aligned-spins) phase where $T < T_c$. The behavior of the spontaneous magnetization (in zero

4 The Curie temperature is named after Pierre Curie (1859–1906), a French physicist who showed that magnetism emerges below a critical temperature. In 1903, he received the Nobel Prize in Physics with his wife, Marie Skłodowska-Curie, and Henri Becquerel for their joint research on radiation phenomena.

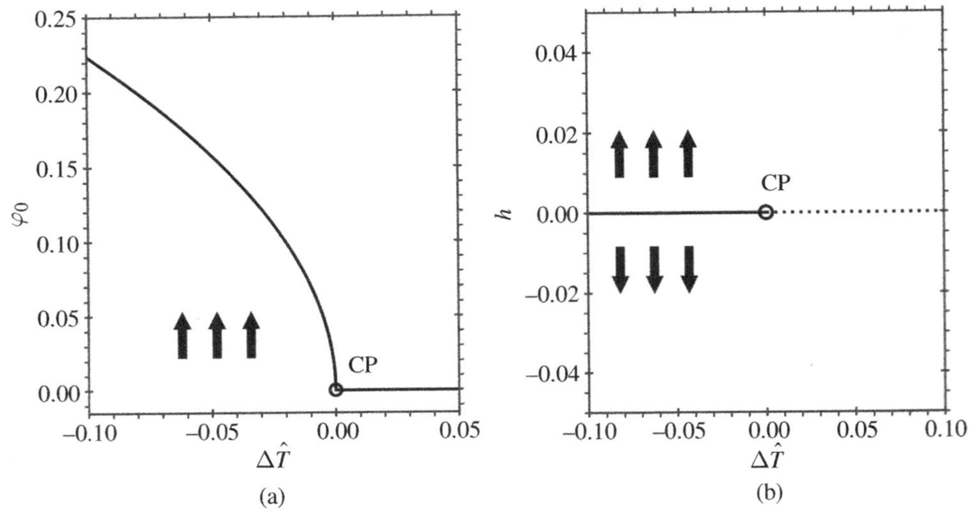

Figure 4.3 (a) The spontaneous order parameter (in zero-magnetic field, $h = 0$), predicted by the Landau theory for the Ising model, as a function of distance to the critical temperature, $\Delta\hat{T} = (T - T_c)/T_c$. The curve is given by Eq. (4.3). The direction of the order parameter is arbitrarily chosen to be positive. (b) Symmetry break in a magnetic field. The solid line ($h = 0$) represents the spontaneous magnetization.

magnetic field) is described by a formula mathematically equivalent to Eq. (2.84), derived for the density of the lattice-gas model,

$$\frac{\langle \hat{M} \rangle}{\hat{T}} = \frac{1}{2} \ln \left(\frac{1 + \langle \hat{M} \rangle}{1 - \langle \hat{M} \rangle} \right) \tag{4.2}$$

and is shown in Figure 4.3a. Therefore, in the Ising, lattice-gas, and regular-solution models, the spontaneous order parameter in the meanfield approximation is universally given by

$$\frac{\varphi_0}{\hat{T}} = \frac{1}{2} \ln \left(\frac{1 + \varphi_0}{1 - \varphi_0} \right) \tag{4.3}$$

Some magnetic materials, such as dysprosium aluminum garnet (DAG), exhibit a phenomenon of antiferromagnetism (introduced by Landau in 1933), where neighboring spins tend to align in opposite directions. The order parameter for the antiferromagnetic phase is the average magnetization of the sublattice of spins with the same orientation. The temperature of a second order transition between the paramagnetic and the antiferromagnetic phase is referred to as Néel temperature,[5] which is analogous to the Curie temperature for ferromagnets.

Alternatively, in a lattice-gas model for a liquid–gas system (Section 2.7.2), the order parameter can be identified as the reduced density difference $\varphi = (\rho - \rho_c)/\rho_c$. The spontaneous order parameter, φ_0, is

$$\varphi_0 = \begin{cases} \rho - \rho_c = 0 & T \geq T_c \\ (\rho_{cxc} - \rho_c)/\rho_c & T < T_c \end{cases} \tag{4.4}$$

where ρ_{cxc} is the density of either the coexisting liquid or gaseous phase below the critical point. Thus, $\varphi_0 = 0$ in the supercritical-fluid phase along the critical isochore and is $\varphi_0 \neq 0$ along two-phase coexistence. Similarly, in an incompressible binary lattice, which exhibits a

5 Louis Eugène Félix Néel (1904–2000) was a French physicist who received a Nobel Prize in Physics in 1970 for his studies of the magnetic properties of solids.

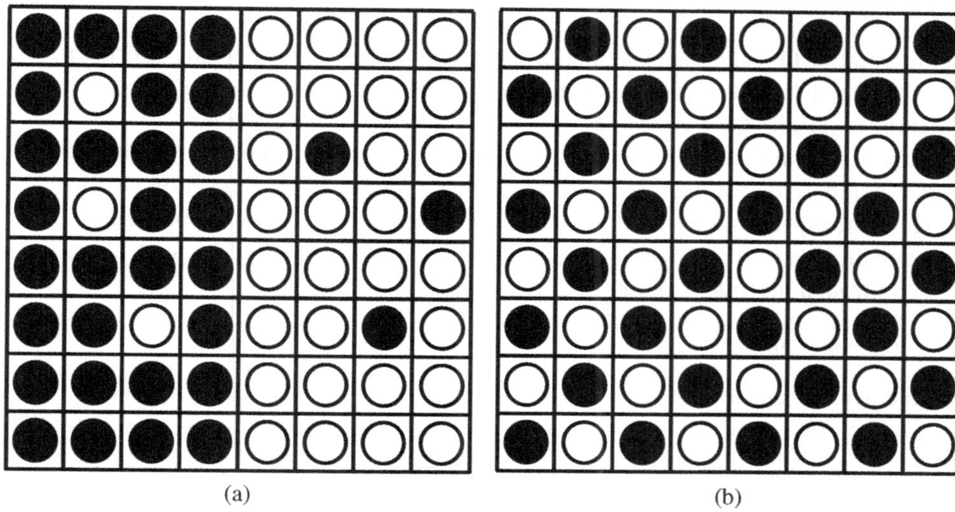

Figure 4.4 Two possible scenarios of phase transitions in binary alloys. (a) Phase separation, similar to an incompressible binary liquid, is illustrated, and (b) alternation of molecular species in a binary alloy (antiferromagnetic-like ordering) is shown.

liquid-liquid phase transition and is described by the regular-solution model (Section 2.8.3), the order parameter can be expressed through the molecular fraction as $\varphi = (x - x_c)/x_c$.

Binary solid solutions could exhibit two alternative phase transitions as illustrated in Figure 4.4. They may exhibit solid–solid phase separation, like binary liquids with a positive deviation from ideality, or, unlike binary liquids, form an ordered binary alloy, in which the neighboring molecules tend to be opposite species. Binary alloys are analogs of antiferromagnets in that the order parameter is expressed through the average concentration of the sublattice containing the molecules of the alternative species. A well-studied example of a binary alloy is β-brass (Cu–Zn).

The analogy between the symmetry break of the alignment of spins in a magnetic system and the spatial separation in a fluid is not just formal. Imposing an arbitrarily small gravitational field, which defines a fixed direction into an isotropic-fluid system, explicitly reveals a decrease in the symmetry at the emergence of two-phase coexistence. This procedure is similar to imposing an external magnetic field onto a ferromagnet to select the direction of magnetization and then reducing the field to an arbitrary small value, see Figure 4.3b.

Landau theory states that the phase transition is of the first order if the order parameter is discontinuous and is of the second order if the order parameter, along with other first derivatives, is continuous across the transition point, while the second derivatives of the free energy are discontinuous. Commonly, the order parameter is a thermodynamic coordinate (Section 2.4), a first derivative of the appropriate thermodynamic potential with respect to the corresponding ordering field (Section 3.5).

Different liquid crystal phases (illustrated in Figure 2.6) are described by different order parameters, a tensor for the nematic phase and two-component vectors for the smectic A (SmA) and the smectic C (SmC) phases. The nematic order parameter is a symmetric second-rank traceless tensor $\hat{Q}_{ij}$. Its matrix is given by

$$\hat{Q}_{ij} = \begin{pmatrix} -\frac{1}{2}\hat{Q} & 0 & 0 \\ 0 & -\frac{1}{2}\hat{Q} & 0 \\ 0 & 0 & \hat{Q} \end{pmatrix} \tag{4.5}$$

where $\hat{Q}$ is the amplitude of the order parameter. For simple rod-like molecules, the amplitude $\hat{Q}$ is the average orientation (a scalar), as given in Eq. (3.60), with respect to the nematic director, $\boldsymbol{n}$, see Figure 3.5. The nematic phase breaks the continuous rotational symmetry of the isotropic liquid phase in that the molecules have some average orientation given by the director $\boldsymbol{n}$.

The structure of the SmA phase consists of layers parallel to $\boldsymbol{n}$ with thickness $\sim\ell$ of the order of the molecular length, so that the continuous translational symmetry in the direction parallel to $\boldsymbol{n}$ is broken. Thus, the order parameter of the SmA phase differs from the nematic phase due to the existence of one-dimensional translational order normal to $\boldsymbol{n}$. The density of the SmA phase may be represented through the order parameter, ψ, as

$$\rho(z) = \overline{\rho}[1 - \psi_{\mathrm{A}}(z)] \tag{4.6}$$

where $\overline{\rho}$ is the average density, z is the spatial coordinate along the director ($\hat{\boldsymbol{n}}$), and $\psi_{\mathrm{A}}(z) = \psi_{\mathrm{A},0}\cos(q_0 z)$. The order parameter, $\boldsymbol{\psi}_{\mathrm{A}} = \psi_{\mathrm{A}}\hat{\boldsymbol{n}}$, is a vector with two components: $q_0 = 2\pi/\ell$, the wavenumber characterizing the average distance between smectic layers, and $\psi_{\mathrm{A},0}$ the amplitude of the density wave. In the absence of smectic layering, fluctuations of the director are unconstrained ("soft"), diverging in the thermodynamic limit ($q \to 0$) as $\propto q^{-2}$ [5]. In the SmA phase, since the layer locally wants to be parallel to $\boldsymbol{n}$, then the soft fluctuations of the nematic director are suppressed by layering.

In some liquid crystals exhibiting SmA order, a further decrease in temperature may result in the transition to the less symmetric SmC phase. In SmC, the uniaxial symmetry is broken due to the emergence of a tilt in the molecular axes with respect to the director, the direction of the density of wave. Also, the molecular orientation could be rotated around $\hat{\boldsymbol{n}}$. Therefore, the density wave for SmC is given by

$$\psi_{\mathrm{C}} = \psi_{\mathrm{A},0}\left(1 + \psi_{\mathrm{C},0}\right)\cos\left(q_0 z\right) \tag{4.7}$$

such that the order parameter for the transition from SmA to SmC is given by

$$\psi_{\mathrm{C},0} = \theta_{\mathrm{C}}\cos\phi \tag{4.8}$$

where θ_{C} is the amplitude of the additional order, oscillating with respect to the director, and ϕ characterizes the tilt (the azimuthal angle). Thus, this order parameter could also be viewed as a two-component vector, like the order parameter for SmA. Like the nematic-SmA transition, the SmA-SmC transition can, in principle, be either first or second order. In all systems studied so far, the SmA–SmC transition was found to be second order. The transitions from the nematic and isotropic phases to SmC are always first order. An example phase diagram of two mixtures of liquid crystals, exhibiting first- and second-order phase transitions between isotropic, nematic, SmA, and SmC phases, is shown in Figure 4.5. Common examples of order parameters in condensed-matter systems are presented in Table 4.1.

More complex types of ordering in liquid crystals could arise. A remarkable example of such structures is the existence of "blue phases" (BP) [8]. If the molecules are chiral or if the liquid crystal is doped with a chiral additive, the preferred molecular orientation is no longer constant but spontaneously forms a helix. This is called the cholesteric or chiral nematic phase, and the distance over which the preferred orientation rotates through 2π radians is called the pitch. If the pitch is greater than about 500 nm, the cholesteric phase undergoes a transition directly to the isotropic liquid phase with increasing temperature. In the case of a smaller pitch, however, the transition from the cholesteric to the isotropic liquid phase is usually via one, two, or three blue phases.

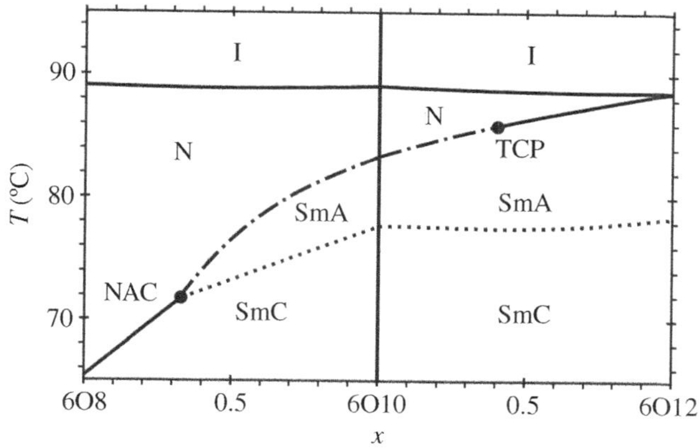

Figure 4.5 Phase diagram for two mixtures of liquid crystals, $\overline{6\text{O}8}$-$\overline{6\text{O}10}$ and $\overline{6\text{O}10}$-$\overline{6\text{O}12}$ [4, 6], showing four phases: isotropic (I), nematic (N), smectic A (SmA), and smectic C (SmC). Solid curves depict first-order transitions, while the dotted curves are second-order transitions. The dash-dotted curve depicts the segment of the nematic-smectic transition line, which is predicted to be second order by the Landau theory, but actually is weakly first order (with a hardly detectable heat of transition) due to the coupling between the fluctuations of the nematic director and smectic ordering (to be discussed in Section 4.4.4). TCP indicates the location of the N-SmA tricritical point predicted by the Landau theory. NAC is the N-SmA-SmC multicritical point.

Table 4.1 Different symmetries of order parameters

Order parameter type	System Examples
Scalar	Fluids, fluid and solid mixtures, order-disorder transition in crystals
Tensor	Nematic liquid crystals
One-component vector	Ising (anisotropic) ferromagnets and ferroelectrics
Two-component vector	Heisenberg ferromagnets, superfluid helium, superconductors, SmA and SmC liquid crystals

These blue phases are quite exotic. One possibility is that the preferred direction of orientation adopts a helical structure, that is quite different from the structure present in the cholesteric phase. Alternatively, a collection of defect lines, called "disclinations," form a dense network throughout the phase. In two of the blue phases (BP_I and BP_{II}), these disclinations form either a simple cubic or body-centered cubic lattice. In the third blue phase (BP_{III}), these disclinations, if they are present at all, form an amorphous structure. A phenomenological theory has demonstrated that the BP_{III}-Isotropic transition is analogous to the liquid–gas transition in fluids, such that there is coexistence of a "liquid of defects" and a "gas of defects" that is terminated at a critical point [9]. The analogy is not trivial as the temperature mainly plays the role of the ordering field, while the entropy mainly plays the role of the order parameter (being the conjugate coordinate to the temperature) since the densities of the coexisting phases are almost equal (similar to a so-called "lattice-liquid" model [10]). Experiments of a pure steroisomer of *S*,*S*-MBBPC in its racemic mixture have shown that the phase transition line between the amorphous blue phase (BP_{III}) and the isotropic liquid terminates at a critical point in the temperature-chiral-fraction plane [7], as illustrated in Figure 4.6.

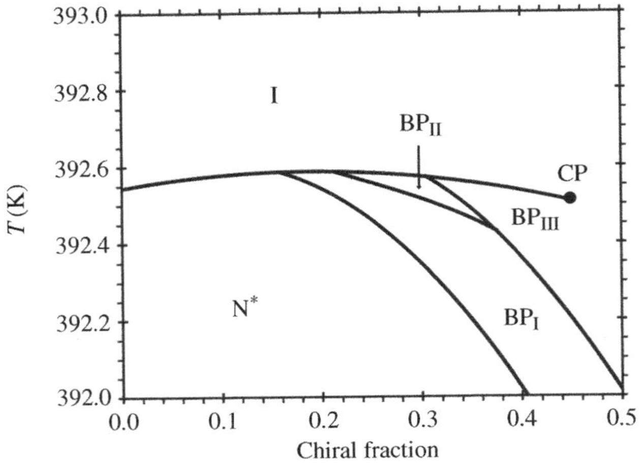

Figure 4.6 Phase diagram of a pure stereoisomer (*S*,*S*-MBBPC), in its racemic mixture. The critical point is denoted by CP, the blue phases I, II, and III by BP$_I$, BP$_{II}$, and BP$_{III}$, the triple point by TP, the isotropic phase by I, and the chiral nematic phase by N*. Source: Adapted from Ref. [7].

The two-component vector order parameter in smectics is analogous to the order parameter for superfluid helium and superconductors, first noticed by de Gennes[6] [5, 11, 12]. In both cases, the transition to the more-ordered phase is characterized by the emergence of a wave function (the order parameter), with a certain amplitude and phase. In particular, in smectics the order parameter is the density wave, while in superfluid helium and superconductors, the order parameter is the quantum wave function that describes the Bose–Einstein condensate through a certain amplitude and phase (see Section 2.12). Note that in smectics, the density wave is a measurable property, while in superfluid helium, the measurable property is the fraction of the superfluid component (the square of the wave function). The order parameter for these systems is also analogous to the order parameter for the "Heisenberg ferromagnet" (also referred to as the XY model), a variant of the Ising ferromagnet, in which the spins are free to rotate in the XY plane (in addition to being oriented up or down).

4.3 Truncated Landau Expansion

The Landau theory assumes that the behavior of various phase transitions could be described by expanding an appropriate thermodynamic potential, referred to as the "free energy," in powers of the order parameter, φ, in the immediate vicinity of the transition point. For example, in the Ising model, the free energy is an analog of the Helmholtz energy, which, in this case, is a function of the temperature and magnetization, $F = F(T, M)$. For the liquid–gas transition, the free energy is the density of the Helmholtz energy, $\rho A = \rho A(T, \rho)$, while for the liquid–liquid transition in weakly-compressible binary mixtures, the appropriate free energy is the Gibbs energy, $G = G(T, P, x)$. In the following material, we will use a general notation for the appropriate dimensionless free energy, $\hat{f}$. For example, for the liquid-gas transition in a single-component fluid, the dimensionless density of the Helmholtz energy is $\hat{f} = \rho A/\rho_c k_B T_c$.

6 Pierre-Gilles de Gennes was a French physicist (1932–2007) and a Nobel Prize laureate in physics (1991) for his outstanding contributions to soft condensed matter physics, in particular into liquid crystals and polymers.

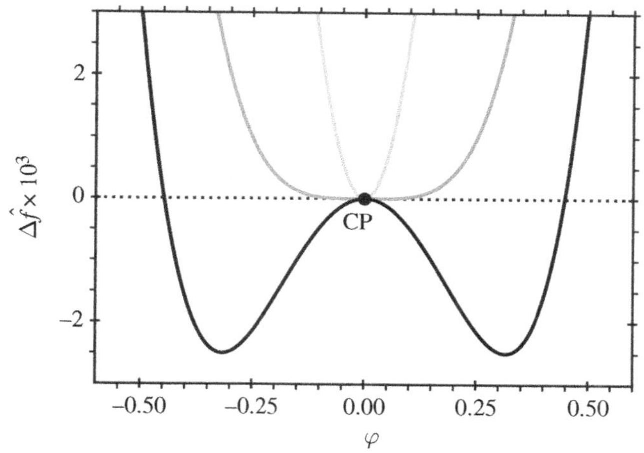

Figure 4.7 The dimensionless free energy density, $\Delta\hat{f}$, for the truncated Landau expansion in the vicinity of the critical point (CP) for supercritical (light gray), critical (dark gray), and subcritical (black) isotherms. Below the critical temperature, $\Delta\hat{T} = (T - T_c)/T_c < 0$, the free energy has two minima, corresponding to the liquid and gas coexisting phases or the direction of the magnetic field produced from a ferromagnet, while above the critical temperature $\Delta\hat{T} > 0$, the free energy has a single minimum at $\varphi = 0$, corresponding to the supercritical phase or the paramagnetic phase of a ferromagnet.

For the simple systems mentioned above, the Landau expansion, truncated to the fourth order, has the form

$$\hat{f} = \hat{f}_{bg} + \frac{1}{2!}a_0\Delta\hat{T}\varphi^2 + \frac{1}{4!}u_0\varphi^4 \tag{4.9}$$

in which a_0 and u_0 are the expansion coefficients, $\Delta\hat{T} = (T - T_c)/T_c$ is the reduced distance to the critical temperature, and $\hat{f}_{bg}$ is a thermodynamic background function. The behavior of this free energy is shown in Figure 4.7. For $T < T_c$, the free energy has two minima corresponding to the two alternative equilibrium values of the order parameter, while for $T > T_c$, the free energy has a single minimum at $\varphi = 0$.

The numerical values of a_0 and u_0 depend on the particular equation of state chosen to represent the system. For example, for the Ising, lattice-gas, and binary-lattice models, $a_0 = 1$ and $u_0 = 2$. Furthermore, for the van der Waals fluid, $a_0 = 6$ and $u_0 = 9$. The function $\hat{f}_{bg}$ for fluids can also be represented by an expansion

$$\hat{f}_{bg} = \hat{f}_c + \hat{S}_c\Delta\hat{T} + \hat{\mu}_c\varphi + \left.\frac{\partial\hat{\mu}}{\partial\hat{T}}\right|_c \Delta\hat{T}\varphi \tag{4.10}$$

where $\hat{f}_c$, $\hat{S}_c$, and $\hat{\mu}_c$ are the Helmholtz free energy, the entropy, and the chemical potential at the critical point. Since the absolute values of the chemical potential and entropy are arbitrary for phase equilibria (see Section 2.14), then, correspondingly, both $\hat{\mu}_c$ and $(\partial\hat{\mu}/\partial\hat{T})|_c = \hat{S}_c - (\partial\hat{P}/\partial\hat{T})|_c$ are also arbitrary. Thus, the background part of the free energy can be considered to be independent of the order parameter, as the terms linear to φ in Eq. (4.10) will not play any role in the description of phase transitions. For this reason, we define the part of the Landau expansion that is relevant to describe the properties of phase-transitions as $\Delta\hat{f} \equiv \hat{f} - \hat{f}_{bg}$.

The field conjugate to the order parameter may be expressed through the derivative of the free energy, as

$$h \equiv \left.\frac{\partial(\Delta\hat{f})}{\partial\varphi}\right|_T = a_0\Delta\hat{T}\varphi + \frac{1}{3!}u_0\varphi^3 \tag{4.11}$$

For example, in a magnetic system, h, is an external magnetic field, while the product $-h\varphi$, is the contribution to the free energy due to the alignment of magnetic spins in the presence of this field. For this reason, h, is often referred to as the "ordering field." The symmetry-breaking effect of an external magnetic field in a magnetic system is shown in Figure 4.3b. For a liquid–gas transition, $h = (\mu - \mu_0)/k_BT_c$, where μ_0 is a function of temperature, given by

$$\mu_0 = \begin{cases} \mu(T, \rho = \rho_c) & T \geq T_c \\ \mu_{cxc}(T, \rho = \rho_{cxc}) & T < T_c \end{cases} \tag{4.12}$$

For a liquid–liquid transition in a weakly-compressible binary mixture, $h = (\mu_{21} - \mu_{21,0})/k_BT_c$, the deviation of the exchange chemical potential from its value along the critical concentration ($T \geq T_c$) or along the liquid–liquid coexistence ($T < T_c$).

The free energy in Eq. (4.9), $\Delta\hat{f} = \hat{f}(\varphi, \Delta\hat{T}) - \hat{f}_{bg}$, depends on the order parameter, φ. An alternative formulation of the free energy is another thermodynamic function, $\Delta\hat{\Psi} = \hat{\Psi}(h, \Delta\hat{T}) - \hat{\Psi}_{bg}$, in which $\hat{\Psi}$ is "field-dependent," as it depends on the conjugate field, h, while $\hat{\Psi}_{bg}$ is a background function similar to $\hat{f}_{bg}$. This free energy is obtained through a Legendre transformation of the order-parameter-dependent free energy as, $\Delta\hat{\Psi} = \Delta\hat{f} - h\varphi$, such that

$$\Delta\hat{\Psi} = \frac{1}{2!}a_0\Delta\hat{T}\varphi^2 + \frac{1}{4!}u_0\varphi^4 - h\varphi \tag{4.13}$$

For example, in fluids, $\hat{\Psi}$, is the dimensionless density of the grand canonical potential, $\hat{\Psi} = \hat{\rho}\hat{\Omega} = -\hat{P}$, which is related to the density of the Helmholtz energy through the relation given by Eq. (2.37a), and via a Legendre transform of the form $\rho\Omega = \rho A - \rho\mu$. In a dimensionless form, the corresponding Legendre transform may be expressed as $-\hat{P} - \hat{\Psi}_{bg} = \hat{f} - \hat{f}_{bg} - h\varphi$.

The condition for equilibrium in the Landau theory is $\partial(\Delta\hat{\Psi})/\partial\varphi|_T = 0$, which corresponds to the definition of the ordering field, h, given by Eq. (4.11). Thus, the equilibrium value of the order parameter is given by $\varphi = -\partial(\Delta\hat{\Psi})/\partial h|_T$. Note that in the limit of zero-ordering field ($h = 0$), both expansions of the free energy, Eqs. (4.9) and (4.13), become equivalent and $\hat{\Psi}_{bg} = \hat{f}_{bg}$.

From both forms of the free energies, one may obtain various thermodynamic properties of the system. For instance, in zero-ordering field ($h = 0$), Eq. (4.11) gives the value of the spontaneous order parameter, $\varphi(h = 0) \equiv \varphi_0$, as

$$\varphi_0 = \begin{cases} 0 & T \geq T_c \\ \pm\sqrt{\frac{6a_0}{u_0}\left(-\Delta\hat{T}\right)} & T < T_c \end{cases} \tag{4.14}$$

yielding two solutions for $T < T_c$ that correspond to either the liquid and gaseous phases in coexistence and to the positive or negative signs of the spontaneous magnetization in the Ising ferromagnet (shown in Figure 4.8). Note that Eq. (4.14) is the linear approximation, near the critical temperature, of the general result for the spontaneous order parameter in the entire temperature range given by Eq. (4.3).

The entropy and heat capacity are determined from the temperature derivatives of the free energy, as

$$\frac{S}{k_B} = -\left.\frac{\partial\hat{f}}{\partial\Delta\hat{T}}\right|_{h=0} = \frac{S_{bg}}{k_B} + \begin{cases} 0 & T > T_c \\ \dfrac{3a_0^2}{u_0}\Delta\hat{T} & T < T_c \end{cases} \tag{4.15}$$

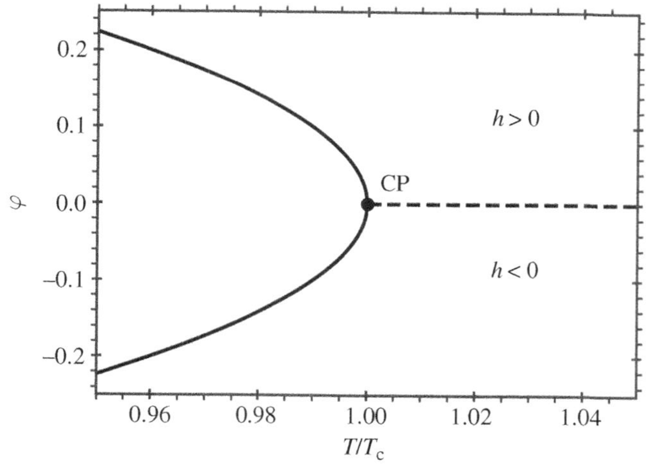

Figure 4.8 The order parameter predicted by the truncated Landau expansion. The spontaneous order parameter, $\varphi = \varphi_0$ (solid curve), given by Eq. (4.14), is obtained from the minima of the free energy, see Figure 4.7. The dashed line corresponds to the line of zero-ordering field ($h = 0$) above the critical temperature.

and

$$\frac{C_{h=0}}{k_B} = \frac{\hat{T}}{k_B}\frac{\partial S}{\partial \Delta\hat{T}}\bigg|_{h=0} = \frac{C_{bg}}{k_B} + \begin{cases} 0 & T > T_c \\ \dfrac{3a_0^2}{u_0}\hat{T} & T < T_c \end{cases} \tag{4.16}$$

where S_{bg} and C_{bg} are the background values of the entropy and heat capacity determined through the first and second temperature derivatives, respectively, of the background free energy $\hat{f}_{bg}$ in Eq. (4.9) or $\hat{\Psi}_{bg}$ in Eq. (4.13). As follows from Eq. (4.16), at the critical point, the heat capacity in zero field, $C_{h=0}$, exhibits a finite discontinuity (see Figure 4.9a). Depending on the physical nature of the phase transition, the heat capacity, $C_{h=0}$, represents different measurable properties of the system. For example, $C_{h=0}$ represents the heat capacity in zero magnetic field for an Ising ferromagnet; the isochoric heat capacity, C_V, at the critical density for the liquid–gas transition in single-component fluids; or the isobaric heat capacity, $C_{P,x}$, at the critical composition for the liquid–liquid transition in weakly-compressible binary liquids.

In addition, the inverse susceptibility, $\hat{\chi}^{-1}$, in zero field is given by

$$\hat{\chi}^{-1} = \frac{\partial^2 \hat{f}}{\partial \varphi^2}\bigg|_{\Delta\hat{T}} = \begin{cases} a_0\Delta\hat{T} & T \geq T_c \\ -2a_0\Delta\hat{T} & T \leq T_c \end{cases} \tag{4.17}$$

The linear behavior of the inverse susceptibility for ferromagnets near the Curie critical temperature (Figure 4.9b) is known as the "Curie-Weiss Law." Like the heat capacity in zero field, the susceptibility, $\hat{\chi}$, has different meanings depending on the nature of the phase transition. For instance, it represents the dimensionless magnetic susceptibility for an Ising ferromagnet, $\hat{\chi} \propto (\partial M/\partial h)|_T$; it is proportional to the isothermal compressibility, $\hat{\chi} \propto \rho^2\kappa_T$ for a single-component fluid Eq. (2.47); or it is proportional to the osmotic compressibility, $\hat{\chi} \propto c^2\kappa_\Pi$ in binary mixtures Eq. (2.100).

The symmetry of the free energy for both the fluid phase transition and the spin orientation in ferromagnets is broken in the presence of the ordering field. The effect of the ordering field on the field-dependent free energy, given by Eq. (4.13), is shown in Figure 4.10. Notice that depending

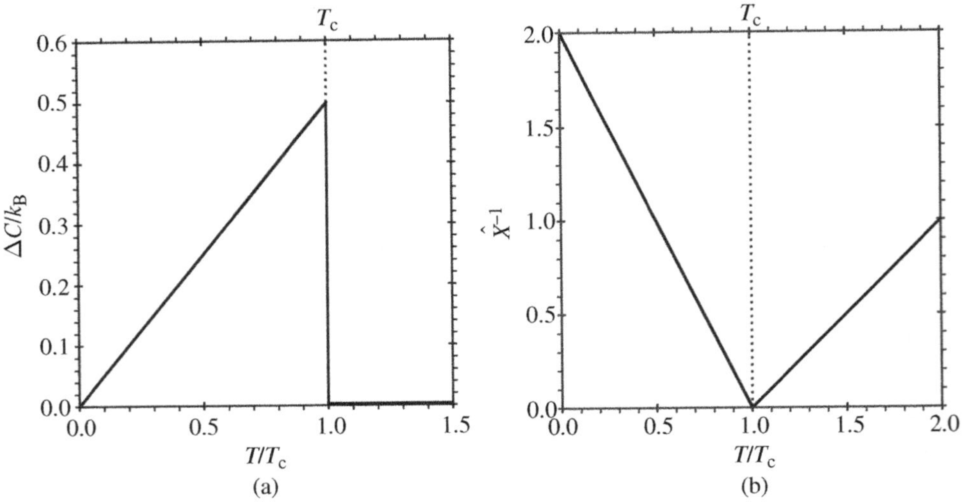

Figure 4.9 (a) The excess heat capacity, $\Delta C/k_B = [C(\varphi_0) - C_{bg}]/k_{bg}$ and (b) the inverse susceptibility predicted by the Landau theory from second derivatives of the free energy. The second-order phase transition is indicated by (a) a "jump" discontinuity in the heat capacity and (b) a discontinuous change in the amplitude of the inverse thermodynamic susceptibility.

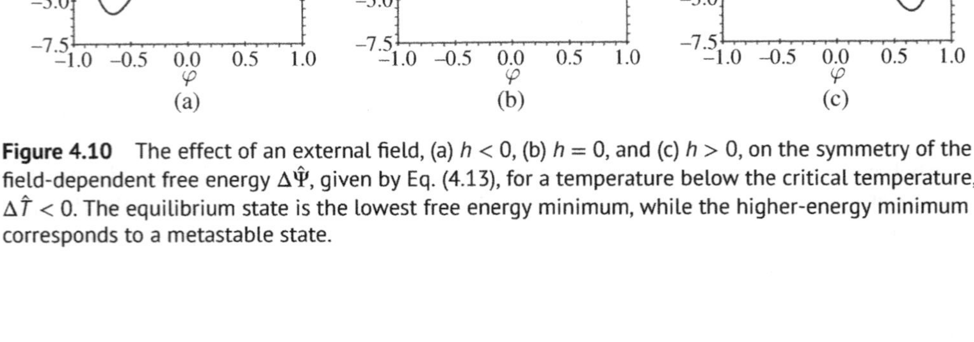

Figure 4.10 The effect of an external field, (a) $h < 0$, (b) $h = 0$, and (c) $h > 0$, on the symmetry of the field-dependent free energy $\Delta\hat{\Psi}$, given by Eq. (4.13), for a temperature below the critical temperature, $\Delta\hat{T} < 0$. The equilibrium state is the lowest free energy minimum, while the higher-energy minimum corresponds to a metastable state.

on the sign of the order parameter (or direction of h) a bias toward one of the alternative states is created and the symmetry of the free energy is broken.

Asymptotic Critical Power Laws in the Landau Theory

The superscript "±" indicates a property above (+) or below (–) the critical point.

- Heat capacity

$$\frac{C_{h=0}}{k_B} = A_0^{\pm}|\Delta\hat{T}|^{\alpha} + \frac{C_{bg}}{k_B} \tag{4.18}$$

where $A_0^+ = 0$, $A_0^- = 9a_0^2/u_0$, and $\alpha = 0$.

- Spontaneous order parameter

$$\varphi_0 = \pm B_0|\Delta\hat{T}|^{\beta} \tag{4.19}$$

where $B_0 = \sqrt{6a_0/u_0}$ and $\beta = 1/2$.

- Susceptibility

$$\hat{\chi} = \Gamma_0^{\pm}|\Delta\hat{T}|^{-\gamma} \tag{4.20}$$

where $\Gamma_0^+ = a_0$, $\Gamma_0^- = 2a_0$, and $\gamma = 1$.

- Ordering field

$$h = D_0|\varphi|^{\delta} \tag{4.21}$$

where $D_0 = (u_0/6)^3$ and $\delta = 3$.

The field-dependence of the order parameter is given through Eq. (4.11), such that the order parameter along the critical isotherm depends on the ordering field as

$$\varphi = \left(\frac{6}{u_0}h\right)^{1/3} \tag{4.22}$$

Therefore, the field-dependent susceptibility along the critical isotherm is given by

$$\hat{\chi} = \frac{1}{3}\left(\frac{6}{u_0}\right)^{1/3} h^{-2/3} \tag{4.23}$$

The field-dependent order parameter for different isotherms is shown in Figure 4.11.

As discussed in this section, the Landau expansion predicts certain power-law values of the temperature and field dependence of the thermodynamic properties asymptotically close to the critical point. Each critical power law is characterized by a so-called, "critical exponent" and "critical amplitude." For example, as shown by Eq. (4.14), the critical power law for the order parameter is given by

$$\varphi_0 = \pm B_0|\Delta\hat{T}|^{\beta} \tag{4.24}$$

where in the Landau theory the critical amplitude is $B_0 = \sqrt{6a_0/u_0}$ and the critical exponent is $\beta = 1/2$. A list of the most common critical exponents and amplitudes predicted by the Landau theory are presented in the insert above.

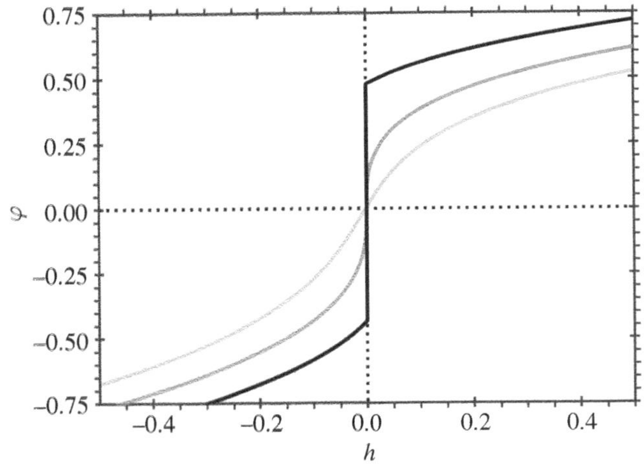

Figure 4.11 The field-dependence of the order parameter predicted by the truncated Landau expansion for subcritical (black), critical (dark gray), and supercritical (light gray) isotherms. Note that below the critical temperature, the order parameter exhibits a jump discontinuity with respect to the ordering field, h, corresponding to a first-order phase transition.

4.4 Extended Landau Expansion

The variety of phase transitions of different physical nature go beyond the simple systems described by the Ising, the lattice-gas, or the binary-lattice models. The behavior of all of the transitions can be addressed by a generalization of the Landau expansion, which for the field-dependent free energy (shown up to the sixth power) reads as

$$\Delta\hat{\Psi} = \frac{1}{2!}a_0\Delta\hat{T}\varphi^2 + \frac{1}{3!}b_0\varphi^3 + \frac{1}{4!}u_0\varphi^4 + \frac{1}{5!}d_0\varphi^5 + \frac{1}{6!}g_0\varphi^6 - h\varphi \tag{4.25}$$

where, in addition to the coefficients a_0 and u_0 in the truncated expansion Eq. (4.9), b_0, d_0, and g_0 are additional Landau coefficients.

The existence and the behavior of the different terms in the generalized Landau expansion, Eq. (4.25), depend on the physical nature of the order parameter. For instance, if the order parameter is a vector (e.g. magnetization or polarization in ferromagnetic or ferroelectric systems) the odd terms are absent and the expansion is symmetric. If the order parameter is a scalar (e.g. the density or concentration in fluids or binary mixtures, see Section 4.2), the symmetry of the expansion is only asymptotic as the coefficient of the cubic term $b_0 \propto \Delta\hat{T} \sim \varphi^2$, such that the cubic and fifth ordered terms are of the same order in φ. Consequently, the odd-ordered terms may be incorporated into a redefined scalar order parameter and only the even-ordered terms will remain.

Landau Theory: Note 1

A key concept in the Landau theory is that all terms in the truncated expansions are of the same order in φ or $\Delta\hat{T}$. Consider the truncated symmetric expansion of the order-parameter-dependent free energy, given by Eq. (4.9). In zero-ordering field, when $\varphi = \varphi_0$, one can show with use of Eq. (4.14) that $\Delta\hat{f}$ may be expressed as

$$\Delta\hat{f}\,(T < T_c) = -\frac{3a_0^2}{u_0}(\Delta\hat{T})^2 + \frac{3a_0^2}{2u_0}(\Delta\hat{T})^2 = -\frac{3a_0^2}{2u_0}\left(\Delta\hat{T}\right)^2 \tag{4.26}$$

such that term proportional to $\Delta\hat{T}\varphi^2$ and φ^4 are both of the order $(\Delta\hat{T})^2$. The same concept may be applied to the cubic and fifth-ordered terms as well as other higher-order terms.

If the order parameter is a tensor (with φ being its absolute value), the cubic term does not vanish at the transition point; thus, such a transition is of the first order. A common example is the transition between the nematic liquid crystal and its isotropic liquid phase (see Figure 4.1). If the coefficient of the fourth-order term, u_0, vanishes and changes its sign at the transition point, the sixth-order term is essential to maintain the stability of the system. If the coefficient u_0 is negative, the transition is first order, while if $u_0 = 0$, this case corresponds to the emergence of a tricritical point. In Sections (4.4.1–4.4.4), based on the generalized Landau expansion, we discuss the asymmetry in fluid-phase behavior, the first-order phase transition, and the emergence of a tricritical point.

4.4.1 Asymmetry of the Order Parameter in Fluids

In real fluids, the symmetry between the liquid and vapor states is broken, resulting in an asymmetric coexistence curve (see, for example, Figure 2.11). In this case, if the order parameter is still defined through $\varphi = (\rho - \rho_c)/\rho_c$, the Landau expansion for the order-parameter-dependent free energy in zero field could be expressed up to the fifth order term, as

$$\Delta\hat{f} = \frac{1}{2!}a_0\Delta\hat{T}\varphi^2 + \frac{1}{3!}b_0\varphi^3 + \frac{1}{4!}u_0\varphi^4 + \frac{1}{5!}d_0\varphi^5 \tag{4.27}$$

where the third-order coefficient, $b_0 = b_1\Delta\hat{T}$, where b_1 is a constant. Since, as the system approaches the critical point, the second- and third-order terms vanish, the condition $u_0 > 0$ maintains the thermodynamic stability and the higher-order terms can be neglected. For this particular case, the ordering field has the form

$$h = \left.\frac{\partial\hat{f}}{\partial\varphi}\right|_{\Delta\hat{T}} = a_0\Delta\hat{T}\varphi + \frac{1}{2!}b_1\Delta\hat{T}\varphi^2 + \frac{1}{3!}u_0\varphi^3 + \frac{1}{4!}d_0\varphi^4 \tag{4.28}$$

In zero field ($h = 0$), the free energy, given by Eq. (4.27), has a single minima above the critical temperature ($\Delta\hat{T} > 0$) at $\varphi = 0$, while below the critical temperature ($\Delta\hat{T} < 0$), there are two minima at

$$\varphi_0 = \left(\frac{\rho - \rho_c}{\rho_c}\right)_{\text{cxc}} = \pm\sqrt{\frac{6a_0}{u_0}(-\Delta\hat{T})} + \left(\frac{b_1}{u_0} - \frac{3a_0d_0}{5u_0}\right)\Delta\hat{T} \tag{4.29}$$

Note that the first term in Eq. (4.29) is the same as for the truncated, symmetric Landau expansion (given by Eq. 4.14), while the second term in Eq. (4.29) contributes to the asymmetry between the two coexisting phases.

Defining the positive solution of Eq. (4.29) to be the liquid state, ρ_ℓ, and the negative solution to be the gaseous state, ρ_g, then the sum of these two quantities represents the asymmetry of the coexistence curve,

$$\rho_\ell + \rho_g = 2\rho_c\left[1 + \left(\frac{b_1}{u_0} - \frac{3a_0d_0}{5u_0}\right)\Delta\hat{T}\right] \tag{4.30}$$

This property, $\Delta\hat{\rho}_d \equiv (\rho_\ell + \rho_g)/2\rho_c$, is known as the *rectilinear diameter*, which is commonly given in the form

$$\Delta\hat{\rho}_d = 1 + D\Delta\hat{T} \tag{4.31}$$

where the slope, D, is expressed through the Landau coefficients as

$$D = \frac{b_1}{u_0} - \frac{3a_0d_0}{5u_0} \tag{4.32}$$

Note that the van der Waals equation of state has a special symmetry when represented in the Landau expansion, such that $b_1 = 0$ and $d_0 = -u_0$ [13, 14]. Thus, since $a_0 = 6$ and $u_0 = 9$, the slope of the rectilinear diameter for a van der Waals fluid is $D = 2/5$, see Figure 4.12.

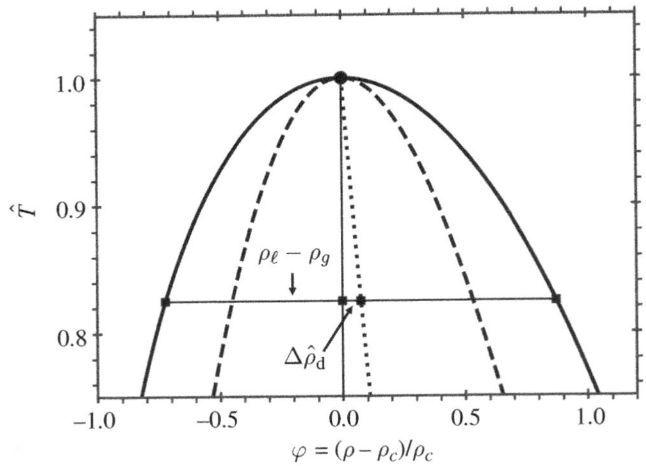

Figure 4.12 The rectilinear diameter, $\Delta\hat{\rho}_d$, emerges from the asymmetric coexistence of the VDW EOS. The diameter only asymptotically becomes linear, with a slope $D = 2/5$ (as predicted by the Landau theory), at the critical point.

The asymmetric terms in the Landau expansion, given by Eq. (4.27), may be eliminated by redefining the order parameter for fluids as $\varphi = (\rho - \rho_c)/\rho_c - D|\Delta\hat{T}|$. Correspondingly, the ordering field, $h = (\mu - \mu_0)/k_B T_c$, will be given by Eq. (4.11) and the free energy will be described by the symmetric Landau expansion, as shown in Figure 4.7.

4.4.2 First-Order Phase Transitions

There are many examples of first-order phase transitions with different physical origins. Boiling-condensation (liquid–gas) transitions, melting-freezing (solid–liquid) transitions, and sublimation-deposition (solid–gas) transitions are common examples of first-order phase transitions. Some phase transitions in liquid crystals, such as the isotropic-nematic transition (see Section 4.2), and solvation-precipitation in solutions are also examples of first-order phase transitions.

Since isotropic liquids and gases are bodies of the same symmetry, the boiling-condensation transition does not exhibit a change in symmetry. Thus, these systems may exhibit both first-order or second-order (critical point) phase transitions. However, this is not the case for solid–liquid or solid–gas transitions, as solids have a fundamentally different symmetry than fluids, such that these transitions are always first order [15]. Consequently, there is no critical point for the melting-crystallization line.

Traditionally, phase transitions are explained through the competition between the energy and the entropy expressed as the change in the Helmholtz energy: $\Delta A_t = \Delta U_t - T\Delta S_t$. Thus, phase transitions are commonly viewed as an energy-driven event, since the energy of attraction between particles dominates the entropy term at low temperatures, causing the emergence of the ordered phase (like in the lattice-gas or Ising models). At high temperatures, the entropy term becomes dominant, reflecting the emergence of the disordered phase.

However, in principle, the transition from fluid to solid does not require attraction between molecules. In this cases, the phase transition is fully entropy-driven [16]. The first example of a purely entropy-driven first-order transition to a nematic liquid was demonstrated in the Onsager

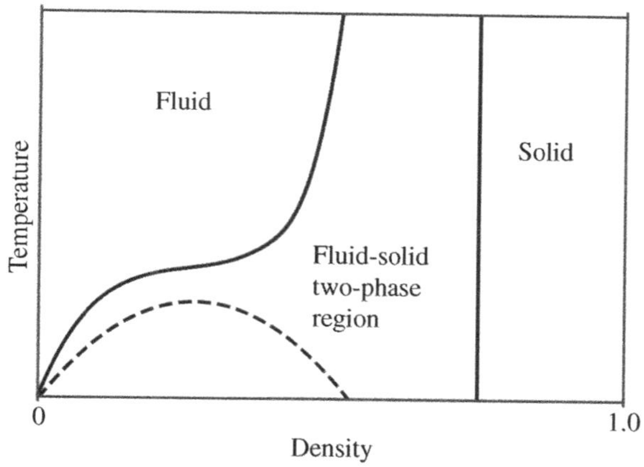

Figure 4.13 Schematic phase diagram of a substance with repulsion between molecules significantly dominating over attraction. In this case, the solid–fluid phase transition is mainly entropically driven. The dashed curve indicates the metastable fluid-fluid transition, located below the temperature of fluid-solid coexistence.

model of non-attracting rods [17]. It has also been shown that the van der Waals model, with an infinitely weak long-ranged attraction ($a \to 0$), exhibits entropy-driven crystallization [18]. More recently, entropy-driven crystallization (without a transition to a condensed fluid phase) has been experimentally and computationally observed for systems of non-attracting colloidal spheres [19] (see, for example Figure 4.13). One may also notice, from the phase diagram of methane (Figure 2.17), that crystallization occurs from the hard-fluid region (above the Zeno line), where the repulsion between molecules dominates over attraction [16].

Lastly, one of the unique characteristics of a first-order transition is the existence of metastability, the ability of the system to remain temporarily in the previously stable phase below or above the transition temperature or pressure [20, 21]. For example, as discussed in Section 2.6.3, if a gas is supercooled below the condensation point (T_{tran}) at constant pressure, the temperature of thermodynamic stability of the gaseous phase is $T^* \equiv T_s < T_{\text{tran}}$, where T_s is the temperature of the stability limit (spinodal), while if a liquid is superheated above the boiling point at constant pressure, the temperature of thermodynamic stability of the liquid phase is $T^{**} \equiv T_s > T_{\text{tran}}$, see Figure 2.14.

4.4.3 Weakly First-Order Phase Transitions

The most sensitive test to determine whether a phase transition is of the first or second order is through a measurement of the latent heat (the change in the enthalpy). For example, slow-scanning adiabatic calorimetry experiments have an extremely high resolution of about 0.001 Joule (because the rate of scanning can be reduced to a microkelvin per second). Such an exceptional resolution can discriminate between truly second-order transitions (with no latent heat) and weakly first-order transitions (with latent heat per molecule $\ll k_B T$) commonly observed in liquid crystals [4, 22].

In order to apply the Landau expansion to any phase transition, the order parameter must be small at the transition point. Consequently, the Landau expansion cannot commonly describe the solid–liquid transition as the order parameter is typically not small. Thus, the Landau expansion can only describe a so-called "weak" first-order phase transition, where the emerging order parameter is small. In these transitions, the enthalpy and volume change of the transition must both be small. For example, in weakly compressible liquid crystals, despite the very large slope of the neumatic-isotropic transition in pressure–temperature coordinates, $(dP/dT)|_{\tilde{\alpha}\tilde{\beta}}$, corresponding to an almost negligible volume change of transition (see the Clapeyron equation, Eq. 2.62), the transition is usually weakly first-order, as the heat of the transition, $\Delta H_{\tilde{\alpha}\tilde{\beta}}$, is of the

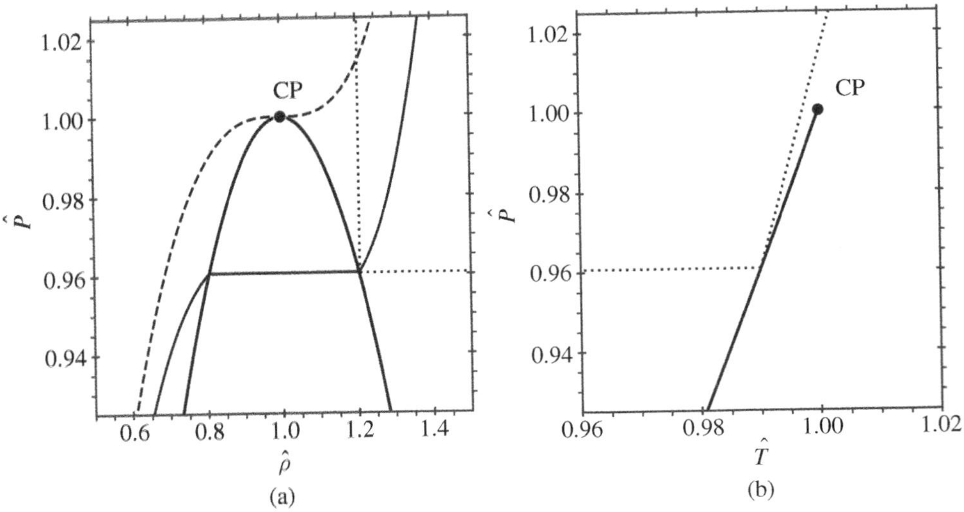

Figure 4.14 Illustration of a weakly first-order boiling transition in (a) density–pressure coordinates and (b) pressure–temperature coordinates. The dotted horizontal line in (a) and (b) correspond to isobaric heating from the liquid phase to the boiling temperature at phase coexistence. The other dotted lines in (a) and (b) correspond to isochoric cooling from the liquid phase to the boiling temperature at phase coexistence.

order $\Delta H_{\tilde{\alpha}\tilde{\beta}} \lessapprox 0.1\, k_B T$ [4], which is compare with the heat of crystallization of water $\Delta H_{\tilde{\alpha}\tilde{\beta}} \sim k_B T$ and the heat of boiling of water $\Delta H_{\tilde{\alpha}\tilde{\beta}} \sim 10\, k_B T$.

The weak first-order nature of phase transitions could be accounted for in the Landau expansion, given by Eq. (4.25), through the specific behavior of different terms in the expansion, namely, the cubic term, the quartic term, and the term containing the ordering field. Each of these three cases is considered in the remainder of this section.

4.4.3.1 Weakly First-Order Boiling-Condensation Transition

The ordering field generates a finite value of the order parameter at any temperature. Changing the temperature, while maintaining this finite value of the order parameter $\varphi = \varphi(T)$, allows one to reach the phase transition temperature at which the ordering field vanishes and the order parameter reaches its spontaneous value, $\varphi = \varphi_0$. This event is depicted for a magnetic ferromagnet in Figure 4.11 and for the boiling-condensation transition in a fluid in Figure 4.14. If the transition is a weakly first-order phase transition, occurring close to the critical point, the effect is described by the Landau expansion as illustrated in Figure 4.10.

At a temperature above the boiling-condensation temperature ($T > T_{\text{tran}}$), the ordering field for a single-component fluid, $h = (\mu - \mu_0)/k_B T_c$ (as defined through Eq. 4.12), generates a larger magnitude of the order parameter than its spontaneous value ($|\varphi| > |\varphi_0|$). Upon cooling along a noncritical isochore (e.g. $\varphi = (\rho - \rho_c)/\rho_c > 0$), when the order parameter reaches its spontaneous value $\varphi = \varphi_0$, the ordering field vanishes and the system exhibits a first-order phase transition into the two-phase region. Upon further cooling, the liquid is supercooled with respect to the gaseous state (a metastable state emerges between $T^* < T < T_{\text{tran}}$). This metastable state is depicted by the higher free energy minimum shown in Figure 4.10, which disappears for $T < T^*$.

4.4.3.2 Weakly First-Order Transition Emerging from the Cubic Term

The existence of the third-order term in the generalized Landau expansion, Eq. (4.25), depends on the symmetry of the order parameter. If the order parameter is a vector, the odd-ordered terms will not exist, but if it is a scalar or tensor (as for nematic liquid crystals), these terms may exist. The well-studied case of a weakly first-order transition is the isotropic-nematic transition in liquid crystals. The application of Landau theory to this transition was made by de Gennes and Prost [5] and is often referred to as the "Landau-de Gennes theory."

In contrast to the behavior of isotropic fluids and fluid mixtures (see Section 4.4.1), the third-order term does not vanish at the isotropic-nematic transition point, thereby making this transition first order. In this case, the truncated Landau expansion of the order-parameter-dependent free energy has the form

$$\Delta\hat{f} = \frac{1}{2!}a_0\Delta\hat{T}^*\varphi^2 + \frac{1}{3!}b_0\varphi^3 + \frac{1}{4!}u_0\varphi^4 \tag{4.33}$$

where $\Delta\hat{T}^* = (T - T^*)/T_{\text{tran}}$ is the temperature distance to the stability limit of the disordered phase, T^*, with $T^* < T_{\text{tran}}$. The condition for a weakly-first order transition emerges if the cubic term in the expansion is small and negative. For the neumatic-isotropic transition in liquid crystals, the free energy is expanded in powers of $\varphi = \hat{Q}$, the magnitude of the tensor order parameter $\hat{Q}_{ij}$, given by Eq. (4.5). In this case, typical values of b_0 range from $b_0 \sim -0.1$ to $b_0 \sim -0.2$ [4]. The free energy for this transition is shown in Figure 4.15a for various temperatures.

For most nematic liquid crystals, the first-order transition from the isotropic phase is weak, with $|b_0| \ll 1$. This is in sharp contrast to the Onsager model of rods, which exhibits a strong first-order transition to the nematic phase. The entropy-driven phase transition in the Onsager model is caused by pure repulsion of rigid rods (see Section 4.4.2). The physical reason for the soft nature of the nematic-isotropic transition in real liquid crystals is that the flexibility and biaxiality of the anisotropic molecules produce more complex molecular interactions. The attraction between different parts of the molecules plays a significant role. Molecular flexibility and biaxiality may renormalize the Landau coefficients, b_0 and u_0, making them small parameters, which is reflected in the isotropic-nematic transition being weakly first order. Consequently, the application of the Landau expansion to the quantitative description of the properties near the phase transition temperature may require accounting for higher-order terms (fifth and sixth order) [4].

The relation between the transition temperature and the temperature of the stability limit of the disordered phase is obtained from the condition $\Delta\hat{f}(T = T_{\text{tran}}) = 0$, giving,

$$T^* = T_{\text{tran}}\left(1 - \frac{b_0^2}{3a_0u_0}\right) \tag{4.34}$$

The finite value of the order parameter emerges at the transition point ($T = T_{\text{tran}}$) with a magnitude $\varphi(T_{\text{tran}}) = -2b_0/u_0$. Typically, for the nematic-isotropic transition in liquid crystals, this value is of the order ~0.1, see Figure 4.1, and correspondingly, $T_{\text{tran}} - T^* \sim 1\,\text{K}$.

The value of the order parameter for other temperatures is obtained by the minimization of the free energy, Eq. (4.33), as

$$\varphi = \begin{cases} 0 & T > T_{\text{tran}} \\ -\dfrac{3b_0}{2u_0}\left(1 + \sqrt{1 - \dfrac{8a_0u_0}{3b_0^2}\Delta\hat{T}^*}\right) & T < T_{\text{tran}} \end{cases} \tag{4.35}$$

such that at the stability limit, $T = T^*$, the order parameter has the value, $\varphi(T^*) = -3b_0/u_0$, which is larger than $\varphi(T_{\text{tran}})$ as $b_0 < 0$.

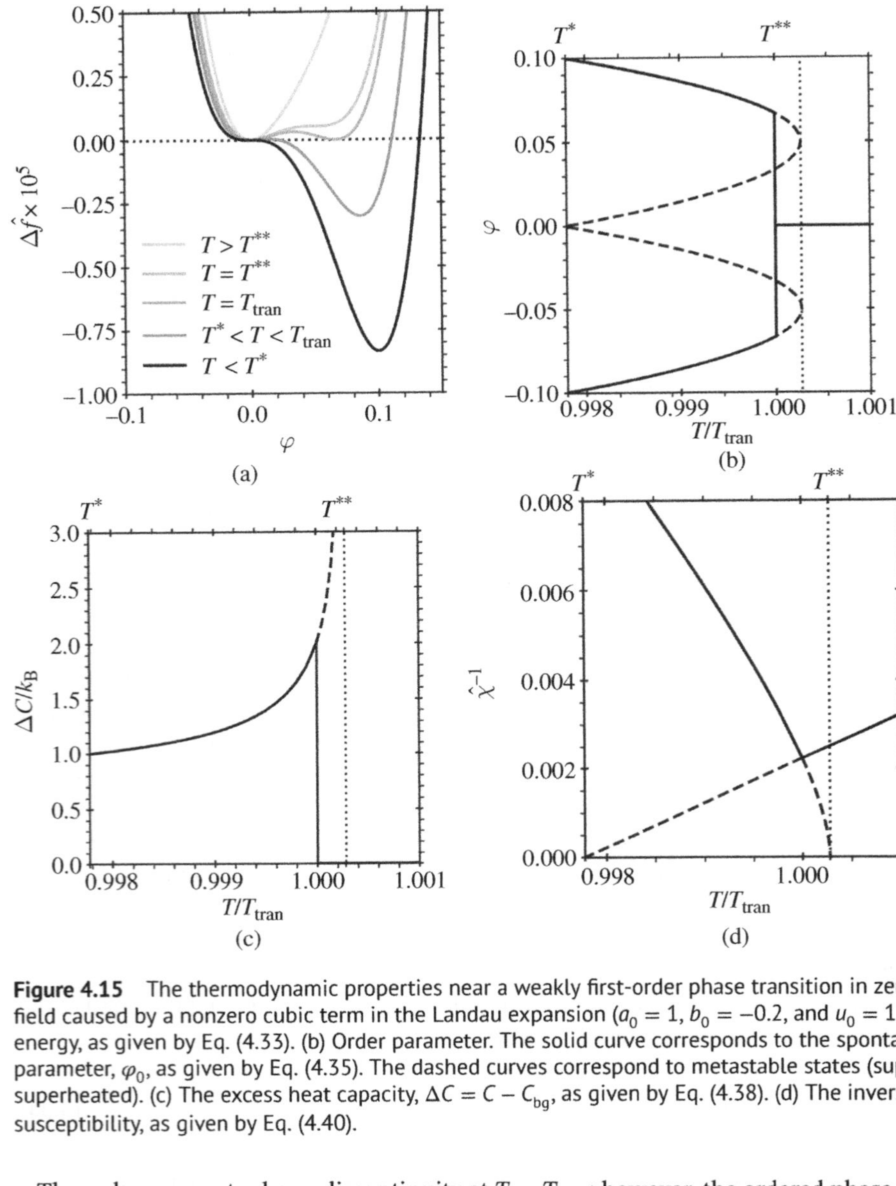

Figure 4.15 The thermodynamic properties near a weakly first-order phase transition in zero-ordering field caused by a nonzero cubic term in the Landau expansion ($a_0 = 1$, $b_0 = -0.2$, and $u_0 = 12$). (a) Free energy, as given by Eq. (4.33). (b) Order parameter. The solid curve corresponds to the spontaneous order parameter, φ_0, as given by Eq. (4.35). The dashed curves correspond to metastable states (supercooled or superheated). (c) The excess heat capacity, $\Delta C = C - C_{\mathrm{bg}}$, as given by Eq. (4.38). (d) The inverse susceptibility, as given by Eq. (4.40).

The order parameter has a discontinuity at $T = T_{\mathrm{tran}}$; however, the ordered phase may be superheated into a metastable state until it reaches the temperature of the stability limit of the ordered phase, T^{**}. Beyond this temperature, Eq. (4.35) has no solution, becoming imaginary. Therefore, the condition $1 - 8a_0u_0\Delta\hat{T}^*/(3b_0^2) = 0$, defines $T = T^{**}$ as

$$T^{**} = T^* + T_{\mathrm{tran}}\frac{3b_0^2}{8a_0u_0} = T_{\mathrm{tran}}\left(1 + \frac{b_0^2}{24a_0u_0}\right) \tag{4.36}$$

Thus, accounting for Eq. (4.36) in Eq. (4.35), one obtains the order parameter in the stable and metastable states for $T < T^{**}$ as

$$\varphi = -\frac{3b_0}{2u_0}\left(1 + \sqrt{\frac{8a_0u_0}{3b_0^2}\Delta\hat{T}^{**}}\right) \tag{4.37}$$

where $\Delta\hat{T}^{**} = (T^{**} - T)/T_{\text{tran}}$. Therefore, at the limit of stability of the ordered phase, the metastable order parameter has a discontinuity, $\varphi(T^{**}) = -3b_0/(2u_0)$. The behavior of the order parameter in the stable and the metastable states is illustrated in Figure 4.15b.

The heat capacity in the disordered phase (including the supercooling condition) and the ordered phase (including the superheating condition) is found as

$$\frac{C_{h=0}}{k_B} = \frac{C_{\text{bg}}}{k_B} + \begin{cases} 0 & T > T_{\text{tran}} \\ \dfrac{3a_0^2}{u_0}\left[1 + \dfrac{b_0}{4}\sqrt{\dfrac{6}{a_0 u_0}}\left(\Delta\hat{T}^{**}\right)^{-1/2}\right] & T < T_{\text{tran}} \end{cases} \tag{4.38}$$

Note that when $b_0 \to 0$, the square-root anomaly of the heat capacity disappears. The heat capacity is illustrated in Figure 4.15c. The discontinuity in the heat capacity is $\Delta C/k_B = (C - C_{\text{Bg}})/k_B = 12a_0^2/u_0$, where ΔC is the excess heat capacity at the transition ($T = T_{\text{tran}}$).

The anomaly of the heat capacity in the ordered phase is generated by the cubic term in the Landau expansion. Experiments confirm the existence of this anomaly in the nematic phase, see Figure 4.16a. However, there is also an excess of the heat capacity in the isotropic phase, although the anomaly is much less pronounced. This excess is caused by fluctuations of the anisotropy (around the average zero value) in the isotropic phase near the weakly first-order NI transition. This effect is not captured by the Landau-de Gennes theory, and it is associated with pretransitional nematic fluctuations of anisotropy, discussed in Section 4.9. Moreover, small impurities, always present in real substances, split the transition into a temperature interval between the beginning and the end of the formation of a new phase, $T_{\text{NI}}^{(1)}$ and $T_{\text{NI}}^{(2)}$, see Figure 4.16b. If there were no impurity, the heat of the transition could be obtained through the integral of the heat capacity, which is a delta function $\tilde{\delta}\left(T - T_{\text{NI}}\right)$ at the transition point. $\tilde{\delta}$ is a function of $T - T_{\text{NI}}$, defined as

$$\tilde{\delta}\left(T - T_{\text{NI}}\right) = \begin{cases} \infty & T = T_{\text{NI}} \\ 0 & T \neq T_{\text{NI}} \end{cases} \tag{4.39}$$

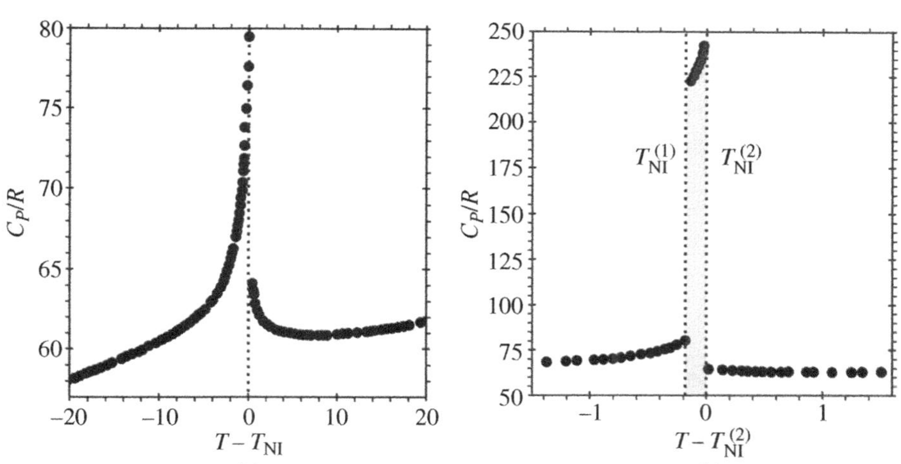

Figure 4.16 Isobaric heat capacity of MBBA in the vicinity of the nematic-isotropic phase transition. (a) The excess heat capacity observed in both phases. (b) A split in the transition temperature, T_{NI}, between $T_{\text{NI}}^{(1)}$ and $T_{\text{NI}}^{(2)}$ (dotted lines), is caused by the existence of small impurities in the MBBA sample, which is found in the immediate vicinity of the transition point [4]. The gray area corresponds to the estimate for the latent heat of the transition.

such that, the integral over all temperatures gives, $\Delta H = \int \tilde{\delta}\left(T - T_{\mathrm{NI}}\right) dT$. If there is an impurity, the heat capacity is no longer a delta function, such that a measurement of the heat capacity within the interval of temperatures, $T_{\mathrm{NI}}^{(1)}$ and $T_{\mathrm{NI}}^{(2)}$, enables one to accurately estimate the heat of the transition as $\Delta H = \int_{T_{\mathrm{NI}}^{(1)}}^{T_{\mathrm{NI}}^{(2)}} C_P dT$, as shown by the gray area in Figure 4.16b.

The susceptibility in the disordered phase (including the supercooling condition) and the ordered phase (including the superheating condition) is found as

$$\hat{\chi}^{-1} = \frac{\partial^2 \hat{\Psi}}{\partial \varphi^2} = \begin{cases} a_0 \Delta \hat{T}^* & T > T_{\mathrm{tran}} \\ \sqrt{\frac{3b_0^2 a_0}{2u_0} \Delta \hat{T}^{**}} + 2a_0 \Delta \hat{T}^{**} & T < T_{\mathrm{tran}} \end{cases} \tag{4.40}$$

The susceptibilities for $T > T^*$ and for $T < T^{**}$ cross at $T = T_{\mathrm{tran}}$ with a finite value of $\hat{\chi}^{-1} = b_0^2/3u_0$. Also note that in the superheated metastable region, $T_{\mathrm{tran}} < T < T^{**}$, the inverse susceptibility does not follow the Curie-Weiss law (see Eq. 4.17), scaling with temperature as $\hat{\chi}^{-1} \sim |\Delta \hat{T}^{**}|^{1/2}$ in the lowest approximation. The behavior of the susceptibility predicted for a weakly first-order phase transition caused by a nonzero cubic term in the Landau expansion is illustrated in Figure 4.15d.

Experimental confirmation of the behavior of the susceptibility in the isotropic phase in BMOAB (near the nematic-isotropic transition) obtained from light-scattering measurements is shown in Figure 4.17. The experimental data also demonstrates small deviations from linearity in the immediate vicinity of the transition temperature. These deviations, like those observed for the heat capacity, may also be attributed to the effect of the pretransitional fluctuations of anisotropy, as the transition is weak, and thus, near-critical.

However, many liquid crystals exhibits a direct transition from the isotropic phase to the smectic A (SmA) phase characterized by one-dimensional translational order. For example, the phase diagram of a mixture of two liquid crystals, $\overline{6}O\overline{6}$–$NO_2$ and $\overline{6}0\overline{10}$, shown in Figure 4.18a, demonstrates either the nematic gap between the SmA and isotropic phases ($\overline{6}0\overline{10}$) or a direct isotropic-SmA transition ($\overline{6}O\overline{6}$–$NO_2$). The nematic phase vanishes at the triple point, $\mathrm{TP}_{\mathrm{NIA}}$.

The isotropic-SmA transition is "moderately" of first order [24], being characterized by a more significant gap between the transition temperature and the temperature of stability of the isotropic phase, T_{A}^*, with respect to the SmA phase. Moreover, the deviations of the inverse susceptibility from linearity are also more significant than for an isotropic-nematic transition. Experiments on many liquid crystals and their mixtures demonstrate that there is a correlation between the

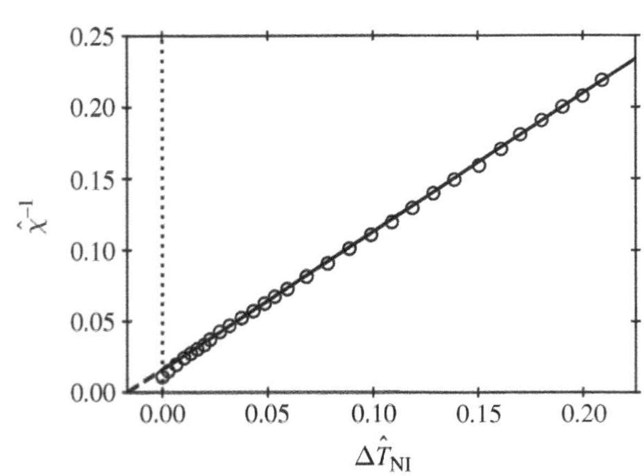

Figure 4.17 Temperature dependence of the inverse susceptibility in the isotropic phase of BMOAB. Source: Adapted from Refs. [6, 23].

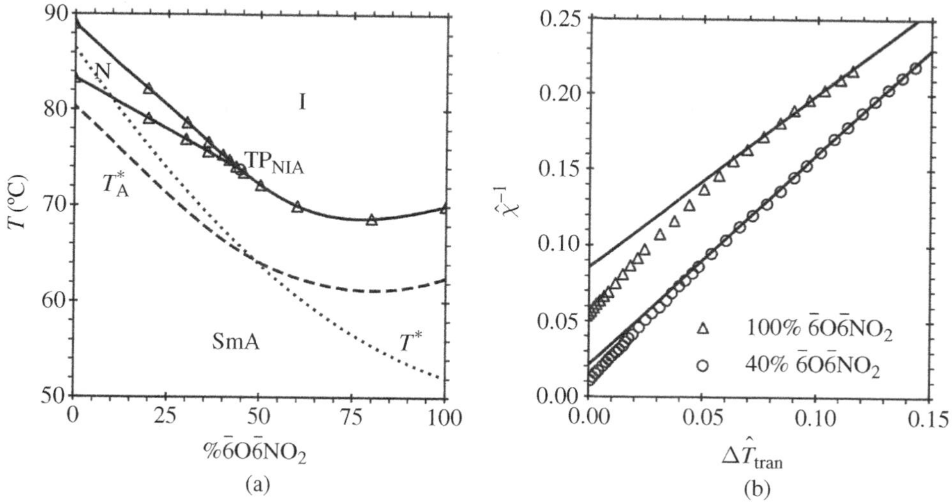

Figure 4.18 (a) Isotropic (I), nematic (N), and smectic A (SmA) phase behavior of a mixture of liquid crystals, $\overline{6}\mathrm{O}\overline{6}\mathrm{NO}_2$ and $\overline{6}\mathrm{O}\overline{10}$ [4, 6]. Solid curves depict first-order phase transitions between these phases, while the dashed and dotted curves are the predictions for the stability limits of the isotropic liquid with respect to the nematic and SmA states, T^*_A and T^*, respectively. TP_NIA indicates the N-I-SmA triple point. (b) Susceptibility, obtained from static light-scattering experiments, for pure $\overline{6}\mathrm{O}\overline{6}\mathrm{NO}_2$ and near the triple point for the mixture $\overline{6}\mathrm{O}\overline{6}\mathrm{NO}_2$–$\overline{6}\mathrm{O}\overline{10}$, indicated by the open circle on (a) [4].

emergence of the smectic order and the temperature difference between the stability of the isotropic liquid with respect to the nematic and smectic phases, $T^* - T^*_\mathrm{A}$. The smaller the gap, the larger the pretransitional enhancement of the orientational susceptibility and the stronger the first-order nature of the transition [6]. This effect is demonstrated in Figure 4.18b and is qualitatively predicted by the theory of "weak crystallization" [25, 26].

According to the theory of weak crystallization, close to the absolute stability limit, $\Delta\hat{T}^*_\mathrm{A} = (T - T^*_\mathrm{A})/T_\mathrm{tran} \ll 1$, the coupling between the nematic and SmA fluctuations of the order parameters contribute to the free energy as $\lambda\delta\hat{Q}|\delta\psi_\mathrm{A}|^2 \propto (\Delta\hat{T}^*_\mathrm{A})^{3/2}$, where λ is a coupling constant. Thus, this coupling enhances the cubic term in the Landau expansion, making the first-order transition less weak. Consequently, the contribution of the entropy due to the coupling of the translational order is proportional to $(\Delta\hat{T}^*_\mathrm{A})^{-1/2}$, while the contributions of this coupling to the isobaric heat capacity and to the susceptibility are proportional to $(\Delta\hat{T}^*_\mathrm{A})^{-3/2}$. In the weak crystallization theory, the contributions from the translational order are small compared to the background properties. In Figure 4.18b, the inverse susceptibility could be described by the function,[7]

$$\hat{\chi}^{-1} \approx a_0\Delta\hat{T}^* - a_1(\Delta\hat{T}^*)^{-1/2} - a_2(\Delta\hat{T}^*_\mathrm{A})^{-3/2} \tag{4.41}$$

where a_1 and a_2 are constants. The first square-root correction is attributed to the fluctuations of the nematic order parameter, which will be discussed in Section 4.9. The second correction is predicted by the weak crystallization theory, where a_2 is proportional to the nematic-SmA coupling constant, $a_2 \propto \lambda$. Note that, according to the theory, both corrections are small with respect to the leading term.

7 In the first order approximation, the function for the susceptibility is $f(x) \approx (a_0\Delta\hat{T}^*)^{-1} + x$, while its inverse $f^{-1}(x) \approx a_0\Delta\hat{T}^* - x$ where $x \equiv a_1(\Delta\hat{T}^*)^{-1/2} + a_2(\Delta\hat{T}^*_\mathrm{A})^{-3/2}$.

In this subsection, the effect of compressibility on a first-order phase transition was neglected for simplicity. In reality, solids and liquids are, at least, weakly compressible. In this case, the transition temperature and the Landau expansion coefficients would depend on pressure. Thus, the equations described in this section would be the same for any given pressure, while the derivatives of the free energy would be taken at constant pressure. In this case, the heat capacity described by Eq. (4.38) would be the isobaric heat capacity.

An interesting scenario could arise if $b_0 = b_0(P)$ changes its sign. Accordingly, the order parameter would change its sign, and the point where $b_0(P) = 0$ would correspond to a second-order phase transition (an "isolated critical point"). This scenario has been predicted in mixtures of rod-like and disk-like liquid crystals (see Ref. [4]).

4.4.3.3 Weakly First-Order Transition Emerging from the Quartic Term

A weakly first-order phase transition could also occur if the fourth-order term in the Landau expansion is small and negative. Examples of first-order phase transitions induced by a negative quartic term include: the nematic-SmA transition close to the isotropic phase, the nematic-SmC transition in liquid crystals, and many transitions in solids, which originate from a coupling between competing order parameters. In this case, the higher-ordered terms in the Landau expansion will be required to maintain stability.

For simplicity, consider the symmetric case of the generalized Landau expansion Eq. (4.25) of the order-parameter dependent free energy, given by

$$\Delta\hat{f} = \frac{1}{2!}a_0\Delta\hat{T}^*\varphi^2 + \frac{1}{4!}u_0\varphi^4 + \frac{1}{6!}g_0\varphi^6 \tag{4.42}$$

where $u_0 < 0$. The free energy is illustrated in Figure 4.19a. The transition temperature is obtained in relation to the temperature of the stability limit of the disordered phase from the condition, $\Delta\hat{f}(T = T_{\text{tran}}) = 0$, giving,

$$T_{\text{tran}} = T^*\left[1 + \frac{5u_0^2}{8a_0g_0}\right] \tag{4.43}$$

such that the value of the spontaneous order parameter at the transition temperature is $\varphi_0^2(T_{\text{tran}}) = -15u_0/g_0$.

The temperature dependence of the spontaneous order parameter is obtained from the minimization of the free energy, Eq. (4.42), as

$$\varphi_0^2 = \begin{cases} 0 & T > T_{\text{tran}} \\ -\frac{10u_0}{g_0}\left(1 + \sqrt{1 - \frac{6a_0g_0}{5u_0^2}\Delta\hat{T}^*}\right) & T < T_{\text{tran}} \end{cases} \tag{4.44}$$

At the stability limit of the disordered phase, $T = T^*$, the spontaneous order parameter has a magnitude of $\varphi_0^2(T^*) = -20u_0/g_0$.

The ordered phase may be superheated into a metastable state until it reaches the temperature of the stability limit of the ordered phase, T^{**}. Beyond this temperature, Eq. (4.44) has no solution, becoming imaginary. Therefore, the condition $1 - 6a_0g_0\Delta\hat{T}^*/(5u_0^2) = 0$, defines $T = T^{**}$ as

$$T^{**} = T^* + T_{\text{tran}}\frac{5u_0^2}{6a_0g_0} = T_{\text{tran}}\left(1 + \frac{5u_0^2}{54a_0g_0}\right) \tag{4.45}$$

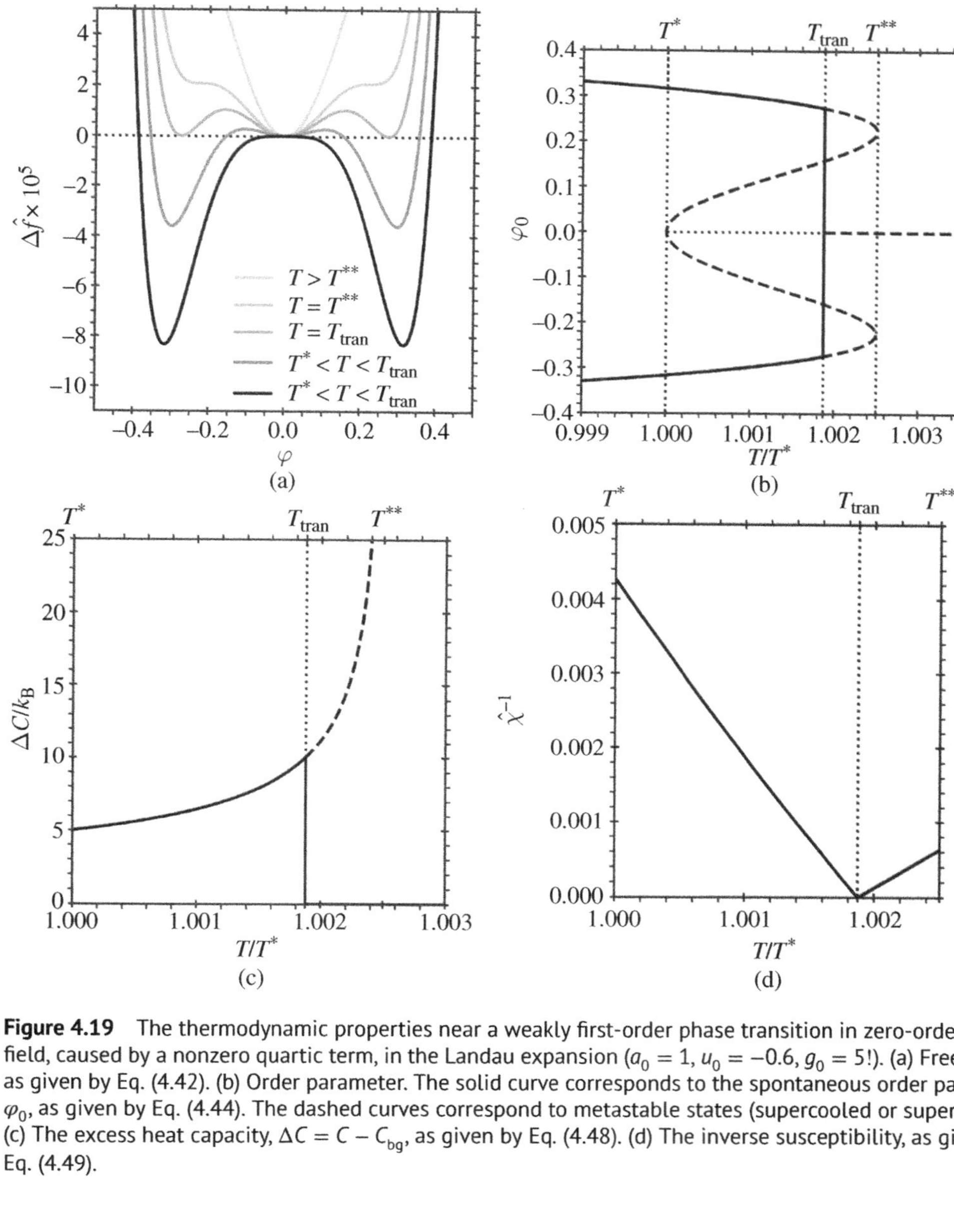

Figure 4.19 The thermodynamic properties near a weakly first-order phase transition in zero-ordering field, caused by a nonzero quartic term, in the Landau expansion ($a_0 = 1$, $u_0 = -0.6$, $g_0 = 5$!). (a) Free energy, as given by Eq. (4.42). (b) Order parameter. The solid curve corresponds to the spontaneous order parameter, φ_0, as given by Eq. (4.44). The dashed curves correspond to metastable states (supercooled or superheated). (c) The excess heat capacity, $\Delta C = C - C_{bg}$, as given by Eq. (4.48). (d) The inverse susceptibility, as given by Eq. (4.49).

Thus, accounting for Eq. (4.45) in Eq. (4.44), one obtains the spontaneous order parameter in the stable and metastable states for $T < T^{**}$ as

$$\varphi_0^2 = -\frac{10u_0}{g_0}\left(1 + \sqrt{\frac{6a_0g_0}{5u_0^2}\Delta\hat{T}^{**}}\right) \tag{4.46}$$

where $\Delta\hat{T}^{**} = (T^{**} - T)/T_{tran}$. Therefore, at the limit of stability of the ordered phase, the metastable order parameter has a discontinuity, $\varphi_0^2(T^{**}) = -10u_0/g_0$. The behavior of the order parameter is illustrated in Figure 4.19b.

The entropy in zero field, $\hat{S} = S/k_B$ is obtained from the derivative of the free energy with respect to temperature, as given in Eq. (4.15), as

$$\hat{S} = \hat{S}_{bg} - \frac{a_0}{2}\varphi_0^2 \tag{4.47}$$

The entropy discontinuity across the transition is $\Delta\hat{S}(T_{tran}) = a_0^2\varphi_0^2(T_{tran})/2 = -15a_0u_0/(2g_0)$. Thus, the latent heat of the first-order transition is $T_{tran}\Delta\hat{S}(T_{tran}) = -15a_0u_0T_{tran}/(2g_0)$. The heat capacity exhibits a discontinuity at the transition temperature, diverging at the stability limit of the order phase as

$$\frac{C}{k_B} = \frac{C_{bg}}{k_B} + \begin{cases} 0 & T > T_{tran} \\ \dfrac{60a_0^2}{g_0}\left[1 + \left(\dfrac{6a_0g_0}{5u_0^2}\Delta\hat{T}^{**}\right)^{-1/2}\right] & T < T_{tran} \end{cases} \tag{4.48}$$

Figure 4.19c demonstrates the behavior of the heat capacity for the weakly first-order phase transition caused by the negative quartic term in the Landau expansion. An example of an experiment that confirmed the existence of this anomaly in the ordered phase (SmC) of a liquid crystal, $\overline{6}O\overline{8}$, near the nematic-SmC transition is illustrated in Figure 4.20. The small pretransitional excess in the heat capacity in the nematic phase is attributed to fluctuations of the SmC order parameter, similar to the pretransitional effects associated with fluctuations of the nematic order parameter in the isotropic phase near the nematic-isotropic transition, as illustrated in Figure 4.16a.

Similarly, the inverse susceptibility is finite at the transition temperature, having a magnitude $15u_0^2/(8g_0)$, vanishing at the stability limit of the disordered phase, T^*. Upon heating, the inverse susceptibility also vanishes at the stability limit of the ordered phase, T^{**}. The susceptibility is given in the lowest-order approximation by

$$\hat{\chi}^{-1} = \begin{cases} a_0\Delta\hat{T}^{*} & T > T_{tran} \\ 4a_0\Delta\hat{T}^{**} - \frac{10u_0^2}{3g_0}\left(1 - \sqrt{\frac{6a_0g_0}{5u_0^2}\Delta\hat{T}^{**}}\right) & T < T_{tran} \end{cases} \tag{4.49}$$

Figure 4.19d shows the behavior of the inverse susceptibility in the vicinity of the phase transition with a negative quartic term as predicted by the Landau theory.

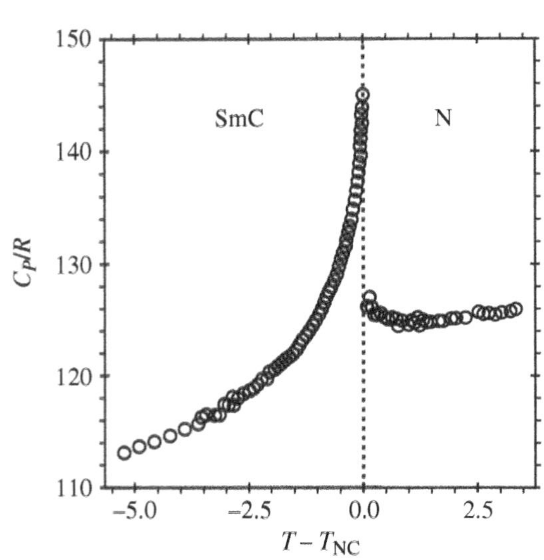

Figure 4.20 Anomaly of the heat capacity in a liquid crystal, $\overline{6}O\overline{8}$, near its weakly first-order nematic-SmC transition ($T_{NC} \approx 65$ °C, see Figure 4.5). Source: Adapted from Ref. [23].

4.4.4 Tricritical Phase Transition

The value of the fourth-order coefficient in the symmetric Landau expansion could be tuned by an additional thermodynamic field, like the pressure or the chemical potential of a solute. The point at which $u_0 = 0$ separates the line of second-order phase transitions from the line of first-order phase transitions. This special point is referred to as a "tricritical" point [15, 27, 28]. Tricritical points have been found in several ferromagnetic, antiferromagnetic, and ferroelectric materials, in superfluid mixtures of helium isotopes, and in some solids exhibiting order-disorder transitions, such as in KH_2PO_4 and NH_4Br as well as other ammonium halides. In all of these cases, the mechanism of the emergence of tricriticality is a result of the coupling between a vector order parameter and an additional order parameter (a scalar or a tensor), conjugated to the additional physical field (see more details in Appendix 4.A).

A schematic phase diagram that illustrates the appearance of a tricritical point along the line of phase transitions in an incompressible ferromagnet upon the addition of a nonmagnetic impurity is shown in Figure 4.21. Depending on the concentration of the impurity, the critical temperature decreases, and at the tricritical point, the line of Curie critical temperatures becomes a line of first-order phase transitions characterized by a discontinuity of the magnetization and the emergence of the two-phase region.

A well-studied example, both experimentally and theoretically, is tricriticality in a mixture of helium isotopes (Figure 4.22). The He-4 isotope exhibits a second-order phase transition (referred to as the "lambda transition") from a normal viscous fluid to a superfluid (with no viscosity) at the temperature, $T_\lambda = 2.17\,\text{K}$. The addition of He-3 isotopes lowers the temperature of the λ-transition until it reaches the tricritical point (at ~67% concentration of He-3) [29, 30].[8] Above

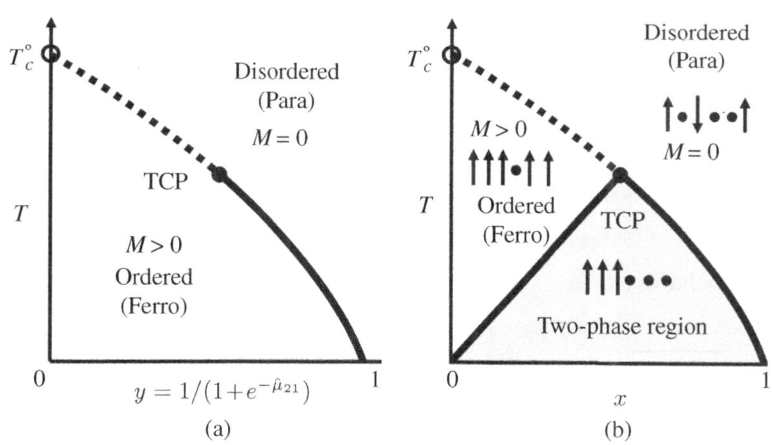

Figure 4.21 Sketches depicting a tricritical transition in ferromagnetic materials in zero magnetic field upon the addition of a nonmagnetic impurity. (a) Temperature vs. a generalized "activity," defined through the exchange chemical of the impurity as $y = 1/(1 + e^{-\tilde{\mu}_{21}})$. (b) Temperature vs. concentration of the impurity. In (a) and (b), the dotted curve is the line of second-order phase transitions ("λ-line"), while the solid curves represent the first-order phase transitions. The tricritical point (TCP) is indicated by the closed circle, while the critical temperature of a pure ferromagnet (T_c°) is indicated by the open circle.

8 In this case, He^3 acts like a "nonmagnetic impurity" as discussed above, since it does not exhibit superfluidity at these conditions. In fact, it was later discovered that He^3 also exhibits boson-condensation, but at extremely low temperatures (~ μK) [29].

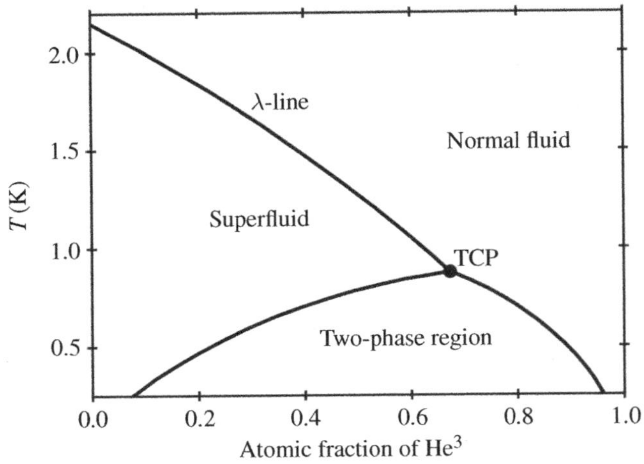

Figure 4.22 Phase diagram for the ^{3}He– ^{4}He mixtures near the lambda line and tricritical point.

this concentration, the transition becomes first order, accompanied by phase separation of the superfluid and the normal fluid phases.

The temperature dependence of the thermodynamic properties at the tricritical point may be described by the Landau theory when $u_0 = 0$. In this case, the field-dependent Landau expansion for the free energy, given from Eq. (4.42), has the form

$$\Delta\hat{f} = \frac{1}{2!}a_0\Delta\hat{T}_{\mathrm{TCP}}\varphi^2 + \frac{1}{6!}g_0\varphi^6 \tag{4.50}$$

where $\Delta\hat{T}_{\mathrm{TCP}} = (T - T_{\mathrm{TCP}})/T_{\mathrm{TCP}}$ is the distance to the tricritical transition temperature, T_{TCP}. The ordering field is given from the condition $h = \partial(\Delta\hat{f})/\partial\varphi$, as

$$h = a_0\Delta\hat{T}_{\mathrm{TCP}}\varphi + \frac{1}{5!}g_0\varphi^5 \tag{4.51}$$

In zero field, the spontaneous order parameter, $\varphi = \varphi_0$, reads as

$$\varphi_0 = \begin{cases} 0 & T > T_{\mathrm{TCP}} \\ \pm\left(5!a_0\Delta\hat{T}_{\mathrm{TCP}}/g_0\right)^{1/4} & T < T_{\mathrm{TCP}} \end{cases} \tag{4.52}$$

The heat capacity in zero field behaves as

$$\frac{C_{h=0}}{k_{\mathrm{B}}} = \frac{C_{\mathrm{bg}}}{k_{\mathrm{B}}} + \begin{cases} 0 & T > T_{\mathrm{TCP}} \\ \dfrac{a_0^{3/2}}{2}\sqrt{\dfrac{30}{g_0\Delta\hat{T}_{\mathrm{TCP}}}} & T < T_{\mathrm{TCP}} \end{cases} \tag{4.53}$$

diverging at the tricritical temperature in the ordered phase. The inverse susceptibility vanishes at the tricritical temperature as

$$\hat{\chi}^{-1} = \begin{cases} a_0\Delta\hat{T}_{\mathrm{TCP}} & T > T_{\mathrm{TCP}} \\ -4a_0\Delta\hat{T}_{\mathrm{TCP}} & T < T_{\mathrm{TCP}} \end{cases} \tag{4.54}$$

The behavior of the physical properties at the tricritical condition is shown in Figure 4.23. Figure 4.24 illustrates the experimental validity of the prediction given by Eq. (4.54) for the inverse susceptibility.

The Landau theory well-describes the behavior of tricriticality in physical systems. However, one may notice, upon comparison between Figures 4.21 and 4.22, a difference; the slope of the lambda

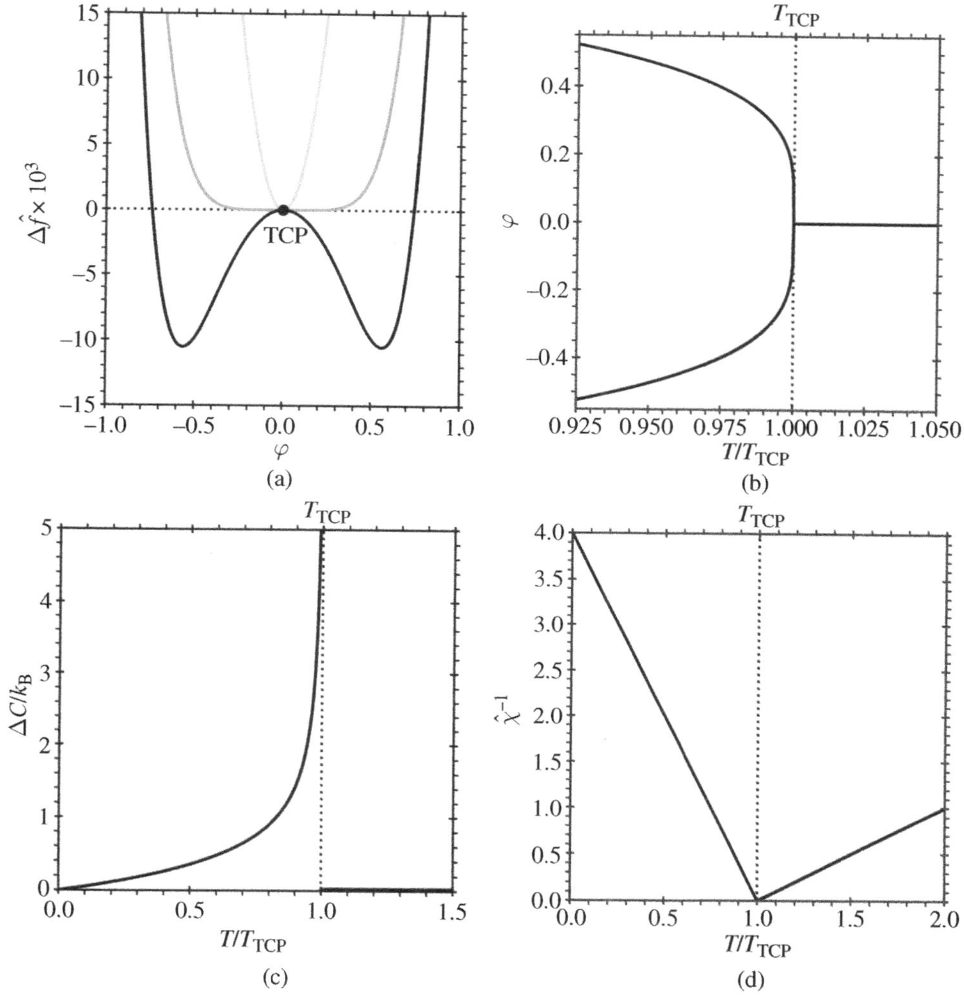

Figure 4.23 The thermodynamic properties near a tricritical phase transition caused by a zero-quartic term in the symmetric Landau expansion ($a_0 = 1$ and $g_0 = 2 \cdot 5!$). (a) Free energy, as given by Eq. (4.50). (b) Order parameter. The solid curve corresponds to the spontaneous order parameter, φ_0, as given by Eq. (4.52). The dashed curves correspond to metastable states (supercooled or superheated). (c) The excess heat capacity, $\Delta C = C - C_{bg}$, as given by Eq. (4.38). (d) The inverse susceptibility, as given by Eq. (4.54).

line changes at the tricritical point as seen in the phase diagram of He^3-He^4 mixtures. This slope is not predicted by Landau theory and is attributed to fluctuations of the order parameter near the tricritical point. The effects of fluctuations will be discussed in Chapter 8.

Another example of a phase transition affected by the sign change of the quartic term in the Landau expansion is the nematic-SmA transition in mixtures of liquid crystals (see Figure 4.5). Like for He^3–He^4 mixtures, the quartic term in the Landau expansion of the SmA vector order parameter, ψ_A, (the density wave describing the one-dimensional translation order) is tuned by the concentration of the component. For this transition, ψ_A is coupled to the tensor order parameter, $\hat{Q}_{ij}$, characterizing the nematic phase. The coupling between these two order parameters ($\sim |\psi_A|^2\hat{Q}_{ij}$) renormalizes the fourth-order term in the landau expansion in powers of ψ_A,

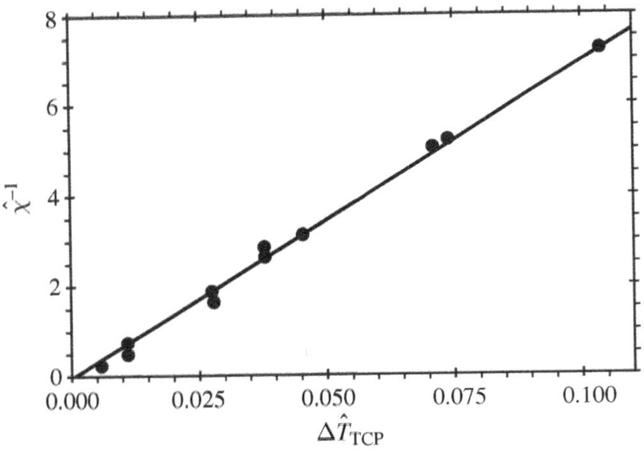

Figure 4.24 Magnetic susceptibility of an antiferromagnet (DAG) in zero magnetic field near its tricritical temperature, $T_{\mathrm{TCP}} \approx 1.81$ K. Source: Adapted from Ref. [31].

changing its sign from positive to negative when fluctuations of $\hat{Q}_{ij}$ become significant upon the approach to the isotropic-nematic phase transition. Consequently, the isotropic-nematic-SmA tricritical point emerges when the quartic term disappears.

However, there is another subtlety in the physics of the nematic-SmA transition, originating from the coupling between ψ_{A} and fluctuations of the nematic director, $\delta\boldsymbol{n}$, of the order $\sim \lambda|\psi_{\mathrm{A}}|^2|\delta\boldsymbol{n}|^2$. Fluctuations of the nematic director scale as $|\delta\boldsymbol{n}|^2 \propto |\psi_{\mathrm{A}}|$, such that an extremely small cubic

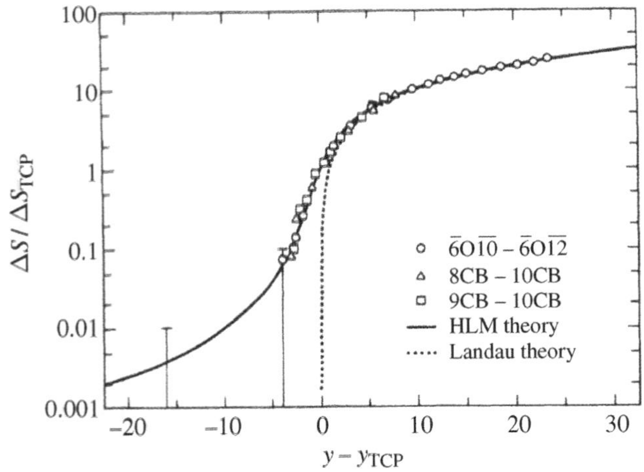

Figure 4.25 Confirmation of the Halperin–Lubenski–Ma (HLM) effect by accurate calorimetry experiments in mixtures of liquid crystals [32]. The vertical axis is the normalized entropy discontinuity of the transition, $\Delta S/\Delta S_{\mathrm{TCP}}$, where ΔS_{TCP} is the entropy discontinuity at the Landau tricritical point emerging from the cubic term in the Landau expansion. The horizontal axis is the distance to the tricritical point, $y - y_{\mathrm{TCP}} = (a'R/\Delta S_{\mathrm{TCP}})(x - x_{\mathrm{TCP}})$, where x is the molecular fraction of a liquid-crystal mixture and a' is a system-dependent constant. When $y - y_{\mathrm{TCP}} \gg 1$, then $y - y_{\mathrm{TCP}} \propto \Delta S/\Delta S_{\mathrm{TCP}}$, while when $y - y^* \ll 0$, then $y - y_{\mathrm{TCP}} \propto -(\Delta S/\Delta S_{\mathrm{TCP}})^{1/2}$, as the transition remains weakly first-order. Source: Reproduced with permission from Ref. [32], © American Physical Society (1990).

term, scaling as $\lambda|\psi_A|^3$, emerges in the free energy regardless of the sign of the quartic term. This phenomenon is known as the "Halperin–Lubenski–Ma" (HLM) effect [33]. This effect is observed when the nematic-SmA transition is near its tricritical point predicted by the Landau theory ($u_0 = 0$), as the small cubic term makes the transition always weakly first-order. This behavior has been confirmed by accurate calorimetry experiments [32, 34] (as shown in Figure 4.25) and studies of interfacial dynamics, which qualitatively differ for first- and second-order transitions [32]. The solid curve in Figure 4.25, describing the behavior of the entropy across the transition, is calculated from the Landau expansion with a small cubic term and a quartic term tuned by the concentration of the mixture. Above the predicted Landau tricritical point, the entropy change is linear with composition, while below the predicted Landau tricritical point, the entropy change is quadratic with composition, as the transition remains weakly first order. However, in the region far below the tricritical point, the entropy of the transition becomes extremely small, being hardly detectable. The error bars in Figure 4.25 show the limits of the calorimetry experiments.

4.5 Meanfield Approximation

The major assumption of the Landau theory is that, in the vicinity of the critical point, the free energy is an analytical function of temperature and the order parameter. The physical reason for the analyticity of the free energy is that the effects of fluctuations of the order parameter on the macroscopic thermodynamic properties are neglected. This approach is known as the "meanfield" approximation. In the meanfield approximation, the complex picture of interactions between molecules is replaced by a mean field (or a mean force) that acts equally on all molecules; such as the van der Waals constant a, see Eq. (2.71), or the interaction parameter ω in the lattice-gas model, see Eq. (2.80), where a single parameter accounts for the uniform interactions between molecules. Meanfield theories also include the Bragg-Williams approximation in the theory of ordering alloys, Curie–Weiss theory of ferromagnetism, and the microscopic theory of superconductivity. Traditional engineering equations of state for fluids are all based on the meanfield approximation.

The meanfield approximation is valid only if the range of molecular interactions is very long, such that all (or nearly-all) molecules are subject to the same interaction force. Indeed, in some phase transitions, like the transition to superconductivity in metals and the transition to the ordered phase in ferroelectrics, the interactions between particles or quasiparticles are long ranged. Thus, these phase transitions could be well-described by the meanfield approximation. However, in many real solid materials and in all fluids, the interactions between molecules, atoms, or magnetic spins are short ranged. As a result, fluctuations of the order parameter may modify the physical properties, such that they cannot be fully described by the Landau theory. Despite this fact, the Landau theory remains as a major guidance in the theory of phase transitions, being a starting point for a qualitative or semi-quantitative description of phase transitions as well as for understanding phase diagrams in a variety of materials.[9]

9 With respect to the significance of the Landau theory, it is worthy to cite a remark of Valeriĭ L. Pokrovskiĭ [35]: "The main achievement of Landau is not the meanfield approximation used for calculations, but the fundamental notion of spontaneous symmetry violation and the order parameter as a measure of this violation. Due to the concept of the order parameter, phase transition theory became a cross-disciplinary branch of science, much like the theory of oscillations."

4.6 Spatial Correlations of Order-Parameter Fluctuations

The meanfield approximation corresponds to the assumption that the fluctuations of the order parameter are small, spatially independent, and Gaussian. This implies that fluctuations between neighboring volumetric domains are noninteracting. Therefore, while the fluctuations depend on the macroscopic thermodynamic properties, like the compressibility, they do not effect these properties.

In the following material, we consider the first correction to the pure meanfield description. In this approximation, the fluctuations may spatially correlate, such that the thermodynamic properties become spatially dependent, but in the thermodynamic limit (see Section 1.2) the macroscopic properties will still be unaffected by the fluctuations. For instance, consider the correlations of density fluctuations in fluids. In this case, the "local" spatially-dependent density is given by

$$\rho(\mathbf{r}) = \frac{\mathrm{d}N(\mathbf{r})}{\mathrm{d}V_{\mathrm{t}}} \tag{4.55}$$

Consequently, the dimensionless spatial-dependent susceptibility has the form

$$\hat{\chi}_{\mathrm{i}}(\mathbf{r}) = \left.\frac{\partial \hat{\rho}(\mathbf{r})}{\partial \hat{\mu}}\right|_{T} = \frac{k_{\mathrm{B}} T_{\mathrm{c}}}{\rho_{\mathrm{c}}} [\rho(\mathbf{r})]^2 \kappa_T(\mathbf{r}) \tag{4.56}$$

The interdependence of densities in neighboring domains, separated by a mesoscopic distance $r = |\mathbf{r} - \mathbf{r}'|$, is described by the spatial correlation function. Consider an isotropic fluid, where the thermodynamic properties are invariant upon rotation and translation. These conditions imply that the spatial correlation function between two fluctuations should only be a function of the distance, r. One can express the correlation function of two fluctuations at the positions $r = 0$ and r through the average fluctuation of the density at each of those locations

$$C(r) = \frac{\langle \delta\rho(0)\, \delta\rho(r) \rangle}{\rho} - \tilde{\delta}(r) \tag{4.57}$$

where $C(r)$ is the correlation function and $\tilde{\delta}(r)$ is the delta function (both expressed in units of density), defined as

$$\tilde{\delta}(r) = \begin{cases} \infty & r = 0 \\ 0 & r \neq 0 \end{cases} \tag{4.58}$$

Thus, the integral over all space, $\int \tilde{\delta}(r)\, \mathrm{d}V_{\mathrm{t}} = 1$. The purpose of the delta function in Eq. (4.57) is to constrain the correlation function such that two molecules cannot occupy the same space, as $C(r = 0) \to \infty$.

Integrating Eq. (4.57) over the volume V_{t}^*, one obtains

$$\int C(r) \mathrm{d}V_{\mathrm{t}}^* = V_{\mathrm{t}}^* \frac{\left\langle (\delta\rho)^2 \right\rangle}{\rho} - 1 = \rho k_{\mathrm{B}} T \kappa_T - 1 \tag{4.59}$$

Note that in Eq. (4.59), after integration over space, one obtains the integrated mean square fluctuation of the density, which is spatially independent and is determined by the macroscopic compressibility, Eq. (3.28). As follows from Eq. (3.31), in an ideal gas of molecules or quasiparticles $V_{\mathrm{t}}^* \left\langle (\delta\rho)^2 \right\rangle / \rho = 1$, and thus, $\int C(r) \mathrm{d}V_{\mathrm{t}}^* = 0$. For the fluctuations of a generalized density, the correlation function would be expressed through the generalized susceptibility, given by Eq. (3.47).

In weakly-compressible liquids, the compressibility is small and the integral of the correlation function, $\int C(r) \mathrm{d}V_{\mathrm{t}}^*$, is close to -1. This means that the correlations between neighboring molecules are restricted by the mutual impenetrability of the molecules (referred to as the "anticorrelation" of particles). In the vicinity of the critical point, the compressibility is extremely large, such that the delta function can be neglected. For a dense fluid, the integral of the correlation function, $\int C(r)\mathrm{d}V_t$, is positive, and the correlation function decays with distance, r, as [15]

$$C(r) \cong \frac{\langle \delta\rho(0)\delta\rho(r)\rangle}{\rho} = \frac{1}{(4/3)\pi\xi_0^3}\frac{e^{-r/\xi}}{r/\xi_0} \tag{4.60}$$

where the characteristic length scale ξ is the *correlation length* of fluctuations and ξ_0 is the range (or "radius") of direct intermolecular interactions. For short-ranged interactions, ξ_0 is on the order of the molecular radius. When the correlation length is small, $\xi \ll r$, the correlation function decays exponentially, while when the correlation length is very large, $\xi \gg r$, the correlation function becomes long-ranged, decaying as $C(r) \sim 1/r$. The behavior of the correlation function is illustrated in Figure 4.26.

The correlation length can be defined as a length scale whose square is spatially averaged with a probability-distribution function, given by the correlation function, as

$$\xi^2 = \int r^2 C(r)\mathrm{d}V_t \tag{4.61}$$

Furthermore, the characteristic distance of short-ranged intermolecular interactions, sometimes referred to as the "bare" correlation length, ξ_0, is obtained by averaging its square with a probability-distribution function, given by the intermolecular potential, as

$$\xi_0^2 = \rho \int r^2 \frac{U(r)}{k_B T}\mathrm{d}V_t \tag{4.62}$$

The function, given by Eq. (4.60), is commonly known as the Ornstein–Zernike correlation function.

The Ornstein–Zernike correlation function characterizes the density-density correlations of a fluid in the region where the correlation length is larger than the range of intermolecular interactions. Outside of this region, the correlation functions becomes oscillatory, like in crystalline solids. In liquids, these oscillations are observed at the scale of the molecular size, decaying over distance. The region of oscillatory behavior and Ornstein–Zernike behavior of the correlation function is

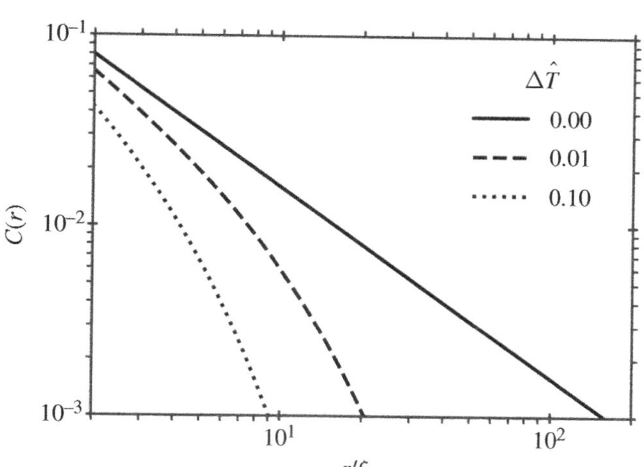

Figure 4.26 Ornstein–Zernike correlation function for the lattice-gas model: relatively away from the critical point (dotted), closer to the critical point (dashed), and at the critical point (solid).

qualitatively separated by the so-called, "Fisher–Widom Line" [36, 37], which is located roughly around the Frenkel line, see Figure 2.31.

4.7 Mesoscopic Susceptibility

As seen from Eq. (4.59), the integral of the correlation function over space is proportional to the macroscopic spatially-independent susceptibility. However, the correlation function itself does depend on the length scale, as follows from Eq. (4.60). This feature of the nature of correlations can be reconciled by representing the fluctuations as a superposition of "waves" in a continuous range of wavenumbers, from $q = 2\pi/\ell$, where ℓ is the size of the system, (practically from $q = 0$) to $q = q_0 \sim v_0^{-1/3}$ (the molecular size cutoff). Such a representation of a complex function by a superposition of harmonic waves is called a "Fourier series." Relations between coordinate-dependent and wavenumber-dependent (momentum-dependent) properties are determined by Fourier transformations.

In particular, the change of the free energy per unit volume, $\delta\hat{f}$, due to a fluctuation of dimensionless density, $\delta\hat{\rho}$, in zero-ordering field is obtained by a Fourier summation over all the possible wave numbers as

$$\delta\hat{f} = \frac{1}{2}\sum_q \hat{\chi}_q^{-1}|\delta\hat{\rho}_q|^2 \tag{4.63}$$

where $\delta\hat{\rho}_q$ is the density fluctuation characterized by a certain wavenumber (at a certain length scale) and $\hat{\chi}_q$ is the susceptibility at this particular length scale. Note that the factor of 1/2 emerges to avoid double counting positive and negative density fluctuations. This susceptibility (defined in momentum space in terms of the characteristic wavenumber, q) is a Fourier component of the coordinate-dependent susceptibility $\hat{\chi}_i(\boldsymbol{r})$ defined in Eq. (4.56).

Landau Theory: Note 2

A key concept in the Landau theory is that the microscopic ("bulk") spatially-independent free energy may be expressed through the inverse susceptibility, $\hat{\chi}^{-1}$. Consider the truncated symmetric expansion of the order-parameter-dependent free energy, given by Eq. (4.9). With use of Eq. (4.17), one can show that $\Delta\hat{f}$ in zero-ordering field may be expressed as

$$\Delta\hat{f} = \frac{1}{2}a_0\Delta\hat{T}\varphi^2 = \frac{1}{2}\hat{\chi}_{q=0}^{-1}\varphi^2 \tag{4.64}$$

Such that, in a general form, the spatially-independent free energy is $\Delta\hat{f} = (1/2)\hat{\chi}_{q=0}^{-1}\varphi^2$, which is also true for higher-ordered Landau expansions (the only difference being the behavior of $\hat{\chi}^{-1}$).

The work of a fluctuation described by the characteristic wavenumber has the same form as the thermodynamic formula Eq. (3.24), but where the fluctuating properties are found at a fixed wavenumber q, as

$$\frac{W_f(q)}{k_B T} = \frac{|\delta\hat{\rho}_q|^2}{2\left\langle|\delta\hat{\rho}|_q^2\right\rangle} = \frac{|\delta\hat{\rho}_q|^2}{2\hat{\chi}_q} \tag{4.65}$$

Hence, $\left\langle|\delta\hat{\rho}_q|^2\right\rangle = \hat{\chi}_q$. The connection between the correlation function in wavenumber space and in coordinate space, with use of Eq. (4.60), is given by the Fourier Transform,

$$\left\langle\left|\delta\hat{\rho}_q\right|^2\right\rangle = \int e^{i\boldsymbol{q}\cdot\boldsymbol{r}}\left(\frac{1}{4\pi\xi_0^2}\frac{e^{-r/\xi}}{r}\right)\mathrm{d}V_t^* = \frac{\xi^2}{\xi_0^2(1+\xi^2q^2)} \tag{4.66}$$

This transform defines the wavenumber-dependent ("local") susceptibility as

$$\boxed{\hat{\chi}(q) = \frac{\xi^2}{\xi_0^2(1+\xi^2 q^2)} = \frac{\hat{\chi}_{q=0}}{1+\xi^2 q^2}} \tag{4.67}$$

where $\hat{\chi}_{q=0}$ is the susceptibility in the limit $q \to 0$ (identical to the macroscopic thermodynamic susceptibility). If the product of the correlation length and wavenumber is small, $q\xi \ll 1$, then the susceptibility does not depend on the length scale. If this product is large, $q\xi \gg 1$, then the susceptibility is independent of the correlation length, being dependent only on the length scale as $\hat{\chi}(q) = \hat{\chi}_{q=0}/(\xi_0^2 q^2)$.

The results for the first correction to the meanfield approximation of the wavenumber-dependent thermodynamic properties presented in this section could also be generalized for the fluctuations of any density-like order parameter, φ, associated with a generalized density, φ_i. For instance, the order parameter for fluids is $\varphi = (\rho - \rho_c)/\rho_c$, for binary-liquid mixtures is $\varphi = (x - x_c)/x_c$, and for ferromagnetic systems is $\varphi = \langle \hat{M} \rangle$.

4.8 Landau–Ginzburg Functional

The simplest way to connect the mesoscopic susceptibility with other thermodynamic properties and introduce heterogeneities into mesoscopic thermodynamics is to express the wavenumber-dependent free energy, given by Eq. (4.63), in coordinate space. A fluctuation of the order parameter, $\delta\varphi$, and its spatial derivative in coordinate space are given by

$$\delta\varphi(\boldsymbol{r}) = \sum_q \delta\varphi_q e^{i\boldsymbol{q}\cdot\boldsymbol{r}} \quad \text{and} \quad \nabla(\delta\varphi) = \sum_q i\boldsymbol{q}\delta\varphi_q e^{i\boldsymbol{q}\cdot\boldsymbol{r}} \tag{4.68}$$

For Gaussian fluctuations, $\delta\varphi_q\delta\varphi_{-q} = |\delta\varphi_q|^2$. Using this result and the fact that the inverse susceptibility, given by Eq. (4.67), may be expressed as

$$\hat{\chi}_q^{-1} = \hat{\chi}_{q=0}^{-1}\left(1+\xi^2 q^2\right) \tag{4.69}$$

Then, the dimensionless excess free energy, $\Delta\hat{F}$, is obtained from the spatial integral of the order-parameter-dependent free-energy change due to a fluctuation $\delta\hat{f}$, given by Eq. (4.63), as

$$\Delta\hat{F} = \int \left(\frac{1}{2}\hat{\chi}_{q=0}^{-1}\varphi^2 + \frac{1}{2}c_0|\nabla\varphi|^2\right) \mathrm{d}V_t \tag{4.70}$$

where the constant, c_0, emerges from the product $\hat{\chi}_{q=0}^{-1}\xi^2$. Additionally, Eq. (4.70) could be represented for a field-dependent free energy as

$$\Delta\hat{\Phi} = \int \left(\hat{f}(\varphi, \Delta\hat{T}) + \frac{1}{2}c_0|\nabla\varphi|^2 - h\varphi\right) \mathrm{d}V_t \tag{4.71}$$

The first term, $\hat{f}$, is the portion of the dimensionless free-energy density per unit volume in the thermodynamic limit, which is independent of the gradients and which can be obtained by integration of the susceptibility in the limit $q \to 0$. The form of $\hat{f}(\varphi, T)$ depends on the nature of the order parameter, φ, and is given by a particular equation of state. Equation (4.71) is known as the *Landau–Ginzburg functional.*[10]

The addition of the gradient term to the free energy in Eq. (4.71) indicates that the local inhomogeneities (caused by fluctuations) are energetically unfavorable since the presence of the gradient

10 A functional may be considered as a "function of functions."

term increases the local free energy. We also note that the lowest coordinate-dependent term is quadratic because the gradient is a vector while the energy is a scalar. The coefficient c_0 of the gradient term is a phenomenological constant associated with the range of intermolecular interactions. For simple molecular fluids this parameter can be estimated as $c_0 \approx \xi_0^2 v_0^{2/3}$.

According to Eq. (4.69), the correlation length naturally emerges from the Landau–Ginzburg functional. In particular, for a fluid, since the susceptibility along the critical isochore ($\varphi = 0$) diverges as $\hat{\chi}_{q=0} = 1/a_0\Delta\hat{T}$, the correlation length diverges at the critical point as

$$\xi = \xi_0(\Delta\hat{T})^{-1/2} \tag{4.72}$$

where $\xi_0 \equiv \sqrt{c_0/a_0}$. This temperature dependence of the correlation length is presented for the isotropic phase of MBBA in Figure 4.27 Below the critical temperature (in the two-phase region), $\hat{\chi}_{q=0} = (-2a_0\Delta\hat{T})^{-1}$, hence the correlation length diverges as

$$\xi = \xi_0(-2\Delta\hat{T})^{-1/2} \tag{4.73}$$

in each coexisting phase.

The behavior of the coefficient c_0 has important physical implications. If the fluctuations are isotropic, then c_0 is independent of direction. However, if the fluctuations depend on a direction, like in the smectic phases of liquid crystals, then c_0 would be anisotropic and the correlation length would have a different magnitude for transverse and longitudinal fluctuations. Additionally, the coefficient c_0 is usually positive (as gradients increase the free energy). However, in some specific cases, this coefficient can vanish, being renormalized by an additional field variable (such as the chemical potential of an impurity). Then, higher-order gradient terms in the Landau–Ginzburg functional are to be accounted for. The significance of higher-order gradient terms is considered in Chapter 7. A form of the Landau–Ginzburg functional that includes such higher-order gradient terms reads as

$$\Delta\hat{\Phi} = \int \left(\hat{f} + \frac{1}{2!}c_0(\nabla\varphi)^2 + \frac{1}{2!}k_0(\nabla^2\varphi)^2 + \frac{1}{3!}l_0\varphi(\nabla\varphi)^2 - h\varphi\right) dV_t \tag{4.74}$$

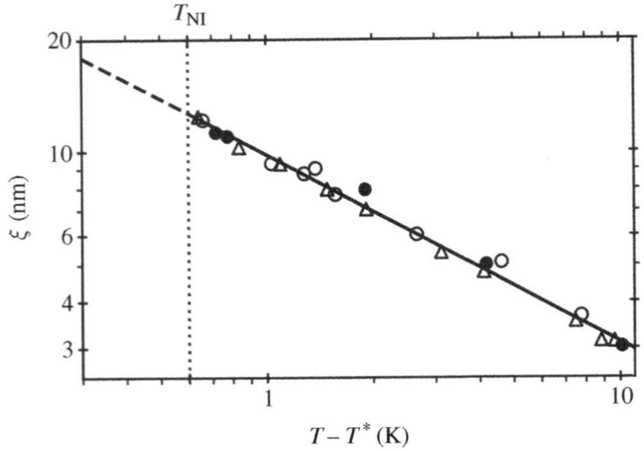

Figure 4.27 Temperature dependence of the correlation length in the isotropic phase of a liquid crystal, MBBA. The symbols represent the experimental data, obtained from light scattering with different polarizations of light [38], while the solid curve is the prediction by the Landau–Ginzburg functional, given by Eq. (4.72). Source: Adapted from Ref. [38].

The $\varphi(\nabla\varphi)^2$ term is asymmetric with respect to the sign of φ (which exists only for a scalar order parameter) and is non-asymptotic (higher order).

4.9 Validity of Landau Theory: Ginzburg Criterion

The meanfield approximation neglects the effects of fluctuations by representing the complex picture of molecular interactions as a single field that acts equally on all particles. While the gradient term in the Landau–Ginzburg functional introduces the correlation length of fluctuations (diverging at the critical point) and the spatial dependence of the thermodynamic properties, these properties, in the thermodynamic limit ($q \to 0$), are the same as in the classical van der Waals theory.

The condition that determines the validity of the meanfield approximation is that the mean square fluctuations of the order parameter must be smaller than the average value of the spontaneous order parameter, $\langle \varphi^2 \rangle \ll \varphi_0^2$. The order parameter fluctuations may be expressed through the general formula for the fluctuations of a generalized density, Eq. (3.46), given in a dimensionless form as

$$\langle \varphi^2 \rangle = \frac{v_0}{V_t^*}\hat{\chi} \tag{4.75}$$

where v_0 is the volume per molecule, adopted as $v_0 = \overline{r}^3$ (where $\overline{r}$ is the average distance between molecules) and $V_t^* \approx \xi^3$, the volume of a fluctuation. With use of Eq. (4.14) for the spontaneous value of the order parameter, the condition for the validity of the meanfield approximation in the ordered phase (below the critical point) reads,

$$\left(\frac{\overline{r}}{\xi}\right)^3 \hat{\chi} \ll -\frac{6a_0}{u_0}\Delta\hat{T} \tag{4.76}$$

Upon substitution of Eqs. (4.17) and (4.73) for the susceptibility and correlation length in the two-phase region, and solving for $\Delta\hat{T}$, one obtains

$$|\Delta\hat{T}| \gg c\frac{u_0^2}{a_0^4}\left(\frac{\overline{r}}{\xi_0}\right)^6 \equiv N_G \tag{4.77}$$

where in this estimate $c = 1/18$, which depends on the method to evaluate fluctuational corrections. The inequality, given by Eq. (4.77), is known as the *Ginzburg criterion* and the combination of the parameters on the right side of the inequality is referred to as the *Ginzburg number*, N_G.[11]

Apart from the numerically small prefactor c that depends on the property considered (such as the order parameter, susceptibility, or heat capacity), there are two physical properties that control the Ginzburg number: the ratio of the average distance between molecules to the range of interactions, $\overline{r}/\xi_0$, and the Landau expansion coefficients (in this case, a_0 and u_0). Note that a_0 is typically of the order of one.

The Ginzburg number becomes zero when the range of intermolecular interactions is infinite or in the special case $u_0 = 0$ (at a tricritical point), such that the meanfield approximation becomes valid. For example, in conventional superconductive materials, the length of the Cooper pairs (see Section 2.12), which represents the range of interactions responsible for the emergence of

11 Vitaly Lazarevich Ginzburg (1916–2009) was a Soviet and Russian physicist, a pupil of Lev Landau, who won a Nobel Prize, together with Alexei Abrikosov and Anthony Leggett, in 2003 for his significant contributions to the theory of superconductivity and superfluidity. Ginzburg was 87 years old at that time.

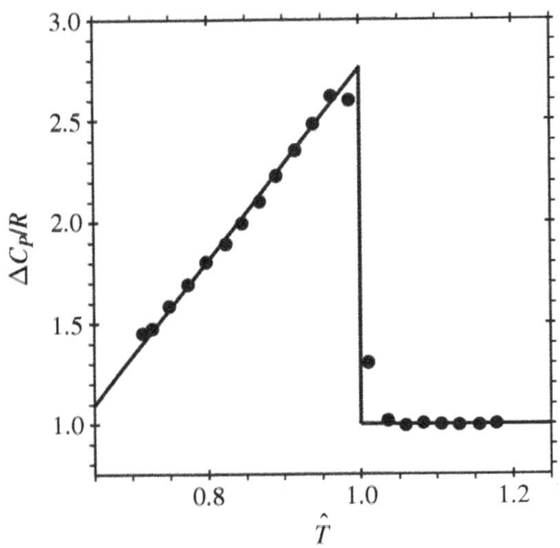

Figure 4.28 The excess heat capacity discontinuity near the second-order phase transition to superconductivity in a metal, FeSe, that accurately follows the predictions of Landau theory, Eq. (4.16). Source: Adapted from Ref. [39].

superconductivity, is on the order of $\xi_0 \sim 10^2 - 10^3$nm, such that the ratio $(\bar{r}/\xi_0)^6 \sim 10^{-12}$–$10^{-18}$ in Eq. (4.77). Thus, the region where the meanfield approximation is invalid becomes practically absent. The meanfield discontinuity in the heat capacity near the second-order phase transition to superconductivity in FeSe demonstrates the validity of the Landau theory, Eq. (4.16), as shown in Figure 4.28.

Another example where the meanfield approximation is largely valid is near the SmA-SmC transition in liquid crystals, see Figure 4.29, and near the weakly first-order isotropic-nematic phase transition, with noticeable pretransitional fluctuation-induced corrections. The physical origin of a small Ginzburg number for these transitions could be related to the strong anisotropy

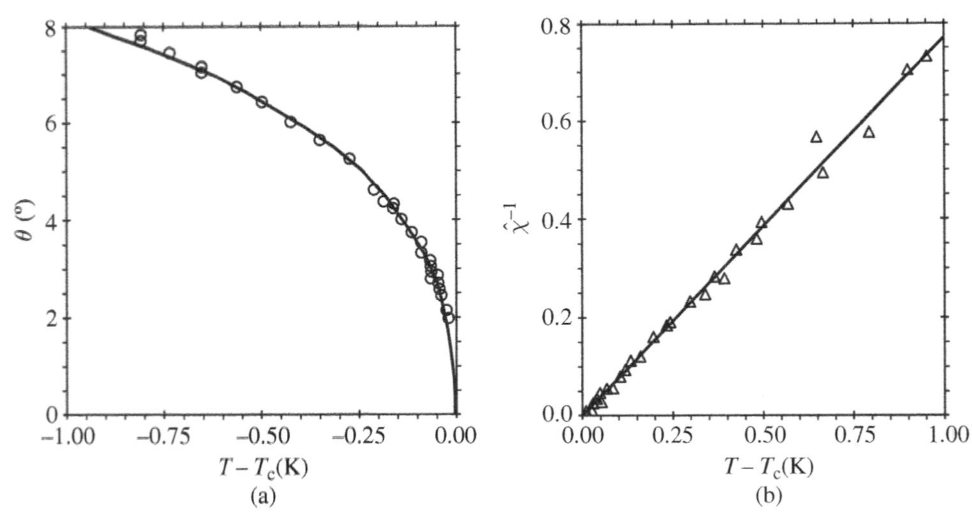

Figure 4.29 Temperature dependence of (a) the mean molecular tilt in the SmC phase and (b) the inverse susceptibility in the SmA phase of a liquid crystal, $\overline{4}O\overline{7}$, obtained by X-ray scattering near the second-order SmA–SmC transition ($T_c \approx 50$ °C) [42]. These properties follow the predictions of Landau theory, given by Eqs. (4.14) and (4.17). Source: Adapted from Ref. [42].

in the molecular shape for a liquid crystal, resulting in a small ratio $\bar{r}/\xi_0 \approx 1/3$–$1/4$, such that $N_G \sim 10^{-4}$–10^{-5}.

In contrast to previous examples, for molecular fluids with short-range interactions ($\bar{r}/\xi_0 \approx 1$), one should not expect the meanfield approximation to be valid in the critical region, as $N_G \gtrapprox 0.1$. The effects of critical fluctuations on the physical properties of fluids will be considered in Chapter 8. There was a long-standing debate in the literature about the possibility of observing meanfield behavior in the critical region of fluids containing ions. It was argued that long-ranged Coulombic forces might allow the meanfield approximation to be valid in such systems. However, phase separation in most fluids containing ions is driven by non-Coulombic forces in solvents with a high dielectric constant and where Coulombic forces are weak. This means that phase separation may occur, even in the absence of ions; examples of such systems include liquid–gas separation of salty water [40] or liquid–liquid separation in solutions of weakly dissociated organic solutes, such as isobutyric acid in water [41]. This effect of segregation where the Ginzburg number is expected to be large is referred to as "solvophobic" phase separation.

On the other hand, in low-dielectric-constant systems with large-sized ions, such as solutions of tetra-*n*-butylammonium picrate in alkanols, phase separation is "Coulombic" (in that it is driven by Coulombic forces between ions). In such systems, the intermolecular potential $U(r)$ determining the correlations between the ions, is screened so that

$$U(r) \sim \frac{e^{-\kappa_D r}}{\varepsilon r} \tag{4.78}$$

where the inverse Debye screening length κ_D is proportional to the dielectric constant, ε [43]. Therefore, the interactions between the ions are effectively short-ranged. It was finally concluded [44] that basically in all ionic fluids, Coulombic or solvophobic, the meanfield approximation is not valid in the critical region.

In the previous examples, the Ginzburg criterion was uniquely defined as each of the systems only had one correlation length. However, for the SmA phase in liquid crystals, the correlation length is different for transverse and longitudinal fluctuations of the order parameter. The ratio $\bar{r}/\xi_0$ and, consequently, the Ginzburg criterion for the validity of the meanfield approximation for the nematic-SmA transition depends on the direction of fluctuations.

In the region where the meanfield approximation is valid, $\Delta\hat{T} > N_G$, the gradient term in the Landau–Ginzburg functional produces small (relative to the homogeneous part of the free energy) fluctuational corrections to the thermodynamic response functions. The corrections grow upon approaching the critical temperature, but always remain small in the region $\Delta\hat{T} > N_G$. The origin of these corrections may be qualitatively understood if one assumes that the correction to the meanfield free energy is proportional to the density of fluctuations, $\sim 1/\xi^3 \propto |\Delta\hat{T}|^{3/2}$, such that the second derivatives (with respect to φ or with respect to $\Delta\hat{T}$) scale as $\sim\xi \propto |\Delta\hat{T}|^{-1/2}$.

The fluctuational correction to the heat capacity was obtained more rigorously via the Landau–Ginzburg functional [45]

$$\frac{\Delta C_{h=0}}{k_B} = \frac{1}{16\pi}\left(\frac{\bar{r}}{\xi_0}\right)^3 (\Delta\hat{T})^{-1/2} \tag{4.79}$$

Phenomenologically, this result was also obtained in a more general form by Mistura and Sette as [46]

$$\frac{\Delta C_{h=0}}{k_B} \propto v_0 \xi \left(\frac{\partial \xi^{-2}}{\partial \hat{T}}\right)_\rho^2 \tag{4.80}$$

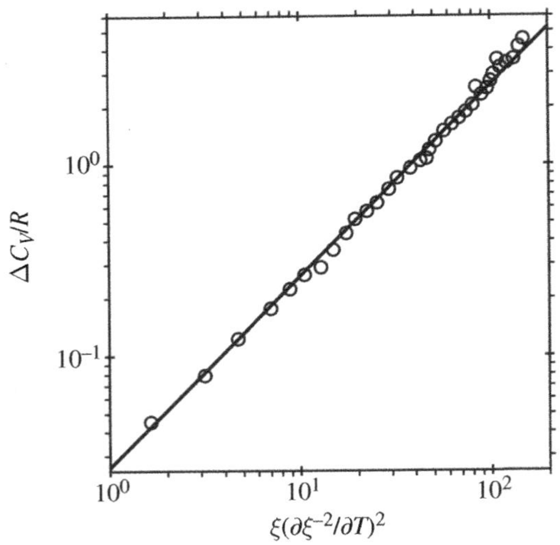

Figure 4.30 Relation between the excess pretransitional heat capacity and the correlation length of fluctuations of anisotropy in the isotropic phase of a liquid crystal, BMOAB. The line corresponds to the theoretical prediction, given by Eq. (4.80). Source: Adapted from Ref. [6].

This correlation was experimentally confirmed for the nematic-isotropic transition in liquid crystals, as demonstrated in Figure 4.30. Fluctuations also modify the susceptibility as [47]

$$\hat{\chi} = \frac{1}{a_0\Delta\hat{T}}\left[1 + \frac{u_0}{8\pi a_0^2}\left(\frac{\overline{r}}{\xi_0}\right)^3 (\Delta\hat{T})^{-1/2}\right] \tag{4.81}$$

The fluctuational corrections to the heat capacity and susceptibility are often referred to as "fluctuation-induced square-root corrections to meanfield." These fluctuational corrections can explain the pretransitional effects observed in the isotropic phase for the heat capacity and susceptibility, as demonstrated in Figs. 4.16 and 4.17. Upon comparing the first correction to the susceptibility in Eq. (4.41) and that in Eq. (4.81), one finds that $a_1 = (u_0/8\pi a_0^2)(\overline{r}/\xi_0)^3$.

One may notice that the coefficients in the square-root corrections to the heat capacity and the susceptibility are proportional to $\sqrt{N_G}$. Consequently, one may obtain alternative forms of the Ginzburg criterion, Eq. (4.77), through the condition that the square-root corrections to the heat capacity and susceptibility must remain small with respect to their regular behavior, a discontinuity in the heat capacity $(3a_0^2/u_0)$ and a diverging susceptibility as $1/\Delta\hat{T}$, given in the meanfield approximation by Eqs. (4.16) and (4.17). These alternative forms of the criteria differ from each other by a numerical factor that depends on the property considered, but are independent of the physics of the phase transition.

4.10 Effects of the Dimensions of Space

In a d-dimensional system, the volume of fluctuations is $V_t^* \approx \xi^d$, the average distance between particles is $v_0 = \overline{r}^d$, and the Ginzburg criterion for the validity of the meanfield regime, given by $\langle \varphi^2 \rangle \ll \varphi_0^2$, reads as

$$\left|\Delta\hat{T}\right|^{(4-d)/2} \gg \frac{\sqrt{2}u_0}{6a_0^2}\left(\frac{\overline{r}}{\xi_0}\right)^d \tag{4.82}$$

For $d = 3$, one obtains the result given by Eq. (4.77), while for $d \geq 4$, the meanfield approximation is always valid. Thus, $d = 4$, known as the "marginal dimensionality," plays a special role in the theory of critical phenomena: it separates meanfield and fluctuation regimes in systems with short-range interactions.

The smaller the dimensionality, the more important the effects of fluctuations. In two-dimensional (planar) systems, the Ginzburg criterion is given by

$$|\Delta\hat{T}| \gg \frac{\sqrt{2}u_0}{6a_0^2}\left(\frac{\bar{r}}{\xi_0}\right)^2 \tag{4.83}$$

Note that in two-dimensional systems, the criterion is three orders of magnitude more severe than that in three-dimensional systems.

In one-dimensional (linear) systems fluctuations are so strong that two-phase coexistence (at a single point) is not possible [15]. Indeed, one could imaging a linear system of size ℓ consisting of alternating sections formed by two different phases. The system can be formally regarded as a "mixture" of the contacting points of each species. In equilibrium (if it would be possible) the number of contacts, n, must decrease, tending toward a single point. The chemical potential of the contact points reads

$$\mu = \left.\frac{\partial G_t}{\partial n}\right|_T = k_B T \ln(n/\ell) + \mu_0(T) \tag{4.84}$$

Since the ratio, n/ℓ, is small, the chemical potential is large and negative. Thus, any increase in n will decrease the Gibbs energy. Therefore, it will be more energetically favorable for the two phases to separate into shorter and shorter sections, and therefore, macroscopic separate phases cannot exist. This statement is known as "Landau–Peierls theorem." The only possibility for linear (one-dimensional) phases to exist at a single point is for the system to possess infinitely long-range interactions, such that fluctuations of the order parameter are completely suppressed.

A special case may arise in a three-dimensional system with one-dimensional translational order, which is realized upon the emergence of the one-dimensional density wave in smectic liquid crystals. According to the Landau–Peierls theorem, true long-ranged order cannot exist in smectic liquid crystals as fluctuations of the phase of the density wave will diverge [5].

4.11 Meanfield Dynamics of Phase Transitions

As discussed in Section 4.2, the Ising model and the lattice-gas model may phenomenologically be described by the Landau theory with an appropriate choice of the order parameter. While these two models show remarkable similarities in thermodynamics, they do not behave the same in dynamics. This difference emerges from the nature of the order parameter for each system. In fluids, the order parameter is conserved, as the total number of particles of each species is conserved, while in the Ising model, the order parameter is nonconserved, as the total number of each alternative spin-orientation is unrestricted, and thus, may change.

This case corresponds to the spatially-independent relaxation of a nonconserved order parameter. In contrast, for a single-component fluid, as discussed in Section 3.11.2, the isobaric entropy fluctuations of a single-component fluid reaches its equilibrium value via spatially-dependent thermal diffusivity with a characteristic rate $\Gamma_T = D_T q^2$ (see Eq. 3.91). This process corresponds to the dynamics of a conserved order parameter.

Another important example of a system with a conserved order parameter is a weakly-compressible binary-liquid mixture. The fluctuations of the concentration of species relax to equilibrium through spatially-dependent mutual diffusion, with a characteristic rate given by $\Gamma_D = Dq^2$ (see Eq. 3.92). The coefficient of mutual diffusion of each species in a binary mixture is given through Eq. (2.176). Asymptotically close to the critical point, one finds, with the use of Eq. (4.20) for the susceptibility, $\chi = (\partial x/\partial \mu_{21})|_T$, that

$$D = \tilde{\alpha}\chi^{-1} = \frac{\rho_c}{k_B T_c}\left(\frac{\tilde{\alpha}}{\Gamma_0^{\pm}}\right)|\Delta\hat{T}| \tag{4.85}$$

such that, in the meanfield approximation, mutual diffusion vanishes at the critical point proportionally to $\Delta\hat{T}$. Expressing the inverse susceptibility through the correlation length, via Eq. (4.72), one obtains,

$$D = \frac{\rho_c}{k_B T_c}\left(\frac{\tilde{\alpha}}{c_0}\right)\xi^{-2} \tag{4.86}$$

Thus, the characteristic rate of diffusion scales as $\Gamma_D \propto |\Delta\hat{T}|q^2$. Note that in the meanfield approximation, the Onsager coefficient, $\tilde{\alpha}$, does not depend on the distance to the critical point. Critical fluctuations in the close vicinity of the critical point may significantly change the prediction given by Eq. (4.86). These effects on the mutual diffusion coefficient will be discussed in Chapter 8.

On the other hand, there are many examples of a nonconserved order parameter, including, but not limited to: magnetization in ferromagnets, polarization in ferroelectrics, the order parameter for superfluid helium, the tensor order parameter for nematic liquid crystals, and the reaction coordinate in an equilibrium chemical reaction. The relaxation time for the fluctuations of anisotropy, given by Eq. (3.63), could be expressed in a general form for any nonconserved order parameter near the phase transition as

$$\tau \approx \tau_0 \hat{\chi}_q \tag{4.87}$$

where $\hat{\chi}_q$ is the susceptibility, given by Eq. (4.69), that accounts for the local fluctuations of the order parameter, as obtained from the Landau–Ginzburg functional.

The nature of relaxation of a nonconserved order parameter was demonstrated through the experimentally observed sound dispersion and sound attenuation near the weakly first-order nematic-isotropic phase transition in liquid crystals. As shown in Figure 4.31, the sound dispersion

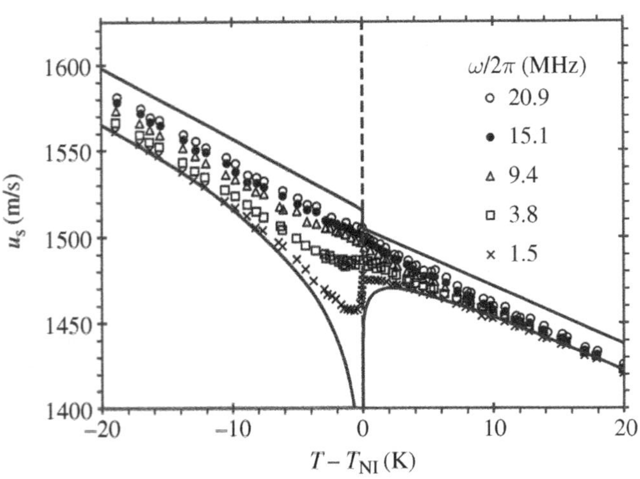

Figure 4.31 Speed of sound in a liquid crystal, MBBA, near its nematic-isotropic transition. The lower and upper curves represent the extrapolated temperature dependencies of $u_s(\omega \to 0)$ and $u_s(\omega \to \infty)$, respectively. The discontinuity of $u_s(\omega \to \infty)$ at the transition point is due to a density discontinuity. Source: Adapted from Ref. [48].

was observed in both phases (nematic and isotropic), though the effects are much more pronounced in the nematic phase. These effects reflect a similar asymmetry observed for the heat capacity, see Figure 4.16, which is associated with the fact that below the transition the order parameter is not zero, while above the transition, only local fluctuations of the order parameter exist. Furthermore, the sound attenuation, as illustrated in Figure 4.32, shows a pronounced maximum at the transition point, which decreases upon an increase of the frequency of sound, ω.

Figures 4.33 and 4.34 demonstrate the universal character of the sound dispersion and sound attenuation for the relaxation of the fluctuations of anisotropy in the isotropic phase of two liquid

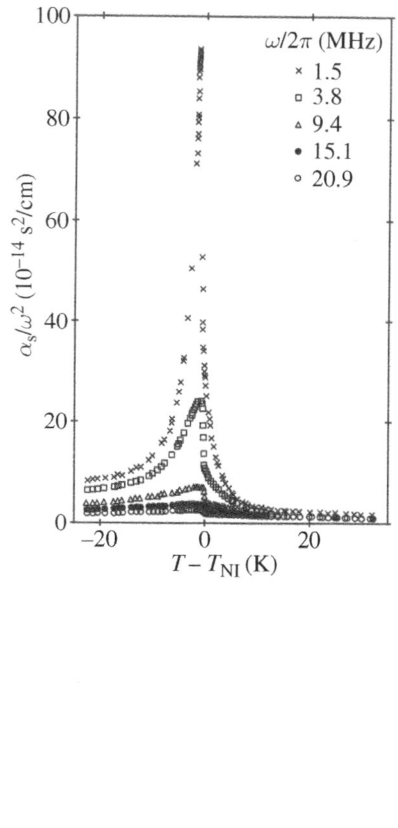

Figure 4.32 Sound absorption in a liquid crystal, MBBA, near its isotropic-nematic phase transition obtained for different frequencies, ω. Source: Adapted from Ref. [48].

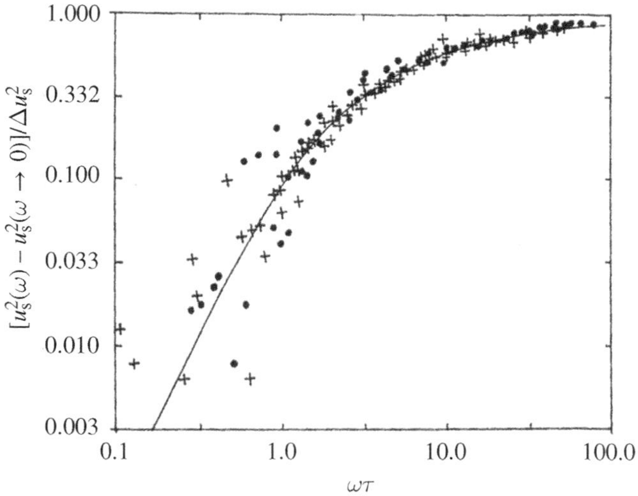

Figure 4.33 The normalized dispersion of the sound velocity in the isotropic phase of two nematic liquid crystals, MBBA (circles) and BMOAB (crosses). The solid curve is the prediction of the relaxation theory developed for the fluctuations of the anisotropy. Source: Adapted from Ref. [49].

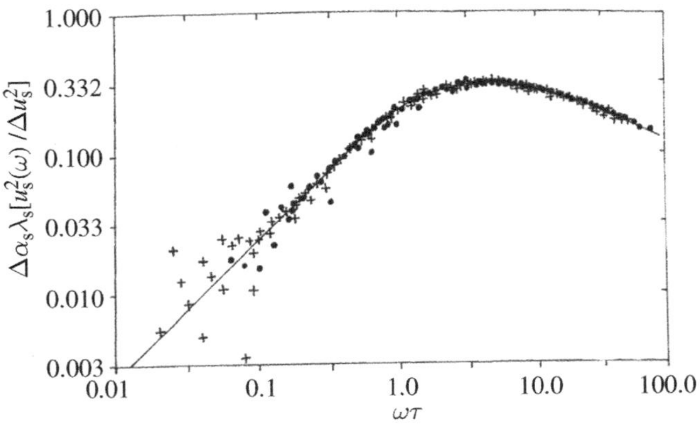

Figure 4.34 Universal function of the normalized excess sound absorption per wavelength, $\Delta\alpha_s\lambda_s[u_s^2(\omega)/\Delta u_s^2]$, for the isotropic phase of two nematic liquid crystals, MBBA (circles) and BMOAB (crosses). The solid curve is the result of the relaxation theory developed for the fluctuations of anisotropy. Source: Adapted from Ref. [49].

crystals as a function of $\omega\tau$ over several orders of magnitude. While the behavior of these functions qualitatively follows the predictions of the classical relaxation theory, presented in Figure 3.11, the exact form of the universal functions for the sound dispersion, f_u, and the sound attenuation, f_α, are different due to the spatial-dependence of the orientational susceptibility in the equation for the relaxation time, Eq. (3.63). The corresponding expressions, derived in the Ornstein–Zernike approximation by Imura and Okana [49], are given by

$$f_u = \sqrt{\frac{2\omega\tau}{\omega\tau + \sqrt{\omega^2\tau^2 + 1}}} \tag{4.88}$$

$$f_\alpha = \sqrt{\frac{2}{\omega\tau}}\left(\sqrt{\frac{1}{\omega\tau} + \sqrt{\frac{1}{\omega^2\tau^2} + 1}} - \sqrt{\frac{2}{\omega\tau}}\right) \tag{4.89}$$

While fluctuations contribute to the sound dispersion and attenuation in the nematic phase with the same order of magnitude as in the isotropic phase, the major contribution to these properties is due to the relaxation of the nematic order parameter. This relaxation is well-described by the classical relaxation theory, see Eqs. (3.97) and (3.98). In particular, the universal function for the sound attenuation could be represented by a linear combination of these two contributions as

$$\Delta\alpha_s\lambda_s\frac{u_s^2}{\Delta u_s^2} \approx (1 - A_m)f_\alpha + A_m f_{cl} \tag{4.90}$$

where $A_m \approx 0.8$ is related to the ratio of the maximum sound dispersion at zero frequency above and below the phase transition, see Figure 4.31. The separation of these two contributions, from the relaxation and fluctuations of the order parameter, is illustrated in Figure 4.35. One may note that at low frequencies, the asymptotic behavior ($\omega\tau \to 0$) for both contributions is the same, $f_\alpha = f_{cl} \to \omega\tau$, corresponding to Onsager's principle (see Section 3.7). However, in the limit of $\omega\tau \to \infty$, the contributions from the relaxation of fluctuations is $f_\alpha \to 1/\sqrt{\omega\tau}$, while the contribution from the relaxation of the order parameter is $f_{cl} \to 1/\omega\tau$.

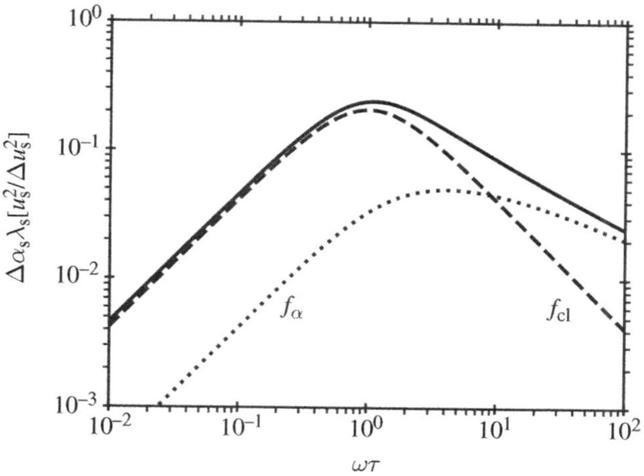

Figure 4.35 Separation of the universal function for the normalized excess sound absorption per wavelength obtained for the nematic phase of a liquid crystal, MBBA. The dashed curve, f_{cl}, is the contribution from the relaxation of the nematic order parameter, as predicted by the classical relaxation theory, Eq. (3.98). The dotted curve, f_α, is the contribution from the fluctuations of the order parameter, Eq. (4.89). The solid curve is the weighted sum of these two contributions as given by Eq. (4.90). Source: Adapted from Refs. [6, 48].

Appendix 4.A Emergence of Tricriticality

In this Appendix, we consider the emergence of tricriticality in the Landau theory via the coupling between two different order parameters: a scalar or tensor, φ, with a vector, $\boldsymbol{\psi}$. An example of this case is the transition to superfluidity in mixtures of helium isotopes, where the vector order parameter for superfluidity is the quantum wave function for boson condensation (see Section 4.2) and the scalar order parameter is the concentration of He^3 in the mixture of He^3-He^4, $\varphi = (x - x_c)/x_c$. The phase behavior of this system is illustrated in Figure 4.22.

Consider a generalized free energy for a system composed of two order parameters, in which one is a scalar, φ, and the other is a vector, $\boldsymbol{\psi}$. The thermodynamic potential may be written in the form

$$\hat{f}(\boldsymbol{\psi}, \varphi) = \hat{f}_\psi(\boldsymbol{\psi}) + \hat{f}_\varphi(\varphi) + \hat{f}_\times(\boldsymbol{\psi}, \varphi) \tag{4.A.1}$$

where $\hat{f}_\psi$ and $\hat{f}_\varphi$ are free energy functions of the vector and scalar order parameter only, while $\hat{f}_\times$ determines the coupling between the two order parameters. In general, the expression for the free energy, dependent on the two-order parameters $\boldsymbol{\psi}$ and ϕ, may be written in the form

$$\hat{f}(\boldsymbol{\psi}, \varphi) = \frac{1}{2!}a_\psi \Delta\hat{T}_\psi |\boldsymbol{\psi}|^2 + \frac{1}{4!}u_\psi |\boldsymbol{\psi}|^4 + \frac{1}{6!}g_\psi |\boldsymbol{\psi}|^6 + \frac{1}{2!}a_\varphi \Delta\hat{T}_\varphi \varphi^2 + \lambda|\boldsymbol{\psi}|^2\varphi \tag{4.A.2}$$

In which a_ψ, u_ψ, g_ψ, and a_φ are the Landau expansion coefficients and $\Delta\hat{T}_\psi$ and $\Delta\hat{T}_\varphi$ are the distances to the critical temperatures, $T_{c,\psi}$ and $T_{c,\varphi}$, corresponding to each order parameter, while λ is a coupling constant. Note that in the vicinity of the phase transition $\Delta\hat{T}_\varphi$ does not vanish and the higher ordered terms of $\hat{f}_\varphi$ may be neglected.

Consider the case when the ordering field conjugate to the vector order parameter, $h_\psi = 0$, while the ordering field conjugate to the scalar order parameter, h_φ, may vary. In this case, Eq. (4.A.2) can be simplified to an expression with only one order parameter via a Legendre transform. $\hat{\Phi}(\boldsymbol{\psi}, h_\varphi) = \hat{f}(\boldsymbol{\psi}, \varphi) - h_\varphi\varphi$, where $h_\varphi = \partial\hat{f}/\partial\varphi$. Minimizing the field-dependent free energy with respect to φ, as $\partial\hat{\Phi}/\partial\varphi = 0$, one finds

$$\varphi = (h_\varphi - \lambda|\psi|^2)/a_\varphi \Delta\hat{T}_\varphi \tag{4.A.3}$$

Using this form of φ and the Legendre transform, the free energy may be written in the form,

$$\hat{f}(\psi, h_\varphi) = \frac{1}{2!}a_\psi \Delta\hat{T}_\lambda|\psi|^2 + \frac{1}{4!}u_\lambda|\psi|^4 + \frac{1}{6!}g_\psi|\psi|^6 \tag{4.A.4}$$

where $\Delta\hat{T}_\lambda$ is the distance to the second-order phase transition line (the λ-line), given by

$$\Delta\hat{T}_\lambda = \Delta\hat{T}_\psi + \frac{2h_\varphi\lambda}{a_\psi a_\varphi \Delta\hat{T}_\varphi} \tag{4.A.5}$$

Along the λ-line $|\psi|^2 = 0, \varphi = h_\varphi/(a_\varphi\Delta\hat{T}_\varphi)$, and the distance to the critical point is renormalized as $\hat{T}_\lambda = \hat{T}_\psi - 2\varphi\lambda/a_\psi$. Thus, the coupling constant λ, determines the shift in the second-order transition temperature,

$$\lambda = -\frac{a_\psi}{2\tilde{T}_{c,\psi}}\frac{d\tilde{T}_{c,\psi}}{d\varphi} \tag{4.A.6}$$

Moreover, the renormalized coefficient of the fourth-order term, u_λ, in Eq. (4.A.5) is given by

$$u_\lambda = u_\psi - \frac{12\lambda^2}{a_\varphi\Delta\hat{T}_\varphi} \tag{4.A.7}$$

as the scalar order parameter increases, such as through the increasing concentration of the impurity, and the distance to the critical temperature associated with the scalar order parameter, $\Delta\hat{T}_\varphi$, decreases, the coefficient of the fourth order term may change its sign. The location of the tricritical point is thus determined from the conditions

$$\Delta\hat{T}_\psi = -\frac{2\lambda\varphi}{a_\psi} \quad \text{and} \quad \Delta\hat{T}_\varphi = \frac{12\lambda^2}{a_\varphi u_\psi} \tag{4.A.8}$$

and the phase diagram has the form of Figure 4.22.

It is important to note that other forms of coupling between the order parameters, not discussed in this Appendix, could occur in the case when $\boldsymbol{\psi}$ and $\boldsymbol{\varphi}$ are both vectors, such that the coupling may take the form, $\lambda|\psi|^2|\varphi|^2$. Such coupling may produce either a "bicritical point" (strong coupling), in which two second-order transition lines merge into one first-order transition line, or a "tetracritical point" (weak coupling), in which two second order transition lines intersect each other [27, 28]. If the system is characterized by more than two order parameters, their coupling could produce a "multicritical point," like the N-SmA-SmC multicritical point in a mixture of liquid crystals, shown in Figure 4.5 [4]. Alternatively, one may find in a multicomponent fluid mixture where ψ and φ are both scalars, such as the density and concentration, a point where three coexisting fluid phases become critical, which is referred to as an "asymmetric tricritical point" [50].

References

1 P. Ehrenfest, "Phasenumwandlungen im ueblichen und erweiterten sinn, classifiziert nach dementsprechenden singularitaeten des thermodynamischen potentiales," *Proc. R. Acad. Amsterdam*, vol. 36, pp. 153–157, 1933. *Eng. Transl. Communications from the Physical Laboratory of the University of Leiden*, Supplement No. 75b.

2 T. T. Haar, ed., *Collected Papers of L. D. Landau*. Gordon and Breach Science Publishers, 1965.

3 P. Kapitsa, "Viscosity of liquid helium below the λ-point," *Nature*, vol. 141, p. 74, 1938.

4 M. A. Anisimov, *Critical Phenomena in Liquids and Liquid Crystals*. New York: Gordon and Breach, 1991.

5 P.-G. de Gennes and J. Prost, *The Physics of Liquid Crystals*. Oxford University Press, 2nd ed., 1993.

6 M. A. Anisimov, "1. Introduction," *Mol. Cryst. Liq. Cryst.*, vol. 162, pp. 1–96, 1988.

7 M. A. Anisimov, V. A. Agayan, and P. J. Collings, "Nature of the blue-phase-III–isotropic critical point: An analogy with the liquid-gas transition," *Phys. Rev. E*, vol. 57, pp. 582–595, 1998.

8 P. J. Collings, *Liquid Crystals: Nature's Delicate Phase of Matter*. Princeton University Press, 2002.

9 T. C. Lubensky and H. Stark, "Theory of a critical point in the blue-phase-III–isotropic phase diagram," *Phys. Rev. E*, vol. 53, pp. 714–720, 1996.

10 C. E. Bertrand and M. A. Anisimov, "Peculiar thermodynamics of the second critical point in supercooled water," *J. Phys. Chem. B*, vol. 115, pp. 14099–14111, 2011.

11 P.-G. de Gennes, "Sur la transition smectique A-smectique C," *C.R. Hebd. Séances Acad. Sci., Ser. B*, vol. 274, pp. 758–760, 1972.

12 P. Pieranski, "Pierre-Gilles de Gennes: Beautiful and mysterious liquid crystals," *C.R. Phys.*, vol. 20, no. 7, pp. 756–769, 2019.

13 A. K. Wyczalkowska, J. V. Sengers, and M. A. Anisimov, "Critical fluctuations and the equation of state of van der Waals," *Physica A*, vol. 334, no. 3, pp. 482–512, 2004.

14 J. Wang and M. A. Anisimov, "Nature of vapor-liquid asymmetry in fluid criticality," *Phys. Rev. E*, vol. 75, p. 051107, 2007.

15 L. D. Landau and E. M. Lifshitz, *Statistical Physics Part 1*, vol. 5 of Course of Theoretical Physics. Oxford: Pergamon Press, 3rd ed., 1980.

16 D. Frenkel, "Entropy-driven phase transitions," *Physica A*, vol. 263, no. 1, pp. 26–38, 1999.

17 L. Onsager, "The effects of shape on the interaction of colloidal particles," *Ann. N.Y. Acad. Sci.*, vol. 51, p. 627, 1949.

18 H. Longuet-Higgins and B. Widom, "A rigid sphere model for the melting of argon," *Mol. Phys.*, vol. 8, no. 6, pp. 549–556, 1964.

19 P. C. Hiemenz and R. Rajagopalan, *Principles of Colloid and Surface Chemistry*. CRC Press, 3rd ed., 2016.

20 V. P. Skripov, *Metastable Liquids*. Wiley, 1974.

21 P. G. Debenedetti, *Metastable Liquids: Concepts and Principles*. Princeton University Press, 1996.

22 M. A. Anisimov and J. Thoen, "Heat capacities in the critical region," in *Heat Capacities of Liquids and Vapours* (E. Wilhelm and T. M. Trevor, eds.), ch. 14, pp. 307–328. Cambridge: Royal Society of Chemistry, 2010.

23 M. A. Anisimov, "Universality of the critical dynamics and the nature of the nematic-isotropic phase transition," *Mol. Cryst. Liq. Cryst.*, vol. 146, pp. 435–461, 1987.

24 P. K. Mukherjee, "Isotropic to smectic-A phase transition: A review," *J. Mol. Liq.*, vol. 190, p. 99, 2014.

25 S. A. Brazovskiĭ, I. E. Dzyaloshinskiĭ, and A. R. Muratov, "Theory of weak crystallization," *Sov. Phys. JETP*, vol. 66, pp. 625–623, 1987.

26 E. Kats, V. Lebedev, and A. Muratov, "Weak crystallization theory," *Phys. Rep.*, vol. 228, pp. 1–91, 1993.

27 M. A. Anisimov, E. E. Gorodetskiĭ, and V. M. Zaprudskiĭ, "Phase transitions with coupled order parameters," *Sov. Phys. Usp.*, vol. 24, p. 57, 1981.

28 I. Lawrie and S. Sarbach, "Theory of tricritical points," in *Phase Transitions and Critical Phenomena* (C. Domb and M. S. Green, eds.), vol. 9. New York City: Academic Press, 1984.

29 D. Vollhardt and P. Wölfle, *The Superfluid Phases of Helium 3.* London: Taylor and Francis, 1990.

30 A. Schmitt, *Introduction to Superfluidity*, vol. 888 of Lecture Notes in Physics. New York: Springer, 2015.

31 N. Giordano and W. P. Wolf, "New method for investigating magnetic tricritical points," *Phys. Rev. Lett.*, vol. 35, pp. 799–803, 1975.

32 M. A. Anisimov, P. E. Cladis, E. E. Gorodetskiĭ, et al., "Experimental test of a fluctuation-induced first-order phase transition: The nematic–smectic-A transition," *Phys. Rev. A*, vol. 41, pp. 6749–6762, 1990.

33 B. I. Halperin, T. C. Lubensky, and S.-K. Ma, "First-order phase transitions in superconductors and smectic-A liquid crystals," *Phys. Rev. Lett.*, vol. 32, pp. 292–295, 1974.

34 C. S. P. Tripathi, P. Losada-Pérez, J. Leys, et al., "Evidence from adiabatic scanning calorimetry for the Halperin-Lubensky-Ma effect at the N-SmA phase transitions in mixtures of 7oCB+heptane with an injected SmA phase," *Eur. Phys. J. E*, vol. 35, p. 54, 2012.

35 V. L. Pokrovskiĭ, "Notes on history of critical phenomena," *Hist. Phys., APS Newsl.*, vol. VII, no. 3, p. 15, 1998.

36 M. E. Fisher and B. Widom, "Decay of correlations in linear systems," *J. Chem. Phys.*, vol. 50, pp. 3756–3772, 1969.

37 C. Vega, L. F. Rull, and S. Lago, "Location of the Fisher-Widom line for systems interacting through short-ranged potentials," *Phys. Rev. E*, vol. 51, pp. 3146–3155, 1995.

38 T. W. Stinson and J. D. Litster, "Correlation range of fluctuations of short-range order in the isotropic phase of a liquid crystal," *Phys. Rev. Lett.*, vol. 30, pp. 688–692, 1973.

39 A. Muratov, A. Sadakov, S. Gavrilkin, et al., "Specific heat of FeSe: Two gaps with different anisotropy in superconducting state," *Physica B*, vol. 536, pp. 785–789, 2018.

40 M. A. Anisimov, J. V. Sengers, and J. M. Levelt Sengers, "Near-critical behavior of aqueous systems," in *Aqueous Systems at Elevated Temperatures and Pressures* (D. A. Palmer, R. Fernández-Prini, and A. H. Harvey, eds.), ch. 2, pp. 29–71. London: Academic Press, 2004.

41 S. C. Greer, "Coexistence curves at liquid-liquid critical points: Ising exponents and extended scaling," *Phys. Rev. A*, vol. 14, pp. 1770–1780, 1976.

42 R. J. Birgeneau, C. W. Garland, A. R. Kortan, et al., "Smectic-A-smectic-C transition: Mean field or critical," *Phys. Rev. A*, vol. 27, pp. 1251–1254, 1983.

43 J.-L. Barrat and J.-P. Hansen, *Basic Concepts for Simple and Complex Liquids.* Cambridge University Press, 2003.

44 M. E. Fisher, "The story of coulombic criticality," *J. Stat. Phys.*, vol. 75, pp. 1–36, 1994.

45 A. Levanyuk, "Contribution to the theory of light scattering near the second-order phase-transition points," *Sov. Phys. JETP*, vol. 36, p. 571, 1959.

46 L. Mistura and D. Sette, "Specific heat near the critical point," *J. Chem. Phys.*, vol. 49, p. 1419, 1968.

47 V. Vaks, A. Larkin, and S. Pikin, "Self-consistent field method for the description of phase transitions," *Sov. Phys. JETP*, vol. 24, p. 240, 1967.

48 M. A. Anisimov, V. P. Voronov, Y. F. Kiyachenko, and et al., "The homogeneous behaviour of absorption and dispersion of ultrasound in the isotropic phase of a nematic liquid crystal," *Mol. Cryst. Liq. Cryst.*, vol. 104, pp. 273–279, 1984.

49 H. Imura and K. Okano, "Theory of anomalous ultrasonic absorption and dispersion of nematic liquid crystals just above the clearing point," *Chem. Phys. Lett.*, vol. 19, pp. 387–390, 1972.

50 C. M. Knobler and R. L. Scott, "Multicritical points in fluid mixtures: Experimental studies," in *Phase Transitions and Critical Phenomena* (C. Domb and M. S. Green, eds.), vol. 9. New York City: Academic Press, 1984.

Exercises

4.1 Explain the concept of the order parameter with regard to the symmetry change and the classification of phase transitions. Give three examples of order parameters and the corresponding systems (scalar, vector, and tensor).

4.2 Determine the truncated Landau expansion for:

a) An incompressible binary mixture, for which the dimensionless Gibbs energy of mixing in the binary regular solution model is,

$$\Delta\hat{G} = \hat{T}[(1-x)\ln(1-x) + x\ln x] + \omega x(1-x)$$

where the interaction parameter ω =constant> 0. Represent the Gibbs energy of mixing in the form of a truncated (two term) Landau expansion in powers of the order parameter near the liquid–liquid critical point of demixing. Find the values of the coefficients: a_0 and u_0. *Hint*: First find the critical parameters (critical temperature and critical composition) and the inverse of the dimensionless osmotic susceptibility, $\hat{\chi}^{-1} = \partial^2(\Delta\hat{G})/\partial\varphi^2$, as a function of $\Delta\hat{T} = (T - T_c)/T_c$ and $\varphi = (x - x_c)/x_c$. The inverse susceptibility will contain the coefficients generally defined by Landau expansion.

b) The van der Waals equation of state. Find the values of the coefficients a_0 and u_0. *Hint*: the inverse susceptibility for this equation of state may be calculated from

$$\hat{\chi}^{-1} = \hat{\rho}k_B\hat{T}\left.\frac{\partial\hat{\rho}}{\partial\hat{P}}\right|_{\hat{T}}$$

where $\hat{\rho} = \rho/\rho_c$, $\hat{P} = P/P_c$, and $\hat{T} = T/T_c$ as a function of $\varphi = (\rho - \rho_c)/\rho_c$. The inverse susceptibility will contain the coefficients generally defined by the Landau expansion.

4.3 Why are critical points of fluid–fluid equilibria equivalent to second-order phase transition? Why should the transitions between fluid phases away from the critical points be defined as first-order phase transitions?

4.4 Based on the truncated symmetric Landau expansion, calculate and plot the behavior of the following properties (adopt $a_0 = 1$ and $u_0 = 2$):

a) The liquid–vapor coexistence curve and the spinodal.

b) The dimensionless isothermal compressibility above and below the critical temperature (along the critical isochore and along the phase coexistence).

c) The discontinuity of the isochoric heat capacity at the vapor–liquid critical point in a single-component fluid across the critical isochore.

4.5 Derive the expression for the diameter of the liquid–gas coexistence in the meanfield theory. For the van der Waals fluid, obtain the rectilinear diameter. Using your results for the binodal (Exercise 2.21), plot the diameter and the binodal on a temperature–density diagram. Show that $D = 2/5$ for small values of $\Delta\hat{T}$.

4.6 Explain why the phase transitions between liquids and crystalline solids (such as the melting of ice) are always of the first order and do not terminate at a critical point?

4.7 Qualitatively explain the difference between the different Landau expansions for first-order transitions in zero-ordering field. Compare the magnitude of the heat capacity discontinuity for an asymmetric Landau expansion with a nonzero negative cubic term and a symmetric Landau expansion with a nonzero negative quartic term near the transition temperature.

4.8 Provide two examples of phase transitions in systems with coupled order parameters.

4.9 Explain the meaning of the density-density correlation function. Plot the correlation function, $C(r)$, for carbon dioxide along the critical isochore at 10%, 1%, and 0.1% above the critical temperature. Assume that the amplitude of the correlation length, ξ_0, is approximately of the order of the molecular radius.

4.10 Explain the meaning of the meanfield approximation and the Ginzburg criterion. Give two examples of systems where the meanfield approximation is practically always valid and explain why.

4.11 Based on the Landau–Ginzburg functional for the local free-energy density,

$$\hat{f} \approx \frac{1}{2}\hat{\chi}_{q=0}^{-1}\varphi^2 + \frac{1}{2}c_0\left(\frac{\partial\varphi}{\partial r}\right)^2$$

obtain the relation between the susceptibility and correlation length in the meanfield approximation.

4.12 If you investigate a two-dimensional condensation of argon on a plain graphite substrate and compare the properties with a classical equation of state: (a) Would you expect larger deviations from the classical equation of state caused by density fluctuations than for a three-dimensional bulk condensation? (b) Would the critical temperature be lower or higher in the two-dimensional case than that in the three-dimensional case?

4.13 Derive the Ginzburg criterion for a d-dimensional system. If the range of interactions (ξ_0) is twice smaller than the average distance between molecules ($\bar{r}$), what would be the difference between the size of the critical region, characterized by the Ginzburg number, N_G, in three and two dimensions?

4.14 Generally, the local (spatially-dependent) free energy is a function of the Cartesian coordinates, x, y, and z. Consider the projection of the Landau–Ginzburg free-energy density on the vertical coordinate:

$$\hat{f}[\varphi(z)] = \frac{1}{2!}a_0\Delta\hat{T}[\varphi(z)]^2 + \frac{1}{4!}u_0[\varphi(z)]^2 + \frac{1}{2}c_0\left(\frac{\partial\varphi}{\partial z}\right)^2$$

where $\Delta\hat{T} = (T - T_c)/T_c$, $\varphi = (\rho - \rho_c)/\rho_c$, and z is the vertical coordinate.

a) Explain the physical meaning of each term. What is the physical difference between the gradient term and the other terms?
b) For the lattice-gas model give a reasonable estimate of the coefficient c_0 expressed through the critical density.

4.15 By comparing the mean-square density fluctuations and the equilibrium difference between vapor and liquid densities, estimate the Ginzburg criterion for carbon dioxide. It is known from light-scattering experiments that for carbon dioxide $\xi_0 = 0.17$ nm (adopt $a_0 = 1$ and $u_0 = 2$).

4.16 Suggest possible routes to change the Ginzburg number in computer simulations of a fluid represented by a lattice model. Construct a system for which the Ginzburg number is on the order $\sim 10^{-6}$ in a computer simulation.

4.17 Formulate the physical difference between the dynamics of conserved and nonconserved order parameters.

4.18 In the meanfield approximation, estimate the magnitude of the diffusion coefficient for a mixture of nitroethane-isooctane at $\Delta\hat{T} = 0.1$ and $\Delta\hat{T} = 0.01$ distance from the critical temperature of demixing at the critical composition. The critical temperature and critical density of the mixture are $T_c = 31$ °C and $\rho_c \approx 0.8\,\text{g/cm}^3$. Adopt the value for the bare correlation length, $\xi_0 = 0.2$ nm. Approximate the molecular mobility through the Stokes–Einstein formula with a temperature-independent viscosity of $\eta \approx 10^{-3}$ Pa · s.

Part II

Applications of Mesoscopic Thermodynamics

5

Polymer Solutions: A Meanfield Description

This chapter includes:

- Behavior of long polymer chains in solution
- Flory–Huggins Theory
- Theta-point tricriticality
- Diffusion of polymer chains in solutions

The thermodynamics of polymer solutions is a rare example in which a mesoscopic length scale, the size of a polymer molecule, explicitly enters the macroscopic equation of state. In this chapter, we restrict the description of polymer solutions to the meanfield approximation. The effects of critical fluctuations near the critical points of demixing will be discussed in Chapter 8.

5.1 Polymer Solutions

We consider solutions of a high-molecular-weight polymer in a low-molecular-weight solvent [1–3]. The polymer molecule under consideration is a long flexible chain consisting of repeated fragments called "monomers." The molecular weight of the polymer, M_p, is related to the molecular weight of the monomer, M_m, by

$$M_p = \mathcal{N} M_m \tag{5.1}$$

where $\mathcal{N}$ is the *degree of polymerization*, the number of monomers comprising the polymer chain.

For example, consider a solution of polystyrene in methyl cyclohexane. The molecular weight of methyl cyclohexane ($C_6H_{11}CH_3$) is $M_{mc} = 98$ g/mol. The molecular weight of styrene ($C_6H_5CHCH_2$) is almost the same, being $M_m = 104$ g/mol. The molecular weight of typical commercial polystyrene samples vary from $M_p = 10{,}000$ ($N = 100$) to $M_p = 30{,}000{,}000$ ($N = 300{,}000$) g/mol. In addition, commercial polymer samples are polydisperse with a certain distribution of molecular weights. The width and the size of the distribution is characterized by a "polydispersity index," which is often the standard deviation in a Gaussian distribution (σ). A sample with a polydispersity index of $\lesssim 1\%$ is considered to be practically "mono-disperse."

Mesoscopic Thermodynamics for Scientists and Engineers, First Edition. Mikhail A. Anisimov and Thomas J. Longo.
© 2024 John Wiley & Sons, Inc. Published 2024 by John Wiley & Sons, Inc.

Mono-Disperse Polymers
Remarkably, examples of mono-disperse polymers are found in biology where the chain length is thoroughly regulated by Nature. For instance,

1) Human DNA has 280 million monomer units (nucleic acids) and a total stretched length of 9.5 cm.
2) Lysozyme is a protein consisting of 129 amino acids.

The solvent in polymer solutions can act either as a "good solvent" or a "poor solvent." Good solvents dissolve a polymer in all proportions. Poor solvents can dissolve only a small portion of the polymer because of the effective repulsion between solvent molecules and polymer chains. However, the ability of a solvent to dissolve polymers depends on the temperature and pressure. The same solvent can be "good" at a high temperature and "poor" at a low temperature. The temperature that separates good-solvent behavior from poor-solvent behavior is called the "Theta temperature". A solvent at the Theta temperature is a *Theta solvent*. A polymer chain at the Theta temperature is "ideal," in that its constituents exhibit random-walk behavior as discussed in Section 5.2. In a good solvent, the polymer chain exhibits self-avoiding-walk behavior (as will be discussed in Section 5.7), while in a poor solvent, the polymer chain is collapsed into a globule.

5.2 Random Walk of a Polymer Chain

At the Theta temperature, each monomer in the polymer chain can exhibit random-walk behavior similar to that of Brownian motion, see Section 3.9. If one considers a polymer chain with degree of polymerization, $\mathcal{N}$, and a monomer length a, then the end-to-end vector $\boldsymbol{R}$, illustrated in Figure 5.1, is found from the sum of $\mathcal{N}$ vectors $\boldsymbol{a}_i$ that correspond to the direction and size of each monomer. In this case, $\boldsymbol{R}$ is found as

$$\boldsymbol{R} = \boldsymbol{a}_1 + \boldsymbol{a}_2 + \cdots = \sum_{i=1}^{\mathcal{N}} \boldsymbol{a}_i \tag{5.2}$$

The average end-to-end distance is obtained as

$$\langle \boldsymbol{R} \cdot \boldsymbol{R} \rangle = \left\langle \left(\sum_{i=1}^{\mathcal{N}} \boldsymbol{a}_i \right) \cdot \left(\sum_{j=1}^{\mathcal{N}} \boldsymbol{a}_j \right) \right\rangle \tag{5.3}$$

Expressing the magnitude of the vector as $R^2 = \boldsymbol{R} \cdot \boldsymbol{R}$, and simplifying Eq. (5.3), one obtains

$$\langle R^2 \rangle = \left\langle \sum_{i=1}^{\mathcal{N}} \sum_{j=1}^{\mathcal{N}} \boldsymbol{a}_i \cdot \boldsymbol{a}_j \right\rangle \tag{5.4}$$

Separating the $\mathcal{N}$ terms when $\boldsymbol{a}_i = \boldsymbol{a}_j$, yields

$$\langle R^2 \rangle = \mathcal{N} a^2 + \left\langle \sum_{i \neq j}^{\mathcal{N}} \boldsymbol{a}_i \cdot \boldsymbol{a}_j \right\rangle \tag{5.5}$$

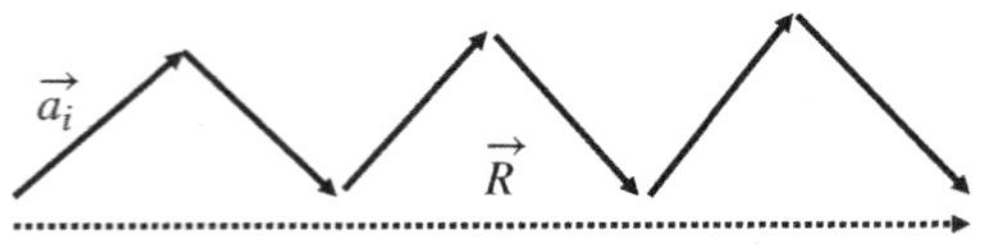

Figure 5.1 Schematic representation of a linear polymer. Each monomer has a length and direction, $\boldsymbol{a}_i$, while the polymer chain is characterized by the end-to-end vector, $\boldsymbol{R}$.

Assuming that each monomer may randomly move in any direction, then upon averaging over all possible movements, one finds

$$\left\langle \sum_{i \neq j}^{\mathcal{N}} \boldsymbol{a}_i \cdot \boldsymbol{a}_j \right\rangle \to 0 \tag{5.6}$$

when a sufficient number of steps are made. Thus, the average end-to-end distance squared is given by

$$\langle R^2 \rangle = \mathcal{N} a^2 \tag{5.7}$$

The root mean square distance is known as the "radius of gyration", R_g, and is given by

$$\boxed{R_g = \sqrt{\langle R^2 \rangle} = a\sqrt{\mathcal{N}}} \tag{5.8}$$

However, one may wonder why these simple connections between monomers have the appearance of being unphysical because successive monomers may not be free to rotate, being constrained by specific bond angles. The inclusion of such short-ranged correlations would not alter the basic random-walk behavior; instead, it simply leads to a different effective step size b, such that the radius of gyrations could be given by

$$R_g = \sqrt{\langle R^2 \rangle} = b\sqrt{\mathcal{N}} \tag{5.9}$$

In this case, one may divide the polymer chain into different subunits with a size, b, that replaces the effective size of the monomers, a. This effect results in a correlation of monomers over a certain distance ℓ_p, known as the "persistence length", which is defined as

$$\ell_p = b^2/a \tag{5.10}$$

Thus, by accounting for the persistence length, one may still consider random-walk behavior of the polymer chain without the restrictions imposed by the chemical bonds between monomers.

The ideal random-walk behavior of the chain can be described by the same Gaussian probability function, Eq. (3.66), used to describe Brownian motion in Chapter 3. This equation can be written in terms of the fluctuations of the end-to-end distance, R, of the polymer chain as

$$w(R^2, \mathcal{N}) = \left(\frac{3}{2\pi \langle R^2 \rangle} \right)^{3/2} \exp\left\{ -\frac{R^2}{2\langle R^2 \rangle} \right\} \tag{5.11}$$

According to Eq. (3.5), this distribution is proportional to the entropy change due to the fluctuations of R, such that $w \propto \exp\left(\Delta S/k_B\right)$. Assuming a completely flexible chain, with $\ell_p = a$, the entropy change is given by

$$\Delta S = -\frac{3}{2}k_B \frac{R^2}{Na^2} = -\frac{3}{2}k_B \frac{R^2}{R_g^2} \tag{5.12}$$

Stretching the ideal chain (at constant temperature and pressure), therefore, increases the Gibbs energy by

$$\Delta G = \Delta H - T\Delta S = -T\Delta S = \frac{3}{2}k_B T \frac{R^2}{R_g^2} \tag{5.13}$$

Therefore, comparing Eqs. (5.11)–(5.13), one observes that the mean square fluctuations of the end-to-end distance of the ideal polymer chain are given by

$$\langle (\delta R)^2 \rangle = \frac{1}{3}R_g^2 \tag{5.14}$$

If the chain is stretched beyond its ideal random-walk configuration, there will be a restoring force proportional to its extension. This restoring force is purely entropic. In contrast to a spring, where the restoring force is due to an increase of the internal energy, for an ideal polymer chain, the restoring force emerges due to the decrease in entropy.

We note that in the ideal chain, the solvent plays no role in the behavior of the chain. When a polymer chain exhibits ideal random-walk behavior, the short-ranged interactions between the monomers and solvent molecules compensate each other. However, this is an exceptional case applicable only to Theta solvents. Throughout the remainder of this chapter, we will discuss the effects of interactions of the monomers and the solvent on the behavior of the polymer chain.

5.3 Flory–Huggins Theory of Polymer Solutions

The thermodynamic description of polymers is primarily associated with the works of Paul Flory.[1] The meanfield equation of state of polymer solutions was developed by Paul Flory, and independently by Maurice Huggins, and it is commonly referred to as the "Flory–Huggins" theory [5–7]. In the simplest version of the Flory–Huggins theory, the polymer solution is represented on a lattice where the polymer-chain fragments are assumed to occupy the same volume as the solvent molecules (Figure 5.2). The polymer volume fraction, φ, is the ratio of the number of cells occupied by the polymer fragments, N_m, to the total number of cells, N, as $\varphi = N_m/N$.

5.3.1 Thermodynamics of Phase Separation

In the Flory–Huggins theory, the entropy of mixing per lattice cell is represented as

$$\frac{\Delta S_{mix}}{k_B T} = -\left[(1-\varphi)\ln(1-\varphi) + \frac{\varphi}{\mathcal{N}}\ln\varphi\right] \tag{5.15}$$

which is similar to that for the lattice-gas model, see Section 2.7.2, except for the division by the degree of polymerization, $\mathcal{N}$, in the second term of Eq. (5.15). This factor accounts for the assembly of the monomers into a chain, and it produces a strong asymmetry in the entropy of mixing with respect to an ideal solution. The enthalpy of mixing per cell is adopted in a symmetric form, $\Delta H_{mix}/k_B T = \varpi\varphi(1-\varphi)$, such that the Gibbs energy per cell can be expressed as

$$\frac{\Delta G_{mix}}{k_B T} = \left[(1-\varphi)\ln(1-\varphi) + \frac{\varphi}{\mathcal{N}}\ln\varphi\right] + \varpi\varphi(1-\varphi) \tag{5.16}$$

where ϖ is the Flory interaction parameter, which is generally a function of temperature and pressure. In the case, $\mathcal{N} = 1$ and $\varpi = \varpi_0/k_B T$ (where ϖ_0 is a constant), Eq. (5.16) describes the regular-solution model, see Section 2.8.3. In the opposite limit, when the polymer chain is composed of an infinite number of monomers ($\mathcal{N} \to \infty$), the second term in the entropy of mixing vanishes. The Flory interaction parameter, ϖ, determines the miscibility of the polymer and the solvent. For $\varpi < 0$, mixing between the polymers and solvent is energetically favorable, while for $\varpi > 0$, mixing will be energetically unfavorable. In the latter case, depending on the temperature and pressure, two liquid phases may coexist: one polymer-rich and the other solvent-rich.

1 Paul Flory (1910–1985) was an American chemist who was awarded a Nobel Prize in 1974 for "his fundamental achievements, both theoretical and experimental, in the physical chemistry of macromolecules" [4].

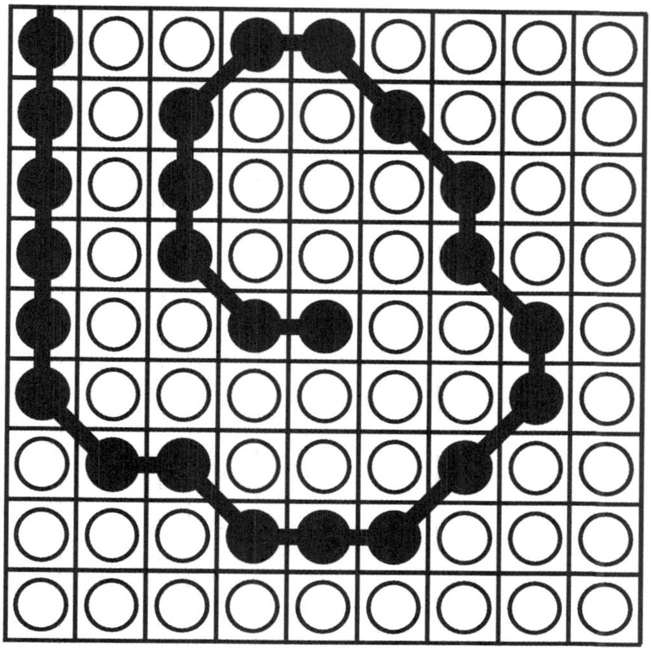

Figure 5.2 A two-dimensional representation of a polymer-chain monomers (black) surrounded by solvent molecules (white) on a lattice as represented in the Flory–Huggins theory of polymer solutions.

By comparing the entropy of mixing in the Flory–Huggins model to the entropy of mixing in an ideal solution, $\Delta S_{\rm id} \equiv \Delta S(\mathcal{N} = 1)$, we note that the excess entropy of mixing, $\Delta S^{\rm E}$, is given by

$$\frac{\Delta S^{\rm E}}{Nk_{\rm B}T} = \frac{\Delta S - \Delta S_{\rm id}}{Nk_{\rm B}T} = -\varphi \ln \varphi \left(\frac{1}{\mathcal{N}} - 1\right) < 0 \tag{5.17}$$

This negative contribution to the entropy of mixing underlines the decrease of the driving force of mixing caused by the reduction in the configurational entropy, due to the existence of the chemical bonds in the polymer chain.

5.3.2 Theta Point

The Gibbs energy per lattice cell (expressed in reduced form as $\hat{G} \equiv G_{\rm t}/Nk_{\rm B}T$) for a polymer solution reads

$$\hat{G} = \hat{\mu}_{\rm m}^{\circ}\varphi + \hat{\mu}_{\rm s}^{\circ}(1 - \varphi) + \Delta\hat{G}_{\rm mix} = \frac{\hat{\mu}_{\rm p}^{\circ}}{\mathcal{N}}\varphi + \hat{\mu}_{\rm s}^{\circ}(1 - \varphi) + \Delta\hat{G}_{\rm mix} \tag{5.18}$$

where $\hat{\mu}_{\rm m}^{\circ}$, $\hat{\mu}_{\rm s}^{\circ}$, and $\hat{\mu}_{\rm p}^{\circ} = \mathcal{N}\hat{\mu}_{\rm m}^{\circ}$ are the reduced (by $k_{\rm B}T$) chemical potentials of the monomer, solvent, and polymer, respectively. The monomer-solvent exchange chemical potential is evaluated from the derivative of the Gibbs energy as

$$\hat{\mu}_{\rm ms} = \left.\frac{\partial \hat{G}}{\partial \varphi}\right|_{T,P} = \hat{\mu}_{\rm m} - \hat{\mu}_{\rm s} = \frac{\hat{\mu}_{\rm p}}{\mathcal{N}} - \hat{\mu}_{\rm s} \tag{5.19}$$

where the chemical potentials of the solvent, $\hat{\mu}_{\rm s}$, and monomer, $\hat{\mu}_{\rm m}$, in solution are given by

$$\hat{\mu}_{\rm s} = \hat{G} - \varphi \left.\frac{\partial \hat{G}}{\partial \varphi}\right|_{P,T} \tag{5.20}$$

$$\hat{\mu}_{\mathrm{m}} = \hat{G} + (1-\varphi)\left.\frac{\partial \hat{G}}{\partial \varphi}\right|_{P,T} \tag{5.21}$$

The critical values of the temperature, T_c, and volume fraction, φ_c, of phase separation are found from the thermodynamic stability conditions,

$$\left.\frac{\partial^2 \hat{G}}{\partial \varphi^2}\right|_{T,P} = \left.\frac{\partial \hat{\mu}_{\mathrm{ms}}}{\partial \varphi}\right|_{T,P} = 0 \tag{5.22}$$

$$\left.\frac{\partial^3 \hat{G}}{\partial \varphi^3}\right|_{T,P} = \left.\frac{\partial^2 \hat{\mu}_{\mathrm{ms}}}{\partial \varphi^2}\right|_{T,P} = 0 \tag{5.23}$$

For the case, $\varpi = \varpi_0/k_B T$ (where ϖ_0 is a constant), these derivatives can be calculated with use of Eq. (5.16); after which, one finds that the critical parameters depend on the degree of polymerization, $\mathcal{N}$, as

$$T_c = \frac{\Theta}{(1+1/\sqrt{\mathcal{N}})^2} \qquad \varphi_c = \frac{1}{1+\sqrt{\mathcal{N}}} \tag{5.24}$$

The parameter, Θ, is known as the "Theta temperature," and is defined as the critical temperature of demixing of a polymer solution with an infinitely long polymer chain at infinite dilution. In the limit of a regular solution ($\mathcal{N} = 1$), $T_c = \Theta/4$, and $\varphi_c = 1/2$, see Section 2.8.3. In the limit of an infinitely long polymer chain, $T_c = \Theta$ and $\varphi_c \to 0$, which is defined as the "Theta-point" limit.

An increase of the degree of polymerization has a dramatic effect on the shape of the phase coexistence curve (binodal) as illustrated on Figure 5.3. The binodal is found by equating the chemical potentials in the two coexisting phases as $\mu_s^{\tilde{\alpha}} = \mu_s^{\tilde{\beta}}$ and $\mu_m^{\tilde{\alpha}} = \mu_m^{\tilde{\beta}}$ as given by Eqs. (5.20) and (5.21). As the degree of polymerization increases, the coexistence curve becomes more asymmetric as the critical volume fraction moves to zero. In the limit of an infinitely long polymer chain, $\varphi_c \to 0$ and the binodal shifts to an "angle-like" shape with a sharp peak at the Theta temperature, $T = \Theta$.

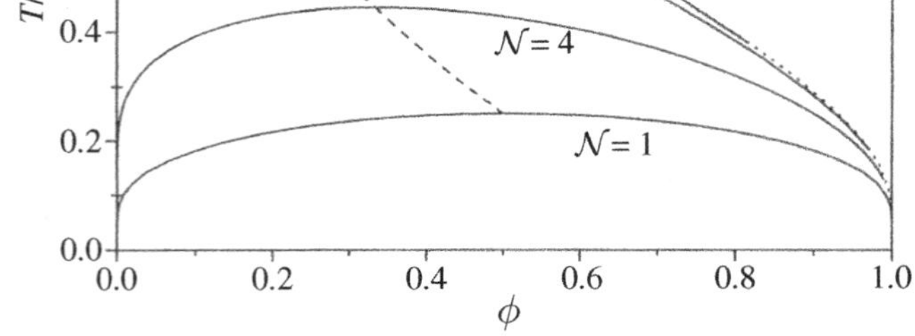

Figure 5.3 Phase coexistence for polymer solutions with different degrees of polymerization, $\mathcal{N}$, calculated from the simplified Flory–Huggins theory ($\varpi = \varpi_0/k_B T$) and plotted as the reduced temperature T/Θ (where Θ is the Theta temperature) vs. volume fraction φ. The dotted line represents the limiting curve of $\mathcal{N} \to \infty$ and the dashed line represents the critical locus. Source: Reproduced with permission from Ref. [8], © Elsevier (1999).

5.4 Liquid–Liquid Critical Point vs. Theta Point

The phase coexistence between two liquid phases in polymer solutions in the Flory–Huggins theory demonstrates two types of behavior [9]. In the vicinity of the liquid–liquid critical point, the shape of the phase coexist is parabola-like (similar to common liquid solutions), while in the vicinity of the Theta point, the coexistence curve adopts an angle-like shape (as shown in Figure 5.3). In this section, these two cases are analyzed through the expansion of the free energy in each region.

5.4.1 Landau Expansion of the Flory–Huggins Free Energy

The Gibbs energy, Eq. (5.16), may be expanded in powers of the order parameter around the critical demixing point in the region relatively far away from the Theta point. That is to say, in the region where the temperature and volume fraction are close to their critical values, while the polymer chain length is large but still finite. Therefore, the polymer solution in the vicinity of the critical demixing point may be described, in the lowest-order approximation, by the two-term truncated Landau expansion in the form of Eq. (4.9).

For demixing polymer solutions, the appropriate order parameter for the Landau expansion is $\phi = \varphi - \varphi_c$, such that the truncated expansion is given by

$$\hat{f} = \frac{G_t}{Nk_B T_c} = \frac{1}{2}\Delta\hat{T}\phi^2 + \frac{1}{12}\sqrt{\mathcal{N}}\phi^4 \tag{5.25}$$

where $\Delta\hat{T} = T/T_c - 1$ is the reduced distance to the critical temperature. For large $\mathcal{N}$, the Landau expansion coefficients, a_0 and u_0, are

$$a_0 = 1 \qquad \text{and} \qquad u_0 = 2\sqrt{\mathcal{N}} \tag{5.26}$$

For a detailed derivation of the Landau expansion of the Flory–Huggins free energy, see Appendix 5.A. One may observe that in the Theta-point limit $\mathcal{N} \to \infty$, u_0 diverges and the Landau expansion is no longer applicable.

From the Landau expansion, all the thermodynamic quantities asymptotically close to the critical point can be obtained (see Section 4.3). First, the equilibrium value of the order parameter along the liquid–liquid coexistence, φ_0, below the critical temperature is given by Eq. (4.14), and for polymer solutions is given by

$$\phi_0 = \pm\sqrt{3}\mathcal{N}^{-1/4}|\Delta\hat{T}|^{1/2} \tag{5.27}$$

Equation (5.27) shows that the critical exponent for the order parameter is $\beta = 1/2$, while its critical amplitude scales with the degree of polymerization as $B_0 \sim \mathcal{N}^{-1/4}$. It was shown by de Gennes and Sanchez [1, 10, 11] that there is a universal relation between the exponent of the $\mathcal{N}$-dependent amplitude and the critical exponent. Defining $B_0 \equiv \pm\sqrt{3}\mathcal{N}^{-b}$ (where $b = 1/4$), then this relation is $b = (1 - \beta)/2$.

Analogous universal relations are observed for other thermodynamic quantities. For instance, consider the heat capacity and susceptibility. In the meanfield approximation, the critical exponent of the heat capacity is $\alpha = 0$, while the critical amplitude is given by Eq. (4.16). For $T > T_c$ (in the homogeneous region), the amplitude is $A_0^+ = 0$, while for $T < T_c$ (in the two-phase region), the amplitude (discontinuity) of the heat capacity is given by

$$A_0^- = \frac{9a_0^2}{u_0} = \frac{9}{2\sqrt{\mathcal{N}}} \tag{5.28}$$

Hence, one may express the exponent of the amplitude as $A_0^- \propto \mathcal{N}^{-a}$, where $a = 1/2$, and therefore, the relation between the amplitude exponent and the critical exponent is $a = (1 + \alpha)/2$. Furthermore, the susceptibility, given by Eq. (4.17), for polymer solution is described by

$$\hat{\chi}^{-1} = \left.\frac{\partial^2 \hat{f}}{\partial \phi^2}\right|_{\Delta \hat{T}} = \begin{cases} \Delta \hat{T} & T > T_c \\ -2\Delta \hat{T} & T < T_c \end{cases} \tag{5.29}$$

such that the amplitudes above and below the critical temperature are both independent of $\mathcal{N}$. Expressing the critical amplitude as $\Gamma_0^{\pm} \propto \mathcal{N}^0 \equiv \mathcal{N}^g$, then the relation between the amplitude exponent and critical exponent is $g = (1 - \gamma)/2$, where the critical exponent, γ, is $\gamma = 1$ in the meanfield approximation, as indicated by Eq. (5.29).

The amplitude of the correlation length is determined by the Landau–Ginzburg free energy functional for the polymer solution in the vicinity of the critical point. In the first-order approximation, the free-energy functional has the form of Eq. (4.71). In polymer solutions, c_0 is proportional to the radius of gyration, $c_0 \sim R_g \approx \sqrt{\mathcal{N}}$ [8, 12]. From the free-energy functional, the correlation length is related to the thermodynamic susceptibility through Eq. (4.69), such that $\xi = \sqrt{c_0 \hat{\chi}_{q=0}}$. In the meanfield approximation, the correlation length is given by Eq. (4.72), where the amplitude of the correlation length depends on the degree of polymerization as

$$\xi_0^+ = \sqrt{\frac{c_0}{a_0}} \propto \mathcal{N}^{1/4} \quad \text{and} \quad \xi_0^- = \sqrt{\frac{c_0}{2a_0}} \propto \mathcal{N}^{1/4} \tag{5.30}$$

Expressing this relation as $\xi_0^{\pm} = \mathcal{N}^{1/4} \equiv \mathcal{N}^n$, then one obtains the relation between the amplitude and critical exponents as $n = (1 - \nu)/2$, where the critical exponent of the correlation length, ν, is $\nu = 1/2$ in the meanfield approximation, see Eq. (4.72).

De Gennes-Sanchez Relations: A Meanfield Summary

Table 5.1 Thermodynamic properties of polymer solutions in the vicinity of the critical point predicted by the Flory-Huggins theory and the de Gennes-Sanchez relations between the amplitude and critical exponents.

Property	Scaling law	Relation
Order parameter	$\phi_0 \sim \pm\mathcal{N}^{-b}\lvert\Delta\hat{T}\rvert^{\beta}$	$b = (1 - \beta)/2$
Heat capacity	$C \sim \mathcal{N}^{-a}\lvert\Delta\hat{T}\rvert^{-\alpha}$	$a = (1 + \alpha)/2$
Susceptibility	$\hat{\chi} \sim \mathcal{N}^{g}\lvert\Delta\hat{T}\rvert^{-\gamma}$	$g = (1 - \gamma)/2$
Correlation length	$\xi \sim \mathcal{N}^{n}\lvert\Delta\hat{T}\rvert^{-\nu}$	$n = (1 - \nu)/2$

Table 5.2 Values for the amplitude and critical exponents, related by the de Gennes-Sanchez relations, as predicted by the Flory-Huggins theory.

Property	Amplitude exp.	Temperature exp.
Order parameter	$b = 1/4$	$\beta = 1/2$
Heat capacity	$a = 1/2$	$\alpha = 0$
Susceptibility	$g = 0$	$\gamma = 1$
Correlation length	$n = 1/4$	$\nu = 1/2$

A summary of the thermodynamic properties predicted by the meanfield Flory-Huggins theory in the vicinity of the critical point is provided in Table 5.1. The relations between the critical exponents and the exponents of the $\mathcal{N}$-dependent amplitudes are known as the "de Gennes–Sanchez relations" (Tables 5.1 and 5.2). At this point, it is important to note that the physical behavior of polymer solutions in the vicinity of the liquid–liquid critical point does not quantitatively agree with the meanfield predictions. As will be demonstrated in Chapter 8, the actual $\mathcal{N}$-dependence of the critical amplitudes and the values of the critical exponents are different, being affected by critical fluctuations, but the de Gennes–Sanchez relations are still remain applicable.

5.4.2 Osmotic-Pressure Virial Expansion

At the Theta point, all of the monomers form an infinitely long single polymer chain ($\mathcal{N} \to \infty$), and the Landau expansion becomes undefined, being unable to appropriately describe the behavior of the phase coexistence of the polymer solution. For a very large (but finite) degree of polymerization, the range of applicability of the Landau expansion becomes smaller as the fourth-order term becomes larger. To resolve this issue, we consider an expansion of the osmotic pressure in powers of the volume fraction, φ, as this expansion is well defined in the Theta-point limit.

The osmotic pressure (expressed in a reduced form as $\hat{\Pi} = \Pi/\rho k_B T$) is related to the Gibbs energy, Eq. (5.16), through a Legendre transform, $\hat{\Pi} = \varphi\hat{\mu}_{ms} - \hat{G}$, and consequently, is expressed through the difference of the chemical potential of the pure solvent, $\hat{\mu}_s^\circ$, and its value in solution, $\hat{\mu}_s$, as $\hat{\Pi} = \hat{\mu}_s^\circ - \hat{\mu}_s$, see Section 2.8.1. The osmotic pressure is given through the expansion of Eq. (5.20) as

$$\hat{\Pi} = \ln(1-\varphi) + \varphi + \frac{\Theta}{2T}\varphi^2 \tag{5.31}$$

where the Theta temperature is related to the Flory interaction parameter by $\Theta = 2\varpi/k_B$. The osmotic pressure can be represented as a virial expansion, expressed in powers of the volume fraction near the Theta point ($\varphi_c = 0$) up to the third order in φ, see Eq. (2.124), yielding

$$\hat{\Pi} = \frac{\Delta\mu_s}{k_B T} = \frac{\varphi}{\mathcal{N}} + B\varphi^2 + C\varphi^3 \tag{5.32}$$

where the second and third virial coefficients are given by

$$B = \frac{1}{2}\left(1 - \frac{\Theta}{T}\right) \quad \text{and} \quad C = \frac{1}{3} \tag{5.33}$$

From the virial expansion, Eq. (5.32), one can define the difference between good, poor, and Theta solvents. Since the second virial coefficient is zero at the Theta point, a Theta solvent is defined where $B = 0$. When $B < 0$ (i.e. where $T < \Theta$), then the attractive part of the virial coefficient dominates the system and mixing of the two phases will be unfavorable – this defines a poor solvent. When $B > 0$ (i.e. where $T > \Theta$), mutual mixing becomes favorable and the solvent is defined as a good solvent.

The coexistence curve between the pure-solvent phase and the polymer-rich phase in the limit of an infinitely long polymer chain has an angle-like shape, which bounds all coexistence curves for shorter chains (see Figure 5.3), and is determined by equating the chemical potentials in each phase of the solvent and monomers, $\mu_s = \mu_s^\circ$ and $\mu_m = \mu_p/\mathcal{N} = 0$. From this condition, $\hat{\Pi} = 0$ and the virial expansion Eq. (5.32) gives two solutions along coexistence,

$$\varphi = 0 \quad \text{and} \quad \varphi = \frac{3}{2}\left(\frac{\Theta - T}{T}\right) \tag{5.34}$$

The first solution corresponds to the pure solvent phase, while the second solution corresponds to the temperature-dependent volume fraction of a polymer in the polymer-rich phase.

5.5 Widom's Crossover Between Meanfield Criticality and Theta-Point Tricriticality

As first demonstrated by Widom[2] [13], the phase coexistence in the Flory–Huggins theory may be represented in a universal way by rescaling the order parameter, $\phi = \varphi - \varphi_c$, and the reduced distance to the critical temperature, $\Delta\hat{T} = (T - T_c)/T_c$, by a factor of $\sqrt{\mathcal{N}}$. One can define the "Widom variable" as

$$\tilde{x} = \sqrt{\mathcal{N}}|\Delta\hat{T}| \tag{5.35}$$

Correspondingly, one can define the rescaled spontaneous order parameter as $\hat{\phi}_0 = (\varphi - \varphi_c)/\varphi_c$, such that the phase coexistence adopts a universal form given for the following limits:

$$\hat{\phi}_0 \simeq \sqrt{\mathcal{N}}\phi_0 = \begin{cases} \pm(3\tilde{x})^{1/2} & (\tilde{x} \to 0), \\ \frac{3}{2}\tilde{x} & (\tilde{x} \to \infty, \text{polymer-rich phase}), \\ -1 & (\tilde{x} \to \infty, \text{solvent-rich phase}) \end{cases} \tag{5.36}$$

where φ_c, in the limit of large $\mathcal{N}$, is $\varphi_c \simeq 1/\sqrt{\mathcal{N}}$ as given by Eq. (5.24). This universal behavior is illustrated in Figure 5.4.

Widom demonstrated that there are two key limits of such scaled behavior. First, asymptotically close to the critical point, where $\Delta\hat{T} \ll 1/\sqrt{N}$ and $\tilde{x} \ll 1$, the order parameter along coexistence is $\hat{\phi}_0 = \sqrt{3\tilde{x}}$, such that the coexistence curve adopts a parabolic shape. This result also corresponds to the behavior of the Landau expansion of the Flory–Huggins free energy. However, Widom's approach also elucidates the behavior of the polymer solution in the alternative limit, where $\Delta\hat{T} \gg 1/\sqrt{N}$ and $\tilde{x} \gg 1$. In this limit, the rescaled order parameter along phase coexistence adopts two solutions: for the polymer-rich phase, $\hat{\phi}_0 = (3/2)\tilde{x}$, while for the solvent-rich phase, $\hat{\phi}_0 = -1$, which corresponds to the "angle-like" shape of the coexistence, Eq. (5.34). Thus, Widom's scaling provides a simple approach, within the Flory–Huggins framework, to describe the behavior of the polymer solution for any value of $\mathcal{N}$.

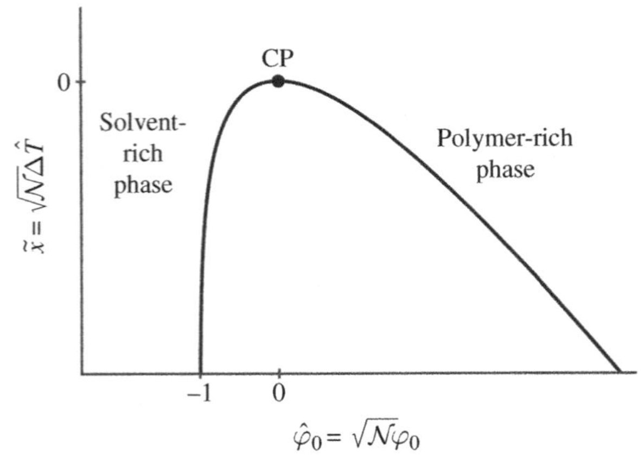

Figure 5.4 The universal phase diagram based on Flory–Huggins theory, as suggested by Widom [13], obtained by rescaling the volume-fraction order parameter, ϕ, and the reduced distance to the critical temperature, $\Delta\hat{T}$, by the square root of the degree of polymerization.

2 Benjamin Widom (born in 1927) is one of the founders of the modern theory of critical phenomena. He made significant contributions to the understanding of the statistical mechanics of fluids, the science of fluid interfaces, and the physics of polymer solutions.

As discussed in Section 4.9, the actual behavior of the phase coexistence asymptotically close to the critical point is affected by critical fluctuations, and cannot be quantitatively described through meanfield theories. The region where the meanfield approximation becomes insufficient corresponds to the limit $\tilde{x} = \sqrt{\mathcal{N}}\Delta\hat{T} \ll 1$ in Widom's scaling. This region can be defined through the Ginzburg criterion, $|\Delta\hat{T}| \ll 1/N_G$, where N_G is the Ginzburg number. The Ginzburg number can be estimated through Eq. (4.77), where for polymer solutions, $a_0 = 1$ and $u_0 = 2\sqrt{N}$, as given in Eq. (5.26), $\bar{r}^3$ is the volume of the monomer, and $\xi_0 \approx \bar{r}N^{1/4}$ is the amplitude of the correlation length, as given by Eq. (5.30), such that

$$N_G \approx \frac{c}{\sqrt{\mathcal{N}}} \tag{5.37}$$

where c is a system-dependent constant of the order of unity. The Ginzburg number indicates the crossover temperature from meanfield behavior away from the critical point to the fluctuation-induced behavior in the asymptotic vicinity of the critical point. It is not accidental that at $\tilde{x} = \sqrt{N}|\Delta\hat{T}| \approx 1$, the Widom crossover indicates the temperature of the crossover from Theta-point behavior to asymptotic critical behavior (either meanfield or fluctuation-induced) in the vicinity of the critical demixing point. In Chapter 8, we will consider the fluctuation-induced critical behavior in more detail.

5.6 Theta Point as a Tricritical Point

In this section, we present thermodynamic arguments to describe the Theta point as a tricritical point. There is a remarkable analogy between the demixing of polymer solutions at an infinite degree of polymerization and a tricritical phase transition, observed in a variety of systems [14] like He^3–He^4 mixtures and some other systems with coupled order parameters (see Appendix 4.A). We illustrate this analogy in Figure 5.5 by comparing the phase behavior in solutions of polystyrene in cyclohexane with a ferromagnet containing a nonmagnetic impurity.

We begin by introducing an appropriate order parameter to describe the Theta-point thermodynamics. De Gennes [1] defined the order parameter characterizing the long polymer chain as a vector, $\boldsymbol{\psi}$, associated with the statistical behavior of the polymer-chain ends. Phenomenologically, if each polymer chain in solution is characterized by an end-to-end vector, as discussed in Section 5.2, then the behavior of the beginning and end of each chain could reflect the symmetry of the polymer. For instance, in the limit $\mathcal{N} \to \infty$ and for temperatures in the good-solvent regime ($T > \Theta$), the probability of the two chain ends meeting is zero. In the same limit, the polymer volume fraction in solution is zero. However, for $T < \Theta$, the polymer chain collapses, expelling the solvent, and, formally, the probability of the two chain ends meeting becomes nonzero. De Gennes proved by integrating over the length of the polymer chain that the square of this probability is equivalent to the polymer volume fraction, $|\psi|^2 = \varphi$ [1]. Therefore, for $T < \Theta$, the polymer-chain order parameter emerges, corresponding to the separation of the system into two phases: one containing the solvent ($|\psi|^2 = \varphi = 0$) and another being the polymer-rich phase ($|\psi|^2 = \varphi > 0$).

Assuming that there is a polymer-chain free energy, $\hat{\Phi}$ (reduced by Nk_BT), that could be expanded in powers of the polymer-chain order parameter, one could express the expansion truncated at the sixth-order term as

$$\hat{\Phi} = \frac{1}{2!}h_2|\psi|^2 + \frac{1}{4!}u_\psi|\psi|^4 + \frac{1}{6!}g_\psi|\psi|^6 \tag{5.38}$$

where the coefficient g_ψ is independent of T and φ, while u_ψ vanishes at the Theta point, defining this point as a tricritical point, see discussed in Section 4.4.4. From Eq. (5.38), one can define the

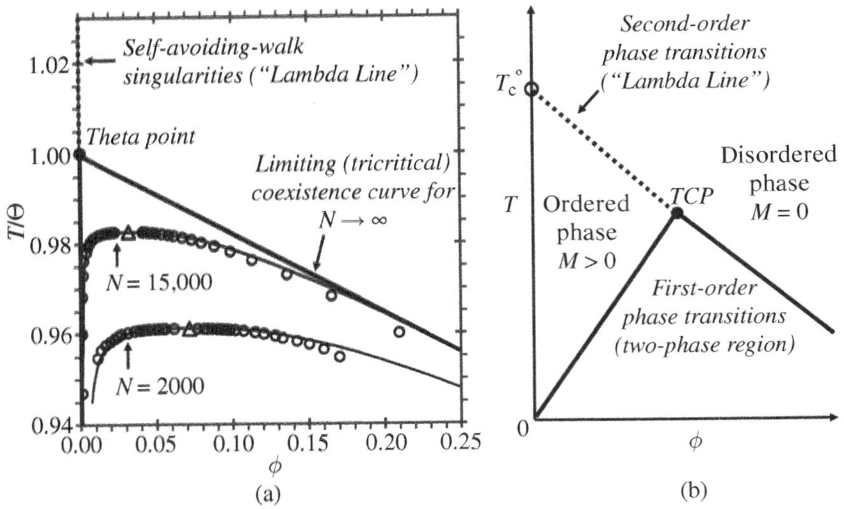

Figure 5.5 Comparison between the Theta-point phase behavior of a polymer solution in the limit of an infinite degree of polymerization ($\mathcal{N}$) and a ferromagnet containing a nonmagnetic impurity in zero magnetic field. (a) Phase behavior of polymer solutions of polystyrene in cyclohexane. As the polymer chain becomes longer, the curved shape of the liquid–liquid coexistence transforms into the "angle-like" shape near the Theta point, where the two branches of the phase coexistence becomes pure solvent and the phase containing the polymer (volume fraction, $\varphi > 0$) [15–19]. The open circles are experimental data, [15–17] the triangles are the critical points of demixing, and the solid black curves are the theoretical predictions [18]. (b) Schematic phase diagram for a ferromagnet (with magnetization, M) containing a nonmagnetic impurity (with concentration, φ). T_c° (open circle) is the Curie temperature of a pure ferromagnet. The tricritical point (TCP) separates the line of second-order phase transitions from the first-order phase transitions, accompanied by phase separation.

"ordering" field $\boldsymbol{h}_1$, conjugate to the order parameter, as $\boldsymbol{h}_1 = \partial\hat{\Phi}/\partial\boldsymbol{\psi}$, which, like the order parameter, is also a vector. The second field, h_2, conjugate to $|\psi|^2 = \varphi$, is expected to play a similar role as the reduced distance to the critical temperature, $\Delta\hat{T}$, in Eq. (5.25).

To connect the expansion coefficients (u_ψ and g_ψ) with physical properties (e.g. φ, T, and $\mathcal{N}$), one must relate the polymer-chain free energy, $\hat{\Phi}$, to the reduced osmotic pressure, $\hat{\Pi}$. This can be done through a Legendre transform of the form, $\hat{\Pi}(T, \boldsymbol{h}_1) = \boldsymbol{\psi} \cdot \boldsymbol{h}_1 - \hat{\Phi}(T, |\psi|)$, which is analogous to the Legendre transform between the osmotic pressure and the Gibbs energy in the Flory–Huggins theory, see the discussion around Eq. (5.31). Performing this Legendre transform, with $\hat{\Phi}$ given by Eq. (5.38), one finds

$$\hat{\Pi} = \frac{1}{2!}h_2|\psi|^2 + \frac{3}{4}u_\psi|\psi|^4 + \frac{5}{6}g_\psi|\psi|^6 \tag{5.39}$$

By definition, Eq. (5.39) is the reduced osmotic pressure, such that the coefficients of the Landau expansion of the polymer-chain free energy in Eq. (5.39) must be related to the coefficients of the virial expansion in Eq. (5.32). Substituting $\varphi = |\psi|^2$ in Eq. (5.32) and matching the corresponding terms, one obtains the following relations. From the coefficients of the $|\psi|^2$ term,

$$h_2 = \frac{2}{\mathcal{N}} - 4B|\psi|^2 - 3C|\psi|^4 \tag{5.40}$$

while from the coefficients of the $|\psi|^4$ and $|\psi|^6$ terms

$$u_\psi = 4B \quad \text{and} \quad g_\psi = 3C \tag{5.41}$$

With use of Eqs. (5.40) and (5.41), the expansion of the polymer-chain free energy can be expressed through the virial expansion coefficients as

$$\hat{\Phi} = -\frac{1}{\mathcal{N}}|\psi|^2 + B|\psi|^4 + C|\psi|^6 \tag{5.42}$$

This expression describes the behavior of the polymer solution in the vicinity of the Theta point. Consequently, the Theta point may be viewed from two perspectives: as in traditional physical chemistry of polymer solutions (based on the Flory–Huggins theory), in which the first two terms in the virial expansion of the osmotic pressure are zero (Section 5.4.2), and in the physics of tricriticality, in which the first two terms of the Landau expansion of the free energy are zero (Section 4.4.4).

One can observe that the polymer-chain free energy, Eq. (5.42), contains all of the features present in the Landau theory of tricritical phase transitions [20] (Section 4.4.4). Phenomenologically, a tricritical point separates lines of second-order phase transitions from first-order phase transitions, as shown in Figure 5.5. In the Landau theory, a tricritical point exists where both the coefficients of the second- and fourth-order terms are zero. The renormalization of the fourth-order term may happen due to the coupling between order parameters (a scalar, ϕ, and a vector, ψ), described by a coupling term in the Landau expansion, in the form of $\lambda\phi|\psi|^2$ (see Appendix 4.A). This coupling term originally exists in the Flory–Huggins Gibbs energy, Eq. (5.16), as $\varpi\varphi^2 = \varpi\varphi|\psi|^2$. Therefore, the Flory–Huggins interaction parameter, ϖ, plays the same role as the coupling constant λ in the theory of tricriticality. The renormalization of the fourth-order term in the polymer-chain free energy, Eq. (5.42), occurs upon approach to the Theta temperature as $B = 1/2 - \varpi = 1/2(1 - \Theta/T)$, in accordance with the Flory–Huggins theory.

As shown in Eq. (5.42), this occurs in polymer solutions when $\mathcal{N} \to \infty$ and $T \to \Theta$, which indicates that the Theta point is a tricritical point. In polymer solutions, states above the Theta temperature, along the line of zero polymer volume fraction $\varphi = 0$ correspond to certain kinds of "critical" states, characterized as self-avoiding-walk singularities [1, 14]. While no real second-order phase transition occurs for these states, the line (along $\varphi = 0$ with $\mathcal{N} \to \infty$) is analogous to the critical (*lambda*) line in other systems with tricritical phase transitions, such as in mixtures of He^3–He^4 [21, 22]. For temperatures below the Theta temperature, the system is phase separated, and thus, can be regarded as a first-order phase transition. In Eq. (5.42), this behavior is observed depending on the sign of the coefficient of the fourth-order term, proportional to B. When $B > 0$, there is no phase separation, while when $B < 0$ the phase transition is first-order.

The value for the spontaneous polymer-chain order parameter, ψ_0, which lies along coexistence for $T < \Theta$ and along the critical volume fraction ($\varphi = 0$) for $T > \Theta$, is given from the minimization of the polymer-chain free energy, $\partial\hat{\Phi}/\partial\psi = 0$, as

$$|\psi_0|^2 = \begin{cases} 0 & \text{for } T > \Theta \\ -\dfrac{B}{2C} \pm \sqrt{\left(\dfrac{B}{2C}\right)^2 + \dfrac{1}{\mathcal{N}}} & \text{for } T \leq \Theta \end{cases} \tag{5.43}$$

Thus, the "angle-like" behavior of the coexistence curve is described in the limit $\mathcal{N} \to \infty$ (for $T \leq \Theta$) as

$$|\psi_0(\mathcal{N} \to \infty)|^2 = \phi_0 = \frac{3}{2}\left(\frac{\Theta - T}{T}\right) \tag{5.44}$$

We note that the same result for the phase behavior is predicted from Flory–Huggins theory, Eq. (5.32), when $\mathcal{N} \to \infty$. Similarly, in the limit $T \to \Theta$, the spontaneous polymer-chain order parameter is given by

$$|\psi_0(T \to \Theta)| = 1/\mathcal{N}^{1/4} = h_2^{1/4} \tag{5.45}$$

The fact that, when $B = 0$, the spontaneous value of the order parameter scales as $h_2^{1/4}$, is another feature common to tricritical phase transitions, see Eq. (4.52).

When the degree of polymerization is large, but still finite, the behavior of the polymer chain is affected by the degree of polymerization. From the polymer-chain free energy, the ordering field for the polymer chain, $h_1(\boldsymbol{r})$, is given by

$$\boldsymbol{h}_1 = \left.\frac{\partial \hat{\Phi}}{\partial \boldsymbol{\psi}}\right|_{h_2} = \boldsymbol{\psi}\left[h_2 + 4B|\psi|^2 + 3C|\psi|^4\right] = \frac{2}{\mathcal{N}}\boldsymbol{\psi} \tag{5.46}$$

The response of the polymer chain to the ordering field is given by the polymer-chain susceptibility defined as, $\hat{\chi}_\psi = (\partial^2\hat{\Phi}/\partial\boldsymbol{\psi}^2)^{-1}$. Along the spontaneous value of the order parameter $|\psi_0|$, and in the limit $\mathcal{N} \gg B$, the inverse susceptibility is given by

$$\hat{\chi}_\psi^{-1} \propto \frac{1}{\mathcal{N}} + B^2 \tag{5.47}$$

The susceptibility scales with the degree of polymerization at the Theta temperature, as $\hat{\chi}_\psi(T \to \Theta) \propto \mathcal{N}$, while in limit $\mathcal{N} \to \infty$, the susceptibility is inversely proportional to the square of the distance to the Theta point, as $\hat{\chi}_\psi(\mathcal{N} \to \infty) \propto B^{-2}$. Moreover, the behavior of the correlation length of fluctuations of the polymer-chain order parameter ξ_ψ, which we refer to as the "de Gennes correlation length," in the vicinity of the Theta point can be determined through a Landau–Ginzburg functional of the form,

$$\Delta\hat{\Phi} = \int \left(\frac{1}{2}\hat{\chi}_\psi^{-1}|\psi|^2 + \frac{1}{2}c_\psi|\nabla\psi|^2\right) dV_t \tag{5.48}$$

which is similar to Eq. (4.71), such that

$$\boxed{\xi_\psi = \sqrt{c_\psi \hat{\chi}_\psi} \propto \frac{\sqrt{\mathcal{N}}}{1 + \sqrt{\mathcal{N}}B}} \tag{5.49}$$

where it is important to note that c_ψ is independent of $\mathcal{N}$, only depending on the size of the monomer. Equation (5.49) has two key limits: at the Theta temperature $\xi_\psi(T \to \Theta) \propto \sqrt{\mathcal{N}}$ and for an infinite degree of polymerization, the de Gennes correlation length is $\xi_\psi(\mathcal{N} \to \infty) \propto B^{-1} \sim (\Theta - T)^{-1}$.

A remarkable feature of tricriticality is the fact that in three dimensions, the system near tricritical points demonstrates essentially meanfield behavior. As discussed in Section 4.10, for Ising-like systems, the marginal dimension of space that separates meanfield criticality and fluctuation-induced criticality is $d = 4$. For tricritical systems, the marginal dimension is $d = 3$ [23]. Therefore, in reality, the crossover between criticality and Theta-point tricriticality in $3d$ systems is essentially a crossover between the fluctuation-induced critical behavior and meanfield tricritical behavior (as will be discussed in Chapter 8).

5.7 Self-Avoiding Walk of a Polymer Chain

In Figure 5.6, for temperatures $T > \Theta$, the line $\varphi = 0$, according to the theory of Theta-point tricriticality, is a line of specific critical states of the infinite polymer chain in a good solvent. Unlike the

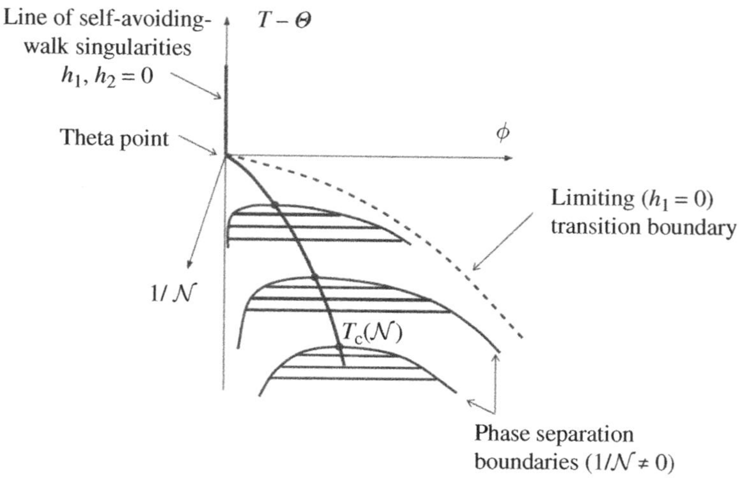

Figure 5.6 Polymer-chain scaling fields and phase separation of polymer solution in the vicinity of the Theta-point (tricritical point). For temperatures $T > \Theta$, the self-avoiding-walk singularity line ("lambda line") for a polymer solution in the limit $\mathcal{N} \to \infty$ is analogous to a line of second-order phase transitions ($h_1 = h_2 = 0$). For temperatures $T < \Theta$, the lambda line transforms into a first-order phase transition line. The line splits into two branches: the polymer branch (dashed curve) and pure solvent (vertical line), where the critical line of demixing at finite $\mathcal{N}$, is shown by the thick solid curve. Source: Reproduced with permission from Ref. [20], © American Institute of Physics (2005).

random-walk behavior of the polymer chain in a Theta solvent (Section 5.2), a polymer chain in a good solvent demonstrates self-avoiding-walk behavior, and, in the limit $\mathcal{N} \to \infty$, this self-avoiding walk exhibits singularities analogous to critical behavior.

To understand this behavior, consider an individual polymer chain in solution. When the monomers of the polymer chain and the solvent molecules repel each other ($B > 0$, a good solvent), the polymer chain swells. This means that the radius of gyration is increasing with degree of polymerization faster than in ideal Theta-solvent conditions. Therefore, one would expect that due to the attraction between solvent molecules and monomers, the radius of gyration scales with the degree of polymerization as $R_g = a\mathcal{N}^{\nu}$, where $\nu > 1/2$. To prove this, one may consider the volume of space that a swollen chain occupies in comparison to an ideal chain. This additional volume is known as the "excluded volume," (see Figure 5.7), and may be obtained by expressing the osmotic pressure, given by Eq. (5.32), through the concentration (partial density), $c = \varphi/\upsilon_0$, as

$$\frac{\Pi}{k_B T} = \frac{c}{\mathcal{N}} + \upsilon_0 B c^2 + \cdots \tag{5.50}$$

where υ_0 is the volume of the monomer. In this form, the excess osmotic pressure with respect to the ideal osmotic pressure, $\Pi_{id} = \Pi(B = 0)$, is

$$\frac{\Delta\Pi}{k_B T} = \frac{\Pi - \Pi_{id}}{k_B T} = \upsilon_0 B c^2 \tag{5.51}$$

Thus, the product $\upsilon_0 B \equiv \upsilon$ defines the excluded volume, υ. The impact of the excluded volume is that adjacent monomers will not overlap with one another and, more generally, any two monomers in the chain repel each other. This restriction on the movement of the monomers within the chain produces the self-avoiding-walk behavior.

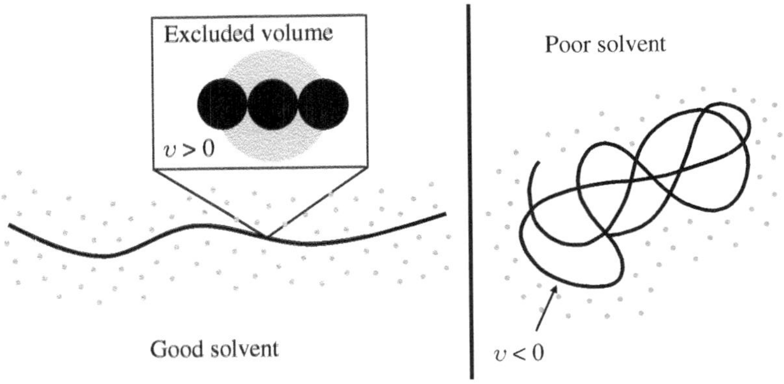

Figure 5.7 Alternative behaviors of a polymer chain in different solvents. In a good solvent, the polymer chain swells, while in a poor solvent, the polymer chain collapses. In a good solvent, the excluded volume, v (eight times the monomer volume, illustrated in gray in the inset), is the space that is inaccessible to a neighboring monomer (black). The larger the excluded volume, the more swollen the polymer chain. The swollen chain exhibits self-avoiding walk behavior. In a poor solvent, the excluded volume is $v < 0$ as the attraction between monomers is stronger than the monomer-solvent attraction. In a Theta-solvent, $v = 0$ and the polymer exhibits ideal random-walk behavior.

The repulsive force between the monomers modifies the Gibbs energy for a single polymer chain, which may be expressed as a sum of the attractive spring-like restoring force that originates from the increase in the entropy when the chain shrinks (Section 5.2), and the repulsive force associated with the excluded volume. The repulsive part of the Gibbs energy, ΔG_{rep}, may be expressed through the excess osmotic pressure and the magnitude of the end-to-end vector, R, as

$$\frac{\Delta G_{\text{rep}}}{k_B T} \approx \frac{\Delta \Pi R^d}{k_B T} \approx v_0 B c^2 R^d \tag{5.52}$$

where d is the dimensionality of space, the concentration of monomers in the chain is $c \approx \mathcal{N}/R^d$, and the effective volume of the chain is $\sim R^d$. Using the expression for the second virial coefficient, given by Eq. (5.33), one may express Eq. (5.52) as

$$\frac{\Delta G_{\text{rep}}}{k_B T} \approx \frac{1}{2} v_0 \frac{\mathcal{N}^2}{R^d}\left(1 - \frac{\Theta}{T}\right) \approx \frac{1}{2} v_0 \frac{\mathcal{N}^2}{R^d} \tag{5.53}$$

For $T \gg \Theta$, $B \approx 1/2$. With use of Eq. (5.13), in which the attractive contribution to the Gibbs energy, $\Delta G_{\text{att}} = \Delta G$, and Eq. (5.53) for the repulsive contribution, the total Gibbs energy, ΔG_{tot}, for a flexible polymer chain in a good solvent is given by

$$\frac{\Delta G_{\text{tot}}}{k_B T} = \frac{\Delta G_{\text{att}} + \Delta G_{\text{rep}}}{k_B T} = \frac{3}{2} \frac{R^2}{\mathcal{N} a^2} + \frac{1}{2} v_0 \frac{\mathcal{N}^2}{R^d} \tag{5.54}$$

The equilibrium value of the average distance between the ends of the polymer chain (radius of gyration) is thus given by the minimization of Eq. (5.54) with respect to R, as

$$R_g = a\mathcal{N}^{3/(d+2)} = a\mathcal{N}^{\nu} \tag{5.55}$$

where the volume of the monomer is given by $v_0 \simeq a^d$. The parameter ν is referred to as the "Flory exponent," and it determines how the radius of gyration increases with chain size [3]. For example, in three dimensions $\nu = 3/5$, which is in agreement with scattering experiments [24]. The behavior of the radius of gyration in a good, Theta, and poor solvent is presented in the table below.

Relations Between the Radius of Gyration and the Degree of Polymerization in $d = 3$ Polymer Systems

1) Good solvent ($B > 0$) – a swollen chain, self-avoiding walk,

$$R_g \propto \mathcal{N}^{3/5} \tag{5.56}$$

2) Theta solvent ($B = 0$) – an ideal chain, random walk,

$$R_g \propto \mathcal{N}^{1/2} \tag{5.57}$$

3) Poor solvent ($B < 0$) – a collapsed chain, globule,

$$R_g \propto \mathcal{N}^{1/3} \tag{5.58}$$

5.8 Phase Separation in Polymer Blends

The treatment of the phase behavior in polymer solutions may be generalized to a mixture of two incompressible polymer species, A and B, with degrees of polymerization $\mathcal{N}_A$ and $\mathcal{N}_B$, respectively [3]. Modifying the Flory–Huggins theory to account for the lengths of each species, the Gibbs energy of mixing of the polymer blend, $\Delta\hat{G}_{pb}$ (reduced by Nk_BT), is given by [5–7]

$$\Delta\hat{G}_{pb} = \left[\frac{\varphi}{\mathcal{N}_A}\ln\varphi + \frac{(1-\varphi)}{\mathcal{N}_B}\ln(1-\varphi)\right] + \varpi\varphi(1-\varphi) \tag{5.59}$$

where the Flory interaction parameter is approximated as $\varpi = \Theta/2T$, see Eq. (5.16). Equation (5.59) has the following two limits: first, in the case when one species is much longer than the other, $\mathcal{N}_A \equiv \mathcal{N} \gg \mathcal{N}_B$ and $\mathcal{N}_B \approx 1$, then Eq. (5.59) reduces to the Gibbs energy of mixing in the Flory–Huggins theory, Eq. (5.16). Hereafter, we refer to this case as the asymmetric limit. Second, in the case when each species is of the same size, $\mathcal{N}_A = \mathcal{N}_B = \mathcal{N}$, then the contribution from the entropy of mixing, the term in brackets in Eq. (5.59), scales with $1/\mathcal{N}$. We refer to this case as the symmetric limit.

The critical volume fraction and critical temperature are determined from the thermodynamic stability conditions, $(\partial^2\hat{G}_{pb}/\partial\varphi^2)|_{T,\mathcal{N}_A,\mathcal{N}_B} = 0$ and $(\partial^3\hat{G}_{pb}/\partial\varphi^3)|_{T,\mathcal{N}_A,\mathcal{N}_B} = 0$, yielding

$$\varphi_c = \frac{\sqrt{\mathcal{N}_B}}{\sqrt{\mathcal{N}_A} + \sqrt{\mathcal{N}_B}} \quad \text{and} \quad T_c = \Theta\left[\frac{1}{\sqrt{\mathcal{N}_A}} + \frac{1}{\sqrt{\mathcal{N}_B}}\right]^{-2} \tag{5.60}$$

In the asymmetric limit, the critical parameters given by Eq. (5.60), reduce to those given by Eq. (5.24). In the symmetric limit, $\varphi_c = 1/2$ and $T_c = \Theta\mathcal{N}/2$, and, when both polymer chains are sufficiently long, such that $\mathcal{N} \gg 1$, then $T_c \to \infty$, and the polymer blend would not mix for any value of the Flory interaction parameter, ϖ.

Just like that given by Eq. (5.25), the truncated Landau expansion may be represented in the form

$$\Delta\hat{G}_{pb} = \frac{1}{2!}\tilde{a}_0\Delta\hat{T}\phi^2 + \frac{1}{4!}\tilde{u}_0\phi^4 \tag{5.61}$$

where the expansion coefficients for the blend are given by

$$\tilde{a}_0 = \left(\frac{1}{\sqrt{\mathcal{N}_A}} + \frac{1}{\sqrt{\mathcal{N}_B}}\right)^2 \quad \text{and} \quad \tilde{u}_0 = \sqrt{\mathcal{N}_A}\sqrt{\mathcal{N}_B}\left(\frac{1}{\sqrt{\mathcal{N}_A}} + \frac{1}{\sqrt{\mathcal{N}_B}}\right)^4 \tag{5.62}$$

The Landau expansion of the Flory–Huggins Gibbs free energy, Eq. (5.25), is recovered in the asymmetric limit. In the symmetric limit, the second- and fourth-order coefficients of Eq. (5.62) are given by $\tilde{a}_0 = 4/\mathcal{N}$ and $\tilde{u}_0 = 32/\mathcal{N}$.

The osmotic pressure for the polymer blend is related to the Gibbs energy through a Legendre transform, $\hat{\Pi}_{pb} = \varphi(\partial\hat{G}_{pb}/\partial\varphi)|_{T,\mathcal{N}_A,\mathcal{N}_B} - \hat{G}_{pb}$, and may be represented as a virial expansion around the Theta point as

$$\hat{\Pi}_{pb} = \frac{\varphi}{\mathcal{N}_A} + \frac{1}{2}\left(\frac{1}{\mathcal{N}_B} - \frac{\Theta}{T}\right)\varphi^2 + \frac{1}{3\mathcal{N}_B}\varphi^3 \tag{5.63}$$

Comparing with the asymmetric limit for the osmotic pressure, Eq. (5.32), one can find that the second and third virial coefficients depend on $\mathcal{N}_B$ as

$$B = \frac{1}{2}\left(\frac{1}{\mathcal{N}_B} - \frac{\Theta}{T}\right) \quad \text{and} \quad C = \frac{1}{3\mathcal{N}_B} \tag{5.64}$$

This dependence on $\mathcal{N}_B$ has important implications for the polymer-chain free energy, Eq. (5.42). Following the derivation in Section 5.6, the polymer-chain free energy of the blend is given by

$$\hat{\Phi}_{pb} = -\frac{1}{\mathcal{N}_A}|\psi|^2 + \frac{1}{2}\left(\frac{1}{\mathcal{N}_B} - \frac{\Theta}{T}\right)|\psi|^4 + \frac{1}{3\mathcal{N}_B}|\psi|^6 \tag{5.65}$$

The condition for Theta-point tricriticality is only observed in the asymmetric limit where $\mathcal{N}_A \gg \mathcal{N}_B \to \infty$. However, in the symmetric limit, the polymer-chain free energy becomes undefined at the Theta point as $T \to \Theta$ and $\mathcal{N} \to \infty$, while $\psi(\boldsymbol{r}) \to 0$, implying that there is no Theta-point tricriticality in a symmetric blend.

The size of the critical region is determined through the Ginzburg criterion, Eq. (5.37), which scales with the degree of polymerization of each species as

$$N_G \sim \frac{1}{\sqrt{\mathcal{N}_A\mathcal{N}_B}} \tag{5.66}$$

Equation (5.66) has the following two limits: in the asymmetric limit, $N_G \sim 1/\sqrt{\mathcal{N}}$, just as predicted by Eq. (5.37), while in the symmetric limit, $N_G \sim 1/\mathcal{N}$. Thus, a symmetric blend, with a large degree of polymerization, has a much smaller critical region than a polymer blend in the asymmetric limit. This result means that the polymer blend would essentially exhibit meanfield behavior, even if the interactions between the monomers are short-ranged. As the polymer blend becomes asymmetric, a tricritical (Theta-point) region would emerge as discussed in Section 5.6.

5.9 Diffusion of Polymer Chains in Solution

In this section, we consider the consequences of the meanfield approach to polymer solutions and polymer blends on the diffusion dynamics. There are two distinct cases for the behavior of the diffusion coefficient. In polymer solutions, the Brownian diffusion of a dilute solution of polymer coils in a low-molecular-weight solvent is described through the Stokes–Einstein equation, given by Eq. (2.184). As shown by Zimm [25], the hydrodynamic radius of the polymer coil, R_h, is proportional to the radius of gyration, such that

$$D \propto \frac{k_B T}{\eta R_g} \tag{5.67}$$

In a Theta solvent, $D \propto \mathcal{N}^{-1/2}$, in a good solvent $D \propto \mathcal{N}^{-3/5}$, and in a poor solvent $D \propto \mathcal{N}^{-1/3}$.

In the region where the Widom variable is $\tilde{x} \ll 1$, the mutual diffusion of a polymer in solution for a large, but finite, degree of polymerization may be considered via the meanfield approach, as given by Eq. (4.86). With use of this equation and Eq. (5.30), for the correlation length in polymer solutions, one finds that the diffusion coefficient asymptotically vanishes at the critical point in the meanfield approximation as $D \propto |\Delta\hat{T}|$.

The prediction for the mutual diffusion coefficient in phase separating polymer blends is quite different. In a symmetric polymer blend, near the critical point of demixing the diffusion coefficient is given through Eq. (4.85) where $\chi^{-1} = \partial^2(\Delta\hat{G}_{pb})/\partial\varphi^2|_T$, and in accordance with Eq. (5.61) for the Landau expansion of the free energy for a polymer blend and Eq. (4.17) for the susceptibility (taking into account that $1/\Gamma_0^+ = \tilde{a}_0 = 4/\mathcal{N}$), the diffusion coefficient scales as $D \propto |\Delta\hat{T}|/\mathcal{N}$, such that $D \propto 1/\mathcal{N}$ as predicted by Rouse [26]. We note that in both polymer solutions and polymer blends, the mutual diffusion (and, correspondingly, the mutual mixing) is severely suppressed for large degrees of polymerization.

As will be discussed in Chapter 8, the actual behavior of the mutual diffusion in polymer solutions is significantly affected by critical fluctuations of concentration. Moreover, for very large degrees of polymerization, the critical dynamics are also affected by a coupling between the dynamics of long polymer chains and the diffusive relaxation of the concentration fluctuations.

5.10 Polymer Chain Near the Critical Point of a Binary Solvent

In addition to the collapse of polymer coils in a poor solvent and the swelling in a good solvent, Brochard and de Gennes have predicted that a polymer coil may collapse and further reswell (to its original size) near the critical point of demixing in a binary solvent [27, 28]. A simplified explanation of the effect predicted by Brochard and de Gennes is as follows. The excluded volume, υ, controls the dimensions of polymer coils in a dilute solution. In a good solvent, the excluded volume is positive, and the coil is swollen. In a poor solvent υ is negative and the coil collapses (see Section 5.7). However, near the critical point of a good binary solvent, the growing correlation length could enhance the preferential adsorption of one of the components, thereby increasing the attraction between the monomers and reducing the excluded volume. In the first-order approximation, the effective excluded volume $\tilde{\upsilon}$, proportional to the second virial coefficient in the Flory theory Eq. (5.51), reads

$$\tilde{\upsilon} = \upsilon(1 - \lambda^2\chi) \tag{5.68}$$

where λ is the strength of the preferential adsorption and $\chi \sim \xi^2 \sim (T_c/T - 1)^{-1}$ is the osmotic susceptibility of the binary solvent.

The Brochard–de Gennes effect can also be phenomenologically explained by a coupling between the polymer-chain order parameter (see Section 5.6) and the order parameter $\phi = x - x_c$ for the near-critical binary solvent. In the expansion for the osmotic pressure in powers of the polymer order parameter, ψ, a coupling term, $\lambda|\psi|^2\phi$, should be added. Like the procedure described in Appendix 4.A, minimization of the osmotic pressure with respect to ϕ could renormalized the

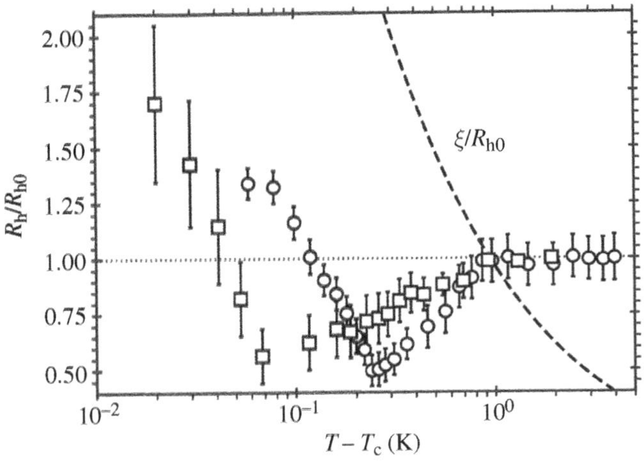

Figure 5.8 Collapse, reswelling, and expansion of polyacrylic acid (M_w = 134,000 g/mol) in a critical 2,6-lutidine + water mixture (squares) and PBMA (M_w = 180,000 g/mol) in a critical nitroethane + isooctane mixture (circles). Data for the reduced hydrodynamic radius, R_h/R_{h0}, are from Refs. [29, 30]. The dashed curve is the reduced correlation length measured independently by DLS [30]. Note that the radius of gyration is related to the hydrodynamic radius as $R_h \approx 0.7R_g$ [31].

fourth-order coefficient in the expansion (the second-order term in the expansion in powers of the polymer volume fraction) as $v[1 - \lambda^2(T_c/T - 1)^{-1}]$.

When the correlation length becomes larger than the radius of gyration, the growth of the susceptibility inside the coil should saturate because the correlation length cannot further grow inside the polymer coil. This effect of partial collapse and reswelling has been qualitatively confirmed computationally and experimentally. However, more accurate studies, especially in a close vicinity to the critical temperature, reveal a more complicated behavior of polymer chains in dilute solutions. Not only does the chain reswell when the correlation length exceeds its size, but it significantly expands beyond its original size – found far away from the critical temperature. This effect is illustrated in Figure 5.8 where the results obtained from two independent studies are presented in reduced scale. The experimental data of [29] were obtained by fluorescent correlation spectroscopy for an individual polymer coil, while the data [30] were obtained for a dilute solution of a polymer (0.4%) in isooctane–nitroethane mixture by DLS. The difference in the refractive indices of species in this mixture is so small ($\Delta n_\lambda = 0.002$) that weak critical scattering does not prevent the measurements of the hydrodynamic radius of the polymer coils.

Appendix 5.A Derivation of Landau Coefficients in the Flory–Huggins Model

Expressing the Gibbs energy of mixing from the Flory–Huggins theory, given by Eq. (5.16), in reduced form $\hat{G}_{\text{mix}} \equiv \Delta G_{\text{mix}}/k_B T_c$, as

$$\Delta\hat{G}_{\text{mix}} = \hat{T}\left[(1-\varphi)\ln(1-\varphi) + \frac{\varphi}{\mathcal{N}}\ln\varphi\right] + \frac{\varpi}{k_B T_c}\varphi(1-\varphi) \quad (5.A.1)$$

where $\hat{T} = T/\hat{T}$ is the reduced temperature, the Flory interaction parameter, ϖ, is related to the Theta temperature through, $\varpi = k_B\Theta/2$, φ is the volume fraction, and $\mathcal{N}$ is the degree of polymerization. The goal is to express Eq. (5.A.1) in the form of the Landau Expansion near the critical point as given by Eq. (4.9) [8],

$$\hat{f} = \frac{1}{2!}a_0\Delta\hat{T}\phi^2 + \frac{1}{4!}u_0\phi^4 \tag{5.A.2}$$

where the reduced distance to the critical point is $\Delta\hat{T} = \hat{T} - 1$ and the order parameter is $\phi = \varphi - \varphi_c$. The critical temperature and critical volume fraction is given by Eq. (5.24), see Section 5.3.2 for details. The reduced chemical potentials of the monomers and the solvent, Eqs. (5.20) and (5.21), are given by

$$\hat{\mu}_s = \hat{G} - \varphi\left.\frac{\partial\hat{G}}{\partial\varphi}\right|_{P,T} = \hat{T}\left[\ln(1-\varphi) + \varphi\left(1 - \frac{1}{\mathcal{N}}\right)\right] + \frac{\varpi}{k_BT_c}\varphi^2 \tag{5.A.3}$$

$$\hat{\mu}_m = \hat{G} + (1-\varphi)\left.\frac{\partial\hat{G}}{\partial\varphi}\right|_{P,T} = \hat{T}\left\{\frac{1}{\mathcal{N}}[\ln\varphi + (1-\varphi)] - (1-\varphi)\right\} + \frac{\varpi}{k_BT_c}(1-\varphi)^2 \tag{5.A.4}$$

The exchange chemical potential, $\hat{\mu}_{ms}$ is found from the first derivative of Eq. (5.A.1) as

$$\hat{\mu}_{ms} \equiv \left.\frac{\partial\hat{G}}{\partial\varphi}\right|_{T,P} = \hat{T}\left[\frac{1}{\mathcal{N}}(\ln\varphi + 1) - (\ln(1-\varphi) + 1)\right] + \frac{\varpi}{k_BT_c}(1-2\varphi) \tag{5.A.5}$$

To determine the Landau expansion of Eq. (5.A.1), one must separate the regular (background) part of the free energy, $\hat{G}_{reg}$, from the critical part of the free energy, $\hat{G}_{cr}$. This is done by expressing the free energy as a sum of $\Delta\hat{G}_{mix} = \hat{G}_{reg} + \hat{G}_{cr}$. This is possible because Landau theory assumes that the free energy is an analytical function of T and φ at the transition point. Since the regular part does not depend on the order parameter, one may express this function as

$$\hat{G}_{reg} = \hat{\mu}_s(\varphi_c) + \varphi\hat{\mu}_{ms}(\varphi_c, \varpi) \tag{5.A.6}$$

Using Eqs. (5.A.3) and (5.A.5), the regular part is determined as

$$\hat{G}_{reg} = \hat{T}\left[(1-\varphi)\ln(1-\varphi_c) + (\varphi-\varphi_c)\left(\frac{1}{\mathcal{N}} - 1\right) + \frac{\varphi}{\mathcal{N}}\ln\varphi_c\right] + \frac{\varpi}{k_BT_c}\left[\varphi_c^2 + \varphi(1-2\varphi_c)\right] \tag{5.A.7}$$

Thus, the critical part is found as $\hat{G}_{cr} = \Delta\hat{G}_{mix} - \hat{G}_{reg}$, and subtracting Eq. (5.A.7) from Eq. (5.A.1) gives

$$\hat{G}_{cr} = \hat{T}\left[(1-\varphi)\ln\left(\frac{1-\varphi}{1-\varphi_c}\right) + \frac{\varphi}{\mathcal{N}}\ln\left(\frac{\varphi}{\varphi_c}\right) + (\varphi-\varphi_c)\left(1 - \frac{1}{\mathcal{N}}\right)\right] - \frac{\varpi}{k_BT_c}(\varphi-\varphi_c)^2 \tag{5.A.8}$$

At the critical point, $\hat{G}_{reg}$ is a constant, while $\hat{G}_{cr}$ vanishes. As such, $\hat{G}_{reg}$ does not contribute to the susceptibility, $\hat{\chi} = (\partial^2\hat{G}/\partial\phi^2)^{-1}$, which diverges at the critical point (see Eq. 4.67).

Using the fact that φ_c and T_c are functions of $\mathcal{N}$ and Θ, given by Eq. (5.24), and using the definition $\varphi = \phi - \phi_c$, then Eq. (5.A.8) can be expressed in the form

$$\hat{G}_{cr} = (\Delta\hat{T} + 1)\left\{\left(\frac{\sqrt{\mathcal{N}}}{1+\sqrt{\mathcal{N}}} - \phi\right)\ln\left(1 - \phi\frac{1+\sqrt{\mathcal{N}}}{\sqrt{\mathcal{N}}}\right) + \frac{1}{\mathcal{N}}\left(\phi + \frac{1}{1+\sqrt{\mathcal{N}}}\right)\ln\left[1 + \phi(1+\sqrt{\mathcal{N}})\right]\right\}$$

$$+ \phi \left(1 - \frac{1}{\mathcal{N}}\right) - \frac{\varpi}{k_B T_c} \phi^2 \tag{5.A.9}$$

The last term, proportional to ϕ^2, can be expressed in terms of the critical temperature as the interaction parameter is related to the temperature as $\varpi = \Theta k_B/2$. Thus, using the relations between T_c and Θ, given by Eq. (5.24), this term can be expressed as

$$\frac{\varpi}{k_B T_c} \phi^2 = \frac{\Theta}{2T_c} \phi^2 = \frac{1}{2} \left(1 + \frac{1}{\sqrt{\mathcal{N}}}\right)^2 \phi^2 \tag{5.A.10}$$

Substituting Eq. (5.A.10) into Eq. (5.A.9) gives the full form of the critical part of the free energy. To determine the Landau coefficients, a_0 and u_0, one must expand this equation in the vicinity of the critical point. This expansion may be performed directly, or (alternatively), one may isolate these coefficients by first determining the susceptibility, as

$$\hat{\chi}^{-1} = \frac{\partial^2 \hat{f}}{\partial \phi^2} = a_0 \Delta \hat{T} + \frac{1}{2} u_0 \phi^2 \tag{5.A.11}$$

The susceptibility may be a useful tool to determine these coefficients as, in the second-order expansion, the temperature is uncoupled from the order parameter – thereby isolating the coefficients a_0 and u_0. Using Eqs. (5.A.10) and (5.A.9), the susceptibility, $\hat{\chi}^{-1} = \partial^2 \hat{G}_{cr}/\partial \phi^2$, is

$$\hat{\chi}^{-1} = (\Delta \hat{T} + 1) \left[\left(\frac{\sqrt{\mathcal{N}}}{1 + \sqrt{\mathcal{N}}} - \phi \right)^{-1} + \frac{1}{\mathcal{N}} \left(\phi + \frac{1}{1 + \sqrt{\mathcal{N}}} \right)^{-1} \right] - \left(1 + \frac{1}{\sqrt{N}}\right)^2 \tag{5.A.12}$$

To determine a_0, one may evaluate the susceptibility along the critical isochore, $\hat{\chi}^{-1}_{\varphi=\varphi_c}$, and compare the leading coefficient of $\Delta \hat{T}$ to that in Eq. (5.A.11). Thus, one obtains

$$\begin{aligned} \hat{\chi}^{-1}_{\varphi=\varphi_c} &= (\Delta \hat{T} + 1) \left[\frac{1 + \sqrt{\mathcal{N}}}{\sqrt{\mathcal{N}}} + \frac{1 + \sqrt{\mathcal{N}}}{\mathcal{N}} \right] - \left(1 + \frac{1}{\sqrt{N}}\right)^2 \\ &= \left(1 + \frac{1}{\sqrt{N}}\right)^2 \Delta \hat{T} \end{aligned} \tag{5.A.13}$$

such that

$$\boxed{a_0 = \left(1 + \frac{1}{\sqrt{\mathcal{N}}}\right)^2} \tag{5.A.14}$$

Evaluating the susceptibility along the critical isotherm gives

$$\hat{\chi}^{-1}_{T=T_c} = \left(\frac{\sqrt{\mathcal{N}}}{1 + \sqrt{\mathcal{N}}} - \phi \right)^{-1} + \frac{1}{\mathcal{N}} \left(\phi + \frac{1}{1 + \sqrt{\mathcal{N}}} \right)^{-1} - \left(1 + \frac{1}{\sqrt{N}}\right)^2 \tag{5.A.15}$$

To determine u_0, one may expand this equation in the vicinity of the critical point to second order in ϕ to get

$$\hat{\chi}^{-1}_{T=T_c} \approx 2 \left(\frac{1 + \sqrt{\mathcal{N}}}{\sqrt{\mathcal{N}}} \right) + \frac{(1 + \sqrt{\mathcal{N}})^4}{N^{3/2}} \phi^2 - \left(1 + \frac{1}{\sqrt{\mathcal{N}}}\right)^2 = \frac{(1 + \sqrt{\mathcal{N}})^4}{N^{3/2}} \phi^2$$

$$= \sqrt{N}\left(1 + \frac{1}{\sqrt{N}}\right)^4 \phi^2 \tag{5.A.16}$$

such that

$$\boxed{u_0 = 2\sqrt{\mathcal{N}}\left(1 + \frac{1}{\sqrt{\mathcal{N}}}\right)^4} \tag{5.A.17}$$

Using Eqs. (5.A.14) and (5.A.17) in the limit when $\mathcal{N}$ is large but finite, one can determine that the truncated Landau expansion of the Gibbs free energy has the form

$$\boxed{\hat{f} = \frac{1}{2}\Delta\hat{T}\phi^2 + \frac{1}{12}\sqrt{\mathcal{N}}\phi^4} \tag{5.A.18}$$

as used in Section 5.4.1.

References

1 P.-G. de Gennes, *Scaling Concepts in Polymer Physics.* Ithaca, NY: Cornell University Press, 1979.

2 R. A. L. Jones, *Polymers*, ch. 5, pp. 73–94. Oxford, UK: Oxford University Press, 2002.

3 M. Rubinstein and R. H. Colby, *Polymer Physics.* Oxford: Oxford University Press, 2003.

4 Nobel Prize Outreach AB, "Paul J. Flory." date accessed: Dec. 2023, https://www.nobelprize.org/prizes/chemistry/1974/flory/.

5 P. J. Flory, "Thermodynamics of high polymer solutions," *J. Chem. Phys.*, vol. 9, p. 660, 1941.

6 M. L. Huggins, "Solutions of long chain compounds," *J. Chem. Phys.*, vol. 9, p. 440, 1941.

7 P. J. Flory, *Principles of Polymer Chemistry.* Ithaca, NY: Cornell University Press, 1953.

8 A. A. Povodyrev, M. A. Anisimov, and J. V. Sengers, "Crossover Flory model for phase separation in polymer solutions," *Physica A*, vol. 264, pp. 345–369, 1999.

9 M. A. Anisimov and C. E. Bertrand, "Thermodynamics of fluids at meso and nano scales," in *Applied Thermodynamics of Fluids* (A. R. Goodwin, C. J. Peters, and J. V. Sengers, eds.), ch. 7, pp. 172–214. Cambridge, UK: Royal Society of Chemistry, 2010.

10 P.-G. de Gennes, "Critical opalescence of macromolecular solutions," *Phys. Lett.*, vol. 26A, no. 7, pp. 313–314, 1968.

11 I. C. Sanchez, "Critical amplitude scaling laws for polymer solutions," *J. Phys. Chem.*, vol. 93, no. 19, pp. 6983–6991, 1989.

12 P. Debye, "Angular dissymmetry of the critical opalescence in liquid mixtures," *J. Chem. Phys.*, vol. 31, p. 680, 1959.

13 B. Widom, "Phase separation in polymer solutions," *Physica A*, vol. 194, no. 1, pp. 532–541, 1993.

14 M. E. Fisher, "The story of coulombic criticality," *J. Stat. Phys.*, vol. 75, pp. 1–36, 1994.

15 J. Kojima, N. Kuwahara, and M. Kaneko, "Light scattering and pseudospinodal curve of the system polystyrene–cyclohexane in the critical region," *J. Chem. Phys.*, vol. 63, p. 333, 1975.

16 M. Nakata, N. Kuwahara, and M. Kaneko, "Coexistence curve for polystyrene–cyclohexane near the critical point," *J. Chem. Phys.*, vol. 62, p. 4278, 1975.

17 M. Nakata, T. Dobashi, N. Kuwahara, et al., "Coexistence curve and diameter of polystyrene in cyclohexane," *Phys. Rev. A*, vol. 18, pp. 2683–2688, 1978.

18 J. S. Hager, M. A. Anisimov, J. V. Sengers, et al., "Scaling of demixing curves and crossover from critical to tricritical behavior in polymer solutions," *J. Chem. Phys.*, vol. 117, p. 5940, 2002.
19 M. A. Anisimov, A. F. Kostko, J. V. Sengers, et al., "Competition of mesoscales and crossover to theta-point tricriticality in near-critical polymer solutions," *J. Chem. Phys.*, vol. 123, p. 164901, 2005.
20 L. D. Landau and E. M. Lifshitz, "Course of theoretical physics," in *Statistical Physics: Part 1*, vol. 5. Oxford, London: Pergamon Press, 3rd ed., 1980.
21 D. Vollhardt and P. Wölfle, *The Superfluid Phases of Helium 3*. London, UK: Taylor and Francis, 1990.
22 A. Schmitt, *Introduction to Superfluidity*, vol. 888 of Lecture Notes in Physics. Cham: Springer International Publishing, 2015.
23 I. D. Lawrie and S. Sarbach, "Theory of tricritical points," in *Phase Transitions and Critical Phenomena* (C. Domb and J. L. Lebowitz, eds.), vol. 9. New York, USA: Academic Press, 1984.
24 D. W. Schaefer and C. C. Han, "Quasielastic light scattering from dilute and semidilute polymer solutions," in *Dynamic Light Scattering: Applications of Photon Correlation Spectroscopy* (R. Pecora, ed.), pp. 181–243. Springer US, 1985.
25 B. H. Zimm, "Dynamics of polymer molecules in dilute solutions: Viscoelasticity, flow birefringence, and dielectric loss," *J. Chem. Phys.*, vol. 24, p. 269, 1956.
26 P. E. Rouse, Jr., "A theory of the linear viscoelastic properties of dilute solutions of coiling polymers," *J. Chem. Phys.*, vol. 21, p. 1272, 1953.
27 F. Brochard and P. D. Gennes, "Collapse of one polymer coil in a mixture of solvents," *Ferroelectrics*, vol. 30, no. 1, pp. 33–47, 1980.
28 C. Williams, F. Brochard, and H. L. Frisch, "Polymer collapse," *Annu. Rev. Phys. Chem.*, vol. 32, no. 1, pp. 433–451, 1981.
29 C. A. Grabowski and A. Mukhopadhyay, "Contraction and reswelling of a polymer chain near the critical point of a binary liquid mixture," *Phys. Rev. Lett.*, vol. 98, p. 207801, 2007.
30 X. Zheng, M. A. Anisimov, J. V. Sengers, et al., "Unusual transformation of polymer coils in a mixed solvent close to the critical point," *Phys. Rev. Lett.*, vol. 121, p. 207802, 2018.
31 K. L. Linegar, A. E. Adeniran, A. F. Kostko, et al., "Hydrodynamic radius of polyethylene glycol in solution obtained by dynamic light scattering," *Colloid J.*, vol. 72, pp. 279–281, 2010.

Exercises

5.1 Obtain, through the work of fluctuations, the expression for the mean-square fluctuations of size for an ideal polymer chain. Explain the physical meaning of the radius of gyration.

5.2 Explain the physical meaning of the excess entropy of mixing for polymer solutions. How does this property depends on the degree of polymerization? Find the excess entropy in the limit of an infinite degree of polymerization.

5.3 The Flory–Huggins model for a solution of a high-molecular-weight polymer in a low-molecular-weight solvent results in the expression for the Gibbs energy of mixing per monomer/solvent site given by Eq. (5.16). Obtain the expressions for the spinodal, the critical temperature, and the critical volume fraction as functions of the degree of polymerization, $\mathcal{N}$.

5.4 Calculate and plot spinodals for polymer solutions in the Flory–Huggins theory with $\mathcal{N} = 1$, 10, and 100.

5.5 Derive the expressions for the chemical potentials: μ_m (monomers) and μ_s (solvent)

5.6 Calculate and plot binodals for polymer solutions in the Flory–Huggins theory with $\mathcal{N} = 1$, 10, and 100. Plot the Gibbs energy of mixing as a function of volume fraction for $\mathcal{N} = 10$ at a temperature below T_c and make a common tangent to obtain the coexisting phases. Compare the result with the corresponding binodal.

5.7 Find the conditions for the Theta point of a polystyrene-methylcyclohexane solution with $\Theta = 344$ K. Derive the virial expansion for the osmotic pressure for this system.

5.8 Explain why the behavior of a polymer chain near the Theta point for an infinite degree of polymerization is analogous to the behavior of a fluid near a tricritical point.

5.9 For $\mathcal{N} = 10, 100$, and 1000, plot the de Gennes correlation length in the vicinity of the Theta point as a function of temperature.

5.10 For the Flory-Huggins model, formulate the conditions of liquid–liquid equilibrium and express the osmotic pressure as a function of the polymer concentration in dilute solution up to the third power. Obtain and plot the equation for the temperature dependence of liquid–liquid coexistence as a function of the volume fraction in the limit of an infinite degree of polymerization adopting $\Theta = 344$ K.

5.11 Plot the excluded volume as a function of temperature for a polystyrene–methylcyclohexane. Explain the physical meaning of the concept of excluded volume.

5.12 Explain the concept of persistence length. Search the internet for the persistence length of human DNA. Explain the difference between the length of a stretched DNA molecule and its radius of gyration.

5.13 Derive the $\mathcal{N}$ dependence of the radius of gyration above the Theta temperature for a d dimensional system. How does the radius of gyration of a polymer chain exhibiting a self-avoiding walk depend on the degree of polymerization in one- and two-dimensional space?

5.14 Derive the power-law asymptotic behavior of the order parameter along coexistence, heat capacity, susceptibility, and correlation length in the Flory–Huggins theory. Explicitly state the $\mathcal{N}$-dependence of the amplitude of each thermodynamic property.

5.15 Derive the power-law asymptotic behavior for a symmetric blend of two polymers, $\mathcal{N}_A = \mathcal{N}_B = \mathcal{N}$, for the same thermodynamic properties as Question (5.14).

6

Surfaces and Interfaces

This chapter includes:

- Thermodynamics of surface tension
- Van der Waals–Landau theory of smooth interfaces
- Wetting transitions
- Homogeneous nucleation and spinodal decomposition

One of the most important applications of the Landau–Ginzburg functional is to interfacial phenomena, the properties of the surface that separates two equilibrium phases. Van der Waals was the first to realize that the thickness of the liquid–gas interface depends on the thermodynamic conditions. In particular, upon approaching the liquid–gas critical point, the interface becomes smooth and thick. In this chapter, the Landau–Ginzburg functional is applied to describe a smooth interface between two fluid phases in the meanfield approximation.

6.1 Surface and Interfacial Tension

At the mesoscale, the interface between phases can significantly affect the behavior of the system. This is because the surface energy is typically smaller than the bulk energy, as the surface-to-volume ratio is $\sim 1/\ell$, where ℓ is the characteristic size of the system. However, for thin films and small droplets, this ratio may be significant, and the contribution of the surface energy cannot be neglected.

Consider the surface illustrated in Figure 6.1, in which the surface area is defined as $\mathcal{A}_\mathrm{t} = \ell z$, where ℓ is the length of the surface and z is the height of a surface. The surface area per molecule is $\mathcal{A} = \mathcal{A}_\mathrm{t}/N$. In general, there are two types of interfaces that may form: *sharp* ("mechanical") interfaces, in which the thickness of the surface is on the order of the molecular size (i.e. the interface between oil and water or liquid and gas far away from the critical point) and *smooth* ("fuzzy") interfaces, in which the thickness is mesoscopic. If the surface is sharp and a gravitational force, $F = -mg$, is applied along the z-direction, then the mechanical work required to alter the surface is given through the integral over the height of the interface as

$$W = \int mg\,\mathrm{d}z = \int \frac{mg}{\ell}\mathrm{d}(\ell z) = \int \sigma \mathrm{d}\mathcal{A}_\mathrm{t} \tag{6.1}$$

where the surface tension is determined as $\sigma = F/\ell = mg/\ell$. The physical origin of the interfacial tension is the attraction between molecules, which forces the surface to shrink, resisting its expansion due to gravity.

Mesoscopic Thermodynamics for Scientists and Engineers, First Edition. Mikhail A. Anisimov and Thomas J. Longo.
© 2024 John Wiley & Sons, Inc. Published 2024 by John Wiley & Sons, Inc.

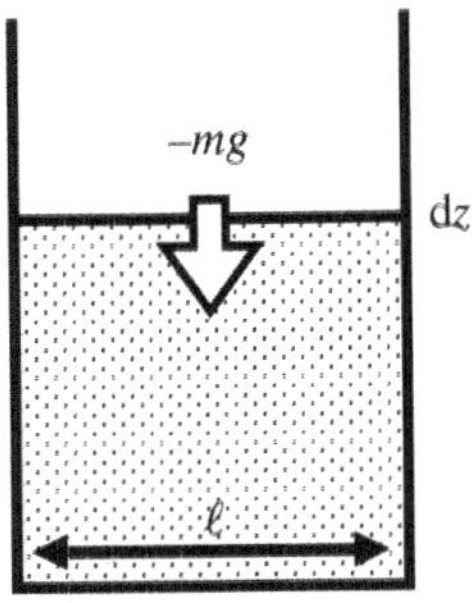

Figure 6.1 A schematic of a film surface that is free to move in the z-direction, but constrained in the other two (like a soap film in a frame). In this example, the surface is influenced by the gravitational force, $F = -mg$.

The interfacial tension between liquid and its vapor, separated by a sharp interface, may be roughly estimated as

$$\sigma \sim \frac{\epsilon}{\mathcal{A}} \sim \frac{k_B T_c}{4(r_0)^2} \tag{6.2}$$

where ϵ is the energy required to move a molecule from the bulk to the surface and r_0 is the radius of the molecule, respectively. For example, this expression gives an order of magnitude estimate of $\sigma \sim 100$ mN/m for the liquid–gas interface in cold water, which is in a reasonable agreement with experiment (see Figure 6.2).

With an increase in temperature, when the contribution to the free energy from the entropy increases, then a sharp interface becomes smooth. This is a key feature of the "thermodynamic" interface. One may characterize the surface tension of such an interface via a variety of thermodynamic potentials (discussed in Section 2.4), such as

$$\sigma = \left.\frac{\partial U_t}{\partial \mathcal{A}_t}\right|_{S_t,V_t} = \left.\frac{\partial A_t}{\partial \mathcal{A}_t}\right|_{T,V_t} = \left.\frac{\partial G_t}{\partial \mathcal{A}_t}\right|_{T,P} \tag{6.3}$$

Each of these potentials are useful when determining the surface tension in different statistical ensembles. Consider the temperature dependence of σ for the liquid–gas interface in water, as illustrated in Figure 6.2. At low temperatures, where the entropic contribution is small, σ is large and the interface is sharp. At high temperatures, where the entropic contribution is large, σ is small and the surface is smooth. The surface tension vanishes at the critical temperature.

The surface energy and the temperature change caused by increasing or decreasing the surface area may be determined explicitly by expressing the interfacial tension through the

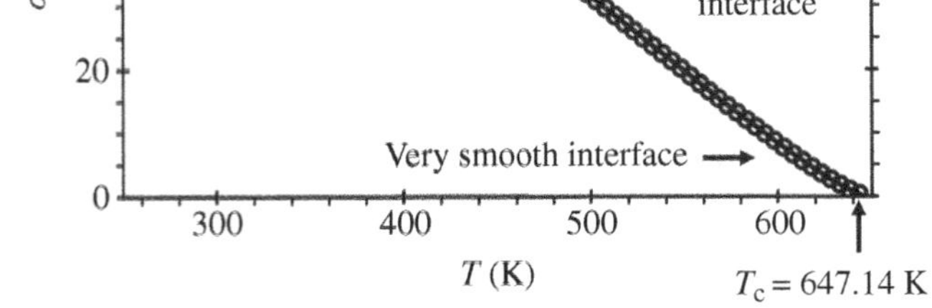

Figure 6.2 Liquid–gas interfacial tension of water from reference data (circles), while the curve is an interpolation between the surface tensions of the mechanical (sharp) and the near-critical (very smooth) interfaces [1]. The solid curve is the interpolation between two limits, given by $\sigma = 235.8|\Delta\hat{T}|^{1.26}(1 - 0.625|\Delta\hat{T}|)$. The difference between the critical exponent of the surface tension predicted by the van der Waals–Landau theory, $\theta = 3/2$ given by Eq. (6.22), and the value obtained by this interpolation, $\theta = 1.26$, is associated with the effects of critical fluctuations (to be discussed in Chapter 8).

thermodynamic potentials. The surface energy is determined by expanding the internal surface energy, $U_t^A = U_t^A(T, \mathcal{A}_t)$, in the form

$$dU_t^A = \left.\frac{\partial U_t^A}{\partial T}\right|_{\mathcal{A}_t} dT + \left.\frac{\partial U_t^A}{\partial \mathcal{A}_t}\right|_T d\mathcal{A}_t \tag{6.4}$$

The second term in Eq. (6.4) may be simplified using the Legendre transform, $U_t^A = A_t^A + TS_t^A$, such that

$$\left.\frac{\partial U_t^A}{\partial \mathcal{A}_t}\right|_T = \left.\frac{\partial (A_t^A + TS_t^A)}{\partial \mathcal{A}_t}\right|_T = \left.\frac{\partial A_t^A}{\partial \mathcal{A}_t}\right|_T + T\left.\frac{\partial S_t^A}{\partial \mathcal{A}_t}\right|_T \tag{6.5}$$

The first term on the right-hand side of Eq. (6.5) is σ, as defined in Eq. (6.3), while the second term may be simplified via the Maxwell relation,

$$\frac{\partial^2 A_t^A}{\partial T \partial \mathcal{A}_t} = -\left.\frac{\partial S_t^A}{\partial \mathcal{A}_t}\right|_T = \left.\frac{\partial \sigma}{\partial T}\right|_{\mathcal{A}_t} \tag{6.6}$$

Therefore, combining Eqs. (6.4) and (6.5), the surface energy per unit area may be expressed through the surface tension and temperature as

$$\left.\frac{\partial U_t^A}{\partial \mathcal{A}_t}\right|_T = \sigma - T\left.\frac{\partial \sigma}{\partial T}\right|_{\mathcal{A}_t} \tag{6.7}$$

The temperature change, due to the adiabatic expansion at the surface, is determined by expressing the entropy at the surface, $S_t^A = S_t^A(T, \mathcal{A}_t)$, as

$$dS_t^A = \left.\frac{\partial S_t^A}{\partial T}\right|_{\mathcal{A}_t} dT + \left.\frac{\partial S_t^A}{\partial \mathcal{A}_t}\right|_T d\mathcal{A}_t = 0 \tag{6.8}$$

For an adiabatic expansion process, where $dS_t^A = 0$, Eq. (6.8) can be solved for the change in temperature per surface area,

$$\left.\frac{\partial T}{\partial \mathcal{A}_t}\right|_{S_t^A} = \left.\frac{\partial \sigma}{\partial T}\right|_{\mathcal{A}_t} \left(\left.\frac{\partial S_t^A}{\partial T}\right|_{\mathcal{A}_t} \right)^{-1} \tag{6.9}$$

where the Maxwell relation given by Eq. (6.6) was used. Integrating from the initial temperature, T_0 and corresponding surface area $\mathcal{A}_t(T_0)$, to the final temperature, T, and surface area, $\mathcal{A}_t(T)$, gives

$$\Delta T = \int_{\mathcal{A}_t(T_0)}^{\mathcal{A}_t(T)} \frac{T}{C_A} \left.\frac{\partial \sigma}{\partial T}\right|_{\mathcal{A}_t} d\left(\frac{\mathcal{A}_t}{n_A}\right) \tag{6.10}$$

where the heat capacity of the interface at constant area is $C_A = T(\partial S_t^A/\partial T)|_{\mathcal{A}_t}$. Thus, upon an adiabatic expansion of a surface, the temperature will decrease by $\Delta T = T - T_0$. Alternatively, adiabatic compression would increase the temperature of the surface by $\Delta T = T - T_0$.

6.2 Van der Waals–Landau Theory of Smooth Interfaces

Van der Waals was the first to consider a continuous variation in the density of a fluid across a smooth liquid–gas interface [2]. The ideas and results of van der Waals could be formulated in a contemporary language through the Landau–Ginzburg free energy functional – introduced in Section 4.8 (see the comments on the history of this field and references in Section 1.3).

Consider the surface tension of a liquid–gas interface, defined by Eq. (6.3), in a simple fluid with a constant total volume in equilibrium at the critical isochore. For this system, the order parameter for each phase is the density, spatially varying along the z-direction as, $\varphi(z) = (\rho(z) - \rho_c)/\rho_c$, see Section 4.2. To construct the Landau–Ginzburg functional for the surface tension, the minimum of the free energy in each phase must be equal. Thus, the appropriate free energy for this functional is the excess density of the Helmholtz energy, defined by Eq. (2.69), as only this part is responsible for the interface between the liquid and gas phases in the system. Taking this into account, the free energy functional has the form

$$\sigma_{LG} = \frac{\partial[\Delta(\rho A_t)]}{\partial A_{LG}} = \rho_c k_B T_c \int \left(\Delta\hat{f}_{cr}(\varphi) + \frac{1}{2} c_0 \left|\frac{d\varphi}{dz}\right|^2 \right) dz \tag{6.11}$$

where the subscript "LG" refers only to the interface between the liquid and gas phases and $\Delta\hat{f}$ is the expansion of the excess density of the Helmholtz energy in the vicinity of the liquid–gas critical point. The excess free energy may be expressed in the form $\Delta\hat{f}_{cr} = \Delta\hat{f} - \Delta\hat{f}(\varphi_0)$, where $\Delta\hat{f}$ is the symmetric critical part of the two-term truncated Landau expansion, given by Eq. (4.9).

The expression for the free energy along coexistence emerges upon the substitution of the spontaneous order parameter, $\varphi = \varphi_0$, see Eq. (4.14), into Eq. (4.9). Upon subsequent factorization of this expression, $\Delta\hat{f}(\varphi_0)$ is obtained as

$$\Delta\hat{f}(\varphi_0) = -\frac{1}{4!} u_0 \varphi_0^4 = -\frac{3a_0^2}{2u_0} (\Delta\hat{T})^2 \tag{6.12}$$

Thus, $\Delta\hat{f}_{cr}$ is given in the form,

$$\Delta\hat{f}_{cr} \equiv \Delta\hat{f} + \frac{1}{4!} u_0 \varphi_0^4 = \frac{1}{4!} u_0 (\varphi^2 - \varphi_0^2)^2 \tag{6.13}$$

Note that $\Delta\hat{f}_{cr}$ has the form of a symmetric double well with minimum values located (for all temperatures) at $\Delta\hat{f}_{min} = 0$.

Therefore, using the excess free energy, Eq. (6.13), the expression for the surface tension may be written as

$$\sigma = \rho_c k_B T_c \int \left(\frac{1}{4!} u_0 (\varphi^2 - \varphi_0^2)^2 + \frac{1}{2} c_0 \left|\frac{d\varphi}{dz}\right|^2 \right) dz \tag{6.14}$$

The Landau–Ginzburg functional contains two parts, the contribution from the bulk thermodynamic part, Eq. (6.13), and the unfavorable energy gain in the free energy due to the spatially-inhomogeneous interface between the two coexisting phases, the gradient term in Eq. (6.14).

The shape of the interfacial profile is obtained from the minimization of the free energy functional. The integrand is minimized when the gradient term is balanced by the quadratic term, as

$$\frac{1}{4!} u_0 (\varphi^2 - \varphi_0^2)^2 = \frac{1}{2} c_0 \left|\frac{d\varphi}{dz}\right|^2 \tag{6.15}$$

Solving for the density gradient along the two coexisting phases gives

$$\frac{d\varphi}{dz} = \sqrt{\frac{u_0}{12c_0}} (\varphi^2 - \varphi_0^2) \tag{6.16}$$

Integrating Eq. (6.16), from one phase to the other, the density profile is obtained as

$$\varphi = \varphi_0 \tanh\left[-z\varphi_0 \sqrt{\frac{u_0}{12c_0}} \right] \tag{6.17}$$

Equation (6.17) may be simplified by using the equilibrium value of the order parameter, Eq. (4.14), such that

$$\varphi_0\sqrt{\frac{u_0}{12c_0}} = \sqrt{\frac{-a_0\Delta\hat{T}}{2c_0}} \equiv \frac{1}{\zeta} \tag{6.18}$$

where ζ is the thickness of the interface. In the two-phase region, for $T < T_c$, the correlation length of concentration fluctuations, ξ, given by Eq. (4.73), is related to ζ as, $2\xi = \zeta$. Thus, far away from the critical point, where the correlation length is small ($\xi \ll 1$), the interface is sharp, while in the vicinity of the critical point, where the correlation length is very large ($\xi \gg 1$), the interface becomes very smooth. The interfacial density profile is given by

$$\boxed{\varphi = \varphi_0 \tanh\left(-\frac{z}{\zeta}\right)} \tag{6.19}$$

Note that at the liquid–gas critical point, the correlation length becomes infinite and the density profile becomes undefined. The universal density profile predicted from the van der Waals–Landau theory is shown in Figure 6.3.

The equilibrium value for the surface tension is obtained when the thermodynamic and spatially-inhomogeneous terms are balanced, Eq. (6.15), such that the surface tension is obtained upon integrating over the entire space as

$$\frac{\sigma}{\rho_c k_B T_c} = \int_{-\infty}^{\infty} \frac{1}{12} u_0(\varphi^2 - \varphi_0^2)^2 dz \tag{6.20}$$

in which the bulk phases are defined to exist at $z = \pm\infty$. Then, upon substitution of the expression for the density profile, given in Eq. (6.19), Eq. (6.20) becomes

$$\frac{\sigma}{\rho_c k_B T_c} = \int_{-\infty}^{\infty} \frac{1}{12} u_0\varphi_0^4 \left(\tanh^2\left(-\frac{z}{\zeta}\right) - 1\right)^2 dz \tag{6.21}$$

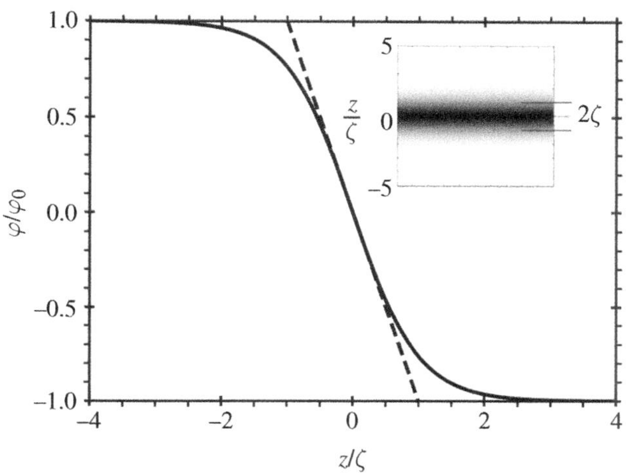

Figure 6.3 The normalized order parameter profile of a smooth interface, as predicted by Eq. (6.19). The inset shows a representation of the order parameter variation of the smooth interface. The intensity of the dark area in the inset corresponds to the variation of the order parameter gradient, given by Eq. (6.16), with a maximum at the center of the interface.

Equation (6.21) gives the interfacial tension as a function of temperature as

$$\frac{\sigma}{\rho_c k_B T_c} = \frac{1}{9}u_0\varphi_0^4\zeta = \frac{2\sqrt{3}}{9}\varphi_0^3\sqrt{c_0 u_0} = 4\sqrt{\frac{2c_0 a_0^3}{u_0^2}}\left(\frac{T_c - T}{T_c}\right)^{3/2} \tag{6.22}$$

where the relation between ζ and φ_0 is given through Eq. (6.18). Thus, one may express the interfacial tension via a critical power law (see Section 4.3) in the form

$$\sigma = \sigma_0 |\Delta\hat{T}|^{\vartheta} \tag{6.23}$$

such that, in the meanfield approximation, the critical exponent is $\vartheta = 3/2$ and the critical amplitude, σ_0, which depends on the Landau expansion coefficients, as $\sigma_0 = 4\rho_c k_B T_c\sqrt{2c_0 a_0^3/u_0^2}$. Note that the same result could be obtained from the gradient term in Eq. (6.14) as

$$\frac{\sigma}{\rho_c k_B T_c} = \int_{-\infty}^{\infty} c_0 \left|\frac{d\varphi}{dz}\right|^2 dz = \frac{\sigma_0}{\rho_c k_B T_c}|\Delta\hat{T}|^{3/2} \tag{6.24}$$

The results of the van der Waals–Landau theory are illustrated in Figure 6.4. These results are applicable to any smooth interface between near-critical phases, such as for the calculation of the density and composition profile in binary and multi-component mixtures. For example, the universal density profile obtained for the ternary solution of water, cyclohexane, and tert-butanol is shown in Figure 6.5, which accurately follows the predictions of the van der Waals–Landau theory, given by Eq. (6.16).

In a polymer solution, the amplitude of the critical power law for the interfacial tension, σ_0, in the critical region, where $\Delta\hat{T} \ll \mathcal{N}^{-1/2}$ (see the criterion in Section 5.5), depends on the degree of polymerization, $\mathcal{N}$. With use of the Landau expansion coefficients, given by Eq. (5.25), $u_0 \sim \mathcal{N}^{1/2}$ and $c_0 \sim \mathcal{N}^{1/2}$, then $\sigma_0 \sim \mathcal{N}^{-1/4}$. Consequently, the critical power law may be formulated as $\sigma \sim \mathcal{N}^{-m}|\Delta\hat{T}|^{\vartheta}$. Sanchez [5] demonstrated that the exponent of the $\mathcal{N}$-dependent amplitude,

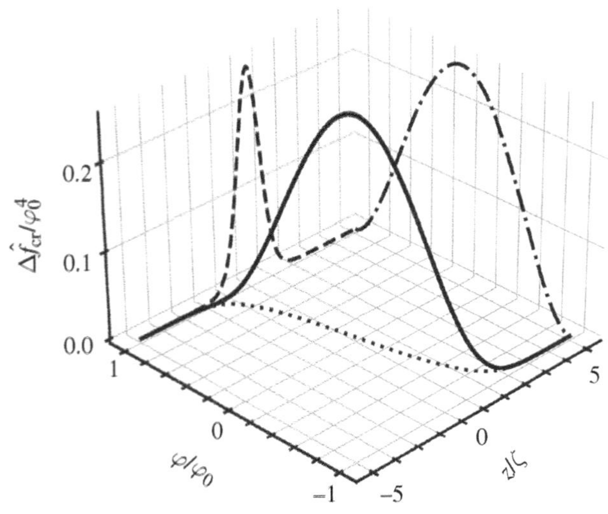

Figure 6.4 Three-dimensional universal plot of the normalized interfacial free energy, $\Delta\hat{f}_{cr}/\varphi_0^4$, given by Eq. (6.21). The solid curve depicts the free energy in three dimensions. The dashdot and dashed curves are projections of the free energy for constant z and constant φ, respectively. The dotted curve depicts the order parameter profile, Eq. (6.19).

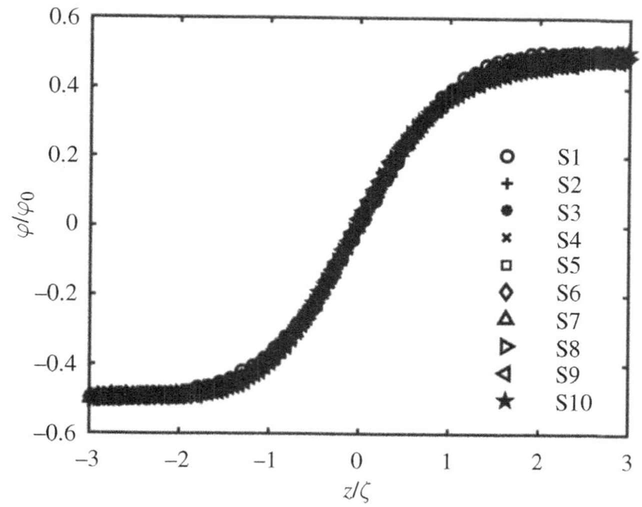

Figure 6.5 The universal density profile ($\varphi/\varphi_0 = (\rho - \overline{\rho})/2\overline{\rho}$), where $\overline{\rho}$ is the average density, obtained by molecular dynamics simulations for 10 solutions (S1–S10) of the ternary mixture of water, cyclohexane, and tert-butanol (TBA) in a broad range of TBA compositions, which have different proximity to the critical point of demixing (along the dotted line in the phase diagram of this system, see Figure 2.25) [3, 4]. Source: Reproduced with permission from Ref. [3], © 2017, American Chemical Society.

m, is related to the exponent of the temperature dependence of the surface tension, ϑ, through $m = (2 - \vartheta)/2$. In the tricritical Theta-point region, given by the criterion $\mathcal{N}^{-1/2} \ll \Delta\hat{T}$, where one phase consists of almost pure solvent, the interface becomes strongly asymmetric, with a thickness proportional to the de Gennes correlation length of a polymer chain, ξ_ψ, described by Eq. (5.49) [6, 7].

6.3 Curved Interfaces and Tolman's Length

Up to this point, we have considered the properties of planar interfaces. However, for a curved interface, like for a droplet or in a capillary tube, the curvature of the surface affects the interfacial profile and the interfacial tension. Consider a droplet of liquid surrounded by vapor. The total energy of the droplet may be expressed through the bulk energy, U_t, of each phase ($\tilde{\alpha}$ and $\tilde{\beta}$) and the surface energy of the interface between them, $U_\text{t}^{\mathcal{A}}$, such that the total energy, U_t^{tot}, is given by

$$U_\text{t}^{\text{tot}} = P_{\tilde{\beta}} V_{\tilde{\beta}} - P_{\tilde{\alpha}} V_{\tilde{\alpha}} + \sigma \mathcal{A}_\text{t} \tag{6.25}$$

Assuming spherical droplets, where $V_{\tilde{\beta}} = V_{\tilde{\alpha}} = (4/3)\pi R^3$, the equilibrium value of the radius of the droplet, R, is obtained from the pressure difference that minimizes the total internal energy of the system. Evaluating, $\partial U_\text{t}^{\text{tot}}/\partial R = 0$, yields

$$P_{\tilde{\alpha}} - P_{\tilde{\beta}} = \frac{2\sigma}{R} \tag{6.26}$$

Equation (6.26) is known as the "Young-Laplace equation." It signifies that the radius, R, is related to the difference in pressure inside and outside of the droplet. For a planar surface, where $R \to \infty$, the pressure difference becomes equivalent between the two phases.

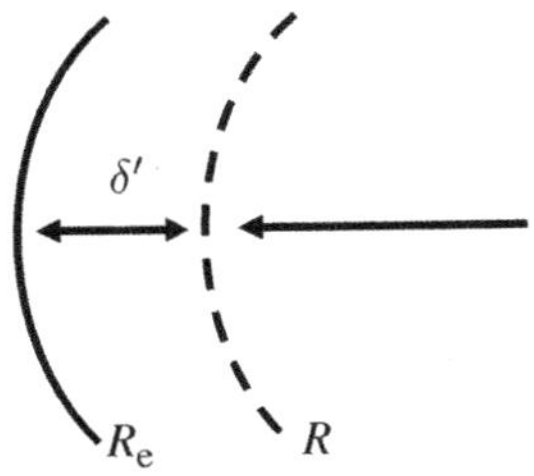

Figure 6.6 Sketch illustrating the concept of Tolman's length (δ') for a droplet of liquid, where R is the radius of the surface of tension and R_e is the radius of the equimolar surface.

Equation (6.26) is accurate for small curvatures, but as the interface becomes more curved, the surface tension may be expressed through an expansion in powers of the curvature, $1/R$. In the first approximation, σ is given in the form

$$\sigma(R) = \sigma_\infty \left(1 - \frac{2\delta'}{R}\right) \tag{6.27}$$

where σ_∞ is the planar surface tension and δ' is the coefficient of the first curvature correction to the surface tension, known as "Tolman's length."[1] One may understand Tolman's length through the difference between the equimolar surface, that separates each phase into an equal number of molecules on each side, and the maximum of the gradient of the density profile, the surface of tension. Thus, Tolman's length is the difference between the equimolar radius, R_e, and the radius of the surface tension, R, such that $\delta' = R_e - R$, illustrated in Figure 6.6. For a sharp interface, Tolman's length is a fraction of the molecular size, while for smooth interfaces, it may become mesoscopic [8, 9].

Depending on the sign of δ', the interface will curve toward one phase or the other. Tolman's length is a thermodynamic characteristic of any interface, but only for curved interfaces, does it affect the surface tension. Regardless of the curvature, Tolman's length is zero only for symmetric fluid coexistence, like in the lattice-gas model; however, all real fluids possess some degree of asymmetric coexistence. Thus, for real fluids, Tolman's length is nonzero.

In the meanfield approximation, the asymmetry in the liquid–gas coexistence is represented by the diameter of the coexistence curve, see the discussion in Section 4.4.1 for more details. With use of the definition of the diameter, given by Eq. (4.31), the asymmetric density profile may be calculated following the same derivation for the symmetric density profile, Eq. (6.19). Thus, one finds that the asymmetric density profile contains a correctional term proportional to the rectilinear diameter,

$$\varphi/\varphi_0 \approx \tanh\left(-\frac{z}{\zeta}\right) + D|\Delta\hat{T}|\tanh^2\left(-\frac{z}{\zeta}\right) \tag{6.28}$$

where D is the slope of the diameter, given by Eq. (4.32). For the van der Waals EOS, $D = 2/5$, as discussed in Section 4.4.1.

Since Tolman's length is associated with the asymmetry of phase coexistence, one may express the pressure difference between the inside of a droplet and its surrounding phase through two contributions, symmetric and asymmetric, as

$$P_{\tilde{\alpha}} - P_{\tilde{\beta}} = (P_{\tilde{\alpha}} - P_{\tilde{\beta}})_{\text{sym}} + (P_{\tilde{\alpha}} - P_{\tilde{\beta}})_{\text{asym}} \tag{6.29}$$

The left-hand side of Eq. (6.29) may be expressed through the Young-Laplace equation Eq. (6.26) with Tolman's correction to the surface tension Eq. (6.27), giving

$$(P_{\tilde{\alpha}} - P_{\tilde{\beta}})_{\text{sym}} + (P_{\tilde{\alpha}} - P_{\tilde{\beta}})_{\text{asym}} \approx \frac{2\sigma_\infty}{R}\left(1 - \frac{2\delta'}{R}\right) \tag{6.30}$$

Since Tolman's correction is associated with the asymmetric part, the symmetric part of the pressure difference must be independent of δ' as

$$(P_{\tilde{\alpha}} - P_{\tilde{\beta}})_{\text{sym}} = \frac{2\sigma_\infty}{R} \tag{6.31}$$

1 Richard C. Tolman (1881–1948) was an American physicist, who made significant contributions to statistical physics and was the first to consider the curvature correction to the interfacial tension.

Meanwhile, the asymmetric part of Eq. (6.30) may be expressed through Tolman's length, such that

$$\frac{2\delta'}{R} \approx -\frac{(\Delta P)_{\text{asym}}}{(\Delta P)_{\text{sym}}} \tag{6.32}$$

where $\Delta P = P_{\tilde{\alpha}} - P_{\tilde{\beta}}$. Therefore, Tolman's length could be estimated through the ratio of the asymmetric to symmetric contributions in the fluid coexistence. It has been shown that the ratio of Tolman's length to the thickness of the interface may be expressed through the ratio of the asymmetric, $\varphi_{0,\text{asym}}$, to the symmetric, $\varphi_{0,\text{sym}}$, contributions to the spontaneous order parameter. The full form of φ_0, containing symmetric and asymmetric parts, is given by Eq. (4.29), in which the first term is $\varphi_{0,\text{sym}}$ and the second term is $\varphi_{0,\text{asym}}$. Thus, Tolman's length may be expressed in the form

$$\frac{2\delta'}{\zeta} \approx \mp\frac{\varphi_{0,\text{asym}}}{\varphi_{0,\text{sym}}} \tag{6.33}$$

where the thickness is related to the correlation length as $\zeta = 2\xi$ (see Eq. (6.18)). With use of the expressions for the symmetric and asymmetric contributions to the order parameter, as discussed in Section 4.4.1, and Eq. (4.72) for the correlation length, one finds that

$$\boxed{\frac{\delta'}{\xi_0} = \mp\frac{D}{\sqrt{6a_0/u_0}}} \tag{6.34}$$

where ξ_0 is the amplitude of the correlation length, defined through Eq. (4.72). For the van der Waals equation of state, all coefficients are given in Section 4.4.1, such that the right hand side of Eq. (6.34) is a constant (equal to 1/5), such that Tolman's length is $\delta' = \xi_0/5$, which is just a fraction of the molecular size. However, in polymer solutions close to the Theta-point ($\mathcal{N} \to \infty$), as shown by Szleifer and Widom [10], the interfacial profile is defined through the de Gennes correlation length, ξ_ψ (see Eq. (5.49)), and the phase asymmetry is so significant that Tolman length becomes mesoscopic, of the order of the interfacial thickness, even in the meanfield approximation, diverging as $\sim \xi_\psi \propto (\Theta - T)^{-1}$ at the Theta point [7].

Generally, in the meanfield approximation, Tolman's length of simple fluids is finite at the critical point. Thus, one may think to conclude that Tolman's length is independent of the temperature and too small to be important for engineering applications. However, as will be discussed in Chapter 8, critical fluctuations significantly affect the properties of real fluids, including the behavior of Tolman's length.

6.4 Wetting Transitions

Molecular interactions between the bulk phase and the surface may affect the fluid's interfacial behavior. If the molecular interaction is attractive, the surface can attract the bulk phase, like dew on glass. Whereas, if the molecular interaction is repulsive, the bulk phase will be repelled, like water beading on a waxed surface or mercury droplets on glass.

Cahn[2] [11] predicted an "interfacial transition" between effectively repulsive and effectively attractive materials (Figure 6.7). This phenomenon is commonly referred to as a "wetting transition." The simplest example of a wetting transition involves the interaction between the bulk liquid–gas phase transition and an inert solid surface or a wall. When the solid surface is exposed

2 John W. Cahn (1928–2016) was an American scientist, who made significant contributions to the equilibrium and nonequilibrium thermodynamics of interfaces and phase transitions.

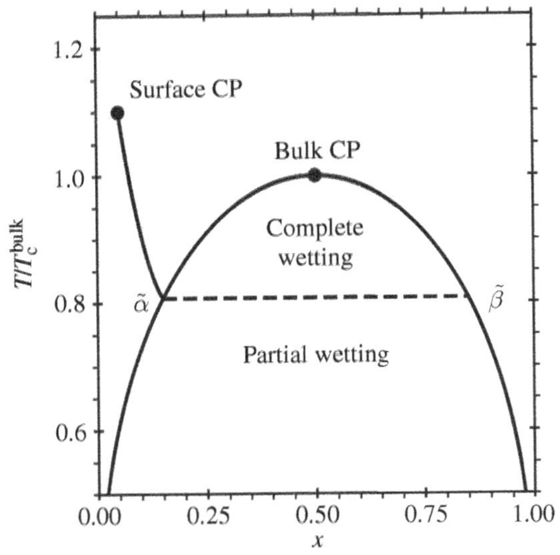

Figure 6.7 Schematic of an interfacial "wetting" phase transition as predicted by Cahn [11]. Complete wetting by the $\tilde{\beta}$ phase occurs only along phase coexistence when both liquid phases are close to their critical demixing point (Bulk CP). In the bulk $\tilde{\alpha}$ phase there is a film of the $\tilde{\beta}$ phase, such that the first-order wetting transition propagates into the bulk, which terminates at a surface critical point (Surface CP). To the left and right of the surface phase transition line are regions of low and high adsorption, respectively.

to the coexisting liquid and gas, two different effects could occur: if the molecules of the liquid interact strongly with the surface, the liquid will spread across the surface as a thick film. In this case, the surface is said to be "wet." Alternatively, if the interactions between the liquid molecules are stronger than their interaction with the surface, then the liquid will clump around itself, and, in this case, the liquid will not wet the surface.

An example of a wetting transition is illustrated in Figure 6.8. If a solid surface is non-wet at a given temperature, raising the temperature will lower the interfacial tension between the liquid and gas, and the solid surface may become wet at a higher temperature. The temperature at which the surface becomes completely wet is the wetting temperature, T_w, and the transition from the non-wet surface to the wet surface, as the temperature is raised, is the wetting transition.

To thermodynamically describe wetting transitions, one may consider three-phase equilibrium between two fluid phases and one solid phase (alternatively, three fluid phases could also produce a wetting transition). Consider a scenario where two fluid phases meet at a surface. The equilibrium condition between any three-phases is referred to as the "three phase line." In equilibrium, the net force on any element of the three-phase line vanishes; meaning that, the surface tensions of each phase balance the forces per length normal to the three-phase line, see the phase triangle shown

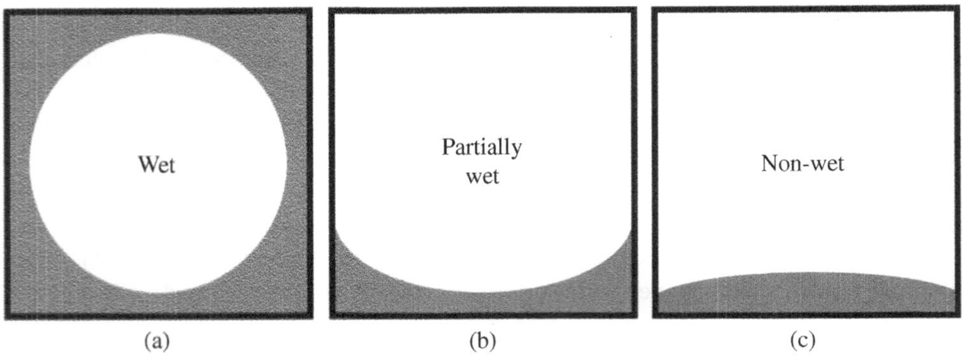

Figure 6.8 Schematic representation of (a) wet, (b) partially wet, and (c) non-wet surfaces.

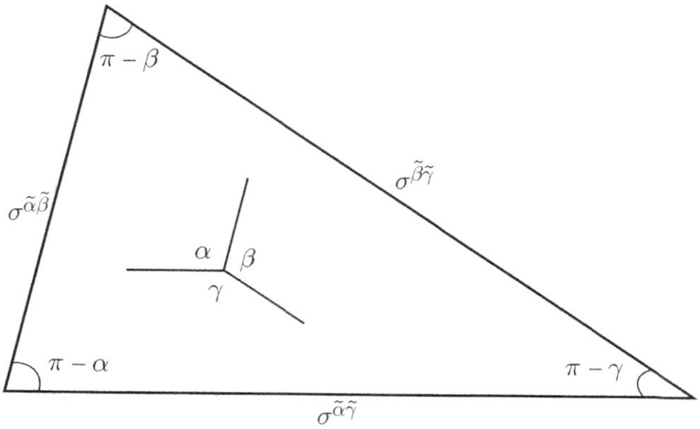

Figure 6.9 Euler's triangle depicting contact angles, α, β, and γ, named after the phases they represent ($\tilde{\alpha}$, $\tilde{\beta}$, and $\tilde{\gamma}$). The triangle relates the contact angles to the surface tensions between each phase.

in Figure 6.9. Each side of this triangle represents a surface tension between two phases. If $\tilde{\alpha}$ and $\tilde{\beta}$ represent liquid phases and $\tilde{\gamma}$ is a solid phase, such that the contact angles, α, β, and γ, are named after each phase. The sum of the contact angles is

$$\alpha + \beta + \gamma = 2\pi \tag{6.35}$$

At equilibrium, for the three phases to be stable, Newton's third law indicates that the surface forces between each species must be balanced. For instance, consider the $\tilde{\gamma}$-phase interface. In this case, the balance of forces acting on the $\tilde{\gamma}$-phase indicates that

$$\sigma^{\tilde{\alpha}\tilde{\beta}} \cos\beta + \sigma^{\tilde{\beta}\tilde{\gamma}} + \sigma^{\tilde{\alpha}\tilde{\gamma}} \cos\alpha = 0 \tag{6.36}$$

When the $\tilde{\gamma}$ phase is a rigid solid, the angle $\gamma = \pi$ and there is only one independent contact angle, $\alpha = \pi - \beta$, such that $\sigma^{\tilde{\beta}\tilde{\gamma}} = \sigma^{\tilde{\alpha}\tilde{\beta}} \cos\alpha + \sigma^{\tilde{\alpha}\tilde{\gamma}}$. This is known as "Young's equation." If $\alpha = \pi$, then $\beta = 0$, and the three surface tensions are related by

$$\sigma^{\gamma\beta} = \sigma^{\alpha\gamma} - \sigma^{\alpha\beta} \tag{6.37}$$

Physically, this corresponds to the $\tilde{\beta}$ phase being spread as a film over the $\tilde{\alpha}$-$\tilde{\gamma}$ interface. This condition is known as "Antonov's rule" [8].

Consider a binary solution (with liquid phases $\tilde{\alpha}$ and $\tilde{\beta}$), in equilibrium with its vapor (phase $\tilde{\gamma}$), which are characterized by three surface tensions, $\sigma^{\tilde{\alpha}\tilde{\beta}}$, $\sigma^{\tilde{\alpha}\tilde{\gamma}}$, and $\sigma^{\tilde{\beta}\tilde{\gamma}}$. In the vicinity of the liquid–liquid critical point, the surface tension between the liquid phases is given in the meanfield approximation by the power law of Eq. (6.23), as

$$\sigma^{\tilde{\alpha}\tilde{\beta}} = \sigma_0^{\tilde{\alpha}\tilde{\beta}} |\Delta\hat{T}|^{3/2} \tag{6.38}$$

The surface tension between the liquid and vapor is determined by the energy of the molecular interactions, such that the difference in surface tensions is proportional to the difference in densities between the two liquid phases as

$$\sigma^{\tilde{\alpha}\tilde{\gamma}} - \sigma^{\tilde{\beta}\tilde{\gamma}} \propto \rho_{\tilde{\alpha}} - \rho_{\tilde{\beta}} \propto |\Delta\hat{T}|^{1/2} \tag{6.39}$$

As predicted by Antonov's rule, Eq. (6.37), the wetting transition temperature is obtained at the point where the surface tension, $\sigma^{\tilde{\alpha}\tilde{\beta}}$, and the difference, $\sigma^{\tilde{\alpha}\tilde{\gamma}} - \sigma^{\tilde{\beta}\tilde{\gamma}}$, become equal. The temperature where these two intersect is indicated by T_w as shown in Figure 6.10.

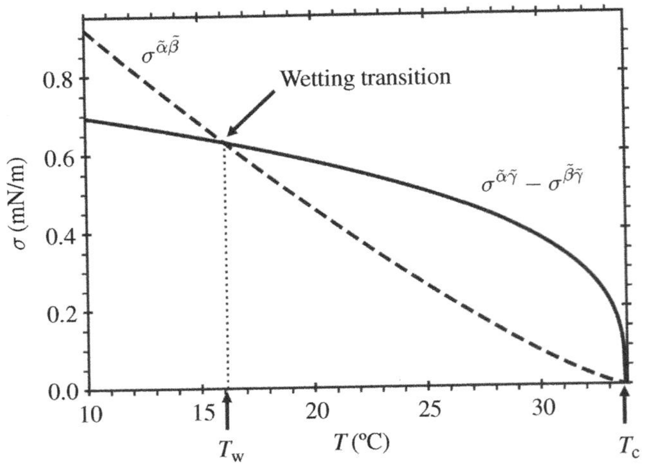

Figure 6.10 The wetting transition predicted for the liquid mixture of methanol and *n*-hexane in coexistence with its vapor at the critical composition. The solid and dashed curves qualitatively follow the meanfield predictions, given by Eqs. (6.38) and (6.39), but with different critical exponent values, 1.26 instead of 3/2 and 0.326 instead of 1/2, as they are affected by critical fluctuations (to be discussed in Chapter 8).

6.5 Nucleation and Spinodal Decomposition

Below the critical temperature, in the two-phase region, there are two different mechanisms for phase formation. If the system is quenched (rapidly cooled) into the metastable region, the region between the spinodal and the coexistence curve (as depicted by quench A in Figure 6.11), then the system may be in a metastable state, in which a new phase may form via the process of "nucleation." Alternatively, if the fluid is quenched inside the unstable (spinodal) region (as depicted by quench B in Figure 6.11), then the system will form a new phase via the process of "spinodal decomposition." The alternative thermodynamics of phase formation due to these two mechanisms is presented in Sections 6.5.1 and 6.5.2.

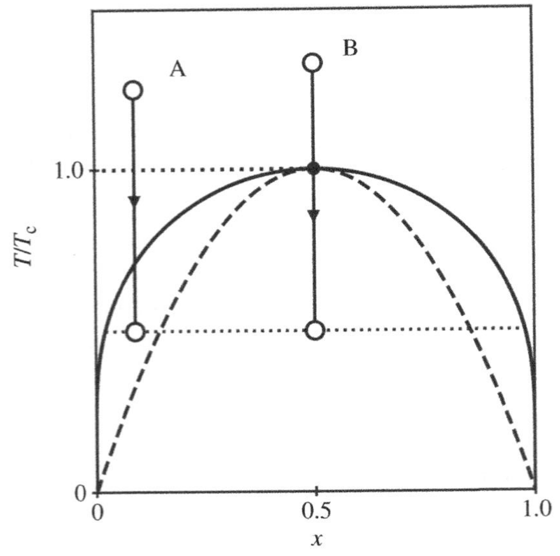

Figure 6.11 Image illustrating a "quench" – when a homogeneous binary mixture is rapidly cooled from a high temperature to a state below the critical temperature of liquid–liquid phase separation (T_c). Quench (A) depicts an off-critical-composition path into the metastable region, while (B) depicts a quench along the critical composition into the unstable region. The binodal (solid) and spinodal (dashed) are plotted for the regular-solution model (see Section 2.8.3).

6.5.1 Gibbs Theory of Homogeneous Nucleation

When a system is quenched into the metastable region between the binodal and spinodal, provided that the quench is fast enough, the system may not immediately separate into two phases. Instead, the transformation must be activated by some perturbation, such as density fluctuations. Initially, density fluctuations will cause small amounts of a new phase to be formed in the originally homogeneous metastable phase. Prior to the formation of a new phase, the system is said to be in a "metastable state," which occurs when supercooling/superheating or, alternatively, over-compressing/over-stretching produce an imbalance in the chemical potential.

For instance, consider the case of a pure fluid over-compressed at constant temperature as presented in Figure 6.12. The metastable phase $\tilde{\beta}$ is characterized by the chemical-potential imbalance, $\Delta\mu = \mu_{\tilde{\beta}} - \mu_{\text{cxc}}$. As follows from Maxwell's equal area rule, two-phase coexistence is established when $\int d\mu = 0$ (see Section 2.6.2). However, in the metastable state, $\Delta\mu = V\Delta P > 0$ (at constant T), where $\Delta P = P - P_{\text{cxc}}$. This imbalance is the driving force for nucleation.

A new phase, $\tilde{\alpha}$, may randomly form within the metastable phase $\tilde{\beta}$ due to density fluctuations. If the substance is perfectly homogeneous (practically speaking: no dust, container's surfaces are clean and neutral, etc.), and, hence, does not have artificial centers of nucleation, then small fluctuation-induced nuclei of phase $\tilde{\alpha}$ may disappear. To continue growing, the nuclei must be large enough to compensate the unfavorable surface energy that occurs when the interface is formed.

Consider a spherical droplet of radius, R, consisting of a new phase $\tilde{\alpha}$ immersed in a uniform metastable phase $\tilde{\beta}$. The formation of phase $\tilde{\alpha}$ will decrease the free energy of the system. The net change in the Gibbs energy may be expressed through the competition between the bulk and surface energies as

$$\Delta G_t = \Delta(N\mu) = -\frac{4}{3}\pi R^3(\Delta P) + 4\pi R^2\sigma \tag{6.40}$$

This competition is illustrated in Figure 6.13. As a result, there is an energy barrier that resists the initial growth of a droplet of phase $\tilde{\alpha}$. The minimum size required for the droplet to grow, R^*,

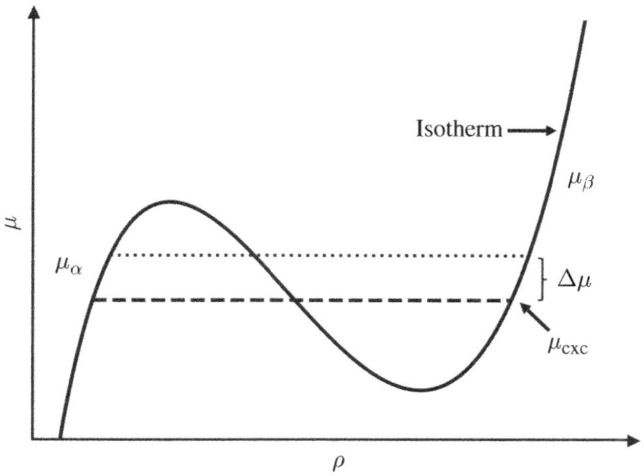

Figure 6.12 Illustration of an imbalance in the chemical potential for a pure fluid over-compressed along an isotherm. This imbalance is the driving force for the formation of a new phase $\tilde{\alpha}$, toward equilibrium phase coexistence.

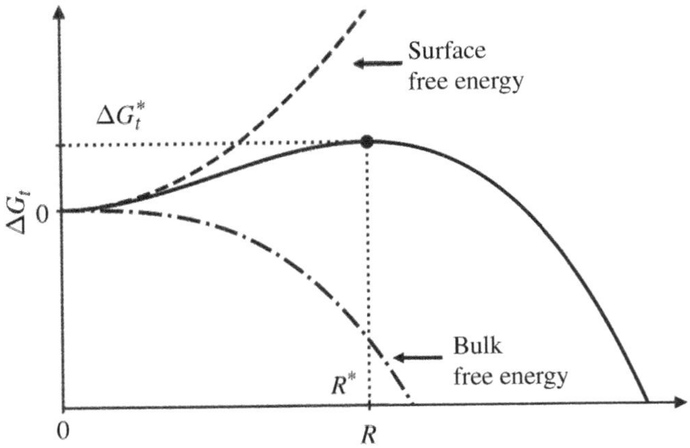

Figure 6.13 Illustration of the competition between the bulk (dashdot) and surface (dashed) contributions to the Gibbs energy of a droplet, with radius R, forming by nucleation in the metastable phase. The total Gibbs energy (solid) is given by Eq. (6.43).

corresponds to the maximum of ΔG_t as

$$R^* = \frac{2\sigma}{\Delta P} \tag{6.41}$$

This condition is similar to the Young–Laplace equation, see Eq. (6.26). Correspondingly, the height of the energy barrier is determined from the maximum of the total free energy at the critical radius as

$$\Delta G_\mathrm{t}^*(R = R^*) = \frac{16\pi}{3}\left[\frac{\sigma^3}{(\Delta P)^2}\right] \tag{6.42}$$

The minimum nucleation radius and the height of the energy barrier are affected by Tolman's length (see Section 6.3), such that

$$\Delta G_\mathrm{t}^*(R = R^*) = \frac{16\pi}{3}\left[\frac{\sigma_\infty^3}{(\Delta P)^2}\right]\left(1 - \frac{2\delta'}{R^*}\right)^3 \tag{6.43}$$

This correction may be noticeable for systems with a small minimum nucleation radius, R^*.

In reality, the nucleation barrier is greatly reduced by the presence of heterogeneous impurities (like dust particles) and the inhomogeneities of the confining surfaces of the container. This phenomenon is referred to as "heterogeneous nucleation." Quantitative descriptions of heterogeneous nucleation depend on the source of the inhomogeneities, and thus, they are highly empirical [12, 13].

Gibbs theory of homogeneous nucleation could be equally applied to a liquid binary mixture quenched under the coexistence along an off-critical-composition path, as illustrated in Figure 6.11. In this case, the role of the density will be played by the concentration, x, and the driving force for the nucleation toward the formation of a new phase $\tilde{\alpha}$ within the metastable phase $\tilde{\beta}$ will be an imbalance in the exchange chemical potential, $\Delta\mu_{21} = \mu_{21,\tilde{\beta}} - \mu_{21,\mathrm{cxc}}$.

6.5.2 Cahn–Hilliard Theory of Spinodal Decomposition

A different phenomenon is observed when the system is quenched into the unstable region, below the spinodal. Consider, for example, a weakly compressible binary liquid solution. When the system is quenched into the unstable region, any local change in composition lowers the free energy. As a result, the effects of composition fluctuations are amplified, leading to the dramatic change of a previously homogeneous solution, which evolves toward demixing and the formation of two separate phases. This profound phenomenon, known as "spinodal decomposition" may be described by the Cahn-Hilliard theory [14, 15], as described in this Section.

An essential feature of spinodal decomposition is that, unlike nucleation in metastable phases, the species flow from regions of low concentration to regions of high concentration. This is the complete opposite of one's usual experience when species mutually diffuse to make the composition uniform. In contrast, spinodal decomposition is driven by a negative mutual diffusion, where the diffusion coefficient, $D = (\hat{\alpha}/\rho)(\partial\mu_{21}/\partial x)|_{P,T}$ is negative as the susceptibility $\partial\mu_{21}/\partial x|_{P,T} < 0$ in the unstable region. Note that this process may become faster or slower depending on the viscosity of the fluid and the proximity to the critical point.

A quantitative description of spinodal decomposition in a binary-liquid mixture in the meanfield approximation is obtained by considering the Landau–Ginzburg free-energy functional, as given by Eq. (4.71), through the critical part of the Gibbs energy per molecule ($\Delta\hat{G} \equiv \Delta G/k_BT_c$), as

$$\Delta\hat{G} = \int \left(\Delta\hat{f}_{cr}(\varphi) + \frac{1}{2}c_0|\nabla\varphi|^2\right) dV_t \tag{6.44}$$

where the spatially-dependent order parameter in this case is $\varphi(\boldsymbol{r}) = x(\boldsymbol{r}) - x_c$. The first term in the integrand of Eq. (6.44) describes the bulk-free energy per unit volume of a mixture, such that $\Delta\hat{f}_{cr} = (1/2)\hat{\chi}^{-1}_{q=0}\varphi^2$, where $\hat{\chi}^{-1}_{q=0} = \partial\hat{\mu}_{21}/\partial\varphi$ as given by Eq. (4.17), and the second term is the energy penalty due to the interaction of local inhomogeneities. The exchange chemical potential between the two phases is given by the variational derivative of the free-energy functional. The variational derivative of any functional, $Y[\{X\}]$ where X is a spatially-dependent function, is given in the form

$$\frac{\delta Y}{\delta X} = \frac{\partial Y}{\partial X} - \nabla \cdot \frac{\partial Y}{\partial \nabla X} \tag{6.45}$$

such that, for the case of Eq. (6.44), the spatially-dependent chemical potential reads

$$\hat{\mu}_{21}(\boldsymbol{r}) = \frac{\partial\Delta\hat{f}_{cr}}{\partial\varphi} - c_0\nabla^2\varphi = \hat{\chi}^{-1}_{q=0}\varphi - c_0\nabla^2\varphi \tag{6.46}$$

The flow of matter is given by the gradient of the exchange chemical potential, Eq. (2.174), as $\boldsymbol{j}_D = -\breve{\alpha}\nabla\hat{\mu}_{21}$, where $\breve{\alpha}$ is the mobility, inversely proportional to the sheer viscosity. Thus, the mutual diffusive flux is given by

$$\boldsymbol{j}_D = -\breve{\alpha}\hat{\chi}^{-1}_{q=0}\nabla\varphi + \breve{\alpha}c_0\nabla^3\varphi \tag{6.47}$$

The time evolution of the composition is obtained through the local conservation of matter,

$$\frac{\partial\varphi}{\partial t} = -\nabla \cdot \boldsymbol{j}_D \tag{6.48}$$

Equation (6.48) is referred to as the "continuity equation," such that the time evolution of the order parameter is given by

$$\frac{\partial\varphi}{\partial t} = \breve{\alpha}\hat{\chi}^{-1}_{q=0}\Delta^2\varphi - c_0\breve{\alpha}\nabla^4\varphi \tag{6.49}$$

The product, $\breve{\alpha}\hat{\chi}^{-1}_{q=0}$, is the mutual diffusion coefficient, D, as given by Eq. (2.176), which is negative as the susceptibility, $\hat{\chi}^{-1}_{q=0}$, is negative in the spinodal region. Moreover, one may notice that the gradient term in the Landau–Ginzburg functional produces a correction to Fick's second law, Eq. (2.174).

With use of Fourier analysis, one may solve Eq. (6.49), yielding the characteristic growth rate, $\omega(q)$, of the local inhomogeneities in the system. Inhomogeneities will grow at a certain rate

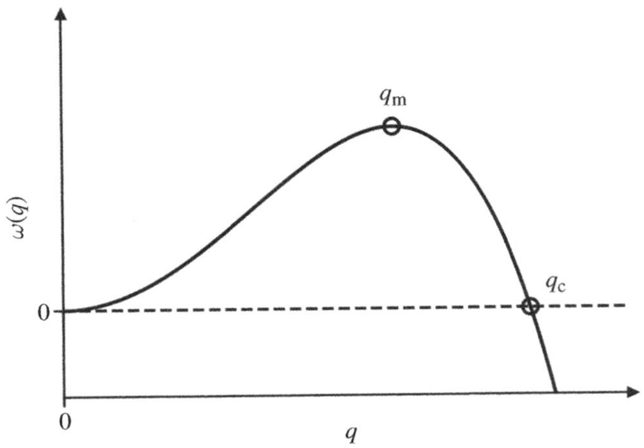

Figure 6.14 The characteristic growth rate of phase formation in the process of spinodal decomposition driven by negative diffusion.

depending on their size, and each size is characterized through a wavenumber, q, inversely proportional to the size of the inhomogeneity. The growth rate has the form

$$\boxed{\omega(q) = -Dq^2(1 - \xi^2 q^2)} \tag{6.50}$$

where the correlation length, $\xi = \sqrt{-c_0\hat{\chi}_{q=0}}$, is defined to be positive. Equation (6.50) is also referred to as the "amplification factor" [14]. A derivation of $\omega(q)$ is provided in Appendix 6.A. The maximum of $\omega(q)$ corresponds to the fastest growing inhomogeneity with a wavenumber $q_m = 1/(\sqrt{2}\xi)$ that designates the characteristic length scale for the growing inhomogeneities. The largest size of inhomogeneities that will grow is designated by the wave number, $q_+ = 1/\xi$. The growth rate is illustrated in Figure 6.14. Note that, in spinodal decomposition, the growth of the phase domain is determined by the correlation length of the fluctuations of the order parameter.

The dynamic behavior described by spinodal decomposition only occurs for short time scales. At longer time scales, once a particular domain, comprised of species A or B, has formed, the domains will slowly increase their size by adopting neighboring molecules of a similar type. This transient behavior is known as "coarsening." The stage of coarsening is manifested by the formation of morphological patterns, being the characteristic feature of spinodal decomposition (Figure 6.15). The nonequilibrium patterns formed during the coarsening process will eventually lead to macroscopic phase separation [17]. However, if the nonequilibrium patterns are "frozen" by a rapid quench, the system will be trapped in this metastable state. This method may be used to develop a well-defined porous material.

6.5.3 Spinodal Decomposition in Mixtures with Interconverting Species

The Cahn–Hilliard theory of spinodal decomposition may be generalized to describe the phase formation in mixtures that exhibit molecular interconversion of species [18]. For simplicity, consider a binary mixture, the same as discussed in Section 6.5.2, except allowing for the alternative species to exhibit simple $A \rightleftharpoons B$ interconversion, where the rate of the forward reaction is equal to that of the reverse reaction. Thus, the interconversion equilibrium constant $K = 1$, see Eq. (2.137), which is assumed to be independent of temperature or pressure. In this case, the Gibbs energy of the reaction, $G_{BA} = -k_B T \ln K = 0$, and there is no preference toward formation of one phase over the other.

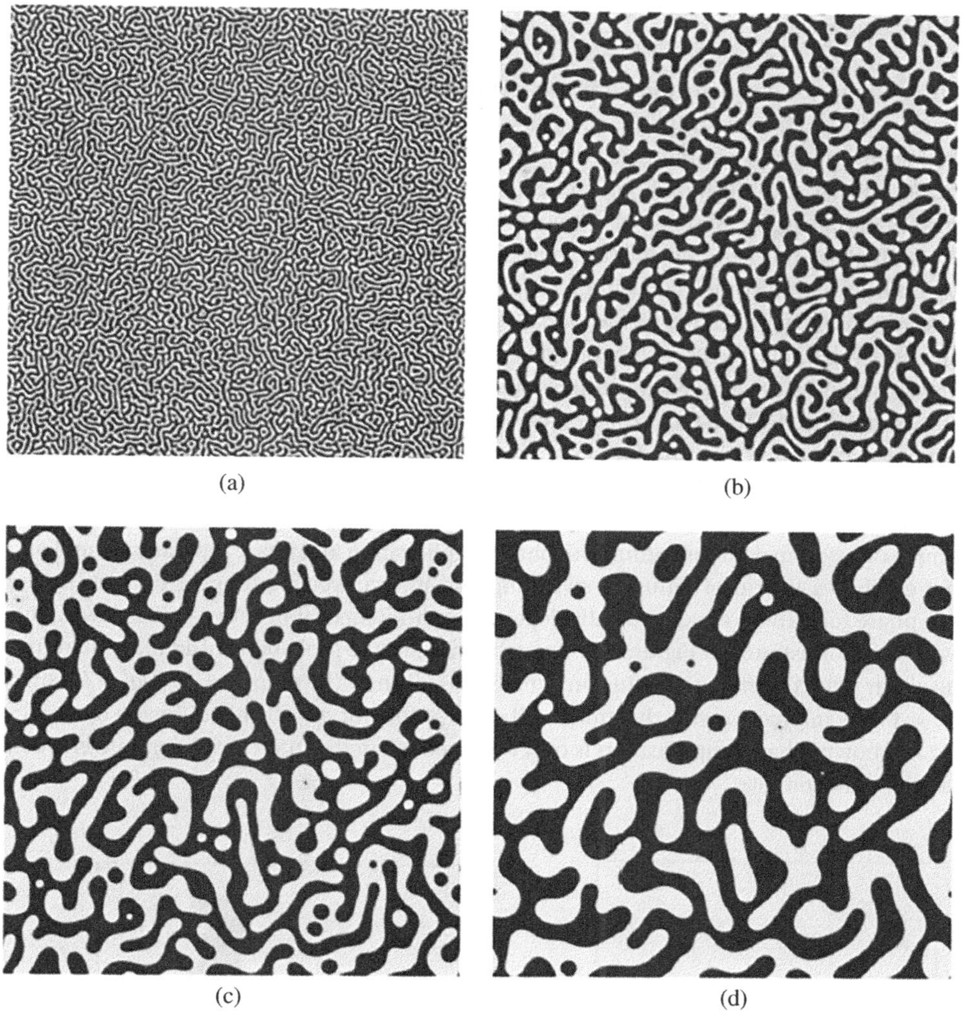

Figure 6.15 Morphological patterns during spinodal decomposition and subsequent coarsening: (a) $t = 100$, (b) $t = 2{,}000$, (c) $t = 10{,}000$, and (d) $t = 30{,}000$ simulation time steps. Source: Reproduced with permission from Ref. [16], © 1999, American Physical Society.

The additional freedom for the species to either diffuse or interconvert manifests in the dynamics of the order parameter, which adopts two-components: a conserved component, associated with the diffusion dynamics; and a nonconserved component, associated with the interconversion dynamics (see discussion in Section 4.11). Consequently, the continuity equation, Eq. (6.48), adopts an additional term, j_{NC}, to account for the nonconserved order-parameter dynamics,

$$\frac{\partial \varphi}{\partial t} = -\nabla \cdot \boldsymbol{j}_D + j_{\mathrm{NC}} \tag{6.51}$$

where the mutual diffusive flux $j_D = -\breve{\alpha}\nabla\hat{\mu}_{\mathrm{BA}}$, as in Eq. (6.47), is associated with the conserved component of the order parameter, while the flux associated with the nonconserved component of the order parameter is $j_{\mathrm{NC}} = -\breve{\delta}\hat{\mu}_{\mathrm{BA}}$, where $\breve{\delta}$ is a kinetic coefficient proportional to the rate of the interconversion reaction. Solving Eq. (6.51) with use of Fourier analysis yields a generalized growth rate in the form

$$\omega(q) = -(Dq^2 + L)(1 - \xi^2 q^2) \tag{6.52}$$

where the interconversion rate is $L = \hat{\chi}_{q=0}^{-1}\check{\delta}$. Note that L has the same units, s^{-1}, as the diffusion rate, Dq^2.

Interconversion yields a new feature in phase formation. If the rate of interconversion is faster than the rate of diffusion, then one phase may grow at the expense of the other, such that the phase transition does not result in equilibrium coexistence of alternative phases. This phenomenon is referred to as "phase amplification," which occurs naturally in such systems as magnetic and ferroelectric materials in the absence of a magnetic or electric field. In these materials, there is no conservation of the number of magnetic spins or electric dipoles with a particular orientation. Similarly, in fluids with fast molecular interconversion, the conservation of the number of alternative molecules may be broken by interconversion, such that phase amplification may occur [19].

In interconverting fluids, the growth of one phase at the expense of another stable phase occurs to avoid the formation of an energetically unfavorable interface, and consequently, two-phase coexistence may be destroyed. Since the formation of the interface is crucial to phase amplification, in macroscopic systems, where the interfacial energy is much smaller than the bulk energy, the system may enter a metastable state in which the energetic benefit of phase amplification is negligible. Thus, the size of the system plays a crucial role in phase amplification.

An example of a system demonstrating both phase amplification and phase separation is an Ising lattice that exhibits a hybrid of diffusion swapping and interconverting flipping dynamics. The outcome of phase formation, which crucially depends on the competition between these two alternative dynamics, is illustrated in Figure 6.16. At each time step, the probability of an individual spin exhibiting either interconversion or diffusion dynamics is given by the ratio $p_r = L/(Dq^2 + L)$, such that for $p_r = 1$ only interconversion dynamics will occur, while for $p_r = 0$ only diffusion dynamics will occur. As shown, phase amplification is only observed when the interconversion rate is much faster than the diffusion rate.

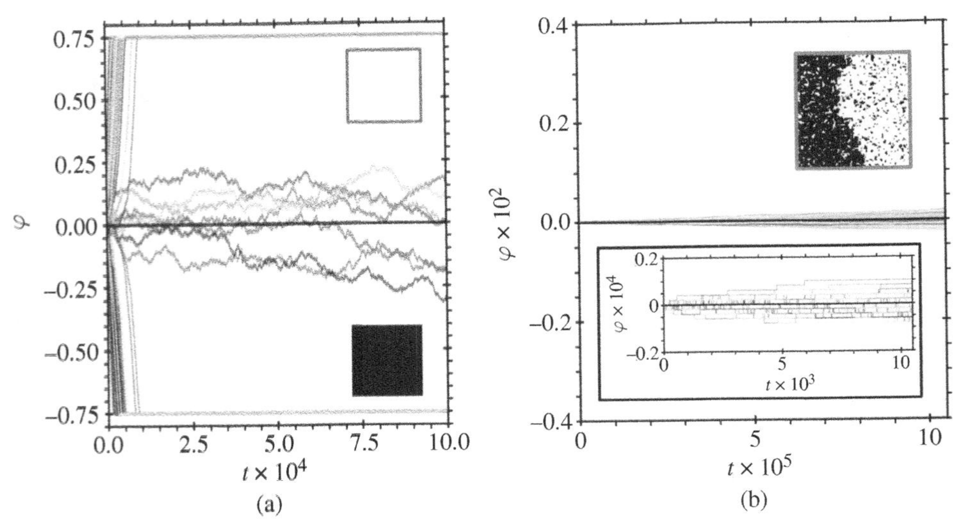

Figure 6.16 Phase amplification in an Ising lattice that exhibits a mixture of diffusion and interconversion dynamics for $N = 100$ realizations of Monte Carlo simulations [19]. (a) At long times, for interconversion probability, $p_r = 1$, one phase grows at the expense of the other. (b) For an extremely small interconversion probability, $p_r = 10^{-9}$, two-phase separation is observed with the sum of alternative values of φ (positive and negative) fluctuating around zero. The inset shows a shorter time scale. The simulation snapshots in (a, b) show the end of the simulation time.

The possibility of phase amplification is particularly relevant for the interpretation of simulations or experiments on substances that exhibit polyamorphism (see Section 2.5), which may originate from interconversion of species. Unlike the idealized symmetric system considered in this section, the volume change and heat of the interconversion reaction may not be zero, which may prevent phase amplification from occurring, depending on the ensemble. For example, in an *NVT* ensemble, phase amplification is restricted by the conservation of matter, such that complete phase separation could be observed. However, in an *NPT* or μVT ensemble, where the size of the simulation box or the number of particles in the system are allowed to fluctuate, then phase amplification could be observed (depending on the size of the system, temperature, and interconversion rate). The possibility of phase amplification has rarely been discussed in studies of polyamorphism. For example, the possibility of phase amplification for liquid–liquid separation in supercooled water models (Figure 2.7) and for fluid–fluid separation in high-pressure hydrogen (Figure 2.9) could be a primary challenge of observing phase separation in these systems [18].

Appendix 6.A Derivation of the Growth Rate in Spinodal Decomposition

Following the discussion in Section 6.5.2, the growth rate equation, Eq. (6.50), is determined from the solution of Eq. (6.49). To verify this solution, we consider a trial function of the form

$$\varphi - \varphi_0 = A(q,t)\cos(\boldsymbol{q}\cdot\boldsymbol{r}) \tag{6.A.1}$$

where A is an unknown amplitude. Note that, due to the symmetry of the system, a choice of cos or sin will yield the same result. Equation (6.A.1) represents plane-wave solutions where φ_0 is the equilibrium value of the order parameter, which is independent of time and position. The left-hand side of the continuity equation, Eq. (6.49), is thus given by

$$\frac{\partial\varphi}{\partial t} = \frac{\partial A(q,t)}{\partial t}\cos(\boldsymbol{q}\cdot\boldsymbol{r}) \tag{6.A.2}$$

while the right-hand side is

$$\breve{\alpha}\hat{\chi}_{q=0}^{-1}\nabla^2\varphi - c_0\breve{\alpha}\boldsymbol{\nabla}^4\varphi = -A(q,t)\breve{\alpha}\left(\hat{\chi}_{q=0}^{-1}q^2 + c_0q^4\right)\cos(\boldsymbol{q}\cdot\boldsymbol{r}) \tag{6.A.3}$$

Equating Eqs. (6.A.2) and (6.A.3), gives

$$\frac{\partial A(q,t)}{\partial t} = -A(q,t)\breve{\alpha}\left(\breve{\chi}_{q=0}^{-1}q^2 + c_0q^4\right) \tag{6.A.4}$$

Equation (6.A.4) is a simple first-order differential equation where, by integrating, the plane-wave amplitude is obtained as

$$A(q,t) = A(q,0)e^{\omega(q)t} \tag{6.A.5}$$

The coefficient, $A(q,0)$, is independent of time. In Eq. (6.A.5), the growth rate of a selected wavenumber, $\omega(q)$, is

$$\omega(q) = -\breve{\alpha}\left(\hat{\chi}_{q=0}^{-1}q^2 + c_0q^4\right) = -\breve{\alpha}\hat{\chi}_{q=0}^{-1}q^2(1 + c_0\hat{\chi}_{q=0}q^2) \tag{6.A.6}$$

In this equation, the prefactor is the negative diffusion coefficient, $D = \breve{\alpha}\hat{\chi}_{q=0}^{-1}$, see Eq. (2.176), while the coefficient of the second term in parenthesis is the correlation length of order parameter fluctuations ξ, see Eq. (4.69). Thus, the growth rate simplifies to

$$\omega(q) = -Dq^2(1 - \xi^2q^2) \tag{6.A.7}$$

With use of Eq. (6.A.5), the expression for the spatial- and time-dependent order parameter is given by

$$\varphi = \varphi_0 + A(q,0)e^{\omega(q)t}\cos(\boldsymbol{q}\cdot\boldsymbol{r}) \tag{6.A.8}$$

As the trial function, Eq. (6.A.1), was defined only for a particular wavenumber, to obtain the complete solution, one must sum over all possible values of q. Thus, the general solution for the spontaneous order parameter in spinodal decomposition is

$$\varphi(\boldsymbol{r},t) = \varphi_0 + \sum_q A(q,0)e^{\omega(q)t}\cos(\boldsymbol{q}\cdot\boldsymbol{r}) \tag{6.A.9}$$

The exact nature of $A(q,0)$ and φ_0 is determined from the initial conditions and the boundary conditions of the system.

References

1 T. Petrova, "Revised release on surface tension of ordinary water substance," *International Association for the Properties of Water and Steam*, 2014.

2 J. D. van der Waals, "Thermodynamic Theory of Capillarity under the Hypothesis of a Continuous Variation of Density," *Verhandel. Konink. Akad. Weten. Amsterdam*, (in Dutch, English Translation by J. S. Rowlinson, *J. Stat. Phys.*, vol. 20, p. 197, 1979).

3 A. A. Novikov, A. P. Semenov, V. Monje-Galvan, et al., "Dual action of hydrotropes," *J. Chem. Phys. C.*, vol. 121, no. 30, pp. 16423–16431, 2017.

4 A. A. Novikov, A. P. Semenov, A. A. Kuchierskaya, et al., "Generic nature of interfacial phenomena in solutions of nonionic hydrotropes," *Langmuir*, vol. 35, no. 41, pp. 13480–13487, 2019.

5 I. C. Sanchez, "Critical amplitude scaling laws for polymer solutions," *J. Phys. Chem.*, vol. 93, no. 19, pp. 6983–6991, 1989.

6 B. Widom, "Phase separation in polymer solutions," *Physica A*, vol. 194, no. 1, pp. 532–541, 1993.

7 M. A. Anisimov and H. St. Pierre, "Diverging curvature correction to the interfacial tension in polymer solutions," *Phys. Rev. E*, vol. 78, p. 011105, 2008.

8 J. S. Rowlinson and B. Widom, *Molecular Theory of Capillarity*. Dover Publications, 2003.

9 M. A. Anisimov, "Divergence of Tolman's length for a droplet near the critical point," *Phys. Rev. Lett.*, vol. 98, p. 035702, 2007.

10 I. Szleifer and B. Widom, "Structure and tension of the interface between dilute polymer solutions," *J. Chem. Phys.*, vol. 90, p. 7524, 1989.

11 J. W. Cahn, "Critical point wetting," *J. Chem. Phys.*, vol. 66, pp. 3667–3672, 1977.

12 V. P. Skripov, *Metastable Liquids*. Wiley, 1974.

13 P. G. Debenedetti, *Metastable Liquids: Concepts and Principles*. Princeton University Press, 1996.

14 J. W. Cahn and J. E. Hilliard, "Free energy of a nonuniform system. I. Interfacial free energy," *J. Chem. Phys.*, vol. 28, p. 258, 1958.

15 J. W. Cahn, "Phase separation by spinodal decomposition in isotropic systems," *J. Chem. Phys.*, vol. 42, pp. 93–99, 1965.

16 J. Zhu, L.-Q. Chen, J. Shen, et al., "Coarsening kinetics from a variable-mobility Cahn-Hilliard equation: Application of a semi-implicit Fourier spectral method," *Phys. Rev. E*, vol. 60, pp. 3564–3572, 1999.

17 J. W. Cahn, "The later stages of spinodal decomposition and the beginnings of particle coarsening," *Acta Metall.*, vol. 14, no. 12, pp. 1685–1692, 1966.
18 T. J. Longo and M. A. Anisimov, "Phase transitions affected by natural and forceful molecular interconversion," *J. Chem. Phys.*, vol. 156, p. 084502, 2022.
19 N. A. Shumovskyi, T. J. Longo, S. V. Buldyrev, et al., "Phase amplification in spinodal decomposition of immiscible fluids with interconversion of species," *Phys. Rev. E*, vol. 103, p. L060101, 2021.

Exercises

6.1 Estimate the liquid–gas surface tension for water at ambient conditions assuming a sharp ("mechanical") interface.

6.2 When the movable boundary separating two phases is not planar, then, at equilibrium, the pressure on both sides of the boundary are no longer equal. For examples, Laplace showed that when two phases are separated by a spherical boundary of radius r, the condition of mechanical equilibrium requires that

$$P_{\text{in}} - P_{\text{out}} = \frac{2\sigma}{r} \tag{6.53}$$

Where P_{in} and P_{out} are the pressures inside and outside the spherical boundary and σ is the interfacial tension between the two phases.
It is possible to cool ultra pure water down to −40°C at 1 atm before crystallization occurs. Under these conditions, the water–ice interfacial tension is 22 dyne/cm, the molar enthalpy of fusion is 6000 J/mole, and the molar volume of ice is 20 cm^3/mole. Estimate the radius of a spherical ice crystal that would form under such conditions. Clearly state and explain any approximations made in the derivation of your result.

6.3 A spherical bubble is enclosed by a thin ethanol film. The bubble is 1 mm in diameter and the film is 10^{-4} mm thick. The bubble is in equilibrium with the surrounding air, which is at 1 bar and 298 K.
For ethanol, the surface tension is believed to be independent of pressure variations but is related to the temperature as

$$\sigma[\text{N/m}] = 0.021\left(1 - \frac{T[\text{K}] - 298}{298}\right) \tag{6.54}$$

The surface heat capacity is approximated as a constant:

$$C_{\mathcal{A}} = T\left.\frac{\partial S}{\partial T}\right|_{\mathcal{A}} = 3\ \text{J/gK} \tag{6.55}$$

The density of ethanol is assumed independent of temperature or pressure and equals 0.79 g/cm^3 and the heat of vaporization of ethanol $\Delta H = 930$ J/g (at 298 K).

(a) Evaluate the surface energy of the ethanol film, $(\partial U^{\mathcal{A}}/\partial \mathcal{A}_t)|_{P,T}$ in J/m^2.
(b) A small tube is inserted into the bubble and air (at 298 K) is blown in so that the bubble rapidly expands to a diameter of 2 mm. Neglect any heat or mass transfer from the air to the liquid film and assume that the temperature dependence of the film is uniform throughout. What is the temperature in this film after the expansion?

6.4 Calculate the critical radius of nucleation for water vapor at 100°C when it is over-compressed by 1 bar. Explain the physical meaning of your result.

6.5 Using the Landau–Ginzburg free-energy functional:
(a) Derive the density profile for a smooth surface. Plot the density profile as a function of the density change along the z-direction.
(b) Using your equation for the density profile, determine the surface tension.

6.6 Explain the origin of Tolman's length (δ'). For what systems is Tolman's length nonzero? Provide two examples.

6.7 Plot the symmetric and asymmetric density profiles on a single plot. What factors affect the asymmetric corrections to the density profile?

6.8 (a) What is a "wetting phase transition?" Explain the thermodynamic origin of wetting transitions. What factors determine the wetting temperature, T_w?
(b) Derive the condition (Antonov's rule) for a wetting transition based on the behavior of surfaces tensions between two critical fluid phases and one noncritical phase in the meanfield approximation.

6.9 Explain the difference between the driving forces for new phase formation in metastable fluids and in thermodynamically unstable fluids (rapidly quenched under spinodal). What are these two processes are called? Give practical examples (substances, conditions) of these alternative processes.

6.10 Consider a simplified model for the critical part of the free energy density, $\Delta\hat{f}_{cr} = k\varphi^2(1 - \varphi^2)$, where k is a constant. Using the variational derivative of the Landau–Ginzburg function with this free energy, obtain the reduced chemical potential for this model. Assuming $k = -1/2$, $c_0 = 1$, plot the reduced chemical potential assuming that the spatially-dependent order parameter has the form, $\varphi = 3\cos(2z)$, where z is a Cartesian coordinate.

6.11 Derive the amplification factor for a system exhibiting spinodal decomposition in the presence of molecular interconversion.

6.12 What system-dependent factors enhance phase amplification in a system that exhibits phase formation and molecular interconversion? Provide at least three examples in your answer.

7

Self-Assembly and Modulated Phases

This chapter includes:

- Equilibrium polymerization
- Thermodynamics of micellization
- Gelation and percolation
- Modulated phases, microemulsions
- Mesoscopic steady-state dissipative structures

The description of phenomena through the Landau–Ginzburg free-energy functional could be extended to the formation of self-assembled and modulated structures in condensed matter. Mesoscopic structure may generally emerge as a result of the interplay between complex inter- and intramolecular interactions. Most notable examples are equilibrium polymerization or the self-assembly of amphiphilic molecules.

7.1 Supramolecular Self-Assembly

One of the most amazing phenomena that occur at the mesoscale is the self-organization of molecules into supramolecular structures. As we have already discussed, in phenomena like nucleation and spinodal decomposition (see details in Section 6.5), new phases can emerge from preexisting ones. However, in equilibrium polymerization, polymer chains are built from a collection of monomers via a chemical reaction. Stable molecular aggregates may also be formed due to strong attractive forces between molecules, such as the formation of mesoscopic (~10–20 nm long) rods from disc-like monomers of cromolyn (disodium cromoglycate) in aqueous solution below room temperature [1, 2].

Similarly, the formation of micelles is also driven by the association of small molecules. However, in this case, these molecules are amphiphilic, such as sodium stearate (common soap) and sodium alcohol sulfates. The substances containing amphiphiles are known as "surfactants" or "detergents." Amphiphilic molecules are unique in that they are made up of both hydrophobic and hydrophilic parts, as illustrated in Figure 7.1. There are two major types of amphiphilic molecules: ionic and nonionic. In ionic amphiphiles, like ordinary soaps, the hydrophobic groups may dissociate in aqueous solutions, forming ions. Thus, the interactions between hydrophobic groups are electrostatic. In contrast, nonionic amphiphiles have covalently bonded hydroxyl groups, which interact with water molecules in solution via hydrogen bonds. Just as in the mixture of oil and water,

Mesoscopic Thermodynamics for Scientists and Engineers, First Edition. Mikhail A. Anisimov and Thomas J. Longo.
© 2024 John Wiley & Sons, Inc. Published 2024 by John Wiley & Sons, Inc.

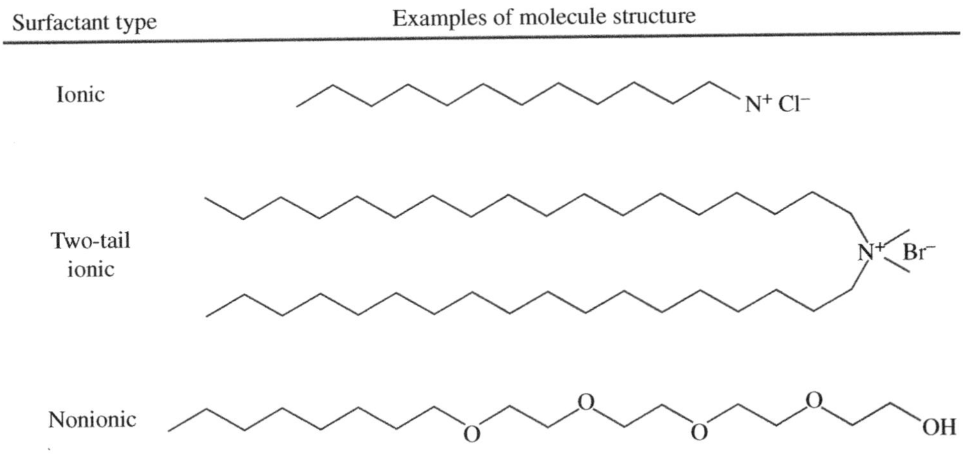

Figure 7.1 Typical examples of amphiphilic molecules: $C_nH_n–NH_3Cl$ (ionic), DDAB (two-tail ionic), and $C_8H_{17}–4\cdot(C_2H_4O)H$ (nonionic).

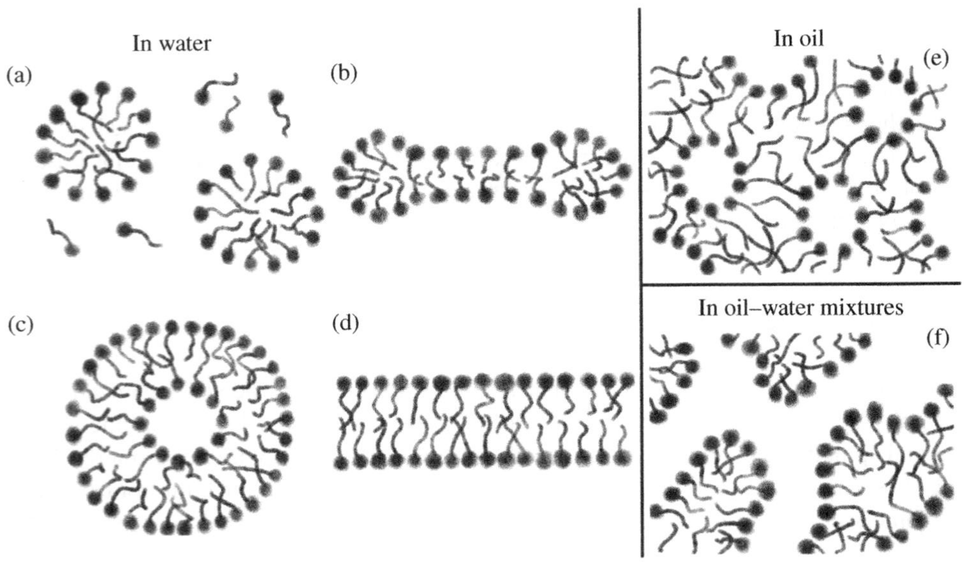

Figure 7.2 Possible supramolecular structures of amphiphiles in solutions: (a) spherical micelles, (b) cylindrical micelle, (c) vesicle, (d) bilayer, (e) inverse spherical micelles, and (f) bicontinuous microemulsion structure. Each amphiphilic structure, originally formed in water or in oil, may correspondingly be solubilized by oil or water.

the hydrophobic groups try to avoid water molecules. When amphiphilic molecules are surrounded by water, the hydrophobic parts may screen the hydrophilic parts from the water, thereby, leading to the formation of supramolecular structures.

In dilute solution, these supramolecular structures adopt a form of spherical aggregates, known as "micelles." Depending on the concentration and chemical structure of the amphiphilic molecules, more complex supramolecular structures may form, such as cylindrical micelles, vesicles, or bilayers, as illustrated in Figure 7.2.

In solutions of amphiphiles in hydrocarbons or lipids ("oil"), the hydrophobic parts of the amphiphilic molecules screen the hydrophilic parts from the solvent, such that inverse micelles may form. If oil is added to an aqueous solution of amphiphiles, then the micelles may solubilize the oil within their hydrophobic cores. In a multicomponent solution of water, oil, amphiphile, and alcohol, a sponge-like phase, known as a "microemulsion," may be formed, also presented in Figure 7.2. The structures formed by amphiphilic molecules are commonly used in pharmaceuticals, food, and drug industries [3].

7.2 Equilibrium Polymerization

One example of self-assembly is the equilibrium polymerization of monomers into a long chain [4]. The transition to polymeric liquid sulfur, at a temperature about ~433 K, is an example of such self-assembly occurring without phase separation. The properties of sulfur near the polymerization transition are completely reversible, similar to a continuous phase transition. A theory by Tobolsky and Eiseberg [5] describes the temperature dependence of the extent of polymerization in terms of a second-order phase transition in the meanfield approximation. The polymerization in sulfur, like in many other real materials, is more complicated because the polymerization of sulfur into its supramolecular structure occurs upon heating, since (below $T \sim 433$ K) liquid sulfur contains octamers that need to be broken before undergoing polymerization [6].

In this section, we consider a simplified version of polymerization into linear flexible chains within the spirit of the Flory–Huggins theory (see Section 5.3). In this case, the degree of polymerization ($\mathcal{N}$), the number of monomers in the chain, depends on the nature of the reacting chemical bonds and the thermodynamic conditions of the reaction, $\mathrm{s} \rightleftharpoons \mathrm{m}$, where s refers to the polymerizing molecules of solvent and m refers to the monomers in the polymer chain. The system could be viewed as a solution of long-polymer chains in a solvent of its "own monomers." Thus, the reaction may be considered as interconversion between the "free monomers" (solvent molecules), s, and the monomers in the polymer chain, m.

In the Flory–Huggins theory, the reduced Gibbs energy per total number of monomers (free and polymerized), $\hat{G} = G_t/Nk_BT$, is given by Eq. (5.18). In this case, the reaction coordinate is the fraction of polymerized monomers. Chemical equilibrium is established when the derivative of the Gibbs energy with respect to the reaction coordinate is zero, see Section 2.9. Applying the chemical reaction equilibrium condition yields,

$$\left.\frac{\partial \hat{G}}{\partial \varphi}\right|_T = \hat{G}_{\mathrm{ms}} - \left.\frac{\partial(\Delta \hat{G}_{\mathrm{mix}})}{\partial \varphi}\right|_T = 0 \tag{7.1}$$

where the reduced Gibbs energy change of the reaction, $\hat{G}_{\mathrm{ms}} = \hat{\mu}^{\circ}_{\mathrm{m}} - \hat{\mu}^{\circ}_{\mathrm{s}}$ is the difference between the chemical potentials of pure monomers (free and polymerized, respectively). This difference defines the chemical reaction equilibrium constant K, as $\hat{G}_{\mathrm{ms}} = -\ln K$. Furthermore, the derivative of the reduced Gibbs energy of mixing is given by

$$\left.\frac{\partial(\Delta \hat{G}_{\mathrm{mix}})}{\partial \varphi}\right|_T = -\ln(1-\varphi) + \frac{1}{\mathcal{N}}\ln\varphi - \left(1 - \frac{1}{\mathcal{N}}\right) + \varpi(1-2\varphi) \tag{7.2}$$

For a solution of a polymer chain in its monomers (a good solvent), one may neglect the enthalpy of mixing, such that $\varpi = 0$. Additionally, in the case of an infinite degree of polymerization ($\mathcal{N} \to \infty$), this equation simplifies to

$$\left.\frac{\partial(\Delta\hat{G}_{mix})}{\partial\varphi}\right|_T = -\ln(1-\varphi) - 1 \tag{7.3}$$

Applying the reaction equilibrium condition, Eq. (7.1), one obtains the relation,

$$\ln K(T,P) = -\left.\frac{\partial(\Delta\hat{G}_{mix})}{\partial\varphi}\right|_T = \ln(1-\varphi) + 1 \tag{7.4}$$

Neglecting the temperature dependence of the heat of reaction, defined through a Gibbs–Helmholtz relation, Eq. (2.30) as, $\hat{H}_{ms} = T^2\partial\hat{G}_{ms}/\partial T$, the equilibrium constant is given as

$$\ln K = \frac{H_{ms}}{k_B}\left(\frac{1}{T_0} - \frac{1}{T}\right) \tag{7.5}$$

where T_0 is the temperature where $\ln K = 0$. Thus, the temperature dependence of the equilibrium fraction of polymerized monomers is given by

$$\ln(1-\varphi) + 1 = \frac{H_{ms}}{k_B}\left(\frac{1}{T_0} - \frac{1}{T}\right) \tag{7.6}$$

A plot of the fraction of polymerized polymers as a function of temperature is shown in Figure 7.3. The fraction of polymerized monomers is zero at the temperature

$$T_p \equiv T_0\left(\frac{k_B T_0}{H_{ms}} - 1\right)^{-1} \tag{7.7}$$

which is the maximum temperature at which the polymer chain may form. Upon cooling below T_p, the fraction of monomers in the polymer chain increases, reaching $\varphi = 1$ at $T = 0$, meaning

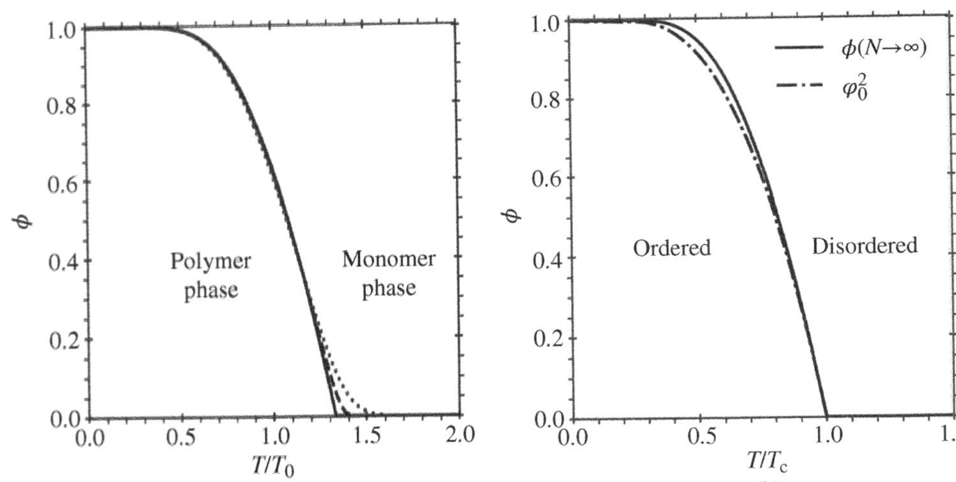

Figure 7.3 (a) Fraction of polymerized monomers as a function of temperature for different degrees of polymerization plotted for $T_0 = 0.5$ and $H_{ms}/k_BT_0 = 4$. Dotted curve corresponds to $\mathcal{N} = 10$, dashed curve to $\mathcal{N} = 30$, and solid curve to $\mathcal{N} \to \infty$. In the limit $\mathcal{N} \to \infty$, the polymerization sharply starts at $T = (4/3)T_0$. (b) Comparison of the sharp polymerization at $\mathcal{N} \to \infty$ with a second-order phase transition in zero field. The fraction of polymerized molecules, $\varphi = |\psi|^2$, where ψ is the order parameter for the polymer chain (solid), while the dash-dotted curve represents the square of the spontaneous order parameter, ϕ_0^2, for the Ising model (see Figure 4.3a). In the linear approximation (near the transition point), the behavior of the two properties become identical.

that, at this temperature, all monomer molecules in the solution have polymerized. For a finite degree of polymerization, the temperature dependence of the equilibrium fraction of polymerized monomers is given by

$$\ln(1-\varphi) - \frac{1}{\mathcal{N}}\ln\varphi + \left(1 - \frac{1}{\mathcal{N}}\right) = \frac{H_{\text{ms}}}{k_{\text{B}}}\left(\frac{1}{T_0} - \frac{1}{T}\right) \tag{7.8}$$

As shown in Figure 7.3, the formation of the polymer chain with a finite degree of polymerization is continuous across T_0. Note that the formation of a polymer with a finite degree of polymerization, like any chemical reaction, is not a thermodynamically defined phase transition.

The behavior of the polymerized monomers with $\mathcal{N} \to \infty$, below T_{p}, changes linearly with $\Delta T = T_{\text{p}} - T$. This is a profound analogy between the sharp polymerization into the infinite polymer chain and a second-order phase transition, as $\varphi = |\psi|^2 \propto T_{\text{c}} - T$, where $T_{\text{c}} = T_{\text{p}}$ and ψ is the de Gennes order parameter characterizing the behavior of the polymer chain (see Section 5.6). The linear behavior of the observable property, the fraction of polymerized monomers φ, near the transition temperature (shown by the solid line in Figure 7.3a), follows from the truncated Landau expansion in powers of the polymer-chain order parameter ψ, given by Eq. (5.38). This analogy is confirmed by a direct comparison with the order parameter of the Ising model in the meanfield approximation for the full range of temperatures from $T = T_{\text{c}}$ to $T = 0$ (see Figure 7.3b).

7.3 Micellization

Unlike the self-assembly of molecules into a long polymer chain, self-assembly through micellization is restricted by the geometry of the amphiphilic molecules and their interactions with the solvent. In the upcoming sections, we describe the thermodynamics of micelle formation (7.3.1), the geometry of the different supramolecular structures (7.3.2), and the effects of fluctuations of the aggregate number, N, the number of amphiphilic molecules in a micelle (7.3.3).

7.3.1 Thermodynamics of Micellization

Consider the formation of mono-disperse micelles with a certain average aggregation number, $\overline{N}$, in a dilute solution. The Gibbs energy per total number of particles of the ternary system has the form

$$G = \mu_{\text{s}}x_{\text{s}} + \mu_{\text{a}}x_{\text{a}} + \mu_{\text{m}}x_{\text{m}} \tag{7.9}$$

where the molecular fractions x_{s}, x_{a}, and x_{m} refer to the fraction of solvent (water or oil), the fraction of free amphiphilic molecules in solution, and the fraction of micelles, respectively. Correspondingly, μ_{s}, μ_{a}, and μ_{m} refer to the chemical potentials of each species.

One may express the Gibbs energy in a simplified form by defining $\varphi_{\text{m}} = \overline{N}x_{\text{m}}$, where φ_{m} is the molecular fraction of amphiphilic molecules aggregated into micelles. In addition, one may express the total concentration of amphiphilic molecules as $x = x_{\text{a}} + \varphi_{\text{m}}$. In this way, the reduced Gibbs energy $\hat{G} \equiv G/k_{\text{B}}T$ may be written as

$$\hat{G} = \hat{\mu}_{\text{s}}^{\circ}x_{\text{s}} + \hat{\mu}_{\text{a}}^{\circ}(x - \varphi_{\text{m}}) + \hat{\mu}_{\text{m}}^{\circ}\varphi_{\text{m}} + \Delta\hat{G}_{\text{mix}} \tag{7.10}$$

where $\hat{\mu}_{\text{s}}^{\circ}$, $\hat{\mu}_{\text{a}}^{\circ}$, and $\hat{\mu}_{\text{m}}^{\circ}$ correspond to the reduced (by $k_{\text{B}}T$) chemical potentials of pure solvent, pure free amphiphilic molecules, and pure aggregated amphiphilic molecules, respectively. The reduced Gibbs energy of mixing, $\Delta\hat{G}_{\text{mix}} \equiv \Delta G_{\text{mix}}/k_{\text{B}}T$ is given by

$$\Delta\hat{G}_{\text{mix}} = x_{\text{s}}\ln x_{\text{s}} + (x - \varphi_{\text{m}})\ln(x - \varphi_{\text{m}}) + \frac{\varphi_{\text{m}}}{\overline{N}}\ln\left(\frac{\varphi_{\text{m}}}{\overline{N}}\right) \tag{7.11}$$

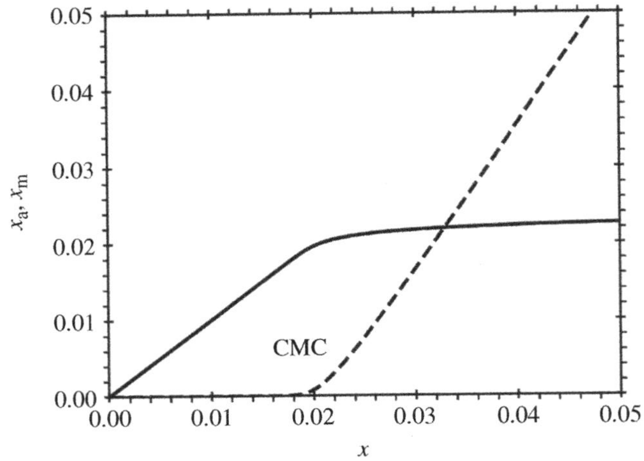

Figure 7.4 The relation between the fraction of free amphiphiles (x_a, solid) and amphiphiles aggregated into micelles (x_m, dashed) as a function of the total fraction of amphiphiles in solution ($x = x_a + x_m$) for a system with $\overline{N} = 30$ and $\hat{\mu}^\circ_{ma} = -3.0$. Below the critical micelle concentration (CMC), $x^* \approx 0.02$, the system consists mostly of free amphiphiles. Above the CMC, almost all of the added amphiphiles are aggregated into micelles. When $\overline{N}$ is large but finite, the micellization is rounded with an uncertainty in the concentration as $\delta x \simeq \ln(x^*/\overline{N})/\overline{N}$.

Note that there is a similarity between the aggregation number, $\overline{N}$, and the degree of polymerization, $\mathcal{N}$, in the Flory–Huggins theory of polymer solutions, see Section 5.3. However, the entropy of mixing in micellar solutions is different, as indicated by the last logarithmic term in Eq. (7.11). The difference is associated with the fact that the entropy of micelles has an additional restriction because the amphiphilic molecules are not free to move within the aggregates (unlike the monomers in an ideal polymer chain). Thus, this term reads as $(\varphi_m/\overline{N})\ln(\varphi_m/\overline{N})$, while in the Flory–Huggins theory, it reads as $(\varphi/\mathcal{N})\ln\varphi$, see Eq. (5.15).

One may define the reduced exchange chemical potential between the free amphiphilic molecules and those in the micelle, $\hat{\mu}_{ma} = \hat{\mu}_m/\overline{N} - \hat{\mu}_a$, as

$$\hat{\mu}_{ma} = \left.\frac{\partial \hat{G}}{\partial \varphi_m}\right|_{T,P} = \frac{\hat{\mu}^\circ_m}{\overline{N}} - \hat{\mu}^\circ_a + \frac{1}{\overline{N}}\ln\left(\frac{\varphi_m}{\overline{N}}\right) - \ln(x - \varphi_m) \tag{7.12}$$

Analogous to the self-assembly by polymerization (see Section 7.2), the process of micelle formation may be viewed as an interconversion between free amphiphilic molecules and those aggregated into micelles. The equilibrium condition of interconversion is $\partial\hat{G}/\partial\varphi_m|_{T,P} = 0$, such that one obtains

$$\hat{\mu}^\circ_{ma} \equiv \frac{\hat{\mu}^\circ_m}{\overline{N}} - \hat{\mu}^\circ_a = -\frac{1}{\overline{N}}\ln\left(\frac{\varphi_m}{\overline{N}}\right) + \ln(x - \varphi_m) \tag{7.13}$$

This condition restricts the number of independent species in solution. In this case, $\hat{\mu}^\circ_{ma}$ is analogous to the Gibbs energy of reaction, as discussed in Section 2.9. Equation (7.13) may be solved for the equilibrium fraction of amphiphilic molecules in the micelles as

$$\boxed{\varphi_m = \overline{N}\left[(x - \varphi_m)e^{-\hat{\mu}^\circ_{ma}}\right]^{\overline{N}}} \tag{7.14}$$

Within this equation, we may define the critical micelle concentration (CMC) as

$$x^* \equiv \frac{c^*}{\rho} = e^{\hat{\mu}^{\circ}_{\mathrm{ma}}} \tag{7.15}$$

Thus, the exchange chemical potential of pure species represents the minimal work required to transfer an amphiphilic molecule from the solution to the micelle, such that the CMC is the minimum concentration of amphiphiles required in solution for micelles to form. When $\hat{\mu}^{\circ}_{\mathrm{m}}/\overline{N} = \hat{\mu}^{\circ}_{\mathrm{a}}$, then $\hat{\mu}^{\circ}_{\mathrm{ma}} = 0$ and $x^* = 1$. This condition indicates that micelles will never form. However, for large micelles when $\overline{N} \gg 1$, then $\hat{\mu}^{\circ}_{\mathrm{m}}/\overline{N} \ll \hat{\mu}^{\circ}_{\mathrm{a}}$, $\hat{\mu}^{\circ}_{\mathrm{ma}} < 0$, and $x^* \ll 1$. This conditions indicates that there is a strong driving force for micellization. In this case, near the onset of micellization (CMC), Eq. (7.14) may be approximated as

$$\varphi_{\mathrm{m}} \cong x - x^* \tag{7.16}$$

The process of micellization is illustrated in Figure 7.4. Note that φ_{m} is rounded across the CMC for finite $\overline{N}$, and only for $\overline{N} \to \infty$ does Eq. (7.16) becomes exact, analogous to the behavior of the fraction of polymerized monomers in the limit of an infinite degree of polymerization (see Figure 7.3). The region around which the linear relation between φ_{m} and x, Eq. (7.16), is valid can be estimated from the expansion in powers of $\delta x = x - x^*$ of Eq. (7.13) as

$$\delta\varphi_{\mathrm{m}} = \varphi_{\mathrm{m}} - \delta x \approx \frac{1}{\overline{N}} \ln\left(\frac{x^*}{\overline{N}}\right) \delta x^2 \tag{7.17}$$

Thus, the formation of micelles with a finite number of amphiphiles $\overline{N}$ is not, strictly speaking, a phase transition, only becoming a thermodynamically defined sharp transition in the limit $\overline{N} \to \infty$.

If the length of the hydrophobic part of an amphiphilic molecule is comparable to the size of the hydrophilic part, the driving force of micellization, $\hat{\mu}^{\circ}_{\mathrm{ms}}$, is weak and, consequently, the aggregation number is small (just a few molecules), while the CMC is large and significantly rounded. Substances consisting of amphiphiles with short hydrophobic tails, such as low-molecular-weight alcohols and amines, are known as "hydrotropes" [7]. Hydrotropes may be completely or significantly water-soluble and can increase the solubility of hydrophobic substances ("oil") in water, serving as co-solvents. Like surfactants, small additions of a hydrotrope reduce the interfacial tension between water and oil. Thus, hydrotropes are effectively used as co-surfactants for stabilization of microemulsions.

While hydrotropes do not form stable micelles in aqueous solutions (due to their small hydrophobic tail), they may form short-lived aggregates. Molecular dynamics (MD) simulations of 5–7 mol% tertiary butanol (TBA) in an aqueous solution have demonstrated the formation of such dynamic noncovalent micellar-like clusters (aggregates) [8]. These clusters have a size on the order of 1 nm and a lifetime from dozens to hundreds of picoseconds, being stabilized by the hydrogen bonds between the hydrophilic parts of the hydrotrope molecules and water [9]. These short-living aggregates can also be viewed as structural or pre-micellar fluctuations in solutions. Such fluctuations of structure are revealed by the fluctuations of entropy (see Section 3.11.2) and are manifested by a sharp peak of the isobaric heat capacity (Eq. 3.34) [10], see Figure 7.5.

The formation of micelles in solution is accompanied by a sharp change in the surface tension between the liquid and gaseous phases or between two liquid phases. Upon the initial addition of amphiphilic molecules, the interface is coated by the amphiphiles. As the amphiphiles accumulate on the interface (a process known as "Gibbs adsorption"), the surface tension decreases proportionally to the amphiphilic concentration. Once the surface has been saturated, then micelles will form in the aqueous phase, and the surface tension will no longer change. This represents an accurate method to find the CMC in experiment, see the qualitative illustration in Figure 7.6.

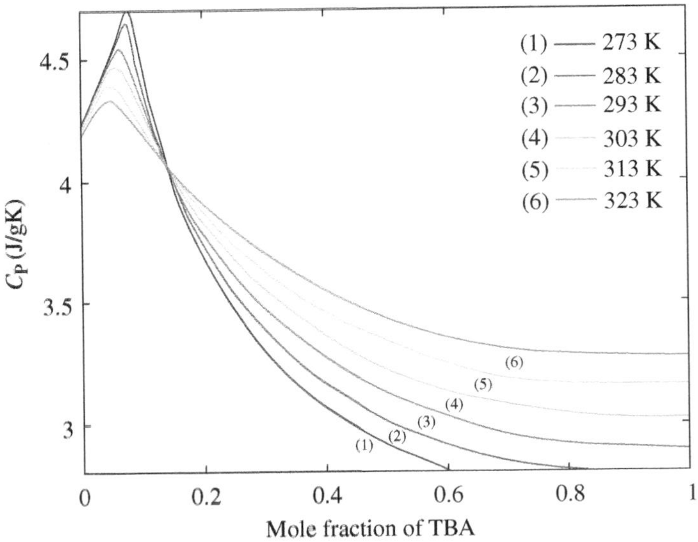

Figure 7.5 The anomaly of the isobaric heat capacity in an aqueous solution of tertiary butanol (TBA) indicating large fluctuations of the entropy. Source: Adapted from Ref. [10].

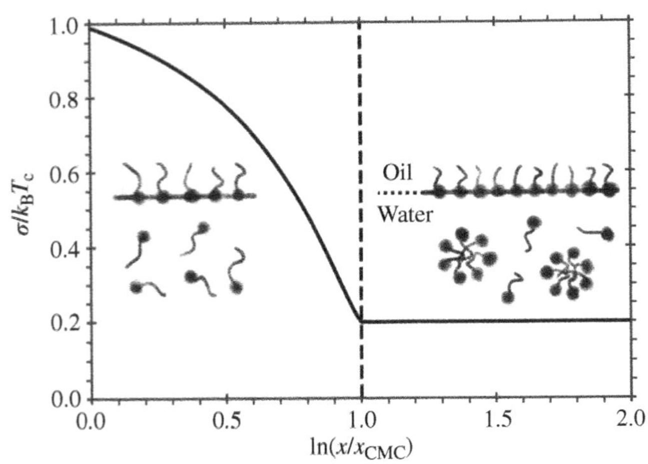

Figure 7.6 The water–oil interfacial tension in a surfactant solution. As the concentration of amphiphilic molecules accumulate along the interface, the surface tension decreases until the CMC is reached, see Eq. (7.19).

The change in the surface tension with respect to the concentration of amphiphiles in solution may be quantitatively obtained from the "Gibbs adsorption isotherm." At constant temperature and pressure, the adsorption isotherm is given by the Gibbs–Duhem equation, Eq. (2.60), as

$$\sum N_i^{\mathcal{A}}\, \mathrm{d}\mu_i + \mathcal{A}_t\, \mathrm{d}\sigma = 0 \tag{7.18}$$

where $N_i^{\mathcal{A}}$ is the number of particles of species i on the surface, defined by the difference of the total number of particles and the number of particles in each of the bulk phases ($\tilde{\alpha}$ and $\tilde{\beta}$) as $N_i^{\mathcal{A}} = N_i - N_i^{\tilde{\alpha}} - N_i^{\tilde{\beta}}$. Assuming an ideal mixing of species, see Section 2.8, the surface concentration of component i, Γ_i, is given by

$$\Gamma_i = \frac{N_i^A}{\mathcal{A}_t} = -\frac{1}{RT}\left(\frac{\partial\sigma}{\partial \ln c_i}\right)_{T,P} \tag{7.19}$$

The change in the surface tension with respect to the concentration of amphiphiles is illustrated in Figure 7.6.

7.3.2 Building a Nanoparticle Through Micellization

In nanotechnology, many processes are based on building nanoparticles through micellization [11]. In this section and the proceeding section, we consider the conditions for which monodisperse micelles may be formed with a stable aggregation number. First, we specify the physical meaning of the driving force for micellization and the effects of the geometry of the amphiphilic molecules on the shape of the self-assembled structures. This can be done with simple geometrical arguments, in which the amphiphilic molecules may be described through an area occupied by the hydrophilic head a_h, the length of the hydrophobic part ℓ, and the volume of the molecule v_0.

The area, a_h, is the area occupied by a single hydrophilic head on the surface of the supramolecular structure. If too many head-groups are forced to be close together, they will repel each other. If they are too far apart, the continuous structure of the surface may be broken since the hydrophobic tails may come into contact with the water molecules [12]. Thus, the chemical potential of the micellized amphiphilic molecules may be approximated by

$$\boxed{\frac{\hat{\mu}_m^o}{\overline{N}} = \frac{\kappa}{a_h} + \sigma a_h} \tag{7.20}$$

where κ is a repulsion constant between head-groups, such as an electrostatic repulsion between equally charged ions in ionic amphiphiles and a hard-core repulsion in nonionic amphiphiles, while σ is the effective interfacial tension between "oil" (hydrocarbons or lipids) and water. This model is depicted in Figure 7.7. The minimum of the chemical potential determines the optimum (equilibrium) area occupied by each amphiphilic head, $\bar{a}_h = \sqrt{\kappa/\sigma}$.

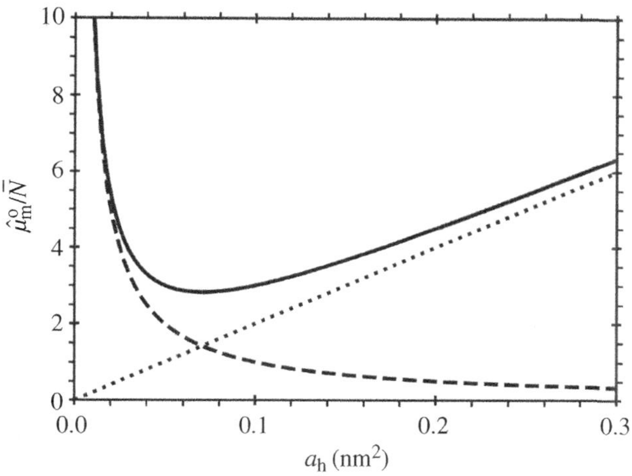

Figure 7.7 The chemical potential of amphiphiles in a micelle (solid), as given by Eq. (7.20). The contribution to the chemical potential due to the repulsion between amphiphilic heads is shown by the dashed curve, while the surface contribution is shown by the dotted curve. The optimum head area, $\bar{a}_h$, is at the minimum of the solid curve.

Possibilities of Supramolecular Shapes

The conditions for the formation of a particular shape of a supramolecular structure may be obtained from the ratio of the volume of the molecule to the area occupied by the amphiphilic head on the surface of the structure. From the volume to surface area ratio, the characteristic length scale of the structure may be treated with respect to the length, ℓ, in one of three categories:

- For spheres:

$$\frac{v}{a_h} \leq \frac{\ell}{3} \tag{7.21}$$

- For cylinders:

$$\frac{\ell}{3} < \frac{v}{a_h} \leq \frac{\ell}{2} \tag{7.22}$$

- For vesicles or flexible bilayers:

$$\frac{\ell}{2} < \frac{v}{a_h} \leq \ell \tag{7.23}$$

- For lamellae (planar bilayers):

$$\frac{v}{a_h} \approx \ell \tag{7.24}$$

- For inverse micelles:

$$\frac{v}{a_h} > \ell \tag{7.25}$$

7.3.3 Fluctuations of Aggregation Number

The equilibrium fraction of surfactant molecules in micelles assumes that every micelle has the same (optimal) aggregation number, $N = \overline{N}$. However, this number could fluctuate as amphiphilic molecules enter and exit the micelle. Consider a simple example of a spherical micelle with radius ℓ (approximately equal to the length of the hydrophobic part of a single amphiphilic molecule), aggregation number N, and surface area per hydrophilic head a_h. Additionally, assuming that the amphiphilic molecule has the geometry of a cone, then the aggregation number is equivalent to the number of monomers that can fit into one micelle, given by

$$N = \frac{\text{Micelle Volume}}{\text{Amphiphile Volume}} = \frac{(4/3)\pi\ell^3}{(1/3)a_h\ell} \tag{7.26}$$

which simplifies to

$$N = \frac{4\pi\ell^2}{a_h} \tag{7.27}$$

Thus, the aggregation number is given by the ratio of the surface area of the micelle to the surface area per hydrophilic head, a_h. The typical value of the head group area is $a_h \approx 0.2\,\text{nm}^2$ and the length of the hydrophobic tail is about $\ell \approx 1\,\text{nm}$, such that the aggregation number is $N \sim 50$.

Since the aggregation number may fluctuate, variations in N are given by $\delta N = N - \overline{N}$. As discussed in Section 7.3.1, the work of a fluctuation of the aggregation number is given by the difference in chemical potentials of pure amphiphiles in the micelle and that of those in the solution,

Eq. (7.13). With use of Eq. (7.27), the expression for the chemical potential for the amphiphilic molecules in the micelle, Eq. (7.20), is given as

$$\frac{\hat{\mu}_{\mathrm{m}}^{\circ}}{\overline{N}} = \frac{\kappa}{4\pi\ell^2}\overline{N} + \sigma\frac{4\pi\ell^2}{\overline{N}} \tag{7.28}$$

The work of a fluctuation of the aggregate number in the lowest approximation (see Eq. 3.42) may be given by

$$\frac{W_N}{k_{\mathrm{B}}T} = \frac{1}{2\overline{N}}\frac{\partial^2(\hat{\mu}_{\mathrm{m}}^{\circ}/\overline{N})}{\partial\overline{N}^2}\delta N^2 = \frac{1}{2\overline{N}}\hat{\chi}_{\mathrm{m}}^{-1}\delta N^2 \tag{7.29}$$

where $\hat{\chi}_{\mathrm{m}}$ is the corresponding susceptibility with respect to the fluctuation of the aggregation number (see the discussion of the generalized susceptibility in Section 3.5), obtained as

$$\hat{\chi}_{\mathrm{m}} = \frac{\overline{N}^3}{8\pi\ell^2\sigma} = \frac{\overline{N}}{2\sigma\bar{a}_{\mathrm{h}}} \tag{7.30}$$

Following the same procedure for obtaining the fluctuations of a generalized density (Section 3.5) and assuming that the length ℓ does not fluctuate, one obtains the relative mean-square fluctuations of the aggregate number as

$$\boxed{\frac{\langle (N-\overline{N})^2\rangle}{\overline{N}^2} = \frac{k_{\mathrm{B}}T}{\overline{N}}\hat{\chi}_{\mathrm{m}} = \frac{k_{\mathrm{B}}T}{8\pi\ell^2\sigma}} \tag{7.31}$$

Thus, the fluctuations of the aggregate number increase for smaller aggregate numbers and smaller oil–water surface tensions. For example, consider an oil–water interfacial tension, $\sigma \approx 30\,\mathrm{mN/m}$ and amphiphilic length, $\ell \approx 1\,\mathrm{nm}$, Eq. (7.31) predicts fluctuations of the aggregate number on the order ~0.5%. In addition to fluctuations of N, if the length of the hydrophobic chain could fluctuate, then the radius of the micelle would also fluctuate. Therefore, to build a robust nanoparticle, without large fluctuations of the aggregate number, the length of the hydrophobic tail and the oil–water interfacial tension must be large. Both parameters are mainly controlled by the chemical and physical properties as well as the geometry of the hydrophobic groups of the amphiphilic molecules.

Note that the Gaussian distribution of the fluctuations of the aggregation number is an approximation only valid for spherical micelles. In the case of cylindrical micelles, the distribution of the aggregation number is not Gaussian. Depending on the geometry of the amphiphilic molecules, their chemical potential in the micelle may not have a minimum at a well-defined aggregation number. Consequently, cylindrical micelles are characterized by a huge and an asymmetric distribution of the aggregation number. For bilayers, there is a qualitatively different situation. The number of individual aggregates is negligible as the bilayer occupies almost the entire space of the system. For more information about the various supramolecular structures, see the extensive literature on this subject, e.g. [13–15].

7.4 Gelation

Gelation is another example of a process of self-assembly, in which molecules form a network. A gel is a material composed of "monomers" (either molecules or molecular clusters) that are able to form a macroscopic network. The monomers are connected by bonds, where the unconnected, or isolated, monomers form a typical liquid. The liquid is comprised of unconnected monomers, referred to as a "sol." When the network structure is formed, the material can be described as being a

Figure 7.8 Schematic representation of a gel network.

soft solid and is referred to as a "gel." A key characteristic of a gel is that if a shear force is temporarily applied (assuming the force is not too strong), the gel will return to its original shape upon the removal of the applied force. In addition to solid-like properties, gels can also have liquid-like properties, in that they are structurally disordered and may contain a considerable fraction of liquid solvent (Figure 7.8).

If a shear modulus appears only at finite frequencies (and disappears in the zero-frequency limit), the gel may be described by the same theory of elasticity as for glasses (see Section 2.5) [13, 16]. Gels with a high density of short rigid segments are likely to be "glassy."

There are two types of gels that may form under equilibrium conditions. A chemical gel is one in which multifunctional monomers form covalent chemical bonds (like an epoxy resin or rubber). Alternatively, a physical gel is one in which the formation of the network depends on the temperature, thereby making the process of physical gelation analogous to a phase transition, such that gelation is thermally reversible. A common example of a physical gel is gelatin, where the network is formed with the hydrogen bonds from hydrolyzed collagen.

The question of the nature of the network may be understood by a microscopic theory, while mesoscopic thermodynamics seeks to answer another question: "what are the general features of the thermo-reversible transition between a sol and a gel?" If this transition is (almost) as sharp as a usual phase transition, one can hope to find similar elements of universality connecting the behavior of apparently different systems. The fact that gelation is often not sharp should not discourage us as we have already considered examples of "rounded" transitions (e.g. micellization or polymerization).

The formation of an infinite gel network is similar to the symmetry breaking that occurs at phase transition points. A model, known as the "percolation theory," can describe the generic features of the sol-gel transition. The simplest version of the percolation model is percolation on a lattice, in which bonds are randomly added between monomers, see Figure 7.9. As more bonds are added, clusters of connected points are formed, which ultimately form an infinite-size cluster spanning the entire system (ultimately, at the gelation point) [17].

In the percolation theory, the fraction of monomers connected to the infinite network, φ, near the percolation threshold is related to the total fraction of interacting monomers, x, as

$$\varphi \propto \begin{cases} 0 & x \leq x_c \\ x - x_c & x > x_c \end{cases} \tag{7.32}$$

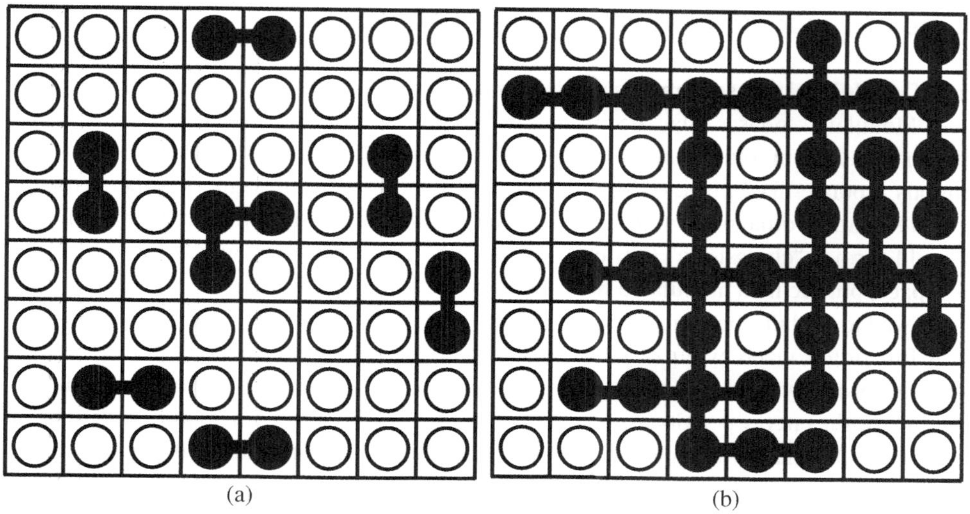

Figure 7.9 Illustration of percolation on a lattice. (a) Cells on a binary lattice are randomly connected with many bonds. (b) As bonds are added, clusters are formed, which eventually span the entire system.

Figure 7.10 The gel fraction, φ, in the classical model of gelation as a function of the fraction of unbonded interacting monomers $(1 - x)$, for a coordination number $z = 3$, given as $\varphi = 1 - [(1 - x)/x]^3$ [13], which, close to the percolation threshold (in the linear approximation), corresponds to Eq. (7.32). Note the striking similarity between the behavior of the gel fraction and the order parameters for polymerization and the magnetization, presented in Figure 7.3b, which indicates that percolation could be viewed as a "second-order phase transition."

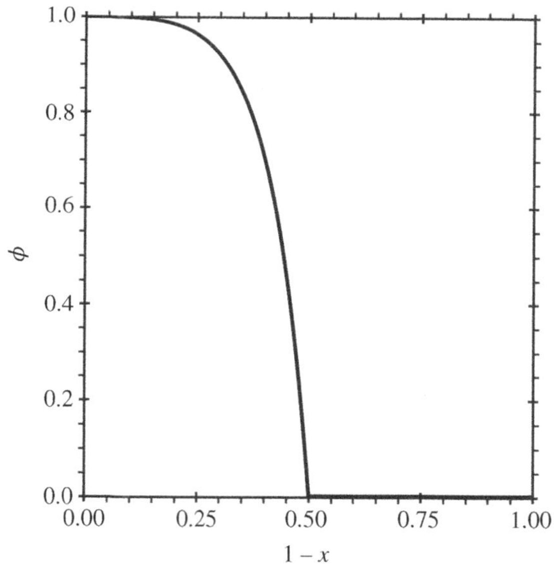

where x_c is the "percolation threshold." Equation (7.32) represents the linear approximation of the behavior of the fraction of monomers in the percolation theory, shown in Figure 7.10.

Similarly, the correlation length for percolation is defined as

$$\xi = \xi_0(x - x_c)^{-\nu} \tag{7.33}$$

where in the meanfield approximation $\nu = 1/2$. Fluctuations of the network change the behavior given by the meanfield approximation. In particular, the actual value for the exponent ν depends on the geometry of the network. Note that the meanfield approximation for the percolation theory is exact for dimensions $d \geq 6$ (Note that the marginal dimension is $d = 4$ for Ising-like systems, see Section 4.10) [13].

7.5 Landau Theory of Modulated Phase Formation

A variety of modulated phases are observed in multicomponent fluids involving amphiphilic molecules. Modulated phases may occur in anisotropic or isotropic systems. For example, amphiphilic substances in solution may form periodic bilayers (lamellar phases) or microemulsions (bicontinuous "sponge-like" phases). Modulated phases with a helical magnetic superstructure are observed in antiferromagnetic materials, like in FeGe [18]. Additionally, some nematic liquid crystals could form a "cholesteric" (or "chiral") nematic liquid phase, in which the director is rotated in space, producing a modulated helical structure.

A general way to explain the formation of a modulated phase is through the Landau–Ginzburg free-energy functional, given by Eq. (4.70). As discussed in Chapters 4 and 6, the inhomogeneities associated with ordinary fluid phase transitions can be described in the meanfield approximation by accounting for the energy penalty associated with the local inhomogeneities (the gradient term with coefficient c_0). Since c_0 is typically positive, fluctuational and interfacial inhomogeneities increased the free energy, such that the system relaxes to its equilibrium homogeneous state or equilibrium two-phase coexistence.

However, an equilibrium "modulated" mesoscopic phase may also exist if the gradient of the order parameter decreases the free energy. If the coefficient of the gradient term, c_0, is tuned by an external parameter, y (associated with the chemical potential of an impurity or any other thermodynamic field), c_0 could change its sign, so that the inhomogeneities would become favorable. To compensate this change in sign and the consequential decrease in the free energy, the next-order gradient term must be added into the free energy functional in Eq. (4.70) to account for thermodynamic stability.

To specify this function, consider a simple example of a weakly-compressible water–"oil"–surfactant mixture. In this case, the order parameter, ϕ, depends on the water–oil ratio and the amount of surfactant, such that $\phi = 0$ corresponds to the critical composition of this ternary system. In common experimental practice, the mutual solubility of water and oil is enhanced by the addition of alcohol ("co-surfactant"). The critical composition, $\phi_c(y)$, depends on the surfactant "activity," $y = 1/(1 + e^{-\hat{\mu}_a})$, where $\hat{\mu}_a = \mu_a/k_B T$ is the reduced chemical potential of the surfactant (amphiphile). In a dilute solution of surfactant, $y \propto x_a$ (where x_a is the molecular fraction of surfactant), and along the critical locus $y = x_a$; the latter is known as the "critical-line" condition, discussed in more detail in Chapter 8. Therefore, for this system, the Landau–Ginzburg free energy functional is given by

$$\Delta\hat{F} = \int \left(\Delta\hat{f}(\phi, \Delta\hat{T}) + \frac{1}{2}c_0(y)|\nabla\phi|^2 + \frac{1}{2}k_0(\nabla 2\phi)^2\right) \mathrm{d}V \tag{7.34}$$

where $\Delta\hat{f}$ is the truncated two-term Landau expansion discussed in Section 4.3.

In the one-phase region ($T > T_c$), $\phi = 0$. In the first-order approximation, the presence of the surfactant smooths the relation between water and oil, such that the critical temperature of water–oil demixing decreases with the surfactant concentration as

$$T_c(x_a) = T_{c,0}\left(1 - \lambda x_a\right) \tag{7.35}$$

where $T_{c,0} \equiv T_c(x_a = 0)$ and $\lambda = (1/T_{c,0})(\mathrm{d}T_c/\mathrm{d}x_a)$. Correspondingly, one may assume that the coefficient of the gradient term in Eq. (7.34) linearly decreases with the constant, λ, as

$$c_0(x_a) = c_{0,x_a=0}(1 - \lambda x_a) \tag{7.36}$$

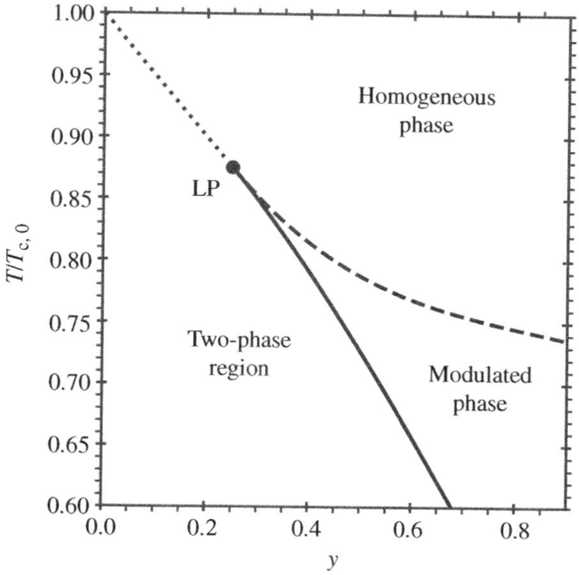

Figure 7.11 A phase diagram of a hypothetical water–"oil"–surfactant mixture at the critical water–oil composition as a function of the "activity," $y = 1/(1 + e^{-\hat{\mu}_a})$, which becomes zero in the limit of zero surfactant concentration, x_a, as the chemical potential, $\hat{\mu}_a \propto \ln x_a \to -\infty$. The curves were obtained from the Landau–Ginzburg functional, Eq. (7.34), with parameters: $a_0 = 1$, $u_0 = 2$, $c_0 = -0.5$, $k_0 = 2.5$, and $\lambda = 4$. The dotted line, given by Eq. (7.35), is the critical line (second-order phase transitions) of water–oil phase separation in the presence of surfactant. The dashed line, given by Eq. (7.42), is the line of second-order phase transitions to the modulated (meso-heterogeneous) phase. The solid line, given by Eq. (7.43), is the line of three-phase coexistence (first-order phase transitions) between oil, water, and modulated phases. The Lifshitz point (LP) corresponds to the intersection of these three phase transitions.

where $c_{0,x_a=0} \equiv c_0(x_a = 0)$. Therefore, there could be a unique surfactant concentration where $c_0 = 0$. This point is known as the "Lifshitz point"[1] and is illustrated in Figure 7.11. In the range of surfactant concentrations where c_0 is negative, a modulated (lamellar or microemulsion) phase may become stable [19–21].

Assuming that the order parameter for the lamellar phase is oscillating in space, like in smectic liquid crystals, as

$$\phi(\boldsymbol{r}) = A_\phi \cos(\boldsymbol{q} \cdot \boldsymbol{r}) \tag{7.37}$$

which is a two-component order parameter comprising an amplitude, A_ϕ, and characteristic wave number of modulation, q_0. With use of this definition for the order parameter, Eq. (7.34), may be expressed as

$$\Delta\hat{F} = \int \left(\frac{1}{2}\hat{\chi}^{-1}_{\phi=0}(q)[\phi(q)]^2 + \frac{1}{64}u_0[\phi(q)]^4 \right) \mathrm{d}V \tag{7.38}$$

where the q-dependent inverse susceptibility, $\hat{\chi}^{-1}_{\phi=0}(q)$, is given by

$$\boxed{\hat{\chi}^{-1}_{\phi=0}(q) = a_0\Delta\hat{T}(x_a) + c_0(x_a)q^2 + k_0 q^4} \tag{7.39}$$

1 Evgeny Mikhailovich Lifshitz (1915–1985) was a Soviet physicist and co-author with Landau of the famous textbook series, "Landau–Lifshitz Course on Theoretical Physics."

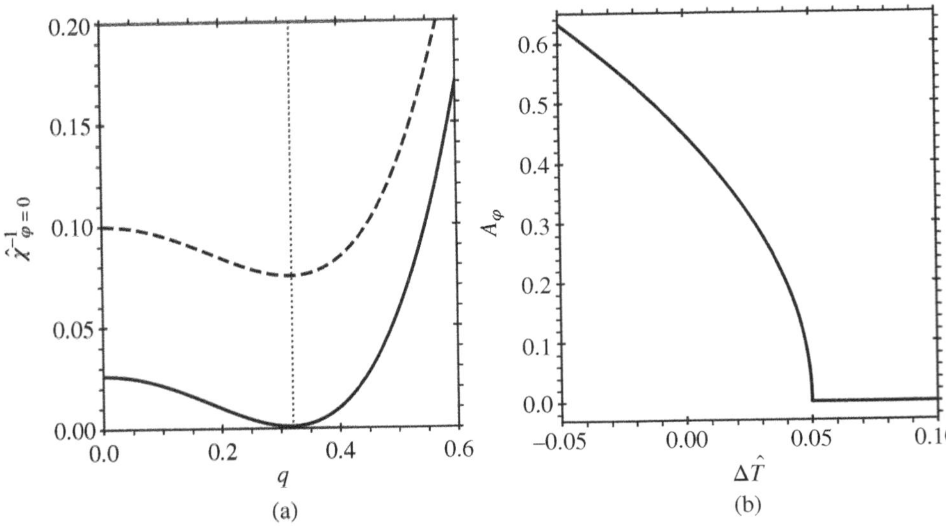

Figure 7.12 (a) Inverse susceptibility, given by Eq. (7.39), for the water–"oil"–surfactant solution ($a_0 = 1$, $u_0 = 2$, $c_0 = -0.5$, and $k_0 = 2.5$). The dashed curve represents a macroscopically homogeneous solution, while the solid curve represents the second-order transition to the modulated phase. The dotted line indicates the equilibrium size of the modulated phase, q_0. (b) Amplitude of the order parameter, given by Eq. (7.37), for the modulated phase of the same system.

where $\Delta\hat{T}(x_a) = \left[T - T_c(x_a)\right]/T_c(x_a)$. The inverse susceptibility is physically related to the generalized osmotic compressibility for a ternary mixture. The behavior of the susceptibility is shown in Figure 7.12a. When $c_0(x_a)$ is negative, $\hat{\chi}^{-1}_{\phi=0}(q)$ has a minimum at the equilibrium wave number of modulation $q = q_0$, inversely proportional to the distance between lamellae, as

$$q_0 = \sqrt{-\frac{c_0}{2k_0}} \tag{7.40}$$

At the Lifshitz point (LP) (where $x_a = x_{LP}$), such that both $\Delta\hat{T}(x_{LP}) = 0$ and $q_0 = 0$. Consequently, the distance between lamellae becomes infinite.

Minimizing Eq. (7.38) gives the amplitude of modulation as

$$A_\phi = \pm\sqrt{\frac{16}{u_0 \hat{\chi}_{\phi=0}(q_0)}} = \sqrt{\frac{8}{u_0}\left(\frac{c_0^2}{2k_0} - a_0\Delta\hat{T}\right)} \tag{7.41}$$

The behavior of the amplitude for a modulated phase is shown in Figure 7.12b. Therefore, both the amplitude and characteristic wave number emerge spontaneously from zero at the LP. Thus, there is a line of second-order phase transitions to the modulated phase, which is characterized by the condition

$$\hat{\chi}^{-1}_{\phi=0}(q_0) = a_0\Delta\hat{T} - \left(\frac{c_0^2}{2k_0}\right) = 0 \tag{7.42}$$

The equation for the line of three-phase coexistence (water–oil–modulated phases) is obtained by equating the free energies of the water and oil in the two-phase region $\hat{F} = -(3/2)(a_0\Delta\hat{T})^2/u_0$ with

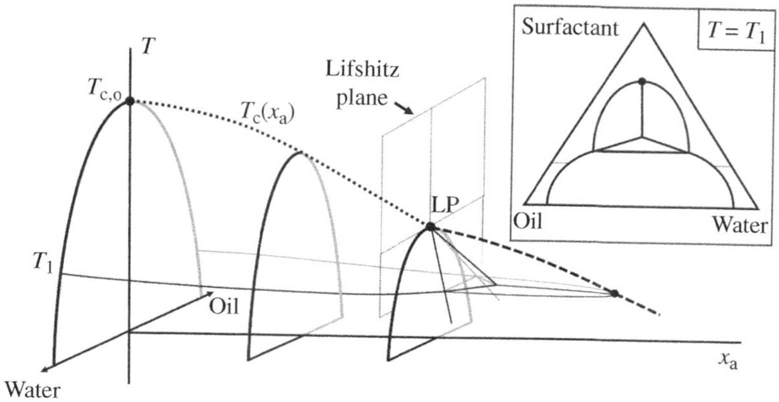

Figure 7.13 Three-dimensional sketch of the phase diagram for a hypothetical "oil"–water–surfactant mixture. It is assumed that the surfactant is equally distributed between the water and oil. The thick dotted curve depicts the critical line of the mixture at the critical oil–water ratio (50/50). The dashed curve shows the second-order transitions between the homogeneous and modulated phases. LP indicates the Lifshitz point. The Lifshitz plane (thin dotted lines) corresponds to the surfactant composition for varying water–oil ratios, at which the coefficient of the gradient term in the Landau-Ginzburg functional, c_0, changes its sign. The solid parabolic curves indicate oil–water phase coexistence (black in front and gray behind). The triangular prism with a corner at LP indicates the three-phase coexistence between water, oil, and modulated phases. The arbitrary isothermal cross-section at $T = T_1$ is shown by the thin solid curves. The inset shows the projection of the three-dimensional diagram along the plane $T = T_1$, in which the solid black curves indicate the first-order phase transitions between water and oil as well as between the microemulsion phase and oil-rich and water-rich homogeneous phases. The horizontal dotted line in the inset indicates the projection of the Lifshitz plane onto the plane $T = T_1$.

the free energy of the modulated phase, $\hat{F} = -4[\hat{\chi}^{-1}_{\phi=0}(q_0)]^2/u_0$. This gives the condition

$$\Delta\hat{T} = \frac{\sqrt{6}-4}{5}\left(\frac{c_0^2}{a_0 k_0}\right) \approx -0.31\frac{c_0^2}{a_0 k_0} \tag{7.43}$$

A phase diagram of a hypothetical water–oil–surfactant mixture at the critical water–oil ratio is shown in Figure 7.11. The three-dimensional phase diagram, which includes the off-critical water–oil ratios, is qualitatively illustrated in Figure 7.13. The Lifshitz plane, which is located at the surfactant concentration where the gradient term changes sign ($x_a = x_{LP}$), indicates a division between two types of homogeneous (macroscopically unmodulated) phase structures. These structures are characterized by two types of correlation functions: one for $x_a < x_{LP}$, given by the Ornstein–Zernike correlation function (Eq. 4.60), and another for $x_a > x_{LP}$, given by an oscillatory correlation function.

In the region, $x_a > x_{LP}$ the modulated state, above the second-order transition line, is metastable, and thus, this phase is macroscopically unmodulated. In practice, the water-rich or oil-rich phases in this region could contain aggregated amphiphiles (solubilized micelles). In this simplified picture of the surfactant–oil–water ternary mixture, the modulated phase (microemulsion) exists only for a 50/50 oil-to-water ratio. As shown in the inset of Figure 7.13, the modulated phase coexists at constant temperature with the water-rich and the oil-rich macroscopically homogeneous phases. Note that the behavior of the ternary mixture at the LP is reminiscent of tricritical behavior (see Section 4.4.4) as the three phases (aqueous, oily, and modulated) become indistinguishable and their coexistence is terminated at one point [23].

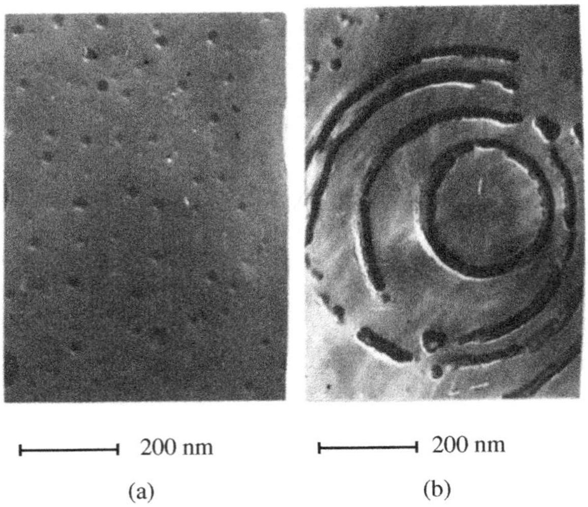

Figure 7.14 Fast-freezing electron microscopy image of aqueous DDAB solutions at two different amphiphile concentrations [22]. (a) 5×10^{-5} mol/L, depicting a nano-vesicle structure, and (b) 8×10^{-5} mol/L, depicting onion-like bilayers. The curvature of the observed lamellae is caused by the cylindrical shape of the frozen sample.

Previously, an assumption that the order parameter exhibits one-dimensional oscillation along a certain direction was made. However, the Landau–Ginzburg functional with a tunable coefficient of the gradient term, c_0, could still be applicable for an order parameter that does not have a preferable direction in modulation. In this case, the predicted mesophase would be isotropic, such that the LP would designate the onset of the formation of a bicontinuous ("sponge-like") microemulsion. In real microemulsion systems, commonly containing co-surfactant and/or salts (in addition to surfactant), the phase diagrams could be much more complicated [24].

In the vicinity of the LP, the van der Waals–Landau theory of interfaces predicts a dramatic decrease in the interfacial tension as it depends on the coefficient of the gradient term as, $\sigma \propto \sqrt{c_0}$ (see Eq. 6.22). Since c_0 becomes very small near the Lifshitz plane, the surface tension significantly decreases in addition to the decrease predicted by the Gibbs adsorption isotherm, Eq. (7.18) [25, 26].

One of the examples of the formation of a lamellar phase was observed in an aqueous solution of didodecyldimethylammonium bromide (DDAB). Since each amphiphilic molecule of DDAB has two hydrophobic tails, it will spontaneously form inverted micelles in hydrocarbons and nano-sized spherical vesicles in dilute aqueous solutions. However, in an aqueous solution, upon a small increase in the concentration, the spherical vesicles self-assemble into a multilayer "onion-like" structure, as shown in Figure 7.14. At the same DDAB concentration, the interfacial tension between the water and the hydrocarbons (octane) almost vanishes, becoming less than 0.1 mN/m. This may suggest that the formation of the lamellar structure occurs along the Lifshitz plane (see Figure 7.13), and, in the presence of oil, this structure may be accompanied by water–oil interfacial instability.

7.6 Dissipative Modulated Structures

The formation of modulated phases in equilibrium systems is a result of the minimization of the equilibrium free energy. Contrary to the mesoscopic structures formed in equilibrium conditions, nonequilibrium spatially-modulated patterns may persist in a steady state in the presence of a continuous energy supply. The structure of the phase domains formed under nonequilibrium conditions resembles equilibrium patterns: modulated phases or bicontinuous microemulsions, as discussed in Section 7.5, and the metastable nonequilibrium structures formed during spinodal decomposition, presented in Figure 6.15. However, the steady-state dissipative structures are fundamentally different as they only exist due to the continuous input of energy.

In this section, we present a case for the formation of nonequilibrium structures existing in steady-state conditions. Consider the system of the hybrid Ising/binary lattice model, which exhibits diffusion and equilibrium interconversion, as discussed in Section 6.5.3. Additionally, we impose an external source of "forced interconversion," which forces the alternative species to be remain at equal compositions, and thus, drives the system away from equilibrium. Consequently, the flux associated with the nonconserved component of the order parameter, j_{NC}, in the continuity equation, Eq. (6.51), adopts an additional term to account for the nonequilibrium order-parameter dynamics (forced interconversion), $\check{\epsilon}\tilde{\mu}_{\mathrm{BA}}$, where $\check{\epsilon}$ is a kinetic coefficient proportional to the rate of forced interconversion and $\tilde{\mu}_{\mathrm{BA}}$ is a nonequilibrium chemical potential. In this case, $j_{\mathrm{NC}} = -\check{\delta}\hat{\mu}_{\mathrm{BA}} + \check{\epsilon}\tilde{\mu}_{\mathrm{BA}}$ and the continuity equation, Eq. (6.51), may be expressed in the expanded form,

$$\frac{\partial \phi}{\partial t} = \check{\alpha}\nabla^2\hat{\mu}_{\mathrm{BA}} - \check{\delta}\hat{\mu}_{\mathrm{BA}} + \check{\epsilon}\tilde{\mu}_{\mathrm{BA}} \tag{7.44}$$

In the lowest-order approximation, the equilibrium chemical potential scales as $\hat{\mu}_{\mathrm{BA}} \sim -\Delta\hat{T}\phi$, which is positive for $T < T_{\mathrm{c}}$, while in the same approximation, the nonequilibrium chemical potential scales as $\tilde{\mu}_{\mathrm{BA}} \sim -\phi$, meaning that the nonequilibrium interconversion always opposes equilibrium interconversion. Consequently, the difference between the equilibrium and nonequilibrium chemical potentials, $\Delta\tilde{\mu}_{\mathrm{BA}} = \hat{\mu}_{\mathrm{BA}} - \tilde{\mu}_{\mathrm{BA}}$, is the driving force for the formation of steady-state structures.

By rearranging Eq. (7.44) to explicitly include this driving force, the dynamics of the equilibrium and nonequilibrium interconversions may be combined into a single term. Redefining the unbalanced chemical potentials as $\tilde{\mu}'_{\mathrm{BA}} = (\check{\epsilon}/\check{\delta})\tilde{\mu}_{\mathrm{BA}}$, a simplified continuity equation may be expressed through two dynamic processes in the form

$$\frac{\partial \phi}{\partial t} = \check{\alpha}\nabla^2\hat{\mu}_{\mathrm{BA}} - \check{\delta}\Delta\tilde{\mu}'_{\mathrm{BA}} \tag{7.45}$$

where $\Delta\tilde{\mu}'_{\mathrm{BA}} = \hat{\mu}_{\mathrm{BA}} - \tilde{\mu}'_{\mathrm{BA}}$, such that the second term describes the effective (equilibrium and nonequilibrium) interconversion dynamics in the system. Solving Eq. (7.45) yields a growth rate in the form

$$\omega(q) = -\check{\epsilon} - (Dq^2 + \check{\delta})(1 - \xi^2 q^2) \tag{7.46}$$

which is illustrated in Figure 7.15 [28].

The results of Monte Carlo simulations of a hybrid Ising/binary lattice model exhibiting diffusion and interconversion of spins in the presence of a nonequilibrium source of spin interconversion, with magnitude $E \propto \check{\epsilon}^2$, are shown in Figure 7.16. The system was simulated on a three-dimensional lattice of size $\ell = 100$ and at temperature $T/T_{\mathrm{c}} = 0.31$. As shown, when $E = 0$, the system exhibits phase amplification as discussed in Section 6.5.3. If $E > E^*$, large enough to overcome equilibrium interconversion, then steady-state periodic structures form with a size inversely proportional to E [27].

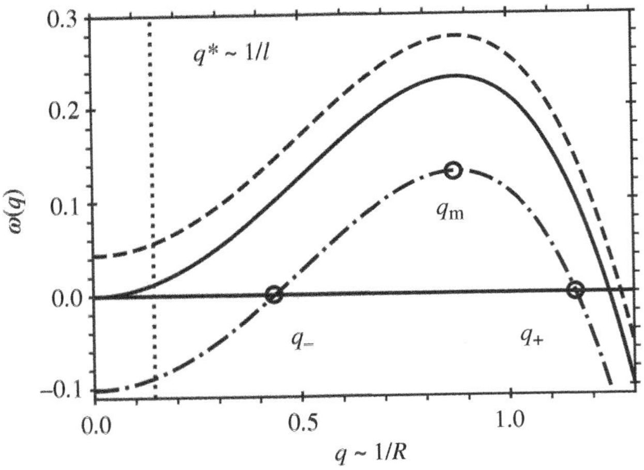

Figure 7.15 Growth rate in systems exhibiting equilibrium diffusion, equilibrium interconversion, and nonequilibrium forced interconversion. Source: Adapted from Ref. [27]. The solid curve depicts the growth rate in a system with only diffusion, as predicted by the Cahn–Hilliard theory, Eq. (6.50), while the dashed curve depicts the growth rate in a system with diffusion and equilibrium interconversion, given by Eq. (6.52), which exhibits phase amplification (see Section 6.5.3). The dash-dotted curve describes the growth rate in a system with a nonequilibrium source of forced interconversion, as given by Eq. (7.46), generating steady-state structures. The size of the microphase domains formed in steady-state conditions is proportional to the inverse wavenumber, $R \sim 1/q_{-}$.

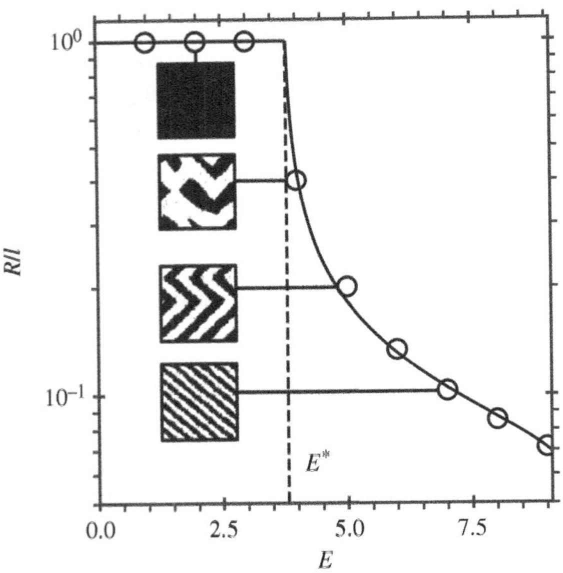

Figure 7.16 Formation of periodic structures in the presence of a continuous nonequilibrium energy source, E, in a hybrid Ising/binary lattice model. The vertical dashed line denotes the onset of steady-state structure formation, E^{*}. The open circles are the results of Monte Carlo simulations (black denotes up-spins while white denotes down-spins), the images are snapshots of the system for selected conditions, and the curves are the theoretical predictions. Source: Adapted from Ref. [27]/National Academy of Science.

References

1 Y. A. Nastishin, H. Liu, S. V. Shiyanovskiĭ, et al., "Pretransitional fluctuations in the isotropic phase of a lyotropic chromonic liquid crystal," *Phys. Rev. E*, vol. 70, p. 051706, 2004.

2 C. Bertrand, K. Linegar, A. Kostko, et al., "Multiscale dynamics of pretransitional fluctuations in the isotropic phase of a lyotropic liquid crystal," *Phys. Rev. E*, vol. 79, p. 041704, 2009.

3 R. Nagarajan, *Amphiphiles: Molecular Assembly and Applications*. Oxford University Press, 2011.

4 N. A. Shumovskyi, T. J. Longo, S. V. Buldyrev, et al., "Modeling fluid polyamorphism through a maximum-valence approach," *Phys. Rev. E*, vol. 106, p. 015305, 2022.

5 A. V. Tobolsky and A. Eisenberg, "Equilibrium polymerization of sulfur," *J. Am. Chem. Soc.*, vol. 81, p. 780, 1959.

6 V. F. Kozhevnikov, W. B. Payne, J. K. Olson, et al., "Physical properties of sulfur near the polymerization transition," *J. Chem. Phys.*, vol. 121, pp. 7379–7386, 2004.

7 W. Kunz, K. Holmberg, and T. Zemb, "Hydrotropes," *Curr. Opin. Colloid Interface Sci.*, vol. 22, pp. 99–107, 2016.

8 D. Subramanian, C. T. Boughter, J. B. Klauda, et al., "Mesoscale inhomogeneities in aqueous solutions of small amphiphilic molecules," *Faraday Discuss.*, vol. 167, pp. 217–238, 2013.

9 A. A. Novikov, A. P. Semenov, V. Monje-Galvan, et al., "Dual action of hydrotropes," *J. Chem. Phys. C.*, vol. 121, no. 30, pp. 16423–16431, 2017.

10 M. A. Anisimov, V. S. Esipov, V. M. Zaprudskiĭ, et al., "An anomaly in the heat capacity and a structural phase transformation of the ordering type in an aqueous solution of t-butanol," *J. Struct. Chem.*, vol. 18, pp. 663–670, 1978.

11 G. Gompper and M. Schick, *Self-Assembling Amphiphilic Systems*. New York: Academic Press, 2nd ed., 1992.

12 J. Israelachvili, *Intermolecular and Surface Forces*. London: Academic Press, 2nd ed., 1992.

13 R. A. L. Jones, *Soft Condensed Matter*. Oxford University Press, 2002.

14 A. P. Shpak and P. P. Gorbyk, *Nanomaterials and Supramolecular Structures Physics, Chemistry, and Applications*. Netherlands: Springer, 2009.

15 J. W. Steed and J. L. Atwood, *Supramolecular Chemistry*. Wiley, 2013.

16 W. H. Wang, "The elastic properties, elastic models and elastic perspectives of metallic glasses," *Prog. Mater Sci.*, vol. 57, no. 3, pp. 487–656, 2012.

17 J. E. Martin, J. Wilcoxon, and D. Adolf, "Critical exponents for the sol-gel transition," *Phys. Rev. A*, vol. 36, no. 4, pp. 1803–1810, 1987.

18 L. D. Landau and E. M. Lifshitz, *Electrodynamics of Continuous Media*, vol. 8 of Course of Theoretical Physics. Oxford: Pergamon Press, 2nd ed., 1984.

19 R. M. Hornreich, M. Luban, and S. Shtrikman, "Critical behavior at the onset of $\vec{k}$-space instability on the λ line," *Phys. Rev. Lett.*, vol. 35, pp. 1678–1681, 1975.

20 A. Michelson, "Phase diagrams near the Lifshitz point. I. Uniaxial magnetization," *Phys. Rev. B*, vol. 16, pp. 577–584, 1977.

21 M. A. Anisimov, E. E. Gorodetsky, A. J. Davydov, et al., "Landau model for self-assembly and liquid crystal formation in surfactant solutions," *Liq. Cryst.*, vol. 11, no. 6, pp. 941–947, 1992.

22 E. V. Zhuravleva, M. A. Anisimov, T. F. Svitova, et al., "Physical chemistry of colloids and interfaces in oil production," in *Proceedings of the 6th IFP Exploration and Production Research Conference* (H. Toulhoat and J. Lecourtier, eds.), p. 133. Paris: Saint-Raphael, 1992.

23 W. Selke, "Spatially modulated structures in systems with competing interactions," in *Phase Transitions and Critical Phenomena* (C. Domb and M. S. Green, eds.), vol. 15. New York: Academic Press, 1992.
24 D. Langevin, *Emulsions, Microemulsions and Foams*. Springer, 2020.
25 M. A. Anisimov, "Interfacial tension of critical liquid mixtures in the presence of a surfactant," *Int. J. Thermophys.*, vol. 13, p. 5, 1992.
26 R. Holyst and M. Schick, "Copolymers as amphiphiles in ternary mixtures: An analysis employing disorder, equimaxima, and Lifshitz lines," *J. Chem. Phys.*, vol. 96, pp. 7728–7737, 1992.
27 T. J. Longo, N. A. Shumovskyi, B. Uralcan, et al., "Formation of dissipative structures in microscopic models of mixtures with species interconversion," *Proc. Natl. Acad. Sci. U.S.A.*, vol. 120, p. e2215012120, 2022.
28 T. J. Longo and M. A. Anisimov, "Phase transitions affected by natural and forceful molecular interconversion," *J. Chem. Phys.*, vol. 156, p. 084502, 2022.

Exercises

7.1 Obtain the relation between the temperature T^* of the onset of polymerization ($\varphi = 0$) and the enthalpy change (heat) of reaction. Approximate the reaction equilibrium constant as

$$\ln K = \frac{\Delta H}{k_B}\left(\frac{1}{T_0} - \frac{1}{T}\right) \tag{7.47}$$

7.2 Derive the expression for the fraction of polymerized monomers as a function of temperature for $\mathcal{N} = 30$ and in the limit of infinite degree of equilibrium polymerization. Plot the corresponding functions. Linearize the function at infinite $\mathcal{N}$ near the onset of polymerization. Adopt the enthalpy of reaction as $\Delta H = 15k_B T_0$ and $T_0 = 300\,\text{K}$.

7.3 Explain the similarity and the difference between the entropy of mixing in polymer solutions and micellar solutions. Comment on the consequences of the differences.

7.4 Derive the expression for CMC given in Eq. (7.15). Prove that at large N the fraction of monomers in micelles behaves as $\varphi \approx x - x^*$.

7.5 The CMC of a nonionic aqueous surfactant solution was found to be 0.1 mol%. Estimate the minimal (thermodynamic) work required to dissociate a micelle with the aggregation number $\overline{N} = 50$ in this solution.

7.6 Derive an expression for the fluctuation of the aggregation number of a spherical micelle in terms of the surface tension and the optimal head area, a_h. Obtain the standard deviation (RMS fluctuation) of a micellar aggregation number, adopting $\sigma = 30\,\text{mJ/m}^2$, $T = 300\,\text{K}$, $\ell = 2\,\text{nm}$, $a_h = 0.1\,\text{nm}^2$, and $\overline{N} = 50$.

7.7 Consider a linear hydrocarbon chain with volume $0.3\,\text{nm}^3$, a critical chain length 0.3 nm, and an amphiphile with an ionic head-group that has an optimum head area in aqueous solution of $a_h = 0.65\,\text{nm}^2$. What supramolecular shape are formed by the amphiphiles? What is the average size and aggregation number of each micelle?

7.8 Explain the similarities and differences between percolation and thermodynamic second-order phase transitions as well as between percolation and polymerization. What is the similarity between the correlation length of critical fluctuations and the size of a percolation cluster?

7.9 In a simple model for $\mu_m/\mathcal{N}$, the free energy change when a specified molecule is taken from the bulk and put in an aggregate of $\mathcal{N}$ molecules, $\mu_m/\mathcal{N}$ is written in terms of the head area, a_h as $\mu_m/\mathcal{N} = \kappa/a_h + \sigma a_h$, where κ is a repulsive constant and σ is an effective interfacial tension.

(a) Derive an expression for the variance in aggregation number of a spherical micelle in terms of σ and the optimum head-group area a_h.

(b) Evaluate the standard deviation of the micelle size for the amphiphilic molecule, assuming that $\sigma = 30\,\text{mN/m}$. Use the volume and critical chain length of a linear hydrocarbon chain from Exercise 7.7.

7.10 Plot the q-dependent inverse mesoscopic susceptibility as a function of q at the temperature of the second-order phase transition (where the system reaches the stability limit to the modulated phase). At which temperature is the transition defined? Adopt the coefficients,

$$a_0 = 1, \quad c_0 = -0.01\,\text{nm}^2, \quad g = 40\,\text{nm}^4 \tag{7.48}$$

Determine the length scale of the phase modulation $\ell = 2\pi/q$.

7.11 Explain the phase diagram of the hypothetical water–"oil"–surfactant mixture at the critical water–oil composition as a function of surfactant concentration, x_a, in the meanfield approximation (Figure 7.11).

(a) Why is the dotted line correspond to the critical line of the oily/aqueous coexistence, while the solid line represents the first-order transition is the transition to the modulated phase is of the second order (explain why or why not)? What phases coexist along the solid line?

(b) What is the physical meaning of the Lifshitz point? With the addition of a surfactant, if the coefficient c_0 becomes negative, where will the metastable line $\Delta\hat{T} = 0$ be located? Why is it metastable?

7.12 Derive the amplification factor for a binary mixture exhibiting spinodal decomposition in the presence of equilibrium and nonequilibrium interconversion. Plot the amplification factor as a function of wavenumber assuming kinetic coefficients: $\breve{\alpha} = 1$, $\breve{\delta} = 0.01$, and $\breve{\epsilon} = 0.02$ in arbitrary units. Compare the steady-state amplification with that predicted by Cahn–Hilliard's theory of spinodal decomposition (without species interconversion).

8

Critical Phenomena

This chapter includes:

- Critical singularities
- Scaling theory
- Effects of critical fluctuations on interfacial phenomena
- Critical phenomena in mixtures
- Finite size scaling
- Critical dynamics

Critical phenomena are one of the most fascinating topics in meso-thermodynamics. In all fluids, simple and complex, in superfluid helium, as well as in many solids, critical anomalies are shaped by mesoscopic fluctuations with diverging length and time scales.

In this chapter, we discuss the nature of fascinating phenomena observed near the critical points (CPs) (second-order phase transitions). The peculiar physics of these phenomena is attributed to the mesoscopic fluctuations of the order parameter (introduced in Section 4.8), which diverge at the critical point. We refer to the region strongly affected by critical fluctuations as the *critical region* and the state of matter in this region as *criticality*. The meanfield description (Landau theory) of the properties in the critical region is generally insufficient. In fact, criticality is characterized by the singular behavior of the thermodynamic and the transport properties. Except for a few notable exceptions, such as second-order phase transitions to superconductive and ferroelectric states, the meanfield description does not yield the correct temperature dependencies of the critical singularities, sometimes even qualitatively. For example, as shown in Figure 8.1, the fluctuation-induced liquid–gas coexistence curve is "flattened," the location of the critical point is shifted, and the rectilinear diameter is curved.

The contemporary theory describing the phenomenology of critical singularities is known as the *scaling theory*. A comprehensive account of scaling concepts and consequences can be found in a review of Fisher[1] [2]. The scaling theory is in excellent agreement with experimental studies of phase transitions in condensed matter and has further penetrated to other areas of natural science. The most accurate experiments on critical phenomena have been performed in a variety of fluids and fluid mixtures (near the liquid–gas and liquid–liquid transitions) as well as near the second-order transition to superfluidity in He^4.

1 Michael E. Fisher (1931–2021) was a British–American physicist who made many seminal contributions to the theory of phase transitions and critical phenomena.

Mesoscopic Thermodynamics for Scientists and Engineers, First Edition. Mikhail A. Anisimov and Thomas J. Longo.
© 2024 John Wiley & Sons, Inc. Published 2024 by John Wiley & Sons, Inc.

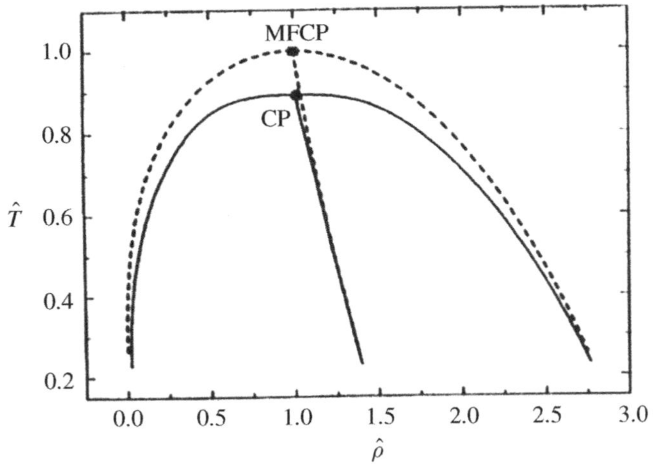

Figure 8.1 Effects of critical fluctuations on the liquid–gas coexistence for the van der Waals fluid, in which it is assumed that the range of intermolecular interactions, ξ_0, is equal to the average distance between molecules, $\bar{r}$, along the critical isochore. The meanfield prediction of the VDW EOS is given by the dashed curve with a meanfield critical point (MFCP), while the fluctuation-induced behavior is given by the solid curve with the actual critical point (CP). Source: Adapted from Ref. [1].

8.1 Challenges in Quantifying Critical Anomalies

Critical phenomena are striking from the first glance. Imagine a liquid mixture in which mutual diffusion of species practically stops, a sound wave is damped after traveling only a few wavelengths, a laser beam is scattered completely on its path through an optical cell, the heat capacity and thermal conductivity diverge, and thermal perturbations do not relax for many hours or even days. All of these anomalies are associated with the critical divergence of an appropriate susceptibility (the magnetic susceptibility in ferromagnets, the isothermal compressibility in pure fluids, or the osmotic compressibility in mixtures, see Section 4.2) and, correspondingly, with the divergence of the correlation length of the order-parameter fluctuations (see Section 4.7 for the meanfield description). Due to these large fluctuations, accurate experiments in the critical region are extremely challenging, as well as their interpretations through idealized theoretical models.

In a typical thermal experiment, an energy perturbation (per molecule), ε, may be neglected if $\varepsilon \ll k_B T$. However, in the critical region, owing to the huge susceptibility, even small perturbations, inevitably associated with any measurement, may lead to a dramatic distortion, resulting in a significant change of the observed anomalies. Thus, the condition $\varepsilon \ll k_B T$ becomes insufficient and should be replaced by a much more severe constraint:

$$\frac{\varepsilon}{k_B T} \ll |\Delta\hat{T}| \tag{8.1}$$

where $\Delta\hat{T} = (T - T_c)/T_c$. Moreover, the time required for the system to relax to equilibrium becomes very long, diverging at the critical point. This phenomenon is known as "critical slowing down." The relaxation time, τ_T, of nonequilibrium gradients of temperature near the liquid–gas critical point of a single-component fluid may be estimated through Eq. (3.91) as

$$\tau_T \approx \frac{\ell^2 \rho C_P}{\kappa} \tag{8.2}$$

where κ is the thermal conductivity and ℓ is a characteristic length scale (e.g. the size of the container). Since the isobaric heat capacity, being proportional to the isothermal compressibility (see Eq. 2.65), diverges at the critical point even in the meanfield approximation, the relaxation time increases upon approaching this point, being limited only by the size, ℓ. To obtain equilibrium data in the critical region, one must be sure that the relaxation time, given by Eq. (8.2), is much smaller than the measurement time. In the interpretation of experimental data, scientists always compare the results with an idealized theoretical model. Consequently, mapping a real (imperfect) system onto an idealized model is one of the major challenges for studies in the critical region.

Therefore, the quality of experimental data collected in the critical region is mainly determined by the distance to the critical point rather than by the resolution and reproducibility of the measurements. Various reproducible distortions of "ideal" critical anomalies, caused by temperature gradients and density inhomogeneities, equilibrium and nonequilibrium ("frozen") impurities, etc., have commonly been sources of misinterpretation and confusion.

For example, due to the anomalously large compressibility, the presence of gravity causes a significant inhomogeneity in the density and strongly affects experiments on fluids in the critical region. Moreover, if the temperature is not uniform in a near-critical fluid, gravity can easily generate convection and convective heat flow [3]. The near-critical convection is the major challenge in obtaining undisturbed thermal-conductivity data in the critical region [4]. The dramatic effect of gravity and critical slowing down on the measurements of the isochoric heat capacity near the critical point of carbon dioxide is illustrated in Figure 8.2a. The effects of gravity were a subject of comprehensive studies in 1970s [5]. The effect of gravity on the calorimetric and viscosity measurements can be significantly reduced by stirring (see Figure 8.2b) and almost totally illuminated in space under microgravity [6]. Alternatively, the effect of gravity may be reduced by selecting liquid mixtures with components of similar densities. Similarly, the major problem in studies of critical

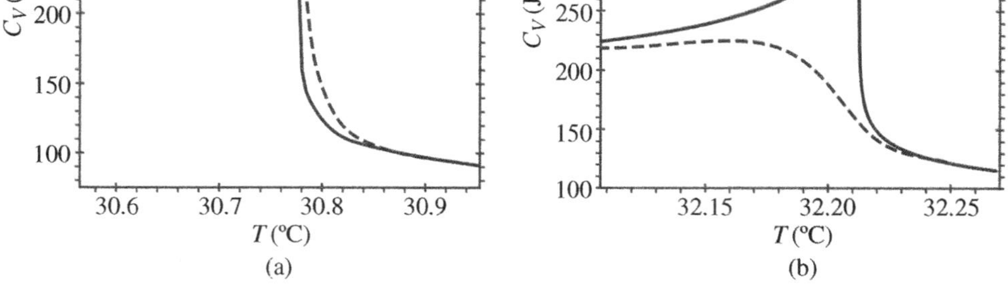

Figure 8.2 (a) The isochoric heat capacity of a 1 mm high (horizontal) disk-shaped sample of CO_2 at the critical density. The effect of gravity is demonstrated by the dashed curve, which shows the results of the same sample tilted, so that its total height is 14 mm [7]. (b) Influence of gravity and equilibration time on the heat-capacity anomaly near the critical point of ethane: With stirring (solid) and without stirring (10 hours equilibration time, dashed). Source: Adapted from Ref. [8].

opalescence, the strong turbidity caused by photons being scattered repeatedly, can be elegantly avoided in binary liquids by selecting components with close refractive indices [3].

It is not surprising that experimental challenges in quantitative studies of the critical anomalies led to the emergence of new methods in experimental physics or a significant modification of already existing ones, such as dynamic light scattering (to study the critical slowing down of fluctuations) or slow-scanning adiabatic calorimetry (to distinguish between second-order transitions and weakly first-order ones). Step by step, experimentalists were able to overcome the major obstacles in obtaining the results suitable for unambiguous interpretation.

Theoretical studies of critical phenomena have also met fundamental challenges. In an early edition of Landau and Lifshitz [9] there was a statement which reads,

> Unlike solids and gases, liquids do not allow a general calculation of their thermodynamic quantities or even their temperature dependence. The reason for this is the presence of strong interactions between the molecules of the liquid without having at the same time the smallness of the vibrations which makes the thermal motions of solids so simple.

While this statement is correct in general, the recent progress in the physics of condensed matter shows that the properties of different liquids demonstrate a remarkable universality in their critical behavior.[2]

For many years, the ideal gas (molecular gas or gas of quasiparticles, such as phonons in solids, see Section 2.12) was the only universal reference state for developing a theory of condensed matter. The effects of interactions between the molecules were often treated as corrections in the form of a virial expansion. Any analytic equation of state (such as the van der Waals equation), which describes fluid condensation and predicts the existence of the critical point, has not changed the situation in principle. As discussed in Section 2.7.1, the van der Waals equation can be expanded as a virial expansion with a second virial coefficient that has a special dependence on temperature [1]. Consequently, the van der Waals equation can be viewed as an extrapolation of the virial expansion to the dense-fluid region. The lack of a sufficient theory of dense fluids has led to a large number of empirical equations of state with many adjustable parameters to be fitted to experimental data. Commonly, in using empirical equations of state, the critical region is usually avoided because such equations could not properly describe the critical singularities.

Alternative to the ideal-gas reference state, the meanfield (Landau) theory of phase transitions is based on the critical-point reference state. In this theory, the complex molecular interactions are approximated by a single mean-force parameter (such as the van der Waals constant, a, or the Flory interaction parameter, ϖ, see Sections 2.7.1 and 5.3). Then, the appropriate free energy is assumed to be an analytical function expandable in powers of the order parameter. The meanfield approach neglects the effects of fluctuations of the order parameter on thermodynamic properties, or, in the next approximation, it considers such effects as small corrections (see Section 4.8). Experimental studies in the critical region undoubtedly show that the shape of the critical singularities cannot be reduced to small corrections within the meanfield approximation. In fact, the actual near-critical free energy is essentially a non-analytical function of the temperature and order parameter.

---

2 In the Richard T. Cox Lecture at the 2006 APS Annual Meeting, Alexander Voronel noted: "Now it is even strange to think that in the 1950s the second-order phase transitions and the liquid–vapor critical point were considered as different fields of physics." In fact, for a large part of the physics community, especially in the United States, and to some degree in Russia, the liquid state was not part of condensed-matter physics at all, actually belonging to engineering or, at best, to physical chemistry.

The adequate theory of near-critical phase transitions resolves the fundamental problem of many strongly interacting degrees of freedom near the stability boundary. Thus, it is not surprising that the development of the theory of critical phenomena, inspired by breakthrough experiments in the critical region, was a major achievement in the physics of condensed matter in the 20th century.

8.2 Critical Singularities

In this section, we present examples of the critical singularities of major thermodynamic and transport properties of individual substances. As illustrated in Figures 8.3–8.9, their behavior is significantly different, sometimes even qualitatively, from the prediction of the Landau (meanfield) theory.

8.2.1 Order Parameter

It had been noticed, since the beginning of the 20th century, but remained unexplained until the 1960s, that the liquid–gas density difference along the phase coexistence near the critical point does not exhibit a parabolic shape, $\Delta\rho \sim \pm(\Delta\hat{T})^{1/2}$, predicted by the van der Waals equation of state and, generally, by the Landau theory of second-order phase transitions Eq. (4.52). As illustrated in Figure 2.10, the densities of liquid–gas coexistence for many fluids is approximately described, in accordance with the law of corresponding states (see Section 2.6.1), by a power law of the from $\Delta\rho \sim \pm(\Delta\hat{T})^{1/3}$, which is much "flatter" than a quadratic parabola but much closer to the prediction of the modern theory.

Figure 8.3 demonstrates accurate simulation results for the Ising/lattice-gas model with a short range of interactions (introduced in Section 6.5.3). Below the critical temperature, this model demonstrates the equivalence of the spontaneous order parameter $\phi_0 = (\rho - \rho_c)/\rho_c$ in the two

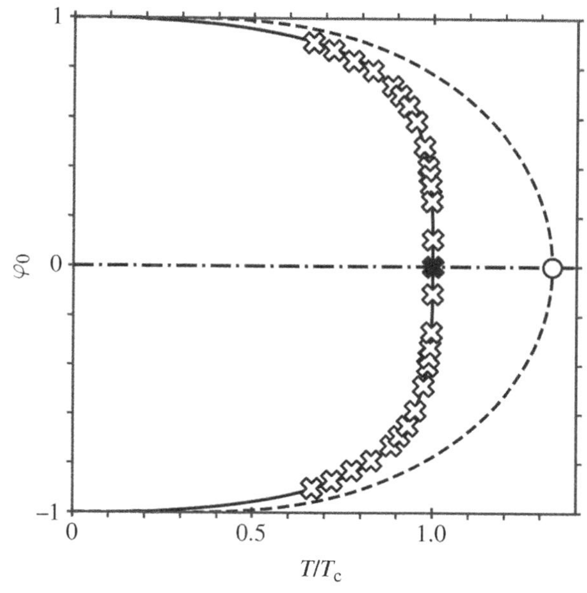

Figure 8.3 Spontaneous order parameter for the lattice-gas model with a short-range of molecular interactions. The crosses are Monte Carlo simulations [10], where the black cross indicates the critical point. The solid curve is the prediction of the scaling theory. The dashed curve is the meanfield approximation, given by the Landau theory (see Eq. 4.3), where the circle indicates the corresponding critical point. Source: Adapted from Ref. [10].

coexisting phases (liquid and gas). Contemporary theory accurately describes the simulation data as

$$\phi_0 = \pm B_0 |\Delta \hat{T}|^{\beta} \tag{8.3}$$

where $\beta \cong 0.326$ and B_0 is a system-dependent amplitude. Critical fluctuations not only flatten the shape of the coexistence curve but also shift the location of the critical point to lower temperatures with respect to the meanfield prediction. This shift can give an estimate of the Ginzburg number, usually between 0.1 and 0.3 for fluids and ferromagnets.

One of the consequences of fluid asymmetry, due to critical fluctuations, is a curvature of the diameter of phase-coexistence, demonstrated by the difference in the shape of the coexistence predicted by the van der Waals EOS (Figure 8.1) and the lattice-gas model (Figure 8.3). Real fluids do not possess the symmetry of the lattice gas (see Section 2.7.2). However, all fluids demonstrate "asymptotic symmetry," meaning that the spontaneous order parameter, ϕ_0 along the phase coexistence, asymptotically close to the critical point, follows the same power law as the lattice gas. This feature is illustrated in Figure 8.4 showing the change of the densities of the coexisting gas and liquid phases of argon upon approaching the critical temperature. It is also seen from Figure 8.4 that the asymptotic behavior of the density of the liquid branch of argon is restricted (within the accuracy of the data) by $\Delta \hat{T} < 10^{-2}$, which is not more than 1–2 K below T_c.

8.2.2 Heat Capacity

The isobaric heat capacity is thermodynamically proportional to the isothermal compressibility, thus strongly diverging at the liquid–gas critical point (see Eq. 2.65), even in the meanfield approximation as $C_P \propto |\Delta \hat{T}|^{-1}$ along the critical isochore (see Eq. 4.16). However, in the meanfield approximation, the isochoric heat capacity does not exhibit any anomaly in the one-phase region. Accordingly, the ratio C_P/C_V should strongly diverge at the critical point, which has been confirmed by experiments.

The Landau (meanfield) theory predicts no critical singularity of the isochoric heat capacity which remains finite at the critical point, exhibiting a discontinuity, given by Eq. (4.16), upon

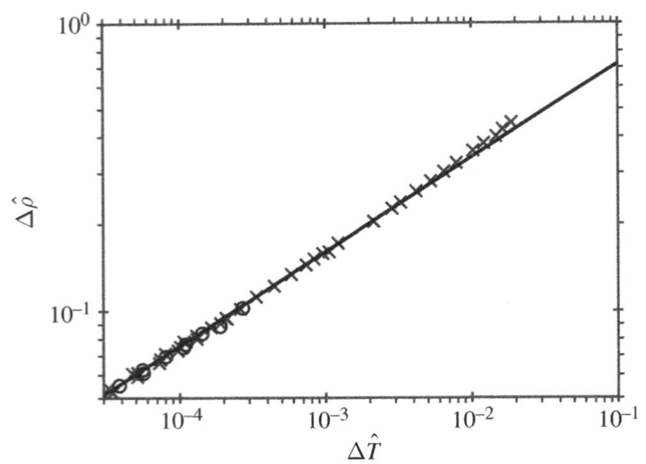

Figure 8.4 Density of liquid–gas coexistence of argon presented in reduced variables where $\Delta\hat{\rho} = (\rho - \rho_c)/\rho_c$ [11]. The liquid branch (crosses) and the gas branch (circles) demonstrate asymptotic symmetry. The solid line corresponds to an asymptotic power law with an exponent $\beta = 0.326$.

crossing the critical temperature along the critical isochore (see Figure 4.9). In fact, the actual behavior of C_V is determined by critical fluctuations. A breakthrough experimental study of argon performed in the early 1960s by Voronel[3] found that the isochoric heat capacity diverges at the critical point, although not as strongly as the isobaric heat capacity [13, 14]. This claim was in striking contradiction with the van der Waals–Landau theory of critical phenomena, which was commonly regarded as "untouchable" at that time. This discovery was a turning point in the development of the contemporary theory of critical phenomena [15], and it was confirmed and further elucidated by other experimental studies.[4]

It has been shown that the isochoric heat capacity of fluids does not follow the meanfield behavior even qualitatively. Instead of exhibiting a finite discontinuity, the heat capacity grows, apparently diverging at the critical point, from both directions, for $\Delta\hat{T} < 0$ and for $\Delta\hat{T} > 0$. Figure 8.5 demonstrates the results of one of the most accurate studies of the isochoric heat capacity of sulfur hexafluoride under microgravity (10^{-6} of Earth's gravity) in a German space mission [6]. This remarkable experiment revealed that the heat capacity diverges in accordance with the power law

$$C_V/k_B = A_0^{\pm}|\Delta\hat{T}|^{-\alpha} + B(T) \tag{8.4}$$

where $\alpha \cong 0.110$, $B(T)$ is a background function of temperature, and $A_0^{\pm}$ is the system-dependent amplitude above (+) and below (−) the critical point. Note that A_0^- is about twice larger than A_0^+.

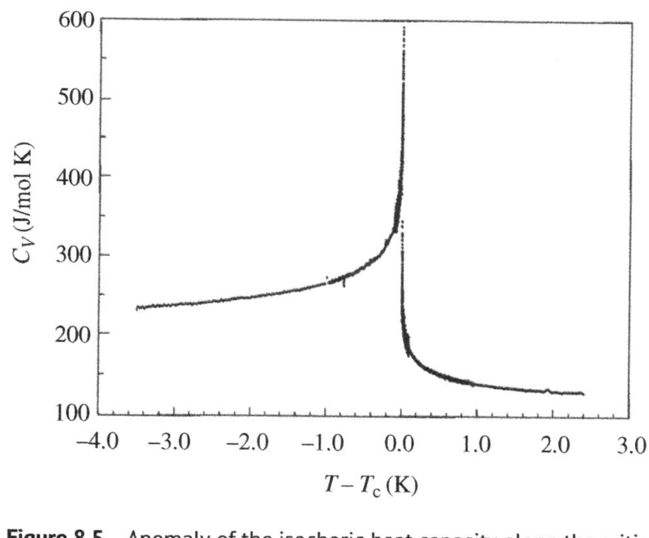

Figure 8.5 Anomaly of the isochoric heat capacity along the critical isochore of sulfur hexafluoride obtained under microgravity. Source: Reproduced with permission from Ref. [6], © (1999), American Physical Society.

3 Alexander V. Voronel is a Soviet and Israeli physicist (born in 1931), currently Emeritus at Tel Aviv University. Voronel made profound contributions to the experimental studies of critical phenomena in the 1960s and 1970s. Voronel was a postdoctoral advisor of one of the authors (M.A.A.) [12].

4 A footnote in the 1964 Russian edition of "Statistical Physics" by Landau and Lifshitz [16], which correctly stated that the isochoric heat-capacity divergence has no grounds in the "classical", meanfield, theory of phase transitions, was wrongly interpreted by many as a fatal criticism of Voronel's conclusion. However, Voronel's discovery was immediately recognized by a bright young theorists: Michael Fisher [17] and Leo Kadanoff [18], who believed in the universal nature of apparently very different phase transitions, such as those in liquid helium, binary alloys, ferromagnetic, and ferroelectric materials, and in classical fluids. By the middle of 1960s, the need for a new theory to adequately explain critical phenomena became obvious for the scientific community [19] (find more details on the breakthrough discoveries in fluid criticality in Ref. [12]).

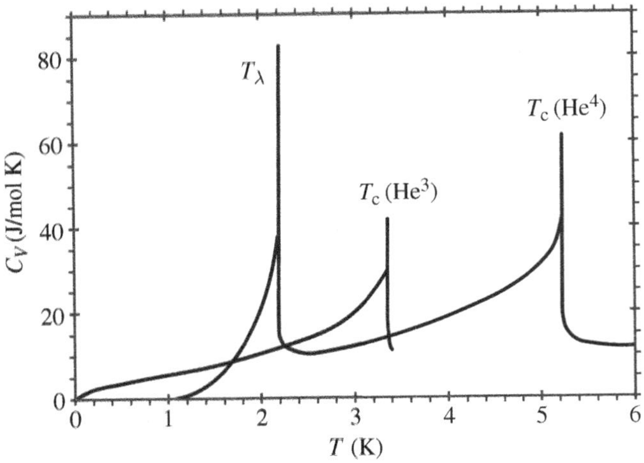

Figure 8.6 Isochoric heat capacity of He^3 and He^4 at their respective critical densities depicted in comparison to the heat capacity of He^4 near the transition to superfluidity (T_λ). Source: With permission from Ref. [20], U.S. Department of Commerce, National Bureau of Standards.

Remarkably, near second-order phase transitions in solids and near the lambda transition in superfluid He^4 (a sharp "cusp" with $\alpha \cong -0.01$ [21]), the isobaric (but not isochoric) heat capacity exhibits an anomaly comparable to the singularity in the isochoric heat capacity near the liquid–gas critical point. Figure 8.6 demonstrates the striking similarities of heat capacities near the lambda transition in He^4 and near the liquid–gas critical points of He^4 and He^3. Further investigations demonstrated that the shape of the heat capacity singularities near the transition to superfluidity and near the critical points of fluids, while being similar, are not identical.[5] The difference in this behavior is explained by the concept of different universality classes in critical phenomena, which will be discussed in Section 8.3.4. However, it was predicted theoretically and confirmed by accurate experiments that in all fluids and fluid mixtures near critical points as well as in anisotropic ferromagnets, like nickel and iron (near the Curie points), the appropriate heat capacity diverges with the same power law.

8.2.3 Susceptibility

In the meanfield approximation, the susceptibility (the magnetic susceptibility in ferromagnets or the isothermal compressibility in fluids, see discussion in Section 4.2) follows $\hat{\chi} \sim |\Delta\hat{T}|^{-1}$, in zero magnetic field or along the critical isochore in fluids (see Eq. 4.17). However, in real fluids and ferromagnets, the divergence of the susceptibility is much stronger. Figure 8.7a demonstrates the divergence of the isothermal compressibility of near-critical carbon dioxide along the critical isochore obtained by accurate light scattering measurements (see Section 3.11.1). In accordance to this experiment and similar experiments on other fluids, along the critical isochore asymptotically close to the critical point, the susceptibility diverges as

$$\hat{\chi} = \Gamma_0^{\pm}|\Delta\hat{T}|^{-\gamma} \tag{8.5}$$

where $\gamma \cong 1.240$ and $\Gamma_0^{\pm}$ is a system-dependent amplitude.

5 Distinguishing between the weak power-law divergence of the heat capacity in fluids and a near-logarithmic singularity in the vicinity of the lambda transition in He^4 is not trivial. Both power-law and logarithmic singularities could be represented by a function $f(x) = (A/\alpha)\,[1 - x^{-\alpha}]$, which, in the limit $\alpha \to 0$, becomes identical to $f(x) = A \ln x$.

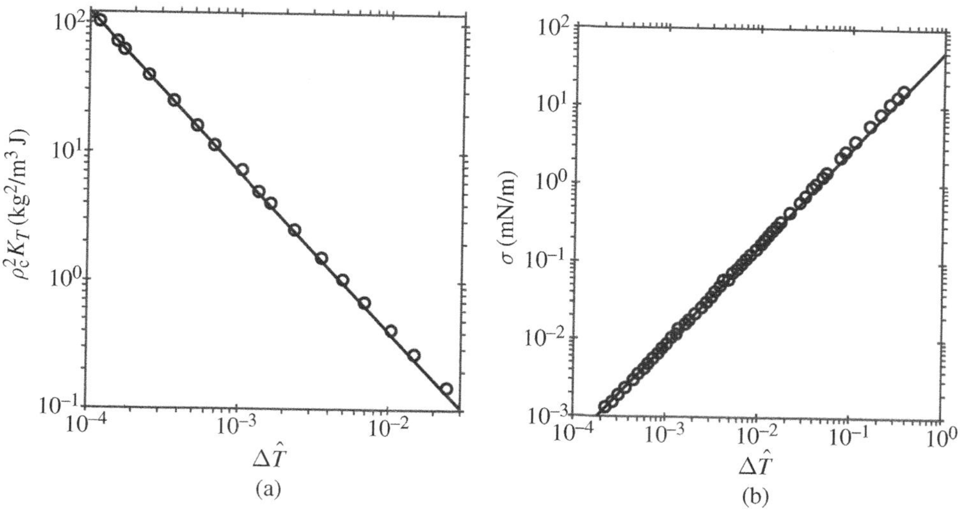

Figure 8.7 (a) Isothermal compressibility of carbon dioxide obtained from bulk light scattering measurements. Source: Adapted from Ref. [22]. (b) Surface tension of xenon obtained by light scattering from the liquid–gas interface in the vicinity of the critical point adapted from Ref. [23, 24]. The solid lines represent fits to power laws with exponents $\gamma = 1.24$ and $\vartheta = 1.26$, respectively.

8.2.4 Interfacial Tension

The van der Waals–Landau theory of the smooth near-critical interface predicts that the liquid–gas interfacial tension vanishes at the critical point as $\sigma \sim |\Delta\hat{T}|^{3/2}$, see Eq. (6.22). Real fluids, however, demonstrate a universal, but significantly different, temperature dependence,

$$\sigma = \sigma_0 |\Delta\hat{T}|^{\vartheta} \tag{8.6}$$

where $\vartheta \cong 1.260$ and σ_0 is a system dependent amplitude. Figure 8.7b shows the outstanding experimental data on the liquid–gas interfacial tension in the two-phase region of xenon obtained by light scattering from the fluctuating interface.

8.2.5 Thermal Conductivity and Shear Viscosity

Transport properties of fluids also exhibit critical singularities. The most remarkable example is the divergence of the thermal conductivity discovered by Sengers[6] in 1962 [27]. This divergence was completely unexpected at that time as it confronted the established classical views on transport properties of fluids in the critical region (see Section 4.11).[7] By eliminating the effects of thermal convection, Sengers proved that the thermal conductivity possesses a pronounced maximum at the critical density. Moreover, the magnitude of this maximum increased upon a gradual approach to the critical isotherm suggesting that the thermal conductivity could tend to infinity at the critical point (Figure 8.8). Later, it was experimentally and theoretically proven that other fluids also exhibit the same divergence of the thermal conductivity in the form of a power law, $\kappa \propto |\Delta\hat{T}|^{0.59}$.

---

6 Jan V. Sengers (born in 1931) is a Dutch–American physicist, currently Emeritus at the University of Maryland, College Park. He made seminal contributions to critical phenomena in condensed matter.

7 Remarkably, at about the same time, when Voronel challenged the classical thermodynamics of the liquid–gas critical point (see footnote 4), Jan Sengers, then a doctoral researcher at the van der Waals laboratory in Amsterdam, confronted the established classical views on transport properties of fluids in the critical region [12].

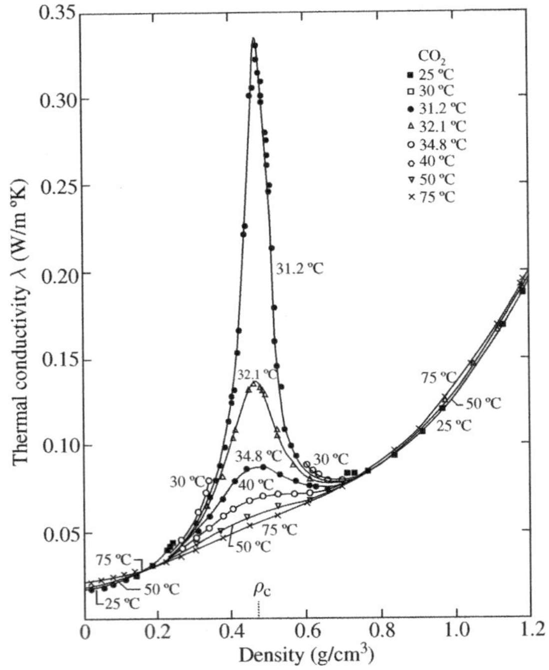

Figure 8.8 Anomaly of thermal conductivity of carbon dioxide in the critical region Ref. [25, 26].

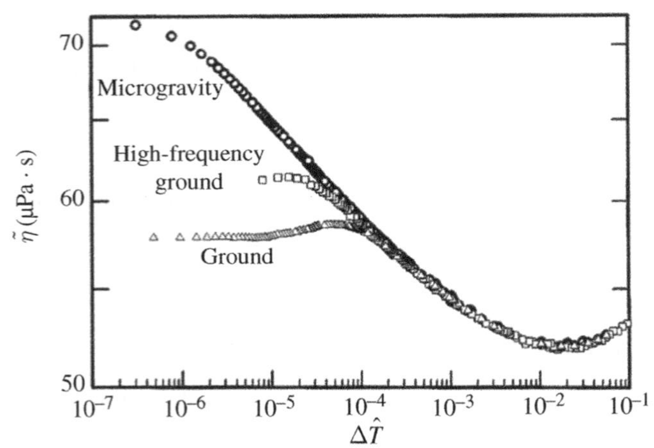

Figure 8.9 Temperature dependence of the viscosity of xenon. Experiments were performed under gravity with and without a high-frequency oscillating viscometer and under microgravity. Source: Adapted from Ref. [28].

Another important transport property of fluids is the shear viscosity, η. For decades, it was believed that the critical enhancement of the viscosity was small, if not negligible. Only at the end of the 20th century, extremely accurate experiments in space, under microgravity, clearly demonstrated that the shear viscosity diverges at the critical point in the form of a power law with an extremely small exponent as

$$\tilde{\eta} \propto |\Delta\hat{T}|^{\psi} \tag{8.7}$$

where $\psi \cong 0.04$.

It is important to note that the universality of the critical singularities of transport properties, unlike that of thermodynamic response functions, do not go beyond fluids and fluid mixtures. Recall from Section 4.11, that the order parameter for fluids is conserved, while the order parameter for magnetic systems is nonconserved. This results in the fundamental difference in their dynamics, and consequently, in their transport properties.

8.3 Nature of Critical Singularities

The origin of the critical singularities could be explained through the diverging fluctuations of the order parameter. In this section, we introduce a simple model to conceptually describe these singularities, referred to as the "fluctuation-cluster" model, and we discuss the phenomenology of critical behavior, known as the "scaling theory."

8.3.1 Divergence of the Correlation Length

The behavior of fluids near critical points, liquid helium near the transition to superfluidity, and many solids near second-order phase transitions (with some meanfield exceptions, such as the transition to superconductivity) can be explained by the divergence of the correlation length of the order-parameter fluctuations.

Figure 8.10 demonstrates three simulation snapshots of the lattice gas in one of the fluid phases (far below the critical temperature), inside the critical region, and in a supercritical fluid (above the critical temperature). In the critical region of fluids, the "size" of the density fluctuations, the correlation length, significantly exceeds the distance between the molecules. The local densities become correlated at long-ranges, while the intermolecular interactions remain short-range (see Section 4.6). Therefore, the system in the critical region becomes mesoscopically heterogeneous, while macroscopically remaining homogeneous. For this reason, the fluid state in the critical region, is sometimes referred to as a "critical meso-phase," even though this state is not thermodynamically a separate phase. In the critical region, the distance to the critical point is significantly smaller than the Ginzburg number Eq. (4.77), and thus, in accordance to the Ginzburg criterion, the meanfield approximation becomes invalid (see Section 4.9).

Accurate light scattering experiments, like those depicted in Figure 8.11, show that the asymptotic temperature dependence of the correlation length in fluids and fluid mixtures along the critical

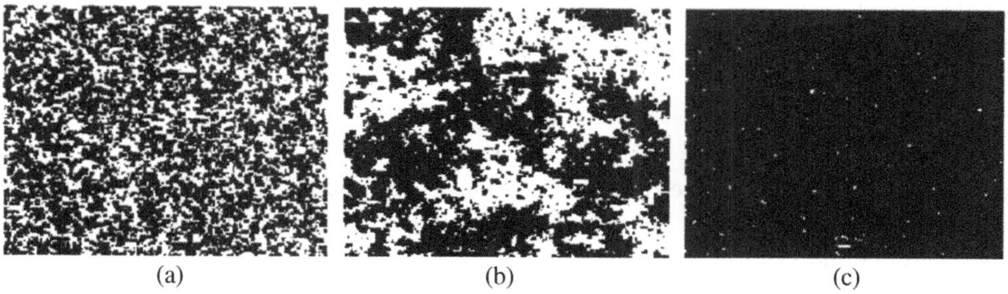

Figure 8.10 Snapshots of Monte Carlo simulations of a two-dimensional lattice gas. The three panels depict the behavior of the system (a) in the supercritical state ($T > T_c$), (b) in the critical region ($T \gtrsim T_c$), and (c) in the liquid phase far away from the critical point ($T \ll T_c$). Source: Reproduced with permission from Ref. [29] © (2008), ELSEVIER.

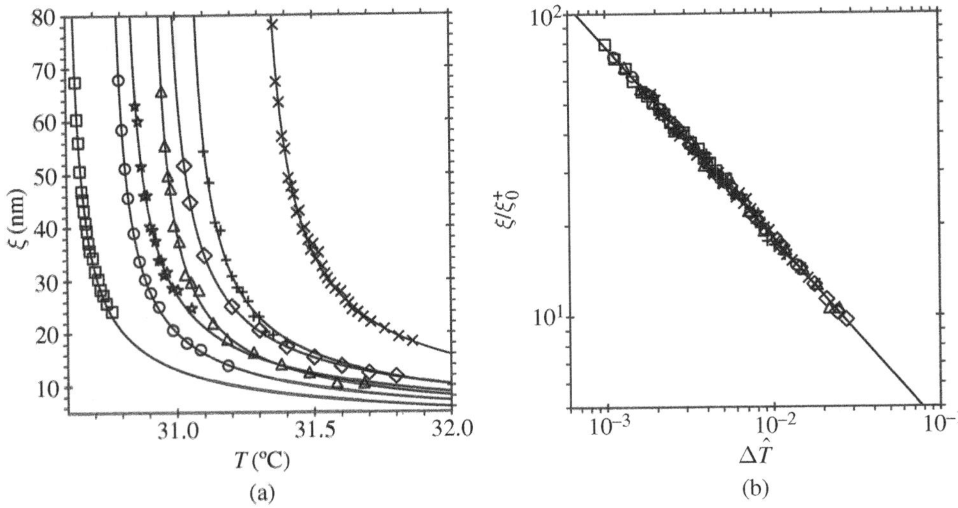

Figure 8.11 (a) Asymptotic divergence of the correlation length near the liquid–liquid critical temperature of nitroethane–isooctane (pure binary solvent, squares) with small additions of polymers: PS-25 (0.05%, circles), PS-25 (0.1%, triangles), PS-50 (0.05%, diamonds), PS-50 (0.1%, crosses), PS-123 (0.01%, stars), PBMA-180 (0.04%, plusses). Source: Adapted from Ref. [30]. (b) The universality of the divergence of the reduced correlation length for all samples described by the same universal power law, given by Eq. (8.8), with $\nu = 0.63$.

isochore or at the critical composition (above the critical point) and along each of the two branches of phase coexistence (below the critical temperature) follows the universal power law,

$$\xi = \xi_0^{\pm} |\Delta\hat{T}|^{-\nu} \tag{8.8}$$

where the critical exponent is $\nu \cong 0.630$ and the critical amplitude $\xi_0^{\pm}$ is the range of direct intermolecular interactions, which is defined by Eq. (4.62) and, typically, is of the order of the molecular size. The critical exponent of the correlation length, ν, is universal for all fluids and fluid mixtures as wells as for anisotropic ferro- and antiferro-magnetic materials, such as nickel and iron. Its value is different from the meanfield prediction, $\nu = 1/2$, which is obtained from the Landau–Ginzburg free-energy functional Eq. (4.72). Most importantly, the asymptotic power-laws, found for other critical anomalies (discussed in Section 8.2), are related to the singularity given by Eq. (8.8). Note that in a typical experiment near the liquid-gas critical point, the existence of gravity would ultimately suppress the divergence of the correlation length. Instead, the correlation length becomes anisotropic, remaining finite in all directions [31].

8.3.2 Fluctuation-Cluster Model

The relation between the divergence of the correlation length and the critical singularities of the thermodynamic properties can be conceptually demonstrated by a "fluctuation-cluster" model [3]. Consider the state of the system in the critical region as a d-dimensional "gas" of fluctuation-induced molecular inhomogeneities ("clusters"), with a half-size being equal to the correlation length. The total d-dimensional volume, V_t, is given by $V_t = (2\xi)^d N_{cl}$, where N_{cl} is the number of clusters. If the interaction between different clusters is neglected, the grand thermodynamic potential of the one-phase system per unit volume in the critical region, $(\Omega - \Omega_c)/V_t = -P + P_c = -\Delta P$, can be expressed as

$$\Delta P = \Delta P_{cr} + \Delta P_{bg} \tag{8.9}$$

where ΔP_{cr} is the fluctuation-dependent "critical part" of the pressure and ΔP_{bg} is a thermodynamic background, which does not depend on the correlation length. For the fluctuation-cluster model, the critical part of the pressure may be approximated as

$$\Delta P_{\mathrm{cr}} \approx \frac{N_{\mathrm{cl}} k_{\mathrm{B}} T_{\mathrm{c}}}{V_{\mathrm{t}}} \approx \frac{k_{\mathrm{B}} T_{\mathrm{c}}}{(2\xi)^d} \tag{8.10}$$

Physically, Eq. (8.10) means that the critical part of the density of the grand thermodynamic potential is proportional to the density of fluctuation-induced inhomogeneities (see discussion on fluctuation corrections to the meanfield behavior in Section 4.9). Therefore, as given by Eq. (8.8), along the critical isochore, $\xi = \xi_0^+ |\Delta \hat{T}|^{-\nu}$, the excess pressure generated by fluctuations is,

$$\Delta P_{\mathrm{cr}} \approx \frac{k_{\mathrm{B}} T_{\mathrm{c}}}{(2\xi_0^+)^d} |\Delta \hat{T}|^{d\nu} \tag{8.11}$$

The excess entropy density due to fluctuations, $\Delta(\rho_{\mathrm{c}} S)$, may be obtained from the Gibbs–Duhem equation Eq. (2.21), $\mathrm{d}P = \rho S \mathrm{d}T + \rho \mathrm{d}\mu$, by evaluating the derivative of the pressure with respect to temperature along the critical isochore. Recall that along this path, the ordering field is $h = (\mu - \mu_0)/k_{\mathrm{B}} T_{\mathrm{c}}$, where μ_0 is the value of the chemical potential along the critical isochore Eq. (4.12), such that in the first-order approximation, $h = 0$. Thus, one finds

$$\Delta(\rho_{\mathrm{c}} S) = \left.\frac{\partial(\Delta P_{\mathrm{cr}})}{\partial T}\right|_{h=0} \approx k_{\mathrm{B}} T_{\mathrm{c}} \frac{d\nu}{(2\xi_0^+)^d} |\Delta \hat{T}|^{d\nu - 1} \tag{8.12}$$

Consequently, one may obtain the diverging (critical) part of the isochoric heat capacity along the critical isochore from the temperature derivative, given by Eq. (4.16), as

$$\frac{\Delta C_V}{k_{\mathrm{B}}} = \frac{1}{k_{\mathrm{B}} T_{\mathrm{c}}} \left.\frac{\partial(\Delta S)}{\partial T}\right|_{h=0} \approx \frac{d\nu(d\nu - 1)}{\rho_{\mathrm{c}} (2\xi_0^+)^d} |\Delta \hat{T}|^{d\nu - 2} \tag{8.13}$$

In accordance with experimental findings (see Section 8.2.2), the excess isochoric heat capacity is given through a power law in the form of Eq. (8.4). Thus, upon comparison of Eqs. (8.4) and (8.13), one obtains the following relations between α and ν as well as between A_0^+ and $\rho_{\mathrm{c}}(\xi_0^+)^d$ as

$$\alpha = 2 - d\nu \tag{8.14}$$

and

$$A_0^+ \rho_{\mathrm{c}} (\xi_0^+)^d = \frac{d\nu(d\nu - 1)}{2^d} \equiv R_\xi \tag{8.15}$$

In three dimensions, adopting $\nu = 0.63$, one obtains $\alpha = 2 - 3\nu = 0.11$ and the amplitude ratio $R_\xi = 0.21$.

Equations (8.14) and (8.15) are examples of the scaling relations between critical exponents and critical amplitudes, and their validity goes well beyond the qualitative fluctuation-cluster model. This exercise demonstrates that the critical singularities of physical properties indeed originates from the divergence of the correlation length. Note that relation (8.14) is exact, yielding the universal value of the critical exponent of the heat capacity, $\alpha = 0.11$, in perfect agreement with modern theory (Sections 8.3.3 and 8.3.4). Relation (8.15) is also universal, although the amplitudes, A_0^+ and ξ_0^+, are system dependent, being affected by specific intermolecular interactions. We also note that the numerical value obtained from the fluctuation-cluster model, $R_\xi = 0.21$, is an approximation associated with the assumed volume of a cluster, $(2\xi)^d$. Generally, fluctuation inhomogeneities are topologically neither cubic cells nor spherical droplets (see Figure 8.10). Contemporary experimental and theoretical studies predict a more accurate universal value, $R_\xi \approx 0.18$. However, it is remarkable that the ratio obtained for the amplitudes of independently measured diverging properties, the heat capacity and the correlation length, is a value that does not depend on the microscopic details.

It is important to note, however, that Eqs. (8.14) and (8.15) explicitly depend on the dimensionality of space. In particular, the meanfield results, $\alpha = 0$ and $\nu = 1/2$ (see Eqs. (4.18) and (4.72), respectively), are valid at $d = 4$. The case $d = 4$ in the theory of critical phenomena is defined as the "marginal dimensionality." For $d \geq 4$, the effects of fluctuations are either marginal (for $d = 4$) or disappear (for $d > 4$).

The fluctuation-cluster model shows that, along the critical isochore in the one phase region, the excess critical part of the pressure is $\Delta P_{\rm cr} \propto A_0^+|\Delta\hat{T}|^{d\nu} \propto \xi^{-d}$, see Eq. (8.11), and with use of the universal relation, Eq. (8.14), one obtains $\Delta P_{\rm cr} \propto A_0^+|\Delta\hat{T}|^{2-\alpha} \propto \xi^{-d}$. Accordingly, one could expect that along fluid-phase coexistence in the two-phase region, $\Delta P_{\rm cr} \propto A_0^-|\Delta\hat{T}|^{2-\alpha} \propto \xi^{-d}$.

8.3.3 Universal Scaling Relations

The fluctuation-cluster model emphasizes the fact that the critical part of the pressure depends on the correlation length of the critical fluctuations, being scaled as ξ^{-d}. One of the major scaling relations between the critical exponents α and ν, see Eq. (8.14), directly follows from this fact. Similar scaling arguments may be used to obtain relations between other critical exponents.

As discussed in Chapter 4 after Eq. (4.13), the critical part of the free energy ($\Delta\hat{f} = \hat{f} - \hat{f}_0$), is related to the critical part of the pressure through a Legendre transform as

$$\Delta P_{\rm cr} = \Delta\hat{f} - h\phi \tag{8.16}$$

In zero ordering field, $\Delta P_{\rm cr} = \Delta\hat{f}$, such that below the critical point

$$\Delta\hat{f} = \frac{1}{2}\hat{\chi}^{-1}\phi_0^2 \propto \xi^{-d} \propto A_0^-|\Delta\hat{T}|^{2-\alpha} \tag{8.17}$$

The spontaneous value of the order parameter, ϕ_0, asymptotically scales as $\phi_0 = B_0|\Delta\hat{T}|^{\beta}$ (see Eq. 8.3), while the susceptibility scales as $\hat{\chi} = \Gamma_0^-|\Delta\hat{T}|^{-\gamma}$ (see Eq. 8.5), such that one obtains another universal relation between the critical exponents α, β, and γ as

$$2 - \alpha = \gamma + 2\beta \tag{8.18}$$

which is exactly satisfied with the best experimental values, $\alpha = 0.11$, $\beta = 0.3255$, and $\gamma = 1.239$. From Eq. (8.17), one may also predict the existence of a universal ratio, R_χ^-, between the critical amplitudes A_0^-, Γ_0^-, and B_0 as

$$R_\chi^- = \frac{A_0^-\Gamma_0^-}{B_0} \tag{8.19}$$

Analogously, it can be shown that, a universal relation, R_χ^+, between A_0^+, Γ_0^+, and B_0, is given by

$$R_\chi^+ = \frac{A_0^+\Gamma_0^+}{B_0} \tag{8.20}$$

implying that the ratios of the corresponding amplitudes above and below the critical temperature are also universal.

Along the critical isobar or the critical isotherm, the susceptibility in nonzero field, may be described by a power law with the critical exponent, δ as

$$\hat{\chi}^{-1} = \delta D_0|\phi|^{\delta-1} \tag{8.21}$$

(see the meanfield power law, given by Eq. (4.21)), while below the critical point in zero-field, the susceptibility may be expressed through the spontaneous order parameter, with use of the same scaling arguments as discussed after Eq. (8.17), as

$$\hat{\chi}^{-1} = \frac{1}{B_0^{\gamma/\beta}\Gamma_0^+}|\phi_0|^{\gamma/\beta} \tag{8.22}$$

Thus, as $h \to 0$, such that $\phi \to \phi_0$, the critical exponents of the susceptibilities given by Eqs. (8.21) and (8.22) should obey the relation

$$\delta - 1 = \gamma/\beta \tag{8.23}$$

while the amplitudes form a universal product

$$R_\phi = D_0 B_0^{\delta-1}\Gamma_0^+ \tag{8.24}$$

One may obtain an approximate relation between the correlation length and the susceptibility, through the Landau–Ginzburg functional, Eq. (4.71), provided that the susceptibility and correlation length are given by their scaling forms, Eqs. (8.5) and (8.8), respectively. One finds, from the Landau–Ginzburg functional, that $\hat{\chi}^{-1}\xi^2 = c_0$, and consequently,

$$\gamma = 2\nu \tag{8.25}$$

This relation is referred to as the Ornstein–Zernike approximation, being exact in the meanfield approximation. However, the actual relation between the exponents γ and ν differs from Eq. (8.25) by a small value, known as the anomalous exponent of the correlation function, η, such that

$$\gamma = \nu(2 - \eta) \tag{8.26}$$

where $\eta = 0.03$, if $\gamma = 1.24$ and $\nu = 0.63$. A good approximation, consistent with Eq. (8.26), of the scaling relation between the susceptibility and correlation length reads

$$\chi = c_0 \xi^{\gamma\nu} \tag{8.27}$$

The universal scaling relations between the critical exponents and the amplitudes are satisfied for both the meanfield values as well as for the fluctuation-induced values (see Table 8.1). Among the critical exponents and their corresponding amplitudes, only two are independent, since they are interrelated by universal relations. Meaning that, only two thermodynamic properties are needed to obtain the complete scaling equation of state in the critical region. Consequently, if new critical singularities are considered, then they will be interrelated to the other properties by additional universal relations.

8.3.4 Classes of Critical-Point Universality

The method to calculate the values of the critical exponents for various phase transitions is known as the "renormalization group" (RG) theory [32, 33], which was pioneered by Wilson.[8] A detailed description of the RG theory is beyond the scope of this book. In this section, we present only the major results of this theory, which have been in excellent agreement with most accurate experimental studies.

In statistical mechanics, the Helmholtz energy is calculated through the canonical partition function Z (see Section 2.10), which may be expressed in the form [34]

$$\frac{A_t}{k_B T} = -\ln Z_A = -\ln \sum_{\phi(\boldsymbol{r})} \exp\left\{-\frac{H[\phi(\boldsymbol{r})]}{k_B T}\right\} \tag{8.28}$$

---

8 Kenneth G. Wilson (1936–2013) was an American theoretical physicist, who was awarded the Nobel in Physics in 1982 for his work on phase transitions.

where $H[\phi(\boldsymbol{r})]$ is the effective Hamiltonian, which represents the local energy of the system, and at mesoscales is assumed to have the same form as the Landau–Ginzburg free-energy functional, given by Eq. (4.71), as

$$\frac{H}{k_B T} = \frac{1}{V}\int\left(\frac{1}{2!}a_0\Delta\hat{T}\phi^2 + \frac{1}{4!}u_0\phi^4 + \frac{1}{2}c_0|\nabla\phi|^2\right)\mathrm{d}V \tag{8.29}$$

If the order-parameter fluctuations are ignored and the order parameter is given by its average value $\phi(\boldsymbol{r}) = \langle\phi(\boldsymbol{r})\rangle$, then the Hamiltonian given by Eq. (8.29) is reduced to the Landau–Ginzburg free-energy functional, Eq. (4.71). RG theory postulates that near the critical point, the correlation length is the only relevant length scale in the system. Thus, the macroscopic properties of fluids can be universally explained and calculated through a single fluctuation-induced mesoscopic length scale.

RG theory reveals that the critical exponents for all systems exhibiting critical phenomena depend only on two physical parameters: the spatial dimensionality, d, and the "dimensionality" (the number of components) of the order parameter. Critical phenomena in systems with the same dimensionality of the order parameter belong to the same universality class. Consequently, all fluids and fluid mixtures, phase separating binary alloys, ammonia halides (orientational order–disorder transitions), and anisotropic ferro- and antiferro-magnets (to list only the most common examples) belong to the same universality class as the Ising and lattice-gas models. For all of these examples, the number of components of the order parameter is $n = 1$, since the order parameter is either a scalar or a single-component vector.

Many other systems belong to different classes of critical-point universality. For example, in superfluid He^4 (theoretically described by the "XY model"), the order parameter is the wave function, which is a two-component vector, $n = 2$, corresponding to the amplitude and phase. In the isotropic Heisenberg ferromagnet, the order parameter is a three-component vector, $n = 3$. The most important result of RG theory is that the critical exponents do not depend on the microscopic details of the interactions between the molecules or quasiparticles, but only depend on the symmetry of the order parameter.

Application of the RG theory to calculate the critical exponents turns out to be especially transparent due to the existence of the marginal dimensionality of space, $d = 4$. At $d = 4$, fluctuations of the order parameter cause only logarithmic corrections to the critical power laws. Thus, the critical exponents remain the same as predicted by the meanfield theory. Correspondingly, the critical exponents for a spatial dimensionality $d < 4$ can be represented by a power series, known as the "epsilon expansion", in terms of the parameter $\varepsilon = 4 - d$.

For a system with arbitrary n, the four most common critical exponents, α, β, γ, and ν are represented as

$$\alpha = \frac{1}{2}\left(\frac{n+4}{n+8}\right)\varepsilon - \frac{1}{4}\frac{(n+2)^2(n+28)}{(n+8)^3}\varepsilon^2 + O(\varepsilon^3) \tag{8.30}$$

$$\beta = \frac{1}{2} + \frac{3}{2}(n+8)^{-1}\varepsilon + \frac{1}{2}\frac{(n+2)(2n+1)}{(n+8)^3}\varepsilon^2 + O(\varepsilon^3) \tag{8.31}$$

$$\gamma = 1 + \frac{1}{2}\left(\frac{n+2}{n+8}\right)\varepsilon + \frac{1}{4}\frac{(n+2)(n^2+22n+52)}{(n+8)^3}\varepsilon^2 + O(\varepsilon^3) \tag{8.32}$$

$$\nu = \frac{1}{2} + \frac{n+2}{4(n+8)}\varepsilon + \frac{(n+2)(n^2+23n+60)}{8(n+2)^3}\varepsilon^2 + O(\varepsilon^3) \tag{8.33}$$

It turns out that the amplitudes in the two lowest terms of the epsilon expansions are numerically small. Thus, although $\varepsilon = 4 - d$ is not a small parameter for $d = 3$, the summation of these

Table 8.1 Critical exponents for systems with different space and order-parameter dimensionalities.

		α	β	γ	δ	ν	η
Meanfield theory		0	1/2	1	3	1/2	0
Planar Ising model ($d = 2$, $n = 1$)		0	1/8	7/4	15	1	1/4
Three-dimensional Ising model ($d = 3$, $n = 1$)	ε expansion[a)]	0.07	0.340	1.244	4.66	0.626	0.037
	RG solution[b)]	0.110	0.326	1.240	4.81	0.63	0.033
Three-dimensional XY model ($d = 3$, $n = 2$)	ε expansion[a)]	−0.02	0.360	1.300	4.46	0.655	0.039
Three-dimensional Heisenberg model ($d = 3$, $n = 3$)	ε expansion[a)]	−0.1	0.377	1.346	4.446	0.678	0.032

a) Calculated up to second order in ε^2, except for the exponent η, which is given with an accuracy to order ε^3.
b) Rounded values in agreement with the best available experimental results.

terms yields values of the critical exponents that reasonably agree with experiments. For the three-dimensional Ising model universality class ($d = 3$ and $n = 1$), the values of α, β, and γ, calculated to the second order in the epsilon expansion are presented in Table 8.1.

The currently adopted (most accurate) values of the critical exponents and universal amplitude ratios are obtained by numerical solutions of the equations of RG theory. These values are presented in Table 8.1. The accurate measurements of the critical exponents in fluids and fluid mixtures strongly support these theoretical values, as illustrated in this chapter.

It is important to note that in the first order of the epsilon expansion $\gamma = 7/6 = 2\nu$, the same as given by the Ornstein–Zernike approximation, Eq. (8.25). Correspondingly, the Landau–Ginzburg functional for the local free-energy Eq. (4.71) is also valid in the first order of the RG epsilon expansion. The difference between γ and 2ν appears only in the second order of the epsilon expansion, which for $n = 1$ reads,

$$\eta = 2 - \frac{\gamma}{\nu} = \frac{\varepsilon^2}{54} + \cdots \approx 0.02\varepsilon^2 \tag{8.34}$$

This numerically very small difference between γ and 2ν reflects a small deviation of the correlation function from the Ornstein–Zernike approximation Eq. (4.67). Correspondingly, asymptotically close to the critical point, at $q\xi \gg 1$, RG theory predicts the following form of the correlation function in momentum space

$$C(q,\xi) \propto q^{1-\eta} e^{-1/(q\xi)} \approx \frac{1}{(q\xi)^{2-\eta}} \tag{8.35}$$

Verification of Eq. (8.35), made possible by light or neutron scattering measurements, is a challenging experimental task because η is very small. However, the most accurate light scattering measurements yield $\eta = 0.03 \pm 0.01$ [35, 36]. Figure 8.12 illustrates these findings by the analysis of the intensity of light scattering in a binary liquid mixture as a function of the correlation length ξ and the scattering wave number, q. If the correlation function followed the Ornstein–Zernike form, the slope of the rescaled data would be zero.

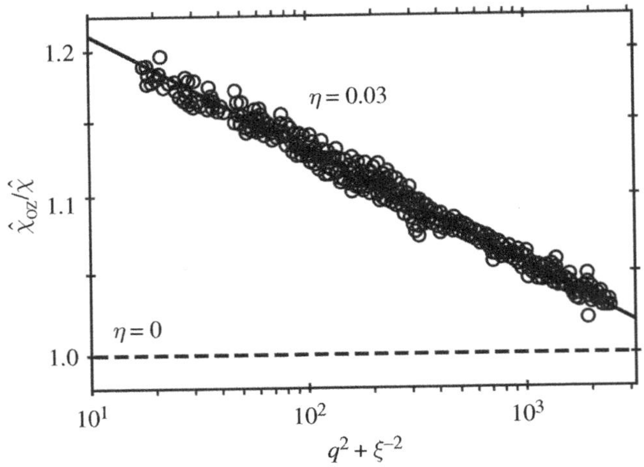

Figure 8.12 Deviation of the q-dependent susceptibility, given by the Fisher–Burford interpolation, Eq. (8.36), from the Ornstein–Zernike approximation ($\hat{\chi}_{OZ}$), Eq. (4.67), near the liquid–liquid critical point of a nitroethane-n-hexane mixture obtained by static light scattering. Source: Adapted from Ref. [35].

In the scaling theory, the spatial-dependent susceptibility is best described by using the following interpolation based on the Fisher–Burford approximation [37, 38],

$$\hat{\chi} \simeq \frac{\hat{\chi}_{q=0}}{(1 + q^2\xi^2)^{1-\eta/2}} \tag{8.36}$$

such that the susceptibility is related to the correlation length by $\hat{\chi}_{q=0} \sim \xi^{2-\eta}$. Thus, the critical exponents γ and ν are universally related through the relation given by Eq. (8.26).

8.4 Phenomenology of Critical Behavior

As discussed in the previous section, critical phenomena exhibit a surprising degree of universality. For instance, the Curie point in the Ising model is thermodynamically equivalent to the liquid–gas critical point in the lattice-gas model (see Section 4.2). The discovery of this universality, also known as the "isomorphism" of critical phenomena, was a profound discovery that brought the behavior of fluids into the realm of condensed-matter physics.[9] In this section, we discuss the phenomenology that describes the critical behavior of single-component fluids in terms of the isomorphism with the Ising model. The concept of isomorphism will be extended to the description of multi-component near-critical mixtures in Section 8.5.

8.4.1 Isomorphism of Critical Phenomena

To introduce the concept of isomorphism, one needs to consider the characteristic properties of the critical behavior in the Ising model through *scaling fields*, *scaling densities*, and *scaling susceptibilities* [40]. In the Ising model, there are two independent scaling fields, h_1, known as the ordering field (or the "strong" field) and h_2, the thermal field (or the "weak" field). A third scaling field, h_3, is a field-dependent thermodynamic potential, such that h_3 is a function of h_1 and h_2. There are

9 The term "isomorphism of critical phenomena" was first introduced by Anisimov, Voronel, and Gorodetskiĭ in 1971 [39].

also two independent scaling densities, ϕ_1 and ϕ_2, defined through the derivatives of the scaling fields as,

$$\phi_1 = \left.\frac{\partial(-h_3)}{\partial h_1}\right|_{h_2} \tag{8.37}$$

$$\phi_2 = \left.\frac{\partial(-h_3)}{\partial h_2}\right|_{h_1} \tag{8.38}$$

The scaling densities are conjugate to their appropriate scaling fields.

In the Ising model, the ordering field is the applied magnetic field, $h_1 = h$, and the thermal field is the reduced distance to the Curie (critical) temperature, $h_2 = \Delta\hat{T}$, while the third scaling field is the density of the "Ising" free energy, $h_3 = \Delta\Phi = \Phi - \Phi_0$ (see Section 4.3). The first scaling density, the order parameter, is the reduced magnetization $\phi_1 = M/M_0$, where M_0 is the magnetization at zero temperature, and the second scaling ("thermal") density, is the reduced entropy, $\hat{S} = S/k_B$.

Equivalently, for the lattice-gas model, the strong field is a combination of the chemical potential and temperature, which in the linear approximation is expressed as

$$h_1 = a_1\Delta\hat{\mu} + a_2\Delta\hat{T} \tag{8.39}$$

where $\Delta\hat{\mu} = \mu/k_B T_c - \hat{\mu}_c$, while a_1 and a_2 are system-dependent coefficients. Absorbing the coefficient a_1 into the strong-field, h_1, one obtains,

$$h_1 = \Delta\hat{\mu} + \frac{a_2}{a_1}\Delta\hat{T} \tag{8.40}$$

In zero-ordering field, one obtains

$$\frac{a_2}{a_1} = \frac{\Delta\hat{\mu}}{\Delta\hat{T}} = \left.\frac{\partial\hat{\mu}}{\partial\hat{T}}\right|_{h_1=0} \tag{8.41}$$

In zero field, one finds that $\partial\hat{\mu}/\partial\hat{T}|_{h_1=0} = S_c - \partial\hat{P}/\partial\hat{T}|_{h_1=0} = 0$, see Eq. (2.185). Therefore, since the critical value of the entropy, S_c, is arbitrary (see Section 2.14), one may select $S_c = \partial\hat{P}/\partial\hat{T}|_{h_1=0}$, such that $a_2/a_1 = 0$.

In the lattice-gas model, the thermal field is $h_2 = \Delta\hat{T}$. Moreover, the third field is the critical part of the grand potential per unit volume, which is the critical part of the pressure, $h_3 = -\Delta\hat{P}_{cr}$, see Eq. (8.16), where $\Delta\hat{P}_{cr}$ is defined as

$$\Delta\hat{P}_{cr} = \Delta\hat{P} - \Delta\hat{\mu} - \hat{S}_c\Delta\hat{T} \tag{8.42}$$

where $\Delta\hat{P} = P/\rho_c k_B T_c - \hat{P}_c$ and $\hat{S}_c = S_c/k_B$. Therefore, the scaling fields in the lattice-gas model are given by

$$h_1 = \Delta\hat{\mu} \tag{8.43}$$

$$h_2 = \Delta\hat{T} \tag{8.44}$$

$$-h_3 = \Delta\hat{P}_{cr} \tag{8.45}$$

In the lattice-gas, the first scaling density is the order parameter, $\phi_1 = (\rho - \rho_c)/\rho_c = \Delta\hat{\rho}$, while the second (thermal) scaling density is the reduced difference in entropy density and critical entropy density, $\Delta(\rho\hat{S}) = \rho\hat{S} - \rho_c\hat{S}_c$ obtained from the derivative, $\phi_2 = \partial(-\Delta P_{cr})/\partial(\Delta\hat{T})|_{\Delta\hat{\mu}} = \Delta(\rho\hat{S})$, with use of Eq. (8.42) for $\Delta\hat{P}_{cr}$. Thus, the scaling densities in the lattice-gas model are

$$\phi_1 = \Delta\hat{\rho} \tag{8.46}$$

$$\phi_2 = \Delta(\rho\hat{S}) \tag{8.47}$$

Therefore, the lattice-gas and Ising models are isomorphic with respect to these scaling fields and scaling densities. Generally, the isomorphism of critical phenomena can be quantified through the concept of "complete scaling," which states that the theoretical scaling fields for any near-critical system can be expressed through the physical field variables (experimentally measurable) characteristic to the phase transition.

8.4.2 Scaling Free Energies and Susceptibilities

Based on the choice of the independent variable, either the ordering field or the order parameter, one could consider two alternative forms of the free energy that would describe the critical behavior of single-component fluids. Asymptotically, the field-dependent free energy, $-h_3 = \Delta\hat{P}_{cr}$, is expressed through h_1 and h_2 as a homogeneous nonanalytic function of the form,

$$\Delta\hat{P}_{cr} = |h_2|^{2-\alpha}\Upsilon^{\pm}\left(\frac{h_1}{|h_2|^{2-\alpha-\beta}}\right) \tag{8.48}$$

where the superscript (±) on the function $\Upsilon^{\pm}$, reflects the difference in Υ for $h_2 > 0$ (above) and $h_2 < 0$ (below) the critical temperature. The actual form of $\Upsilon(z)$, which is a function of the single argument $z = h_1/|h_2|^{2-\alpha-\beta}$, is only known through an explicit equation of state, but asymptotically, the properties of $\Upsilon(z)$ are generally known from the scaling theory,

$$\Upsilon^{\pm}(z) = \begin{cases} z \to 0\ (h_1 \to 0) & \dfrac{A_0^{\pm}}{(2-\alpha)(1-\alpha)} \\ z \to \infty\ (h_2 \to 0) & \pm\dfrac{\delta}{D_0(\delta+1)} z^{(\delta+1)/\delta} \end{cases} \tag{8.49}$$

where the critical amplitudes, $A_0^{\pm}$ and D_0, are defined in Table 8.2. Consequently, in these limits, $\Delta\hat{P}_{cr}$ is given by

$$\Delta\hat{P}_{cr} = \begin{cases} z \to 0\ (h_1 \to 0) & \dfrac{A_0^{\pm}}{(2-\alpha)(1-\alpha)}|h_2|^{2-\alpha} \\ z \to \infty\ (h_2 \to 0) & \pm\dfrac{\delta}{D_0(\delta+1)}|h_1|^{(\delta+1)/\delta} \end{cases} \tag{8.50}$$

Table 8.2 Asymptotic critical power laws for scaling properties.

Property	**Scaling power law**
Order parameter ($h_2 \le 0$, $h_1 = 0$)	$\phi_1 = \pm B_0|h_2|^{\beta}$
Second scaling density ($h_2 \le 0$, $h_1 = 0$)	$\phi_2 = (1-\alpha)A_0^{\pm}|h_2|^{1-\alpha}$
Ordering field ($h_2 = 0$)	$h_1 = \pm D_0|\phi_1|^{\delta}$
Strong susceptibility ($h_1 = 0$)	$\chi_1 = \Gamma_0^{\pm}|h_2|^{\gamma}$
Weak susceptibility[a)] ($h_1 = 0$)	$\chi_2 = A_0^{\pm}|h_2|^{-\alpha}$
Cross susceptibility ($h_2 \le 0$, $h_1 = 0$)	$\chi_{12} = B_0|h_2|^{\beta-1}$

a) In addition to the diverging part of the weak susceptibility, the scaling theory also predicts a fluctuation-induced constant, known as the "critical background." This constant compensates the diverging part away from the critical point (as $h_2 \to 1$), and it is on the order of the meanfield discontinuity of the heat capacity [41].

such that the asymptotic scaling law for the isochoric heat capacity along the critical isochore is recovered for $C_V \sim \partial^2(\Delta\hat{P}_{cr})/\partial h_2^2|_{h_1=0} = A_0^{\pm}|h_2|^{-\alpha}$ and the order parameter along the critical isotherm for $\phi_1 = \partial(\Delta\hat{P}_{cr})/\partial h_1|_{h_2=0} = \pm D_0^{-1}|h_1|^{1/\delta}$.

As the scaling fields are conjugate to the scaling densities, one may generally express the differential of h_3 through the scaling densities as

$$d(-h_3) = \phi_1 \; dh_1 + \phi_2 \; dh_2 \tag{8.51}$$

such that, one may define three scaling susceptibilities: a "strong" susceptibility χ_1, a "weak" susceptibility χ_2, and a "cross" susceptibility χ_{12}, through second derivatives of $-h_3$ as

$$\chi_1 = \left.\frac{\partial\phi_1}{\partial h_1}\right|_{h_2} \tag{8.52}$$

$$\chi_2 = \left.\frac{\partial\phi_2}{\partial h_2}\right|_{h_1} \tag{8.53}$$

$$\chi_{12} = \left.\frac{\partial\phi_1}{\partial h_2}\right|_{h_1} = \left.\frac{\partial\phi_2}{\partial h_1}\right|_{h_2} \tag{8.54}$$

where ϕ_1 and ϕ_2 are defined by Eqs. (8.37) and (8.38). The physical response functions are asymptotically proportional to these three susceptibilities. For example, in zero-ordering field ($h_1 = 0$), the isothermal compressibility and volumetric expansivity are $\kappa_T \propto \alpha_P \propto \chi_1 \sim |\Delta\hat{T}|^{-\gamma}$, the derivative of the density with respect to temperature for each branch along the liquid–gas coexistence is $\partial\rho/\partial T|_{\rho_{cxc}} \propto \chi_{12} \sim |\Delta\hat{T}|^{\beta-1}$, and isochoric heat capacity is $C_V \propto \chi_2 \sim |\Delta\hat{T}|^{-\alpha}$. Note that above the critical point, in zero field ($\rho = \rho_c$), $\chi_{12} = 0$, whereas $\chi_1, \chi_2 \neq 0$.

Alternatively, while the free energy, given by Eq. (8.48), is a function of the ordering field, one may also define an order-parameter-dependent free energy through a Legendre transform of the form, $\Delta\hat{f} = \Delta\hat{P}_{cr} - \phi_1 h_1$, where $\Delta\hat{f}$ is the critical part of the density of the Helmholtz energy (see Section 4.3). Asymptotically, in the scaling theory, this function has the following form

$$\Delta\hat{f} = |h_2|^{2-\alpha}\Theta^{\pm}\left(\frac{\phi_1}{|h_2|^{\beta}}\right) \tag{8.55}$$

Just like $\Upsilon^{\pm}$, given by Eq. (8.49), $\Theta^{\pm}(y)$ is also a function of the single argument $y = \phi_1/|h_2|^{\beta}$, such that $\Theta^{\pm}$ follows the limiting behavior

$$\Theta^{\pm}(y) = \begin{cases} y \to 0 \begin{pmatrix} \phi_1 \to \phi_0 \\ \phi_1 \to 0 \end{pmatrix} & \dfrac{A_0^{\pm}}{(2-\alpha)(1-\alpha)} \\ y \to \infty \; (h_2 \to 0) & \dfrac{D_0}{(\delta+1)} y^{\delta+1} \end{cases} \tag{8.56}$$

where $\phi_1 \to \phi_0$ for $h_2 < 0$ (below) and $\phi_1 \to 0$ for $h_2 > 0$ (above) the critical point. Consequently, in these limits, $\Delta\hat{f}$ is given by

$$\Delta\hat{f} = \begin{cases} y \to 0 \begin{pmatrix} \phi_1 \to \phi_0 \\ \phi_1 \to 0 \end{pmatrix} & \dfrac{A_0^{\pm}}{(2-\alpha)(1-\alpha)}|h_2|^{2-\alpha} \\ y \to \infty \; (h_2 \to 0) & \pm\dfrac{D_0}{(\delta+1)}|\phi_1|^{\delta+1} \end{cases} \tag{8.57}$$

Note that in zero field, ($h_1 \to 0$) the field-dependent, $\Delta\hat{P}_{cr}$, and order-parameter-dependent, $\Delta\hat{f}$, free energies are equivalent.

8.4.3 Linear Model of the Scaled Equation of State

To obtain the explicit form of the scaling functions, $\Upsilon(z)$ and $\Theta(y)$, one needs to formulate a complete equation of state in the critical region. This may be done by replacing the two independent scaling fields, h_1 and h_2, with two parametric variables: a variable r, which measures a "distance" from the critical point and an angular variable θ, which indicates the location on a contour of constant r. This formulation, known as the "parametric scaled equation of state" [43], is given by

$$h_1 = ar^{2-\alpha-\beta}\theta(1-\theta) \qquad \text{and} \qquad h_2 = r(1-b^2\theta^2) \tag{8.58}$$

The simplest form of the parametric equation of state is known as the "linear model," in which the order parameter is given by

$$\phi_1 = kr^{\beta}\theta \tag{8.59}$$

In these equations, b^2 is a universal constant, while a and k represent the two system-dependent coefficients that are related to the critical amplitudes (Table 8.3). The meaning of the parametric variables is illustrated in Figure 8.13. The variable θ varies from $\theta = 0$ at $\phi_1 = 0$ and $h_2 > 0$ to $\theta = \pm b$ at $h_2 = 0$ to $\theta = \pm 1$ at $\phi_1 = \pm\phi_0$ and $h_2 < 0$.

The universal constant b^2 can be approximated as

$$b^2 = \frac{\gamma - 2\beta}{\gamma(1-2\beta)} \tag{8.60}$$

For the three-dimensional Ising model universality class, $b \approx 1.36$, and this scenario is referred to as the "restricted linear model." In this model, the expression for the critical part of the isochoric heat capacity becomes independent of θ as

$$C_V \propto \chi_2 = \frac{ak\gamma(\gamma-1)}{2\alpha b} r^{-\alpha} \tag{8.61}$$

The linear model also predicts values for the universal ratios or products of critical amplitudes (Table 8.4). For example, the amplitude ratio of the heat capacity above and below the critical point is given by

$$\frac{A_0^-}{A_0^+} = 4\left(\frac{\beta}{\gamma}\right)^2\left[\frac{\gamma(1-2\beta)}{2\beta(\gamma-1)}\right]^{\gamma+2\beta} \tag{8.62}$$

Figure 8.14 illustrates that the linear model reasonably describes the relation between this amplitude ratio and the values of the critical exponents, as supported by experiments. Moreover, it has been shown that the predictions of the linear model are consistent with the second-order epsilon

Table 8.3 Critical amplitudes the restricted linear model.

Amplitude	Restricted linear model
Γ_0^+	k/a
Γ_0^-	$0.206k/a$
A_0^+	$0.984ak$
A_0^-	$1.859ak$
B_0	$1.394k$
D_0	$0.477a/k^{\delta}$

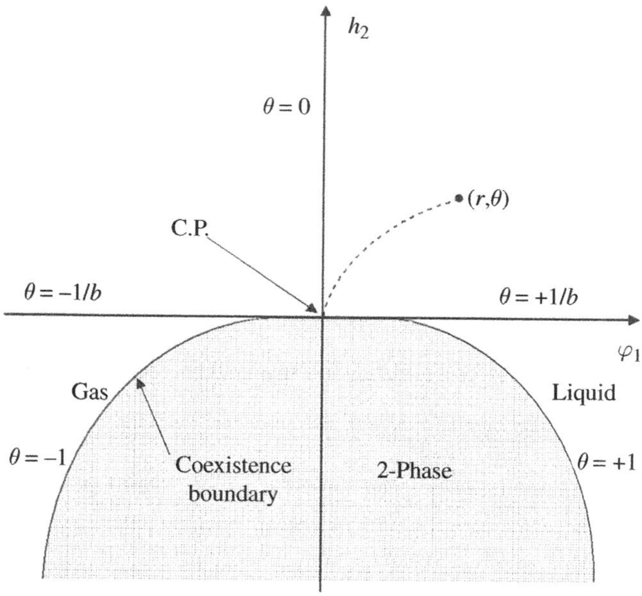

Figure 8.13 Phase diagram for the linear model of a parametric equation-of-state. For the thermal field, $h_2 > 0$, the ordering field $h_1 = 0$ only when the order parameter, $\phi_1 = 0$. For $h_2 < 0$, $h_1 = 0$ everywhere in the two-phase region, i.e. for $h_2 < 0$, $\phi_1 \leq \phi_0$. Source: Reproduced with permission from Ref. [42], © 2010, Royal Society of Chemistry.

Table 8.4 Universal ratios of the amplitudes of scaling powers laws.

Ratio	3D Ising system	Restricted linear model
A_0^+/A_0^-	0.523	0.529
Γ_0^+/Γ_0^-	4.950	4.854
$A_0^+\Gamma_0^+/B_0^2$	0.527	0.506
ξ_0^+/ξ_0^-	1.960	—
$\rho_c A_0^+(\xi_0^+)^3$	0.171	—

expansion in the RG theory [8]. More sophisticated and more accurate parametric equations have also been considered in the literature.

The linear model, supplemented with nonasymptotic corrections, was successfully used to formulate a scaled equation of state for simple fluids – see, for example, Ref. [8] and chapters in books [4, 42]. The linear model, in both its meanfield and scaling versions, was also applied to describe criticality in various physically different systems: ferromagnets [44], blue phase III of liquid crystals [45], supercooled water [46, 47], and for the description of extrema lines around liquid–liquid critical points [48].

8.4.4 Corrections to Asymptotic Scaling Behavior

The scaling laws with universal critical exponents are only valid asymptotically close to the critical point. Thus, in practice, it is necessary to introduce corrections to the asymptotic scaling laws to

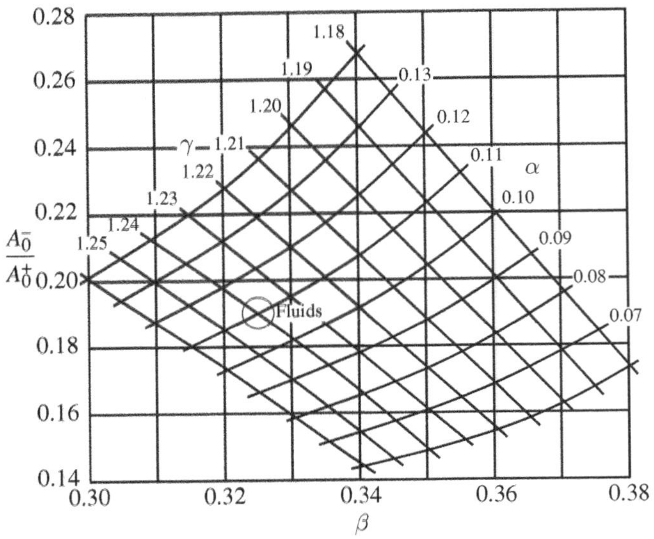

Figure 8.14 A nomogram illustrating the ratios between the critical exponents and the critical amplitudes in the restricted linear model of the parametric equation of state. Source: Adapted from Ref. [8]/AIP Publishing.

describe systems in a broader vicinity of the critical point. There are two types of corrections that will be discussed in this section: symmetric, existing for all Ising-like systems, and asymmetric corrections, specific for fluids.

8.4.4.1 Symmetric Corrections

The asymptotic scaling power laws discussed in Section 8.2 may be expanded to higher-order polynomials. These expansions are often referred to as "Wegner expansions." For instance, consider the first-order Wegner expansion of the spontaneous order parameter below the critical point,

$$\phi_1 = \phi_0 \approx \pm B_0 |h_2|^{\beta} \left[1 + B_1 |h_2|^{\Delta_s} + \cdots\right] \tag{8.63}$$

where $\Delta_s = 0.5$, as obtained in the first-order RG epsilon expansion, is an additional independent universal scaling exponent. The first-order corrections to the asymptotic power laws for various scaling properties are presented in Table 8.5.

The first-order system-dependent Wegner amplitudes, B_1, $A_1^{\pm}$ and $\Gamma_1^{\pm}$, are interrelated through universal relations, like for the zeroth-order (asymptotic) amplitudes. In fact, it has been proven that all of the first-order Wegner amplitudes are related to the Ginzburg number as

Table 8.5 First-order Wegner corrections to the asymptotic scaling powers laws in zero-ordering field ($h_1 = 0$).

Critical power law
$\phi_1 \approx \pm B_0 \lvert h_2\rvert^{\beta} \left[1 + B_1 \lvert h_2\rvert^{\Delta_s} + \cdots\right]$
$\phi_2 \approx \frac{A_0^{\pm}}{1-\alpha} \lvert h_2\rvert^{1-\alpha} \left[1 + A_1^{\pm} \lvert h_2\rvert^{\Delta_s} + \cdots\right]$
$\chi_1 \approx \Gamma_0^{\pm} \lvert h_2\rvert^{-\gamma} \left[1 + \Gamma_1^{\pm} \lvert h_2\rvert^{\Delta_s} + \cdots\right]$

$B_1 \approx A_1^{\pm} \approx \Gamma_1^{\pm} \approx c_W N_G^{-\Delta_s}$, where c_W is a system-dependent constant. Therefore, the Wegner expansion for the spontaneous order parameter may be given in the form

$$\phi_1 = \phi_0 \approx \pm B_0 |h_2|^{\beta} \left[1 + \left(c_W \frac{|h_2|}{N_G} \right)^{\Delta_s} \cdots \right] \tag{8.64}$$

such that one could define the asymptotic critical region by the condition $|h_2|/N_G \ll 1$. The opposition condition ($|h_2|/N_G \gg 1$) is the Ginzburg criterion for the validity of the meanfield approximation, see Eq. (4.77).

8.4.4.2 Asymmetric Corrections for Fluids: Complete Scaling

So far, we have considered systems that are perfectly symmetric, like the Ising model and the lattice gas. However, real systems are generally not symmetric; for example, all real fluids always demonstrate some degree of asymmetry in their phase coexistence. The approach to introduce the fluid-phase asymmetry requires a further specification of the relations between the theoretical scaling fields and the physical field variables. This approach, in the most general form, was formulated by Fisher and coworkers and is known as "complete scaling" [15, 49].

The complete scaling theory predicts that the scaling fields are related to the physical field variables in the linear approximation through a coefficient matrix of the form

$$\begin{pmatrix} h_1 \\ h_2 \\ -h_3 \end{pmatrix} = \begin{pmatrix} a_1 & a_2 & a_3 \\ b_2 & b_1 & b_3 \\ c_2 & c_3 & c_1 \end{pmatrix} \begin{pmatrix} \Delta\hat{\mu} \\ \Delta\hat{T} \\ \Delta\hat{P} \end{pmatrix} \tag{8.65}$$

This equation for a single-component fluid, may be simplified, as was done in Section 8.4.1 for the lattice-gas model,[10] through the following three steps:

- First, with appropriate normalization of coefficients ($a_1 = b_1 = c_1 = 1$ and $c_2 = -1$).
- Second, with an appropriate selection of the value of the critical entropy as $\rho_c S_c = \partial P/\partial T|_{h_1=0}$, allowing for the definition $a_2 = -a_3 \rho_c S_c$.
- Third, with the definition of $\Delta\hat{P}_{cr}$, given by Eq. (8.42).

Additionally, analysis of the physical properties affected by the pressure mixing terms, indicates that $b_3 \Delta\hat{P}$ does not contribute to the first order non-asymptotic corrections to scaling criticality, so, for simplicity, we assign $b_3 = 0$. These simplifications yield,

$$\begin{pmatrix} h_1 \\ h_2 \\ -h_3 \end{pmatrix} = \begin{pmatrix} 1 & 0 & \hat{a}_3 \\ b_2 & 1 & 0 \\ 0 & 0 & 1 \end{pmatrix} \begin{pmatrix} \Delta\hat{\mu} \\ \Delta\hat{T} \\ \Delta\hat{P}_{cr} \end{pmatrix} \tag{8.66}$$

where $\hat{a}_3 = a_3/(1 + a_3)$ and b_2 are new system-dependent constants, which are responsible for the fluid asymmetry. We note that the pressure mixing terms are significant only for fluids with

10 Note that in the complete scaling theory, the leading coefficients are located along the principle diagonal of the coefficient matrix. For example, the relation between field variables in the lattice-gas model, see Eqs. (8.43)–(8.45), is represented in the complete scaling theory as

$$\begin{pmatrix} h_1 \\ h_2 \\ -h_3 \end{pmatrix} = \begin{pmatrix} 1 & 0 & 0 \\ 0 & 1 & 0 \\ -1 & \hat{S}_c & 1 \end{pmatrix} \begin{pmatrix} \Delta\hat{\mu} \\ \Delta\hat{T} \\ \Delta\hat{P} \end{pmatrix} = \begin{pmatrix} 1 & 0 & 0 \\ 0 & 1 & 0 \\ 0 & 0 & 1 \end{pmatrix} \begin{pmatrix} \Delta\hat{\mu} \\ \Delta\hat{T} \\ \Delta\hat{P}_{cr} \end{pmatrix}$$

Since the coefficients are typically constant, one may normalize the scaling fields by the leading coefficients without loss of generality, such that, $a_1 = b_1 = c_1 = 1$ in the coefficient matrix given in Eq. (8.65).

complex molecules exhibiting molecular volume fluctuations, in addition to the number-density fluctuations. In many simple fluids, these additional fluctuations can be neglected, such that $\hat{a}_3 \approx 0$. This scenario is referred to as "revised scaling," in which the asymmetry is controlled only by the mixing term $b_2\Delta\hat{\mu}$ in h_2.

Using the definitions: $-dh_3 = \phi_1 dh_1 + \phi_2 dh_2$ and $\hat{\rho} = \partial\hat{P}/\partial\hat{\mu}|_{\hat{T}}$, one obtains the major result of complete scaling, that the density is expressed through combinations of the scaling densities, ϕ_1 and ϕ_2, as

$$\hat{\rho} = \frac{1 + \phi_1 + b_2\phi_2}{1 - \hat{a}_3\phi_1} \tag{8.67}$$

Expanding the density to second order in ϕ_1 and first order in ϕ_2, gives

$$\hat{\rho} \approx 1 + \phi_1 + \hat{a}_3\phi_1^2 + b_2\phi_2 \tag{8.68}$$

Correspondingly, the order parameter is obtained as

$$\phi_1 = \frac{\Delta\hat{\rho} + b_2\Delta(\rho\hat{S})}{1 + a_3\Delta\hat{\rho}} \tag{8.69}$$

In the scaling theory, the line corresponding to $h_1 = 0$ above the critical point is referred to as the "Widom line" [48], which is a smooth continuation of the coexistence curve in P–T coordinates. For the lattice-gas model, the line $h_1 = 0$, coincides with the critical isochore, given by $\phi_1 = 0$. However, note that in the theory of complete scaling, the line corresponding to the critical isochore, $\Delta\hat{\rho} = 0$, does not exactly coincide with $\phi_1 = 0$ as $\phi_1 \approx b_2\Delta(\rho\hat{S})$ as indicated by Eq. (8.69). However, this line merges asymptotically with the critical isochore with the same slope at the critical point.

8.4.4.3 Consequences of Asymmetry

The existence of the coefficients b_2 and a_3 leads to important experimental consequences. First, the measure of fluid-phase asymmetry, the diameter of the coexistence curve, is singular. As discussed in Section 4.4.1, in the meanfield approximation, the fluid-phase asymmetry is expressed through the rectilinear diameter, $\Delta\hat{\rho}_d = (\rho_V + \rho_L)/2 = D|\Delta\hat{T}| + 1$. Near the critical point, this linear dependence breaks down due to critical fluctuations. Thus, the diameter of the coexistence curve becomes singular. The diameter is obtained from the substitution of the asymptotic scaling behavior for the two coexisting branches below the critical point, yielding

$$\Delta\hat{\rho}_d = \frac{\rho_V + \rho_L}{2} - 1 \approx \hat{a}_3\phi_1^2 + b_2\phi_2 \tag{8.70}$$

such that upon the substitution of ϕ_1 and ϕ_2 (given by Table 8.2), in the lowest approximation, gives

$$\Delta\hat{\rho}_d \approx D_a|h_2|^{2\beta} + D_b|h_2|^{1-\alpha} + D_c|h_2| \tag{8.71}$$

where D_a and D_b are coefficients given by

$$D_a = \hat{a}_3 B_0^2 \quad \text{and} \quad D_b = b_2\frac{A_0^-}{1-\alpha} \tag{8.72}$$

and $D_c \propto b_2$ is a system-dependent coefficient of the linear (higher-order) term. In principle, the scaling fields, Eq. (8.65), may be expressed through higher-order terms that would also contribute to the diameter. A demonstration of the singularity of the diameter of the liquid–gas coexistence for trichlorotrifluoroethane, known as Freon-113, is shown in Figure 8.15.

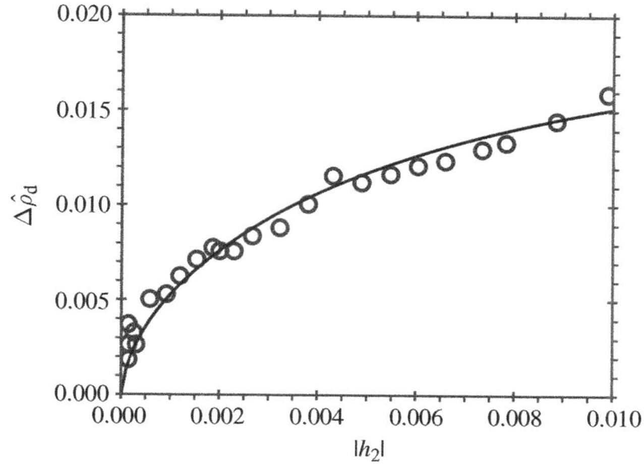

Figure 8.15 The singularity of the diameter for the liquid–gas coexistence in Freon-113. The solid curve corresponds to the nonanalytic terms in Eq. (8.71), where $h_2 = \Delta\hat{T}$. Source: Adapted from Ref. [50].

Second, in complete scaling the isothermal compressibility, $\kappa_T = \hat{\rho}^{-2}\hat{\chi}/(k_B T_c \rho_c)$ where $\hat{\chi}$ is the dimensionless susceptibility defined as $\hat{\chi} = \partial\hat{\rho}/\partial\hat{\mu}|_{\hat{T}}$, is affected by the mixing terms, with coefficients a_3 and b_2 in Eq. (8.65). It was shown, upon calculation of the derivative $\partial\hat{\rho}/\partial\hat{\mu}|_{\hat{T}}$, that the susceptibility is composed of a combination of the three scaling susceptibilities as [8]

$$\hat{\chi} \approx (1 + a_3)^2(1 + 3a_3\phi_1)\chi_1 + b_2^2\chi_2 + 2b_2(1 + a_3)\chi_{12} \tag{8.73}$$

Since, along the critical isochore in the one phase region ($T > T_c$), $\phi_1 = b_2\phi_2$ (as follows from Eq. 8.69), and, in the first-order approximation, $\chi_{12} \approx 0$, then upon the substitution of the scaling susceptibilities given in Table 8.2, one finds

$$\hat{\chi} \approx \Gamma_0^+|h_2|^{-\gamma}\left[1 + \Gamma_\alpha^+|h_2|^{1-\alpha} + \Gamma_\beta^+|h_2|^{\gamma-\alpha}\right] \tag{8.74}$$

where the correction amplitudes are $\Gamma_\alpha^+ \propto b_2^2 A_0^+$ and $\Gamma_\beta^+ \propto a_3 B_0$. Note that the correction terms vanish at the critical point. Since $1 - \alpha < \gamma - \alpha$, one finds that the Γ_α^+ term dominates the asymmetry in the susceptibility.

Lastly, the complete scaling relations, Eq. (8.65), also predict a Yang–Yang anomaly, a divergence in the second temperature derivative of the chemical potential along coexistence below the critical point and along the critical isochore above the critical point, $\partial^2\hat{\mu}/\partial\hat{T}^2|_{\hat{\rho}} \sim -\hat{a}_3|\Delta\hat{T}|^{-\alpha}$. This "anomaly" derives its name from the Yang–Yang relation,

$$\frac{C_V}{k_B} = \left.\frac{\partial^2\hat{P}}{\partial\hat{T}^2}\right|_{\hat{\rho}} - \left.\frac{\partial^2\hat{\mu}}{\partial\hat{T}^2}\right|_{\hat{\rho}} \tag{8.75}$$

Thus, complete scaling predicts that the divergence of the isochoric heat-capacity is generally shared by both $\partial^2\hat{P}/\partial\hat{T}^2|_{\hat{\rho}}$ and $\partial^2\hat{\mu}/\partial\hat{T}^2|_{\hat{\rho}}$, the latter being determined by the pressure-mixing asymmetry coefficient, $\hat{a}_3$. We note that experimental evidence for the Yang–Yang anomaly has been reported for several fluids; however, the experimental evidence is sometimes ambiguous, since the measurements may be strongly affected by small impurities [51].

A rigorous analysis of complete scaling based on an asymmetric effective Hamiltonian for fluids may be found in Ref. [52]. Complete scaling was also extended to spatially inhomogeneous fluids by introducing an additional scaling field associated with the gradient term in the Landau–Ginzburg functional [53].

8.4.5 Crossover Between Meanfield and Scaling Critical Behavior

The theory of critical phenomena, as presented in Section 8.4.4, describes the asymptotic critical behavior. It is possible to extend this theory to describe the thermodynamic behavior in the region where $\Delta\hat{T}$ is still small, but the effects of critical fluctuations are gradually diminishing upon departure from the critical point toward the validity of the meanfield approximation. This procedure, referred to as the "crossover Landau expansion," although being quite complex, may be illustrated in a simplified way, as follows. In the critical region, the order-parameter-dependent free energy, $\Delta\hat{f} = -(h_3 - \phi_1 h_1)$, in the meanfield approximation, has the form of the truncated Landau expansion, see Section 4.3, as

$$\Delta\hat{f} \approx \frac{1}{2!}a_0 h_2 \phi_1^2 + \frac{1}{4!}u_0 \phi_1^4 \tag{8.76}$$

where the thermal field, $h_2 = \Delta\hat{T}$.

The main idea of the crossover theory is that the variables h_2 and ϕ_1, in Eq. (8.76), are rescaled through appropriate crossover functions. For simple fluids, the crossover is controlled by a single parameter, the ratio of the thermal field and the Ginzburg number, defined as $h_2^{\times} \equiv h_2/N_{\mathrm{G}}$. Note that, as explained in Section 4.9, the Ginzburg number is extremely sensitive to the range of interactions, such that the meanfield approximation becomes valid for long-ranged interactions. The Landau expansion of the free energy $\Delta\tilde{f}$, renormalized by fluctuations, may be expressed in the form

$$\Delta\tilde{f} \approx \frac{1}{2!}a_0 \tilde{h}_2 \tilde{\phi}_1^2 + \frac{1}{4!}u_0 \tilde{\phi}_1^4 \tag{8.77}$$

where $\tilde{\phi}_1 = \phi_1 Y_1$ and $\tilde{h}_2 = h_2 Y_2$ represent the order parameter and thermal field rescaled by the crossover functions, Y_1 and Y_2, which depend on $h_2^{\times}$.

The crossover functions contain the following two limits,

$$Y_1\left(h_2^{\times}\right) = \begin{cases} 1 & h_2 \gg N_{\mathrm{G}} \\ |h_2^{\times}|^{(2\beta-1)/2} & h_2 \ll N_{\mathrm{G}} \end{cases} \tag{8.78}$$

$$Y_2\left(h_2^{\times}\right) = \begin{cases} 1 & h_2 \gg N_{\mathrm{G}} \\ |h_2^{\times}|^{(\gamma-1)} & h_2 \ll N_{\mathrm{G}} \end{cases} \tag{8.79}$$

such that one recovers the scaling power laws for the thermodynamic properties in each limit. In particular, for the order parameter and strong susceptibility, one obtains, $\tilde{\phi}_1(h_2^{\times} \ll 1) = \pm B_0|h_2|^{\beta}$ and $\hat{\chi}_1(h_2^{\times} \ll 1) = \Gamma_0^{\pm}|h_2|^{-\gamma}$, where the critical amplitudes for these properties vary with the Ginzburg number as $B_0 \propto N_{\mathrm{G}}^{(1-2\beta)/2}$ and $\Gamma_0^{\pm} \propto N_{\mathrm{G}}^{1-\gamma}$. An additional effect of critical fluctuations is the shift of the actual critical temperature with respect to the meanfield prediction, also being proportional to the Ginzburg number, see Figures 8.1 and 8.3. This approach was used to develop a crossover model for the parametric scaled equation of state [55].

The crossover from meanfield criticality to scaling criticality is confirmed by accurate simulations of the three-dimensional Ising model where the range of interactions ξ_0 (and correspondingly, the Ginzburg number) was tuned by changing the coordination number of interacting spins. The variable $h_2^{\times}$ changed by nine orders of magnitude (see Figure 8.16), such that within the critical region $(h_2 \ll 1)$, the full crossover from asymptotic meanfield behavior to asymptotic scaling behavior

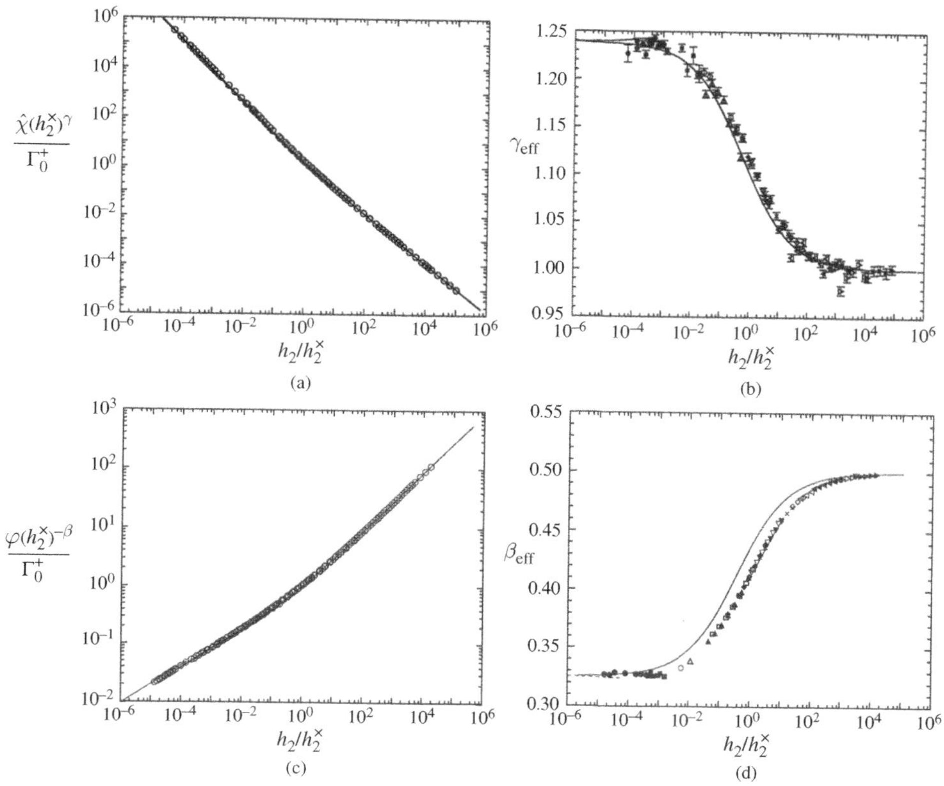

Figure 8.16 Crossover between asymptotic scaling behavior to meanfield behavior, predicted by the Landau expansion, for a three-dimensional Ising model in zero-ordering field with a tunable range of interactions. (a) Normalized susceptibility and (b) the corresponding effective critical exponent, γ_{eff}. (c) Normalized order parameter and (d) the corresponding effective critical exponent, β_{eff}. The solid curves represent the crossover theory. Source: Adapted from Ref. [54]/with permission of Springer Nature.

was recovered. Figure 8.16 illustrates the change in the behavior of the susceptibility and the order parameter, and their corresponding critical exponents throughout this crossover.

Real fluids do not exhibit such long-ranged molecular interactions as presented for the three-dimensional Ising model with a tunable coordination number. Consequently, away from the critical point, the Landau expansion is no longer applicable. Instead, fluids exhibit a crossover from asymptotic scaling behavior to the behavior described by a particular meanfield equation of state away from the critical point.

A similar crossover procedure, which was used for the Landau expansion, was also applied to the full equations of state of the van der Waals fluid and the lattice-gas model. This procedure reproduced the crossover behavior in the entire range of temperatures for real fluids. In particular, it was shown that the isochoric heat capacity of the van der Waals fluid exhibits a crossover from a finite, meanfield discontinuity to a weak singularity upon a decrease of the range of interactions (see Figure 8.17). In the entire temperature range, the crossover behavior of the order parameter for the lattice-gas model with a short range of interactions is shown in Figure 8.3, where the critical exponent of this behavior, β_{eff}, exhibits a non-monotonic crossover to $\beta_{eff} = 0$ at zero temperature (see Figure 8.18). A similar crossover procedure was successfully applied to popular engineering equations of state to describe thermodynamic properties of fluids in a broad range of temperatures and pressures (see, for example, Ref. [56]).

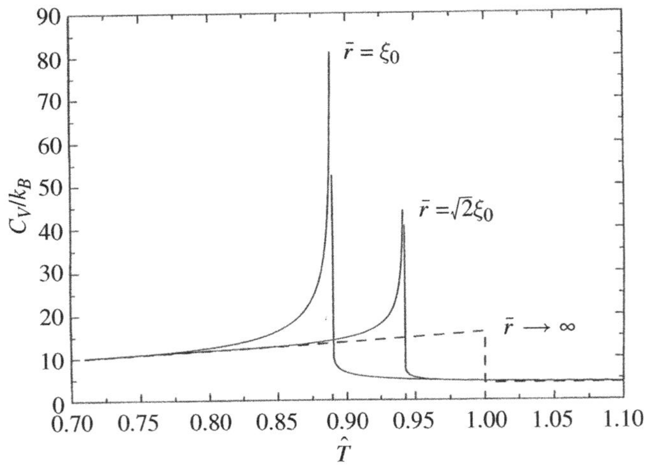

Figure 8.17 Isochoric heat capacity of the van der Waals fluid with three different ranges of molecular interactions: short ranged, medium ranged, and infinitely long ranged (corresponding to the classical behavior). Source: Reproduced with permission from Ref. [1], © 2004, Elsevier.

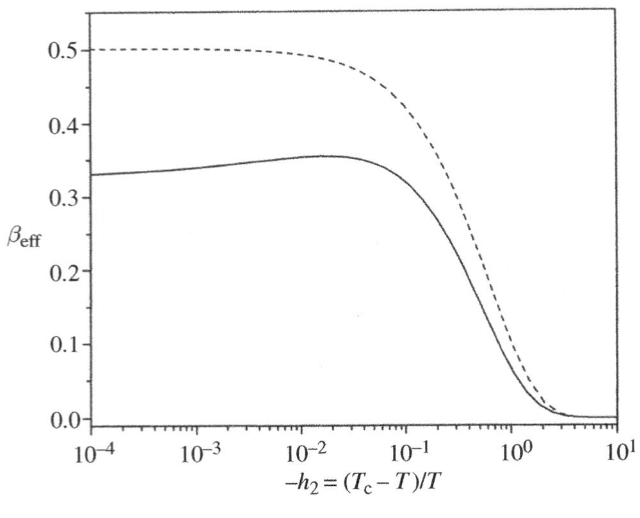

Figure 8.18 The effective critical exponent for the order parameter in the lattice-gas model. The solid curve depicts the crossover from asymptotic scaling behavior to meanfield noncritical behavior upon departure from the critical point, while the dashed curve depicts the crossover from meanfield critical behavior, predicted by the Landau expansion near the critical point, to meanfield behavior far away, as the temperature tends to zero. Source: Reproduced with permission from Ref. [57], © 1999, Elsevier.

8.5 Critical Phenomena in Fluid Mixtures

The critical properties and the critical-phase behavior of fluid mixtures are of significant interest to scientists and engineers because of their broad spectrum of applications, particularly, for supercritical-fluid technology [58, 59]. In this section, we focus on critical phenomena in binary fluids. The phase behavior of binary fluids may exhibit complex phase behavior and different kinds of critical phenomena (see the classification of P–T diagrams in Figure 2.18), bridging the gap between the behaviors of single-component fluids and multicomponent fluids. The variety of phase behavior and ambiguity in the definition of the appropriate order parameter could create an illusion of inapplicability of the critical-point universality to fluid mixtures. For example, the isochoric heat capacity and thermal conductivity diverge at the critical point of single-component fluids, but, generally, remain finite in fluid mixtures.

The generalization of complete scaling to fluid mixtures resolves this controversy by establishing thermodynamic relations between the universal scaling singularities and the observed anomalies in the physical properties of near-critical fluid mixtures. With appropriate thermodynamic transformations between the experimentally observable properties and the theoretical scaling functions, critical-point universality is revealed. Moreover, among all the experiments that were used to prove the fluctuation-induced nature of critical phenomena, the most accurate ones were performed near the liquid–liquid critical points of weakly compressible binary mixtures.

Examples of experimentally observed critical singularities near liquid–liquid critical points in binary mixtures are illustrated in Figures 8.19 and 8.20. These examples, as well as many other experimental studies performed for various mixtures, show that the spontaneous order parameter $\phi_0 = (x - x_c)/x_c$ and isobaric heat capacity at constant composition follow the same 3d-Ising class of universal power laws as were observed for the spontaneous order parameter $\phi_0 = (\rho - \rho_c)/\rho_c$ and isochoric heat capacity in single-component fluids. It was also demonstrated that other thermodynamic properties, like the osmotic compressibility (κ_Π) and the liquid–liquid interfacial tension, follow the same power laws as observed for their counterparts in single-component fluids, like the isothermal compressibility and the liquid–gas interfacial tension. All of these singularities originate from the divergence of the correlation length of the concentration fluctuations (see Figure 8.11).

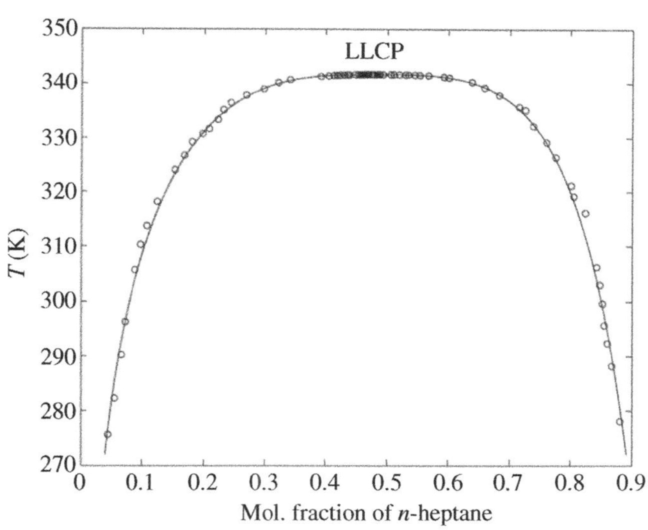

Figure 8.19 Liquid–liquid coexistence curve for a binary solution of n-heptane and acetic anhydride. The solid line is given by Eq. (8.96) and LLCP indicates the liquid–liquid critical point. Source: Reproduce with permission from Ref. [60], © 2005, Walter de Gruyter GmbH.

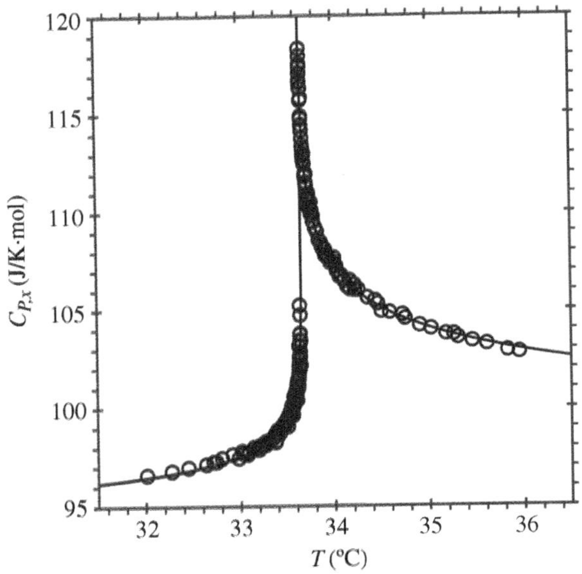

Figure 8.20 Isobaric heat capacity at constant composition near the lower liquid–liquid critical point of a binary mixture of lutidine and water. The divergence of $C_{P,x}$ is described by the scaling power law given by Eq. (8.98). Source: Adapted from Ref. [61].

8.5.1 Complete Scaling for Binary Fluids

The behavior of binary fluids, unlike single-component substances, requires the introduction of an additional variable, x, the molecular fraction of one of the components. Consequently, many fluid mixtures exhibit a continuous line of liquid–gas critical points at various concentrations with or without additional liquid–liquid immiscibility (Type I and II, see Figure 2.18a,b). In more complex cases, the critical line may be interrupted at the critical endpoints, in which a noncritical phase (liquid or gas) emerges (Type III and IV, see Figure 2.18c,d). In type IV behavior, one may observe a continuous crossover from liquid–gas critical phenomena starting from the critical point of single-component B (CP_B) to liquid–liquid critical phenomena at the lower critical end point (LCEP), see Figure 2.18d.

The principle of isomorphism asserts that the thermodynamic behavior in mixtures can still be described in a universal way in terms of three scaling fields: h_1, h_2, and h_3 (see Section 8.4). The different types of critical phenomena observed experimentally emerge from a different relation between the theoretical scaling fields and the observable physical field variables.

In binary mixtures, there are four physical field variables: $\Delta\hat{P}$; $\Delta\hat{T}$; $\Delta\hat{\mu}_1 = (\mu_1 - \mu_{1,c})/k_B T_c$, which replaces $\Delta\hat{\mu}$ in complete scaling of a single-component fluid; and a new field, $\Delta\hat{\mu}_{21} = (\mu_{21} - \mu_{21,c})/k_B T_c$, the chemical potential difference between the chemical potentials of each species, conjugate to x. In the linear approximation, instead of Eq. (8.65), one obtains

$$\begin{pmatrix} h_1 \\ h_2 \\ -h_3 \end{pmatrix} = \begin{pmatrix} a_1 & a_2 & a_3 & a_4 \\ b_2 & b_1 & b_3 & b_4 \\ c_2 & c_3 & c_1 & c_4 \end{pmatrix} \begin{pmatrix} \Delta\hat{\mu}_1 \\ \Delta\hat{T} \\ \Delta\hat{P} \\ \Delta\hat{\mu}_{21} \end{pmatrix} \tag{8.80}$$

In general, the coefficients a_4, b_4, and c_4 are proportional to $x_c(1 - x_c)$, such that Eq. (8.80) reduces to Eq. (8.65) in the limit of a single-component substance ($x_c = 0$ or 1). Equation (8.80) may be simplified by applying similar arguments to those presented in Section 8.4.4 for the complete scaling of single-component fluids. Thereby yielding the simplified form of complete scaling for binary mixtures as

$$\begin{pmatrix} h_1 \\ h_2 \\ -h_3 \end{pmatrix} = \begin{pmatrix} \hat{a}_1 & 0 & a_3 & \hat{a}_4 \\ b_2 & b_1 & 0 & b_4 \\ 0 & 0 & 1 & 0 \end{pmatrix} \begin{pmatrix} \Delta\hat{\mu}_1 \\ \Delta\hat{T} \\ \Delta\hat{\Pi}_{\text{cr}} \\ \Delta\hat{\mu}_{21} \end{pmatrix} \tag{8.81}$$

where $\hat{a}_1 = a_1 + a_3$ and $\hat{a}_4 = a_4 - a_3 c_4$, while the scaling fields, h_1 and h_3, are expressed through the critical part of the grand potential per unit volume, which, by analogy with Eq. (8.42), can be defined as the critical part of the reduced osmotic pressure,

$$\Delta\hat{\Pi}_{\text{cr}} = c_1 \Delta\hat{P} - \Delta\hat{\mu}_1 - \hat{S}_{\text{c}} \Delta\hat{T} + c_4 \Delta\hat{\mu}_{21} \tag{8.82}$$

For binary mixtures, the coefficients a_i, b_i, and c_i as well as the critical parameters: T_{c}, P_{c}, $\mu_{1,\text{c}}$, and $\mu_{21,\text{c}}$ depend parametrically on the actual position along the critical locus. Thus, the coefficients a_1 and c_1 cannot be arbitrarily normalized by an appropriate choice of units (as in Section 8.4). Instead, these coefficients are subject to the following conditions: In the limit of liquid–gas criticality in a single-component substance, like at CP_{B} on Figure 2.18d,

$$a_1, b_1, c_1 \to 1 \qquad \text{while} \qquad a_4, b_4, c_4 \to 0 \tag{8.83}$$

and in the limit of weakly-compressible liquid–liquid criticality, like at LCEP on Figure 2.18d,

$$a_1, c_1 \to 0 \qquad \text{while} \qquad \begin{matrix} a_4 \to 1 \\ c_4 \to x_{\text{c}} \end{matrix} \tag{8.84}$$

The physical meaning of the parametric dependence of these coefficients is a transformation of the scaling fields h_1 and h_3 from liquid–gas to liquid–liquid criticality. In the limit of liquid–gas criticality in a single-component fluid, $h_1 = \Delta\hat{\mu} + \hat{a}_3 \Delta\hat{P}_{\text{cr}}$ and $-h_3 = \Delta\hat{P}_{\text{cr}}$, just as given in Eq. (8.66), while in the limit of liquid–liquid criticality,

$$\begin{pmatrix} h_1 \\ h_2 \\ -h_3 \end{pmatrix} = \begin{pmatrix} 0 & 0 & \hat{a}_3 & 1 \\ b_2 & b_1 & 0 & b_4 \\ 0 & 0 & 1 & 0 \end{pmatrix} \begin{pmatrix} \Delta\hat{\mu}_1 \\ \Delta\hat{T} \\ \Delta\hat{\Pi}_{\text{cr}} \\ \Delta\hat{\mu}_{21} \end{pmatrix} \tag{8.85}$$

where, in this limit, $\Delta\hat{\Pi}_{\text{cr}} = -\Delta\hat{\mu}_1 - \hat{S}_{\text{c}} \Delta\hat{T} + x_{\text{c}} \Delta\hat{\mu}_{21}$.

Order Parameter in Binary Mixtures

- For an infinitely dilute solution in the case when the limit $x \to 0$ is a symmetric fluid (lattice-gas model), $b_2 = 0$ and $\hat{a}_1 = 1$,

$$\phi_1 = \Delta\hat{\rho} \tag{8.86}$$

- For a compressible symmetric binary mixture, $b_4 = 0$ and $\hat{a}_4$ is finite,

$$\phi_1 = \frac{1}{\hat{a}_4} \Delta\hat{\rho}_2 \tag{8.87}$$

- In the limit of an incompressible symmetric mixture (lattice-liquid model), $b_4 = 0$, $\hat{a}_4 = 1$, and $\hat{\rho} = 1$,

$$\phi_1 = \Delta\hat{x} \tag{8.88}$$

If the incompressible binary liquid is perfectly symmetric (molecules of equal sizes), then additional simplifications like $a_3 = 0$, $b_2 = 0$, and $b_4 = 0$ persist. This scenario is referred to as the "lattice-liquid," in which the scaling fields, with the choice of $\hat{S}_c = \partial\hat{\mu}_1/\partial\hat{T}|_{h_1=0}$, are given by

$$\begin{pmatrix} h_1 \\ h_2 \\ -h_3 \end{pmatrix} = \begin{pmatrix} 0 & 0 & 0 & 1 \\ 0 & 1 & 0 & 0 \\ 0 & 0 & 1 & 0 \end{pmatrix} \begin{pmatrix} \Delta\hat{\mu}_1 \\ \Delta\hat{T} \\ \Delta\hat{\Pi}_{cr} \\ \Delta\hat{\mu}_{21} \end{pmatrix} = \begin{pmatrix} 1 & 0 & 0 \\ 0 & 1 & 0 \\ 0 & 0 & 1 \end{pmatrix} \begin{pmatrix} \Delta\hat{\mu}_{21} \\ \Delta\hat{T} \\ \Delta\hat{\Pi}_{cr} \end{pmatrix} \tag{8.89}$$

Note that the lattice-liquid is fully isomorphic with the lattice-gas, see Eqs. (8.43)–(8.45).

For binary mixtures, the Gibbs–Duhem relation, Eq. (2.21), yields

$$d\hat{P} = \hat{s}\ d\hat{T} + \hat{\rho}\ d\hat{\mu}_1 + \hat{\rho}_2\ d\hat{\mu}_{21} \tag{8.90}$$

such that, with use of $-dh_3 = \phi_1\ dh_1 + \phi_2\ dh_2$, one obtains for the density and partial density, $\hat{\rho}_2 = \hat{\rho}\hat{x}$, that

$$\hat{\rho} = \frac{1 + \hat{a}_1\phi_1 + b_2\phi_2}{1 - a_3\phi_1} \quad \text{and} \quad \hat{\rho}_2 = \frac{1 + \hat{a}_4\phi_1 + b_4\phi_2}{1 - a_3\phi_1} \tag{8.91}$$

as well as the molecular fraction, that

$$\hat{x} = \frac{1 + \hat{a}_4\phi_1 + b_4\phi_2}{1 + \hat{a}_1\phi_1 + b_2\phi_2} \tag{8.92}$$

Consequently, for fluid mixtures, the order parameter, ϕ_1, is generally a combination of $\hat{\rho}$ and $\hat{\rho}_2$ as

$$\phi_1 = \frac{b_2}{\hat{a}_4 b_2 - \hat{a}_1 b_4}\Delta\hat{\rho}_2 - \frac{b_4}{\hat{a}_4 b_2 - \hat{a}_1 b_4}\Delta\hat{\rho} \tag{8.93}$$

The limits of Eq. (8.93) are given in Eqs. (8.86)–(8.93).

For fluid mixtures, like for all other systems exhibiting critical phenomena, there are two independent scaling fields, two independent scaling densities, and three diverging scaling susceptibilities: χ_1, χ_2, and χ_{12}. The experimentally measurable response functions, such as the osmotic compressibility (κ_Π) and, at constant composition: $\kappa_{T,x}$, $\alpha_{P,x}$, $C_{P,x}$, and $C_{V,x}$, are generally combinations of these three scaling susceptibilities (Table 8.6) expressed through the coefficients of the scaling fields, Eq. (8.80). An example of such relations for κ_Π and $\kappa_{T,x}$ near the liquid–gas critical locus is given by Eqs. (8.A.10) and (8.A.11). Since the coefficients in the scaling fields, h_1, h_2,

Table 8.6 Response functions diverging at the critical point of a binary fluid.

Strong ($\propto \chi_1 \propto \|h_2\|^{-\gamma}$)	**Weak ($\propto \chi_2 \propto \|h_2\|^{-\alpha}$)**
$C_{P,\mu_{21}}$	$C_{P,x}$
$\left.\frac{\partial x}{\partial \mu_{21}}\right\|_{P,T}$	$\left.\frac{\partial x}{\partial \mu_{21}}\right\|_{\rho,T}$
$\left.\frac{\partial x}{\partial T}\right\|_{P,\mu_{21}}$	$\left.\frac{\partial x}{\partial T}\right\|_{\rho,\mu_{21}}$
$\left.\frac{\partial \rho}{\partial P}\right\|_{T,\mu_{21}}$	$\left.\frac{\partial \rho}{\partial P}\right\|_{T,x}$
$\left.\frac{\partial \rho}{\partial T}\right\|_{P,\mu_{21}}$	$\left.\frac{\partial \rho}{\partial T}\right\|_{P,x}$

and h_3, parametrically depend on the position along the critical loci, the behavior of the response functions correspondingly changes. The principle of isomorphism accounts for all specific cases of critical phenomena, such as: dilute solutions, critical azeotropes, incompressible liquids, and critical temperature or critical pressure extrema.

8.5.2 Weakly Compressible Liquid Mixtures

Since liquid mixtures are commonly weakly compressible, the action of pressure could be introduced as a field variable, "hidden" in the critical parameters. For instance, the pressure dependence on the critical temperature and composition may be expressed in the linear approximation as

$$T_c(P) \simeq T_c^\circ + \frac{dT_c}{dP}\Delta\tilde{P} \quad \text{and} \quad x_c(P) \simeq x_c^\circ + \frac{dx_c}{dP}\Delta\tilde{P} \tag{8.94}$$

where $\Delta\tilde{P} = P_c - P_c^\circ$, the change in the critical pressure along the critical loci, while T_c° and x_c° are constants, defined by the reference critical pressure, $T_c^\circ = T_c(P_c = P_c^\circ)$ as $x_c^\circ = x_c(P_c = P_c^\circ)$.

For instance, consider this effect on a compressible lattice-liquid model. The formulation of the incompressible lattice-liquid scaling fields, Eq. (8.89), may be used with Eq. (8.94) since $\mu_{1,c} = \mu_{1,c}(\Delta\tilde{P})$ and $\mu_{21,c} = \mu_{21,c}(\Delta\tilde{P})$, such that the model becomes weakly dependent on the pressure. Therefore, in this case, $h_1 = h_1(\Delta\tilde{P})$, $h_2 = h_2(\Delta\tilde{P})$, and $h_3 = h_3(\Delta\tilde{P})$, each being dependent on the pressure only through the critical parameters.

Most experiments on liquid–liquid criticality in binary mixtures have been performed near the critical end points, at which the liquid–liquid critical line is terminated by the liquid–gas coexistence, as commonly observed in Type II critical behavior, see Figure 2.18b. Since, away from the liquid–gas critical temperature, the liquid–gas saturation pressure weakly depends on temperature, then the experimental conditions practically correspond to constant pressure [3]. Also note that the liquid–liquid critical line in P–T coordinates is almost vertical (as shown on Figure 2.18b), indicating that the critical parameters weakly depend on the critical pressure.

At constant pressure in a weakly compressible liquid ($\hat{a}_4 = 1$), the molecular fraction, given by Eq. (8.92), may be expanded in powers of the scaling densities to yield

$$\hat{x} \approx 1 + (1 - \hat{a}_1)\phi_1 - \hat{a}\phi_1^2 + (b_4 - b_2)\phi_2 + \cdots \tag{8.95}$$

where $\hat{a} = \hat{a}_1/(1 - \hat{a}_1)$. Taking the difference in the molecular fractions of each component in a binary mixture, such that the non-asymptotic terms, $\hat{a}\phi_1$ and $(b_4 - b_2)\phi_2$ are canceled, the coexistence curve follows the symmetric scaling relation,

$$\frac{\hat{x}_{L1} - \hat{x}_{L2}}{2} \approx (1 - \hat{a}_1)\phi_1 = \pm(1 - \hat{a}_1)B_0|\Delta\hat{T}|^\beta \tag{8.96}$$

The coexistence curve for a system of n-heptane and acetic anhydride following this relation is shown in Figure 8.19.

In addition, for weakly compressible mixtures, the diverging parts of the major thermodynamic response functions (κ_Π, $C_{P,x}$, $\alpha_{P,x}$, and $\kappa_{T,x}$) exhibit the following asymptotic singularities:

$$\kappa_\Pi \propto \chi_1 = \Gamma_0^\pm|\Delta\hat{T}|^{-\gamma} \tag{8.97}$$

$$\Delta C_{P,x} \propto \chi_2 = A_0^\pm|\Delta\hat{T}|^{-\alpha} \tag{8.98}$$

$$\Delta\alpha_{P,x} \propto \chi_2\frac{dT_c}{dP} = A_0^\pm\frac{dT_c}{d\hat{P}}|\Delta\hat{T}|^{-\alpha} \tag{8.99}$$

$$\Delta\kappa_{T,x} \propto \chi_2\left(\frac{dT_c}{dP}\right)^2 = A_0^\pm\left(\frac{dT_c}{dP}\right)^2|\Delta\hat{T}|^{-\alpha} \tag{8.100}$$

Usually, for liquid–liquid transitions dT_c/dP is a very small parameter; for example, $dT_c/dP \approx$ 2 mK/bar for a mixture of nitroethane-isooctane [3]. Therefore, in this case, only the divergences of κ_Π and $C_{P,x}$ may be easily observable, while detecting the divergences of $\alpha_{P,x}$ and $\kappa_{T,x}$ may be challenging.

8.5.2.1 Singular Diameter of Phase Coexistence in Liquid Mixtures

For a weakly compressible binary mixture at constant pressure, the diameter of liquid–liquid coexistence curve follows from Eq. (8.95), as

$$\Delta\hat{x}_d = \frac{x_{L1} + x_{L2}}{2} - 1 \approx -\hat{a}\phi_1^2 + (b_4 - b_2)\phi_2 \tag{8.101}$$

which, upon the substitution of ϕ_1 and ϕ_2, as given in Table 8.2, yields,

$$\Delta\hat{x}_d \approx D_{x,a}|\Delta\hat{T}|^{2\beta} + D_{x,b}|\Delta\hat{T}|^{1-\alpha} + D_{x,c}|\Delta\hat{T}| \tag{8.102}$$

Adopting the same form as for single-component substances Eq. (8.72), where the coefficients for the binary mixture are given in the form

$$D_{x,a} = -\hat{a}B_0^2 \quad \text{and} \quad D_{x,b} = -(b_4 - b_2)\frac{A_0^-}{1-\alpha} \tag{8.103}$$

where $D_{x,c} \propto b_4 - b_2$ is the coefficient of the linear (higher-order) term. The leading singular term (as $2\beta < 1 - \alpha$) in the diameter is $D_{x,a}$, which becomes dominant when the molecular size of the two components are significantly different. An example of the singular behavior of the diameter in various solutions of alkanes in nitrobenzene is presented in Figure 8.21.

8.5.2.2 Closed-Loop Phase Diagrams

Certain liquid mixtures may also possess a closed-loop phase boundary exhibiting an upper and a lower critical point. Figure 8.22 shows an example of such a closed loop for a system of tetrahydrofurane (THF) and water. In some systems with a miscibility gap inside a closed-loop phase boundary,

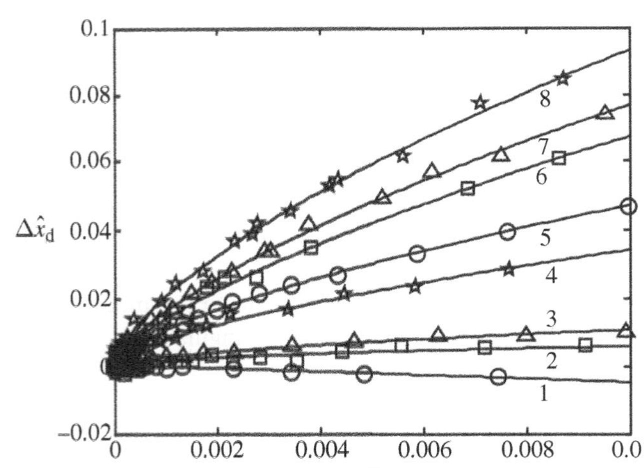

Figure 8.21 The singular diameters of the liquid–liquid coexistence for solutions of alkanes in nitrobenzene: *n*-pentane (1), *n*-heptane (2), *n*-octane (3), *n*-decane (4), *n*-undecane (5), *n*-tridecane (6), *n*-tetradecane (7), and *n*-hexadecane (8) as a function of $h_2 = \Delta\hat{T}$. The solid curves are given by Eq. (8.102). Source: Adapted From Ref. [62].

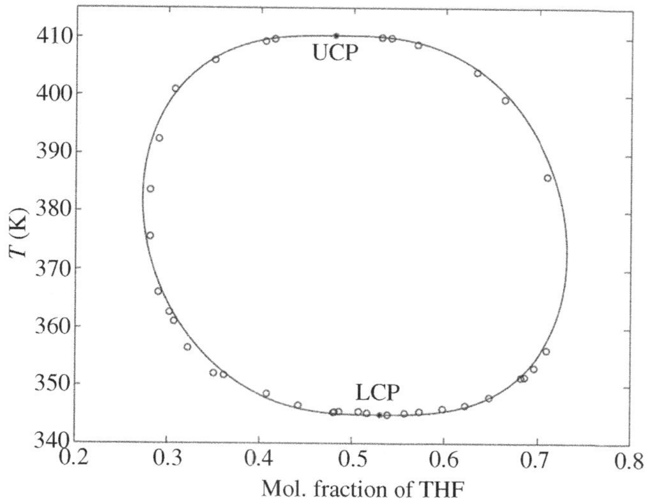

Figure 8.22 Closed-loop liquid–liquid phase coexistence for a binary solution of tetrahydrofurane in water. The solid curve is given by Eq. (8.102), with the inclusion of symmetric and asymmetric corrections to asymptotic scaling behavior as discussed in Section 8.4.4. Source: Reproduce with permission from Ref. [60], © 2005, Walter de Gruyter GmbH.

the miscibility gap may shrink as a function of an external variable (such as the pressure) or due to the addition of a third component, so that the upper and lower critical points may converge to a double-critical point. Equation (8.96) can be used to separately describe the behavior either near the upper, T_c^u, and lower, T_c^l, critical points. However, a formulation of the scaling fields that goes beyond the linear relations with the physical field variables, also allows for a universal description of closed-loop criticality.

A simplified representation of the scaling fields for the lattice-liquid approximation exhibiting closed-loop phase boundaries, may be given by expressing the thermal field, h_2, as

$$h_2 = \Delta\hat{T} + b_5|\Delta\hat{T}|^2 \tag{8.104}$$

where, for the closed loop system, all points are taken in reference to the lower critical point, such that $\Delta\hat{T} = (T - T_c^l)/T_c^l$, in which the location of the critical points weakly depend on pressure, as discussed in Eq. (8.94). At the critical point, all scaling fields must vanish, such that, for the closed-loop system, h_2 is zero not only when $T = T_c^l$, but also, when $T = T_c^u$. Thus, by incorporating these two limits, h_2 may be expressed in the form

$$h_2 = \Delta\hat{T}_{ul} = \frac{(T - T_c^u)(T - T_c^l)}{T_c^u T_c^l} \tag{8.105}$$

Thus, with use of Eq. (8.96), the order parameter of a symmetric closed loop in a binary liquid at constant pressure is given as

$$\Delta\hat{x} \approx \pm B_0|\Delta\hat{T}_{ul}|^{\beta} \tag{8.106}$$

The effect of pressure on some systems may cause the miscibility gap to shrink. This effect is accounted for by the pressure dependence of the two critical amplitudes and critical temperatures. The effect of increasing the pressure on the closed-loop phase diagram of 2-butanol and water is shown in Figure 8.23. Note that at a critical value of pressure, the miscibility gap closes, leading to a double critical point. At the double critical point, the thermal field scales as $h_2 \propto |\Delta\hat{T}|^2$, thus the critical exponents for the thermodynamic properties are doubled, e.g. $\Delta\hat{x} \propto |\Delta\hat{T}|^{2\beta}$.

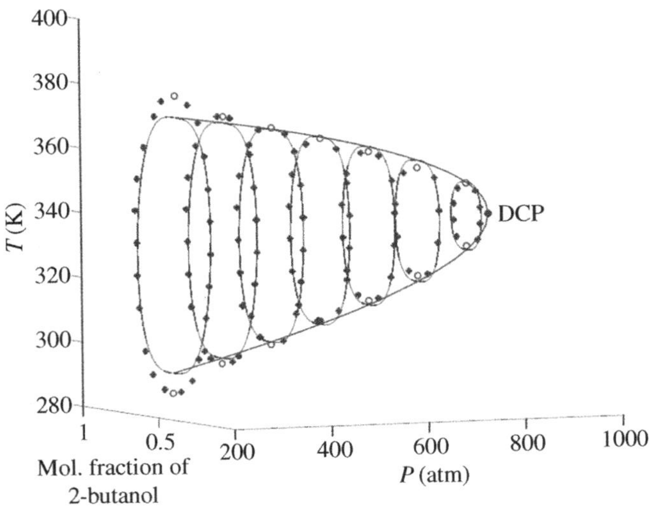

Figure 8.23 Closed loop liquid–liquid phase equilibria in a solution of 2-butanol and water. The symbols indicate experimental phase-boundary data [63], the circular lines are phase coexistence curves given by Eq. (8.106), and the parabolic boundary is the critical locus, which terminates at the double-critical point (DCP). Source: Reproduce with permission from Ref. [60], © 2005, Walter de Gruyter GmbH.

8.5.3 Application to Liquid-Gas Critical Loci

In the previous sections, we introduced the isomorphism of critical phenomena through the complete scaling theory. In Sections 8.4.1 and 8.5.1, we will apply these concepts to describe the critical behavior near the liquid–gas critical line, referred to as the "critical locus," in binary fluids.

8.5.3.1 Critical Line Condition

In general, near-critical binary fluids may be described using the full form of the complete scaling relations, Eq. (8.80), which depends on all physical field variables, including the reduced exchange chemical potential, $\hat{\mu}_{21}$, conjugate to the concentration. In engineering practice, to formulate the scaled equation of state for fluid mixtures, use of $\hat{\mu}_{21} = \mu_{21}/k_B T$ as a field variable is often not convenient, since it diverges in the two pure one-component limits. As suggested by Leung and Griffiths, an alternative definition of the corresponding field variable is a generalized "activity" [64], given by,

$$y = \frac{1}{1 + e^{-\hat{\mu}_{21}}} \tag{8.107}$$

The advantage of this choice, given by Eq. (8.107), is that y goes from 0 to 1, as the mole fraction x varies from 0 to 1. We note that the definition of y is not unique due to the arbitrary choice of the zero values of entropy and energy for each species. Therefore, an additional constraint, $y_c = x_c$, referred to as the "critical-line condition" [41] further simplifies the approach. It must be emphasized that the thermodynamic behavior of mixtures is isomorphic with that of one-component fluids at constant y, but not at constant x, as illustrated in Figure 8.24. The introduction of the activity, y, is conceptually similar to the use of fugacity in engineering thermodynamics to describe fluid-phase equilibria.

8.5.3.2 Peculiar Points on the Liquid–Gas Critical Loci

For mixtures, the scaling susceptibilities χ_1, χ_2, and χ_{12} are not directly measurable quantities. However, experimentally accessible second derivatives of the thermodynamic potentials can be

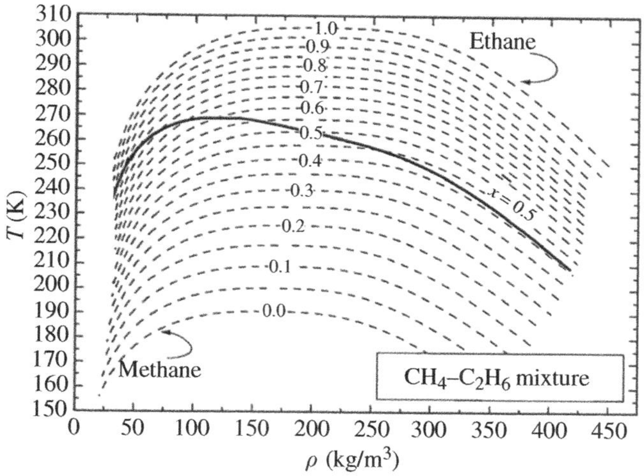

Figure 8.24 Liquid–gas phase boundaries for mixtures of methane and ethane in the temperature-density plane. The dashed curves represent liquid–gas coexistence for different critical values of the effective "activity" y, given by Eq. (8.107), including the coexistence curves of pure methane ($y = 0$) and pure ethane ($y = 1$). At constant temperature, the values of $y = y(T)$ are the same for two coexisting phases. The solid curve represents a cross section at constant composition $x = 0.5$, known as the "dew-bubble" curve [65]. Source: Reproduced with permission from Ref. [66], © 2010, Elsevier.

expressed in terms of these susceptibilities by applying appropriate thermodynamic transformations to the complete scaling relations (see discussion in Section 8.5.1).

As discussed in Appendix 8.A, there are two characteristic parameters, $\hat{\mathscr{T}}_1$ and $\hat{\mathscr{T}}_2$, which are determined by the shape of the critical locus and are responsible for the difference between the critical behavior of a single-component fluid and a binary mixture (see details in Refs. [67, 68]),

$$\hat{\mathscr{T}}_1 = \left[\frac{\Gamma_0^{\pm} \mathcal{K}_1^2}{x_c(1-x_c)}\right]^{1/\gamma} = \left[\Gamma_0^{\pm} x_c(1-x_c)\mathcal{K}^2\right]^{1/\gamma} \tag{8.108}$$

$$\hat{\mathscr{T}}_2 = \left[\frac{A_0^{\pm} \mathcal{K}_2^2}{x_c(1-x_c)}\right]^{1/\alpha} = \left[A_0^{\pm} x_c(1-x_c)\left(\frac{\partial \hat{T}_c}{\partial x_c}\right)^2\right]^{1/\alpha} \tag{8.109}$$

where the critical temperature is reduced by the critical temperature of the solvent, $\hat{T}_c = T_c(x_c)/T_{c1}$, and

$$\mathcal{K} = \frac{d\hat{T}_c}{dx_c}\left[\frac{d\hat{P}_c}{d\hat{T}_c} - \left.\frac{\partial \hat{P}}{\partial \hat{T}}\right|_{h_1=0,\hat{\mu}_{21,c}}\right] \tag{8.110}$$

is the generalized Krichevskiĭ parameter, given by Eq. (2.105), which resolved the critical dilute solution dilemma, see Eq. (2.111).

The first parameter, $\hat{\mathscr{T}}_1$, determines the range of temperatures, $\Delta\hat{T} \gg \hat{\mathscr{T}}_1$ along the critical isochore, $\rho = \rho_c(y)$, in which the response functions: the isothermal compressibility, $\kappa_{T,x}$, and the isobaric heat capacity, $C_{P,x}$, diverge like the strong susceptibility, $\hat{\chi}^{-1}$, while the osmotic compressibility, κ_Π, remains finite, scaling as $\kappa_\Pi \sim x$. The second parameter, $\hat{\mathscr{T}}_2$, determines the range, $\hat{\mathscr{T}}_1 \gg \Delta\hat{T} \gg \hat{\mathscr{T}}_2$, where the strong singularities in $\kappa_{T,x}$ and $C_{P,x}$ are suppressed, while the strong singularity in κ_Π emerges. In the range $\hat{\mathscr{T}}_2 \gg \Delta\hat{T}$, closer to the critical locus or with an increase of the solute concentration, the response functions: $\kappa_{T,x}$ and $C_{P,x}$, diverge weakly, like the weak susceptibility, $\hat{\chi}_2$. The parameter $\hat{\mathscr{T}}_2$ is also responsible for the renormalization of critical exponents at

$\rho = \rho_c$ and constant x, as discussed in Section 8.5.4. It is important to note that $\hat{\mathcal{T}}_2$, unlike $\hat{\mathcal{T}}_1$, does not exist in the meanfield approximation where the critical amplitude of the weak susceptibility, $A_0^{\pm}$, disappears.

Below, we consider the most peculiar points of the liquid–gas critical locus: the critical points of pure components, the critical azeotrope, and the extrema of the critical temperature and critical pressure.

- **Dilute Solutions and Pure-Component Limits**
  Both $\hat{\mathcal{T}}_1$ and $\hat{\mathcal{T}}_2$ vanish in the pure component limits. In a dilute solution near the critical point of the solvent ($x_c \ll 1$),

$$\hat{\mathcal{T}}_1 = (\Gamma_0^{\pm} x_c \mathcal{K}^2)^{1/\gamma} \tag{8.111}$$

$$\hat{\mathcal{T}}_2 = \left[A_0^{\pm} x_c \left(\frac{\mathrm{d}\hat{T}_c}{\mathrm{d}x_c} \right)^2 \right]^{1/\alpha} \tag{8.112}$$

The origin of these characteristic functions may be seen from Eq. (2.111) as

$$\left. \frac{\partial x}{\partial \hat{\mu}_{21}} \right|_{T,P} \approx x_c + x_c^2 \mathcal{K}^2 \Gamma_0^{\pm} |\Delta \hat{T}|^{-\gamma} \tag{8.113}$$

and from the condition, $\Delta \hat{T} < \hat{\mathcal{T}}_2 < \hat{\mathcal{T}}_1$, closer to the critical locus

$$\left. \frac{\partial x}{\partial \hat{\mu}_{21}} \right|_{T,\rho} \approx x_c + A_0^{\pm} \left(\frac{\mathrm{d}\hat{T}_c}{\mathrm{d}x_c} \right)^2 |\Delta \hat{T}|^{-\alpha} \tag{8.114}$$

Figure 8.25 indicates the three regions of mixture behavior. Away from the critical point, for $\Delta \hat{T} > \hat{\mathcal{T}}_1$, one observes pure-fluid critical behavior. For $\hat{\mathcal{T}}_1 > \Delta \hat{T} > \hat{\mathcal{T}}_2$, the strong divergence of the isothermal compressibility (as $\sim |\Delta \hat{T}|^{-\gamma}$) disappears and is replaced by the strong divergence of the osmotic compressibility, while for $\Delta \hat{T} < \hat{\mathcal{T}}_2$, $C_{P,x}$ and $\partial x / \partial \mu|_{T,\rho}$ diverge weakly (as $\sim |\Delta \hat{T}|^{-\alpha}$). Moreover, in the latter region, the isochoric heat capacity at constant composition, $C_{V,x}$, exhibits a cusp (see discussion in Section 8.5.4). Note that this region does not exist in the meanfield approximation (as $\alpha = 0$).

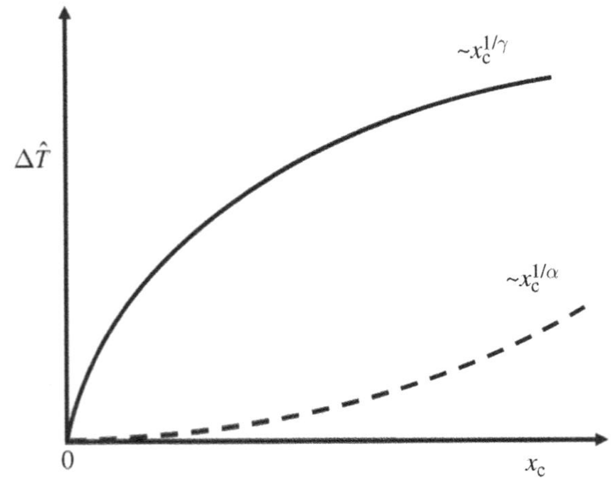

Figure 8.25 Temperature boundaries for observing the singularities of the response functions in dilute solutions near the liquid–gas critical locus.

- **Critical Azeotrope**
  Usually, $\hat{\mathcal{T}}_2 \gg \hat{\mathcal{T}}_1$ (since $\alpha \ll \gamma$), except at the critical azeotrope where $\partial\hat{P}/\partial\hat{T}|_{h_1=0,\mu_{21,c}} = \mathrm{d}\hat{P}_c/\mathrm{d}\hat{T}_c$, such that $\mathcal{K} = 0$, see Eq. (8.110), and $\hat{\mathcal{T}}_1 = 0$, see Eq. (8.108). Consequently, the isothermal compressibility and the isobaric heat capacity both strongly diverge (like in a pure fluid), while the isochoric heat capacity and the osmotic susceptibility diverge weakly because, in this case, $\partial x/\partial\mu_{21}|_{T,P} = \partial x/\partial\mu_{21}|_{T,\rho}$ (see Appendix 8.A).
- **Critical-Line Extrema**
  At the temperature extremum of the critical locus $\partial\hat{T}_c/\partial x_c = 0$, such that $\hat{\mathcal{T}}_2 = 0$. Hence, to satisfy the stability conditions, $\partial\mu/\partial x|_{P,T} = 0$, it follows that $\kappa_{T,x}$ and $\partial x/\partial\mu|_{T,\rho}$ now remain finite at the critical point. As a result, both $C_{P,x}$ and $C_{V,x}$ diverge weakly.
  Another example of critical-line extrema is encountered when $\mathrm{d}\hat{P}_c/\mathrm{d}\hat{T}_c = 0$. This point is sometimes referred to as a "reentrant" critical point. Below the reentrant critical pressure, one phase re-emerges upon lowering the temperature at constant pressure. At the reentrant critical point, $C_{V,x}$ and $C_{P,x}$ remain finite, while the isothermal compressibility diverges weakly. The reentrant critical point is a double critical point, such that along an isobar, the critical exponents are doubled, as discussed in Section 8.5.2.

8.5.4 Renormalization of Critical Exponents

Experimental conditions may or may not correspond to the isomorphic path for which the universal scaling singularities can be observed. For example, in a binary fluid, the thermodynamic path of zero-field ($h_1 = 0$) along the critical isochore is isomorphic upon a change in temperature only if at least one other physical field is kept constant (like μ_{21} or P), see Eq. (8.89). However, along the critical isochore, the experimental path for binary fluids is commonly performed at constant composition, x, while μ_{21} may vary. In the meanfield approximation, the difference between isomorphic and non-isomorphic paths only results in a change of the corresponding amplitudes and, consequently, the "activity," y.

In contrast, in scaling theory, due to the fluctuation-induced divergence of the weak susceptibility, $\chi_2 = A_0^{\pm}|h_2|^{-\alpha}$, the connection between μ_{21} and x becomes non-analytic as $\partial x/\partial\mu_{21}|_{T,\rho} \propto \chi_2 \propto |\Delta\hat{T}|^{-\alpha}$, where $\alpha = 0.11$ for $3d$-Ising systems. Correspondingly, the critical singularities, observed along these alternative isochoric paths, $h_2 = \Delta\hat{T}_x$ (the experimental path at constant x and ρ), and $h_2 = \Delta\hat{T}_{\mu_{21}}$ (the isomorphic path at constant μ_{21} and ρ), are different. Thermodynamic analysis [4] proves that the relation between these paths can be approximated as

$$\frac{\Delta\hat{T}_x}{\hat{\mathcal{T}}_2} = \left(\frac{\Delta\hat{T}_{\mu_{21}}}{\hat{\mathcal{T}}_2}\right)^{1-\alpha}\left[1 + \left(\frac{\Delta\hat{T}_{\mu_{21}}}{\hat{\mathcal{T}}_2}\right)^{\alpha}\right] \tag{8.115}$$

where $\hat{\mathcal{T}}_2$ is defined by Eq. (8.109), see Appendix 8.A for details. Equation (8.115) indicates that the critical region is divided into two parts. When $(\Delta\hat{T}_{\mu_{21}}/\hat{\mathcal{T}}_2)^{\alpha} \gg 1$, then $\Delta\hat{T}_x = \Delta\hat{T}_{\mu_{21}}$. Alternatively, when $(\Delta\hat{T}_{\mu_{21}}/\hat{\mathcal{T}}_2)^{\alpha} \ll 1$, then this relation becomes non-analytic as, $\Delta\hat{T}_x = \hat{\mathcal{T}}_2^{-\alpha}|\Delta\hat{T}_{\mu_{21}}|^{1-\alpha}$. Therefore, the asymptotic critical exponents of the diverging properties will be renormalized when they are obtained along the experimental path. For example, the thermodynamic properties would be observed to diverge as $\kappa_{\Pi} \propto |\Delta\hat{T}_x|^{-\gamma/(1-\alpha)}$, $\xi \propto |\Delta\hat{T}_x|^{-\nu/(1-\alpha)}$, and $C_{P,x} \propto \kappa_{T,x} \propto |\Delta\hat{T}_x|^{-\alpha/(1-\alpha)}$. This effect was first predicted by Fisher [69] in Ising ferromagnets "decorated" by an impurity and is commonly referred to as "Fisher renormalization" of critical exponents.

A more striking consequence of the non-analytic relation between μ_{21} and x at constant density is the suppression and renormalization of the isochoric heat capacity, $C_{V,x}$. Thermodynamically, this property, being the second derivative with respect to temperature, cannot diverge along the critical locus except at special points, such as at the critical azeotrope or in the limits of pure species (known as "Pippard's rule"[11] [3, 70]). Similarly, C_V cannot diverge at the line of the lambda transition in helium or along the Curie line of compressible ferromagnets [71, 72]. The crossover behavior for the singular part of $C_{V,x}$ can be represented by a Padé approximant in the form

$$\frac{\Delta C_{V,x}}{k_B} = \frac{A_0^{\pm}}{\hat{\mathcal{T}}_2^{\,\alpha}}\left[1+\left(\frac{\Delta\hat{T}_{\mu_{21}}}{\hat{\mathcal{T}}_2}\right)^{\alpha}\right]^{-1} \tag{8.116}$$

such that in the limit $(\Delta\hat{T}_{\mu_{21}}/\hat{\mathcal{T}}_2)^{\alpha} \gg 1$, then the singularity of the heat capacity will be seen as $\Delta C_{V,x}/k_B = A_0^{\pm}|\Delta\hat{T}_{\mu_{21}}|^{-\alpha}$. Alternatively, under the condition, $(\Delta\hat{T}_{\mu_{21}}/\hat{\mathcal{T}}_2)^{\alpha} \ll 1$, then the singular part of the isochoric heat capacity will be suppressed, approaching a final value at the critical point, $\Delta C_{V,x}/k_B = A_0^{\pm}/\hat{\mathcal{T}}_2^{\,\alpha}$. Expanding Eq. (8.116) when $(\Delta\hat{T}_{\mu_{21}}/\hat{\mathcal{T}}_2)^{\alpha} \ll 1$ and applying the non-analytic relation between $\Delta\hat{T}_x$ and $\Delta\hat{T}_{\mu_{21}}$, given by Eq. (8.115), one obtains that along the experimental path, at constant composition, the heat capacity exhibits a finite cusp with a renormalized exponent $\alpha/(1-\alpha)$ [4],

$$\frac{\Delta C_{V,x}}{k_B} \approx \frac{A_0^{\pm}}{\hat{\mathcal{T}}_2^{\,\alpha}}\left[1-\left(\frac{\Delta\hat{T}_{\mu_{21}}}{\hat{\mathcal{T}}_2}\right)^{\alpha}\right] = \frac{A_0^{\pm}}{\hat{\mathcal{T}}_2^{\,\alpha}}\left[1-\left(\frac{\Delta\hat{T}_x}{\hat{\mathcal{T}}_2}\right)^{\alpha/(1-\alpha)}\right] \tag{8.117}$$

The renormalization of the singularity in $C_{V,x}$ has been confirmed by accurate experimental studies of dilute solutions of ethane in n-heptane near the liquid–gas critical locus (see Figure 8.26). In this system, the critical locus $T_c(x)$ is very steep, with a large value for the parameter $\hat{\mathcal{T}}_2$; thus, the renormalization of the isochoric heat capacity can be observed even in relatively dilute solutions (1–3%).

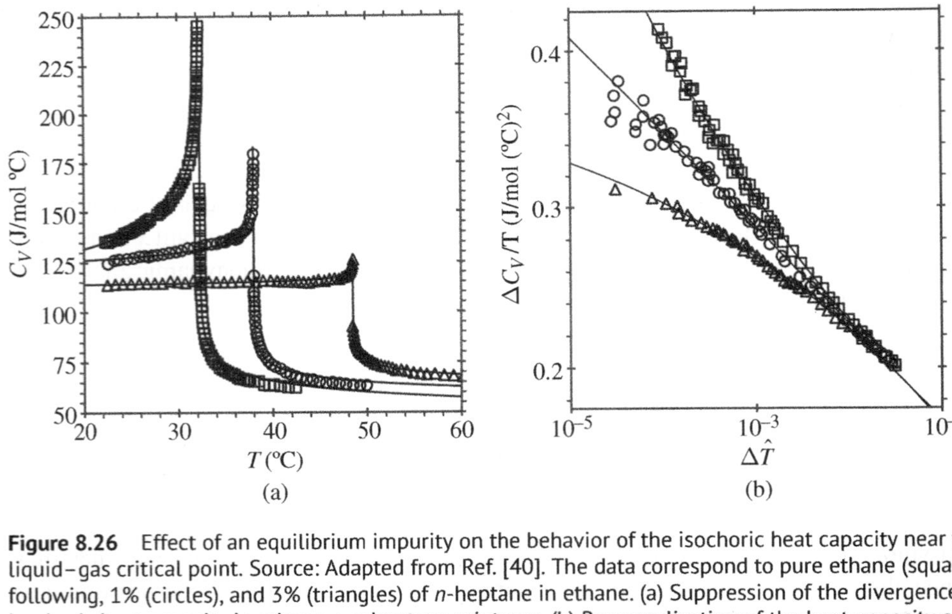

Figure 8.26 Effect of an equilibrium impurity on the behavior of the isochoric heat capacity near the liquid–gas critical point. Source: Adapted from Ref. [40]. The data correspond to pure ethane (squares) following, 1% (circles), and 3% (triangles) of n-heptane in ethane. (a) Suppression of the divergence of the isochoric heat capacity in ethane + n-heptane mixtures. (b) Renormalization of the heat-capacity critical exponent from $\alpha = 0.11$ to $-\alpha/(1-\alpha)$ for ethane + n-heptane mixtures in the one-phase region. The solid curves correspond to $\alpha = -0.11$ (upper), $\alpha = 0$ (middle), and $\alpha = 0.12$ (lower).

11 Sir Alfred Brian Pippard, (1920–2008), was a British physicist, who introduced the notion of coherence length in superconductors.

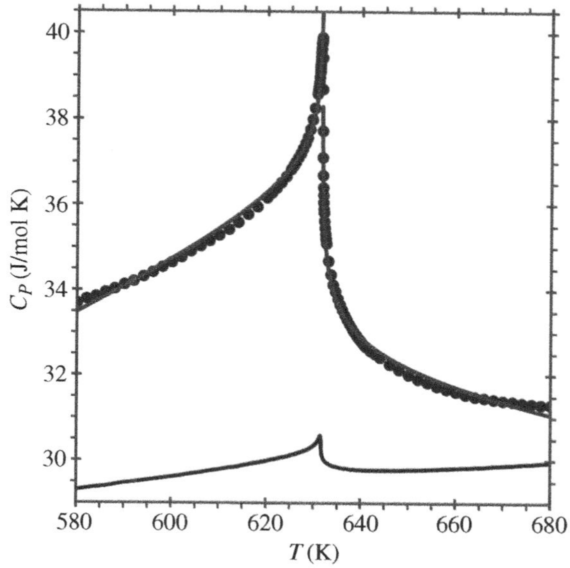

Figure 8.27 The diverging isobaric heat capacity near the Curie point of nickel. The circles indicate experimental data [73], while the solid curve is the theoretical prediction of $C_P \propto \chi_2 = A_0^{\pm}|\Delta\hat{T}|^{-\alpha} + B(T)$, as given in Table 8.2, with $\alpha = 0.11$ and a linear temperature-dependent background. The curve below indicates the difference $C_P - C_V^*$ where C_V^* is the magnetic contribution to the isochoric heat capacity. This difference is proportional to the isothermal compressibility, which weakly diverges with the same critical exponent as C_P, while C_V^* will reach a finite cusp with a renormalized exponent as $C_V \sim \Delta\hat{T}^{\alpha/(1-\alpha)}$. Source: Adapted from Ref. [73].

For liquid–liquid critical points in weakly compressible mixtures, the pressure is a hidden field to be kept constant along the isomorphic path, $\Delta\hat{T}_P$ (the isobar at constant composition), see Section 8.5.2. If both the density and the composition are kept constant along this path, then $\Delta\hat{T}_\rho$ and the isomorphic path, $\Delta\hat{T}_P$, are interrelated via the same construction as the relation between $\Delta\hat{T}_{\mu_{21}}$ and $\Delta\hat{T}_x$, given by Eq. (8.115). In this case, the characteristic function for path renormalization, $\hat{\mathscr{T}}_2$, is given by

$$\tilde{\mathscr{T}}_2 \approx \left[A_0^{\pm}(k_B\rho_c)^2 \left(\frac{dT_c}{dP_c} \right)^2 \right] \tag{8.118}$$

In many liquid mixtures and solids, the function $\tilde{\mathscr{T}}_2$ is too small for experimental detection of the renormalization of critical exponents. Measurements of the isobaric and isochoric heat capacity in nickel (see Figure 8.27) demonstrate that C_P weakly diverges at the Curie point, while C_V exhibits a cusp, reaching a finite maximum value. The weak divergence of the isobaric heat capacity at the Curie point could be revealed only by accounting for the difference $C_P - C_V \propto \kappa_T$, which weakly diverges [4]. Also, the addition of an impurity into a weakly compressible binary-liquid mixture (or a ferromagnet) will renormalize the weak divergence of $C_{P,x}$ (or C_P), respectively, from $-\alpha$ to $\alpha/(1-\alpha)$.

8.6 Effects of Fluctuations on Near-Critical Interfaces

The interface between two phases in the critical region is also affected by order-parameter fluctuations. Near the critical point, the liquid–gas interface becomes smooth with a width extending, in practice, from nanometers to microns. As accurate experiments show (see Figure 8.7b), the surface tension asymptotically vanishes at the critical point as

$$\sigma = \sigma_0 |\Delta\hat{T}|^{\vartheta} \tag{8.119}$$

where the power $\vartheta = 1.26$ is the universal critical exponent for the surface tension and σ_0 is the system-dependent amplitude. Recall that in the meanfield van der Waals–Landau theory, $\vartheta = 3/2$, see Eq. (6.22).

As discussed in Section 6.2, the width of the interface is proportional to the correlation length (see Eq. 6.19), and since the correlation length diverges according to Eq. (8.8) as $\xi \sim |\Delta\hat{T}|^{-\nu}$, then the width of the interface also diverges. In the spirit of the fluctuation-cluster model (Section 8.3.2), for an interface with a width of 2ξ, one may obtain a scaling estimate for the surface tension, based on the fact that the surface tension is the force per unit length (see Eq. 6.1), as

$$\sigma \approx -2\xi \Delta P_{\rm cr} \tag{8.120}$$

With the critical part of the pressure, expressed in Eq. (8.10), the surface tension for a d-dimensional system may be written as

$$\sigma \approx \frac{k_{\rm B}T_{\rm c}}{(2\xi_0^+)^{d-1}}|\Delta\hat{T}|^{(d-1)\nu} \tag{8.121}$$

Comparing Eqs. (8.119) and (8.121), the critical exponent for the surface tension is obtained as $\vartheta = (d-1)\nu$ with an amplitude $\sigma_0 = k_{\rm B}T_{\rm c}/(2\xi_0^+)^{d-1}$. Note that the meanfield result, $\vartheta = 3/2$, is recovered for $d = 4$, while the experimental result, $\vartheta = 2\nu = 1.26$, is recovered for three-dimensional Ising systems, see Eq. (8.6).

The surface tension may also be estimated from the surface contribution to the Landau–Ginzburg functional, see Section 6.2, as $\sigma \sim \phi^2/\xi$. As discussed in Section 8.3.4, the Landau–Ginzburg functional is exact in the first-order epsilon expansion, where the critical exponents are given by: $\gamma = 2\nu$, $\beta = \nu/2$, $\eta = 0$. Thus, in this approximation, one finds that the surface tension follows the same scaling prediction for the critical power law, $\sigma \sim |\Delta\hat{T}|^{2\nu}$.

In addition, critical fluctuations also modify the shape of the density (or concentration) profile. It has been proven in the first-epsilon expansion of the RG theory that the expression for the order-parameter profile near the critical point is given by [74]

$$\frac{\phi}{\phi_0} = \tanh\left(-\frac{z}{2\xi}\right)\left[1 + \frac{2a}{3+a}{\rm sech}^2\left(-\frac{z}{2\xi}\right)\right]^{-1/2} \tag{8.122}$$

where $a = (\sqrt{3}/6)\pi \approx 0.91$ for a three-dimensional system. As shown in Figure 8.28, the fluctuations additionally smooth the density or concentration profile, but the corresponding correction is relatively small with respect to the meanfield result, Eq. (6.19).

Wetting transitions are also affected by critical fluctuations. Consider three-phase equilibrium, between two fluid phases and a solid phase or between three fluid phases, as in Section 6.4. Due to critical fluctuations, the corresponding surface tensions between the three phases should be expressed through scaling relations. In particular, the surface tension between the two near-critical fluid phases ($\tilde{\alpha}$ and $\tilde{\beta}$, not to be confused with the critical exponents), are modified according to Eq. (8.119) as

$$\sigma^{\tilde{\alpha}\tilde{\beta}} = \sigma_0^{\tilde{\alpha}\tilde{\beta}}|\Delta\hat{T}|^{2\nu} \tag{8.123}$$

and the difference in surface tensions between the respective near-critical fluid phase and the non-critical phase ($\tilde{\gamma}$) is related to the difference in densities (or concentrations) of the two fluid phases, such that asymptotically close to the critical point

$$\sigma^{\tilde{\alpha}\tilde{\gamma}} - \sigma^{\tilde{\beta}\tilde{\gamma}} \sim \rho^{\tilde{\alpha}} - \rho^{\tilde{\beta}} \approx B_0|\Delta\hat{T}|^{\beta} \tag{8.124}$$

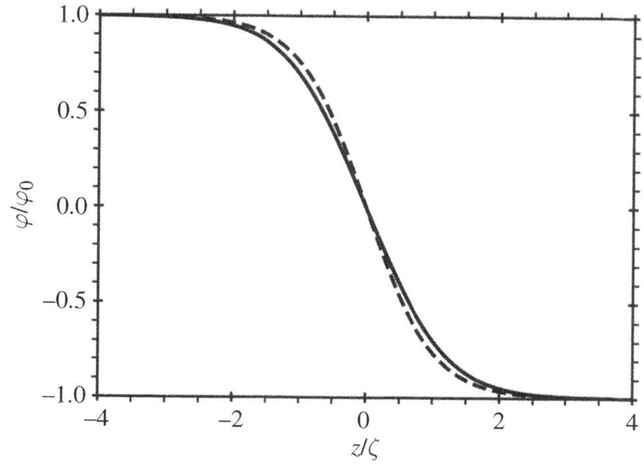

Figure 8.28 Predictions of the density profile in the meanfield approximation (Eq. 6.19, dashed) and in the first-order epsilon expansion of the RG theory (Eq. 8.122, solid).

Correspondingly, the wetting transition temperature, T_w, defined by the condition of Antonov's rule Eq. (6.37) and illustrated in Figure 6.10, will also be shifted by critical fluctuations with respect to the meanfield prediction.

8.7 Divergence of Tolman's Length

A specific consequence of the effects of critical fluctuations on the fluid interface is the divergence of Tolman's length (δ'), the first curvature correction to the surface tension (see Section 6.3). This divergence is purely a fluctuation-induced effect, which does not exist in the meanfield theory.

As discussed in Section 6.3, Tolman's length originates from the fluid-phase asymmetry, as given by Eq. (6.33). For phase equilibria between two fluid phases, $\tilde{\alpha}$ and $\tilde{\beta}$, as discussed in Section 4.4.1, $\phi_{0,\mathrm{asym}} = \Delta\hat{\rho}_\mathrm{d}$ and $\phi_{0,\mathrm{sym}} = \rho_{\tilde{\alpha}} - \rho_{\tilde{\beta}}$, such that Eq. (6.33) reads

$$\frac{\delta'}{\zeta} = \mp c_\delta \frac{\Delta\hat{\rho}_\mathrm{d}}{\rho_{\tilde{\alpha}} - \rho_{\tilde{\beta}}} \tag{8.125}$$

where c_δ is a constant. Critical fluctuations alter this equation in three ways. First, the diameter of the coexistence curve, $\Delta\hat{\rho}_\mathrm{d}$, is transformed from rectilinear to singular, see Eq. (8.71). Second, critical fluctuations modify the temperature-dependence of the difference in fluid phases, such that $\rho_{\tilde{\alpha}} - \rho_{\tilde{\beta}} = B_0|\Delta\hat{T}|^\beta$, where $\beta = 0.326$ for $3d$-Ising systems. Third, they modify the thickness of the interface $\zeta = 2\xi$, as $\xi = \xi_0^-|\Delta\hat{T}|^{-\nu}$, where $\nu = 0.630$ for $3d$-Ising systems. Due to these three effects, Tolman's length is given by

$$\delta' \approx \mp 2c_\delta\xi_0^- \left(\hat{a}_3 B_0|\Delta\hat{T}|^{\beta-\nu} - b_2\frac{A_0^-}{(1-\alpha)}|\Delta\hat{T}|^{1-\alpha-\beta-\nu}\right) \tag{8.126}$$

where $2c_\delta\xi_0^-$ is a universal constant. Note that both terms in Eq. (8.126) will diverge at the critical point. A comparison of the exponents for $3d$-Ising systems shows that the first term diverges as $\beta - \nu = -0.304$, and the second term diverges as $1 - \beta - \alpha - \nu = -0.065$. Therefore, the first term exhibits a much stronger divergence than the second term.

Experimental and computational observations of the divergence of Tolman's length face tremendous challenges. To reliably obtain the curvature correction to the surface tension, one needs to study, experimentally or by simulations, a nano-sized droplet of a highly asymmetric fluid in the critical region ($\Delta\hat{T} \ll 1$). Tolman's length is approximately proportional to the square root of the thickness of the interface, scaling with the correlation length, which diverges at the critical point Eq. (6.19). However, the thickness of the interface, and correspondingly the correlation length, must be significantly smaller than the size of the droplet. Thus, the requirement for the droplet size to be small (large curvature) conflicts with the requirement of a large correlation length (significant critical fluctuations).

At the temperature range of about 1% T_c, the correlation length is typically ~10 nm. Under this condition, accounting for the fact that the value of the leading asymmetry coefficient, a_3, rarely exceeds 0.3 [50], Tolman length is estimated to be of the order of 1 nm. Thus, for a droplet with radius of ~30 nm one could expect a curvature correction to the surface tension of about 3%. On the other hand, for these conditions, in accordance with Eq. (8.119), the surface tension is about 100 times smaller than that far away from the critical temperature; thus, such a small correction to a vanishing surface tension would be difficult to detect.

8.8 Critical Phenomena in Polymer Solutions

The asymmetry of fluid-phase coexistence is most profound in polymer solutions near liquid–liquid demixing. The meanfield description of these systems was presented in Chapter 5. In this section, we consider the effects of critical fluctuations on the Widom crossover from fluid criticality near the demixing points to Theta-point tricriticality.

8.8.1 Ising Criticality and de Gennes–Sanchez Relations

The thermodynamic behavior of polymer solutions near their critical demixing transitions, like all fluids and fluid mixtures, belongs to the three-dimensional Ising-model universality class. Asymptotically close to the liquid–liquid critical point, the thermodynamic properties of polymer solutions (e.g. phase coexistence, osmotic susceptibility, correlation length, etc.) are described by power laws with universal critical exponents, as given in Table 8.1. This scaling behavior has been confirmed in a variety of systems, including polystyrene in cyclohexane [75–83], polystyrene in methylcyclohexane [84], and in polymethylmethacrylate in 3-octanone [85, 86]. In this section, we focus on light-scattering measurements in solutions of polystyrene in cyclohexane. The intensity of static light scattering is proportional to the osmotic susceptibility, $\hat{\chi} \propto \varphi \partial\varphi / \partial\hat{\Pi}$. The intensity, measured at different angles, reveals the wavenumber dependence of the susceptibility, and thus, enables one to recover the correlation length, see Eq. (8.36). For instance, as shown in Figure 8.29, accurate light-scattering measurements in a solution of polystyrene ($M_w = 195,900$ g/mol) in cyclohexane, performed by Jacob et al. [78], have confirmed the universal Ising-model behavior of the correlation length, ξ, and of the osmotic susceptibility, $\hat{\chi}$.

While the critical exponents in polymer solutions, like in all fluids and fluid mixtures, belong to the Ising-model universality class, the critical amplitudes, unique to polymer solutions, diverge or vanish with the degree of polymerization, $\mathcal{N}$, as given in Table 8.7. Static light-scattering measurements of solutions of polystyrene in cyclohexane, from $M_w = 195.900$ g/mol to $M_w = 11.4$ million g/mol [81], have confirmed that the amplitude of the correlation length and the amplitude of the susceptibility satisfy the predictions of de Gennes and Sanchez [87–89], both in the vicinity of the critical point and in the meanfield regime far away from the critical point and in the meanfield

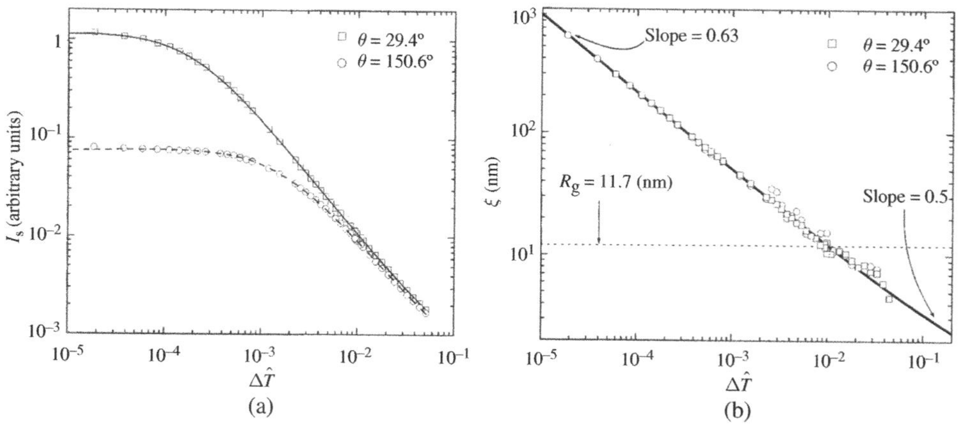

Figure 8.29 Demonstration of Ising scaling behavior in polymer solutions. (a) Light-scattering intensity (proportional to the osmotic compressibility) for a solution of polystyrene (M_w = 195, 900 g/mol) in cyclohexane as a function of the distance to the critical temperature at the critical volume fraction ($\varphi_c \cong 0.07$). (b) Correlation length for the same system. The symbols are the experimental data [78], while the solid curves represent the theoretical prediction [78, 81].

Table 8.7 De Gennes–Sanchez scaling relations for various thermodynamic properties in polymer solutions [87–89] given for both meanfield and scaling.

Scaling law	de Gennes–Sanchez relations	Meanfield amplitude	Scaling amplitude
$\phi \sim \pm\mathcal{N}^{-b}\lvert\Delta\hat{T}\rvert^{\beta}$	$b = (1-\beta)/2$	$b = 0.25$	$b = 0.34$
$C \sim \mathcal{N}^{-a}\lvert\Delta\hat{T}\rvert^{-\alpha}$	$a = (1+\alpha)/2$	$a = 0.5$	$a = 0.555$
$\hat{\chi} \sim \mathcal{N}^{g}\lvert\Delta\hat{T}\rvert^{-\gamma}$	$g = (1-\gamma)/2$	$g = 0$	$g = -0.12$
$\xi \sim \mathcal{N}^{n}\lvert\Delta\hat{T}\rvert^{-\nu}$	$n = (1-\nu)/2$	$n = 0.25$	$n = 0.185$
$\sigma \sim \mathcal{N}^{-m}\lvert\Delta\hat{T}\rvert^{\mu}$	$m = (2-\mu)/2$	$m = 0.25$	$m = 0.37$

The amplitude exponents are given by their meanfield and three-dimensional Ising-model forms.

regime far away from the critical point (see Figure 8.30). The $\mathcal{N}$-dependent nature of the amplitudes raises a fundamental question about the critical behavior of extremely high-molecular-weight polymer solutions close to the limit of an infinite degree of polymerization.

Typically, the critical behavior of simple fluids is controlled by a single mesoscopic length scale, the correlation length of the concentration fluctuations, ξ [4, 15, 90, 91]. However, in polymer solutions, there is a second competing length scale, namely, the size of the polymer coil (as defined through the radius of gyration, R_g) that depends on the degree of polymerization, $\mathcal{N}$. With increasing values of $\mathcal{N}$, upon approaching the limit $\mathcal{N} \to \infty$, the critical concentration (volume fraction), φ_c, decreases to zero, while the critical temperature, T_c, approaches the Theta temperature (see Figure 5.5). In the asymptotic vicinity of the critical temperature, where the correlation length is larger than the radius of gyration, the range of Ising-like critical behavior decreases as R_g increases. The radius of gyration in the polystyrene-cyclohexane system, obtained by neutron-scattering experiments [92, 93], follows the scaling behavior $R_g \sim \mathcal{N}^{1/2}$ along the critical line, see Figure 8.30b. With an increase of R_g, the alternative de Gennes correlation length, ξ_ψ given by Eq. (5.49), emerges. Figure 8.30b verifies the theoretical prediction $\xi_\psi \propto R_g$ (see

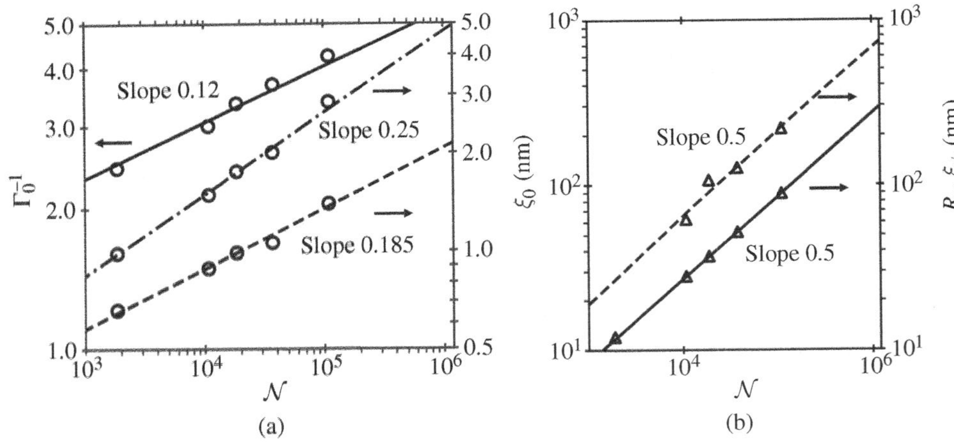

Figure 8.30 (a) Experimental confirmation of the de Gennes–Sanchez scaling relations (Table 8.7). The left axis presents the amplitude of the susceptibility, Γ_0 (solid), while the right axis presents the amplitudes of the correlation length of concentration fluctuations ξ_0, in the critical region (dashed) and away from the critical point in the meanfield region (dotdashed). Source: Adapted from Ref. [81]. The slopes of the lines correspond to the values predicted by de Gennes and Sanchez. (b) The radius of gyration R_g, as a function of the degree of polymerization $\mathcal{N}$, from neutron-scattering measurements [92, 93]. The solid line corresponds to the scaling law, $R_g \sim \mathcal{N}^{1/2}$, along the critical line. Also depicted is the correlation length associated with the polymer-chain fluctuations, ξ_ψ, obtained from the coupling between diffusive and viscoelastic dynamic modes. Source: Adapted from Ref. [83]. The dashed line corresponds to the theoretical prediction $\xi_\psi \propto R_g$ for large $\mathcal{N}$, see Section 5.6 for details.

Section 8.8.2 for details). Thus, the asymptotic amplitude of the correlation length scales with the degree of polymerization as $\xi_0 \sim \mathcal{N}^{0.185}$ in the Ising-like critical region (where $\xi > \xi_\psi$), while it scales as $\xi_0 \sim \mathcal{N}^{0.25}$ in the meanfield region (where $\xi < \xi_\psi$), see Figure 8.40a. Such shrinking of the scaling critical regime corresponds to the behavior of the coexistence curve (see Figure 5.5), leading to the "angle-like" shape in the limit of an infinite degree of polymerization.

8.8.2 Crossover Between Ising Criticality and Theta-Point Tricriticality

The Widom procedure for the crossover between the meanfield critical regime and the Theta-point tricritical regime in polymer solutions (Section 5.5) was generalized to include the effects of critical fluctuations in the vicinity of the critical-demixing point [57, 79–82]. We refer to this approach as the "generalized Widom crossover," to Chapter 12 in [94]. And in this section, we present a simplified version of this crossover, only emphasizing the essential physics of this phenomenon.

First, in the vicinity of the Theta point, since the marginal dimension for tricriticality is $d = 3$, there are logarithmic corrections to the meanfield behavior, induced by fluctuations, that must be accounted for [79, 81, 95]. In this case, the classical chain-length dependence of the critical parameters, φ_c and T_c, as given by Eq. (5.24), are modified by tricritical fluctuations. The modified critical parameters are [79]

$$\varphi_c = \frac{1}{1 + a\sqrt{\mathcal{N}}(1 + \ln \mathcal{N})^{-1/2}} \quad \text{and} \quad \frac{\Theta - T_c}{\Theta} \approx \frac{1}{(\ln \mathcal{N})^{3/22}\sqrt{\mathcal{N}}} \tag{8.127}$$

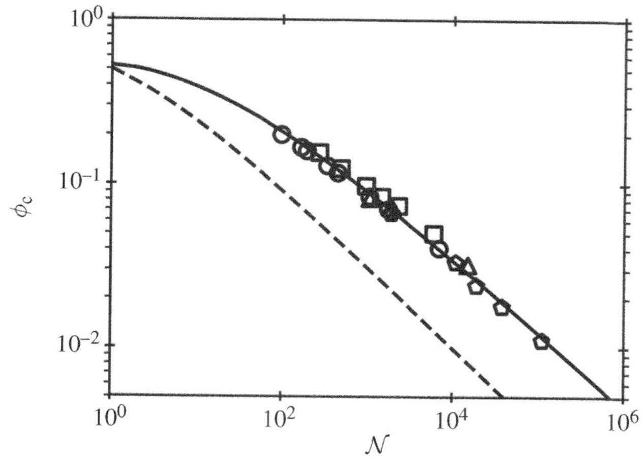

Figure 8.31 Critical volume fraction, φ_c of solutions of polystyrene in cyclohexane measured as a function of the degree of polymerization, $\mathcal{N}$, by Kojima and Nakata et al. [75–77] (triangles), polystyrene in cyclohexane by Anisimov et al. [81] (pentagons), polystyrene in methylcyclohexane by Dobashi et al. [84] (circles), and PMMA in 3-octanone by Xia et al. [85, 86] (squares). The solid curve represents Eq. (8.127), with coefficient $a = 0.95$, while the dashed curve represents the prediction from the Flory–Huggins theory, Eq. (5.24).

where the coefficient a is on the order of unity. Note that there is a significant effect of logarithmic fluctuations on the critical volume fraction, as shown in Figure 8.31, as compared with the prediction of the Flory–Huggins theory, Eq. (5.24). Meanwhile, the effect of logarithmic corrections on the critical temperature is relatively small but still consequential.

In the generalized Widom crossover, as a result of the logarithmic correction to the Theta-point tricritical behavior, a more accurate representation for the Widom variable, x, is $\tilde{x} = (T - T_c)/(\Theta - T_c)$, where the critical temperature is measured experimentally and the Theta temperature is obtained from extrapolation [79–82]. In the Flory–Huggins theory, $\tilde{x} = x$, as $(\Theta - T_c)/T_c \approx 1/\sqrt{\mathcal{N}} = \varphi_c$ for large $\mathcal{N}$. Moreover, the reduced order parameter, $\hat{\phi} = (\varphi - \varphi_c)/\varphi_c$, is rescaled with a critical volume fraction determined from experiment. From here, a simple interpolation function, with a crossover parameter given through the Widom variable, is utilized to modify the universal variables ($\tilde{x}$ and $\hat{\phi}$) to match the scaling limits in the asymptotic Ising-like critical regime. The crossover parameter responsible for the transition from Ising-like criticality to Theta-point tricriticality is given as the ratio $z = \xi/R_g$ [57, 79–82, 96]. We note that in the Flory–Huggins theory, the crossover variable reduces to the Widom variable as $z = \tilde{x}^{-1/2}$. Therefore, the crossover behavior of the phase coexistence in polymer solutions can be represented by a simple generalization of the Widom crossover as

$$\hat{\phi} \approx \hat{\phi}_{\text{MF}}\left(1 + z^2\right)^{(1-2\beta)/4\nu} \tag{8.128}$$

where the amplitude function, $\hat{\phi}_{\text{MF}}$, is predicted from the meanfield Flory–Huggins-like theory. The Ising critical exponent is $\beta = 0.326$ (see Table 8.1). Equation (8.128) has the limits: $\hat{\phi}(z \ll 1) \to \hat{\phi}_{\text{MF}}$ and $\hat{\phi}(z \gg 1) \to B_0|\Delta\hat{T}|^\beta$, where $B_0 \sim \mathcal{N}^{-0.337}$ (see Table 8.7).

In a similar form to the phase coexistence behavior Eq. (8.128), the crossover expressions for the susceptibility and correlation length are given by [81, 82]

$$\hat{\chi}^{-1} \approx \hat{\chi}_{\text{MF}}^{-1}(1+z^2)^{(\gamma-1)/2\nu} \tag{8.129}$$

$$\xi \approx \xi_{\text{MF}}(1+z^2)^{(2\nu-1)/4\nu} \tag{8.130}$$

where the amplitude functions are the susceptibility and correlation length predicted by the mean-field Flory–Huggins-like theory, $\hat{\chi}_{\text{MF}}^{-1} = \overline{\Gamma}_0|\Delta\hat{T}|$ and $\xi_{\text{MF}} = \overline{\xi}_0|\Delta\hat{T}|^{-0.5}$ (where $\overline{\Gamma}_0 \sim \mathcal{N}^0$ and $\overline{\xi}_0 \sim \mathcal{N}^{-0.25}$ are the meanfield amplitudes), see Eqs. (4.20) and (4.72). Asymptotically close to the critical point (for $z \gg 1$), Eqs. (8.129) and (8.130) satisfy the critical power laws, $\hat{\chi} = \Gamma_0|\Delta\hat{T}|^{-\gamma}$ and $\xi = \xi_0|\Delta\hat{T}|^{-\nu}$ (see Table 8.1). For example, consider the phase behavior of the eight different solutions of polystyrene in methylcyclohexane with different degrees of polymerization [79, 84] presented in Figure 8.32a. The universal crossover behavior of this system is shown in Figure 8.32b.

By rescaling the thermodynamic properties, we may represent the crossover phenomenon in a universal form as

$$\tilde{\phi} \equiv \frac{\hat{\phi}|\Delta\hat{T}|_{\times}^{-\beta}}{B_0} \approx \frac{(1+z^2)^{(1-2\beta)/4\nu}}{(\Delta\hat{T}_{\times}/\Delta\hat{T})^{1/2}} \tag{8.131}$$

$$\tilde{\chi} \equiv \frac{\chi|\Delta\hat{T}|_{\times}^{\gamma}}{\Gamma_0} \approx \frac{(1+z^2)^{(1-\gamma)/2\nu}}{\Delta\hat{T}/\Delta\hat{T}_{\times}} \tag{8.132}$$

$$\tilde{\xi} \equiv \frac{\xi|\Delta\hat{T}|_{\times}^{\nu}}{\xi_0} \approx \frac{(1+z^2)^{(2\nu-1)/4\nu}}{(\Delta\hat{T}/\Delta\hat{T}_{\times})^{1/2}} \tag{8.133}$$

Figure 8.33 shows the universal susceptibility (a) and correlation length (b) for five different degrees of polymerization in a polystyrene–cyclohexane solution [75–77, 81, 84–86]. In their rescaled forms, the thermodynamic properties exhibit the middle of their crossover at the Widom crossover point, $\tilde{x} = \Delta\hat{T}/\Delta\hat{T}_{\times} = 1$.

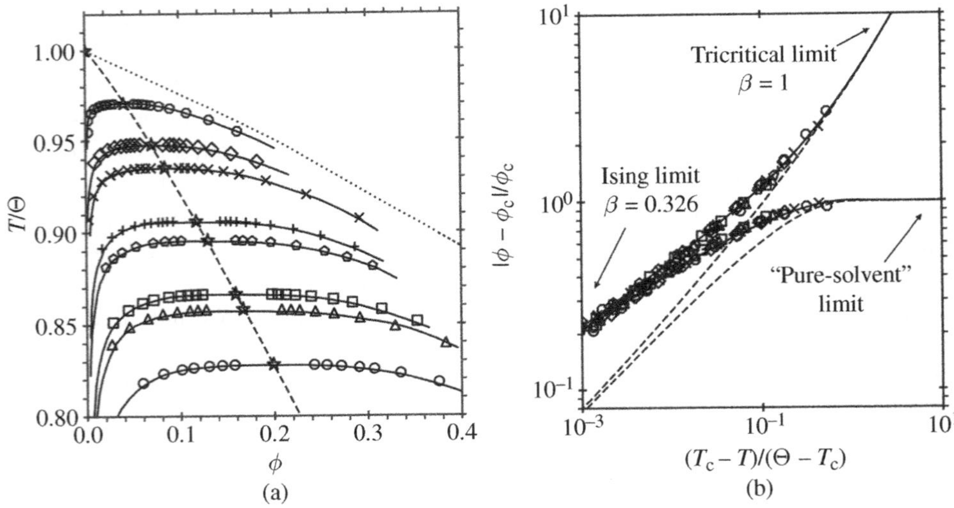

Figure 8.32 Crossover phase behavior in a solution of polystyrene in methylcyclohexane. (a) Phase coexistence for polymer solutions with different molecular weights: from M_w = 10.200 g/mol (lowest) to M_w = 719.000 g/mol (highest). The different types of points are experimental data [84], while the curves are the theoretical prediction [96]. (b) Universal scaled crossover behavior. The polymer-rich branch exhibits a crossover from the asymptotic Ising-like critical region to the tricritical Theta-point region, while the solvent-rich branch exhibits a crossover from the asymptotic Ising-like region to the limit of pure solvent.

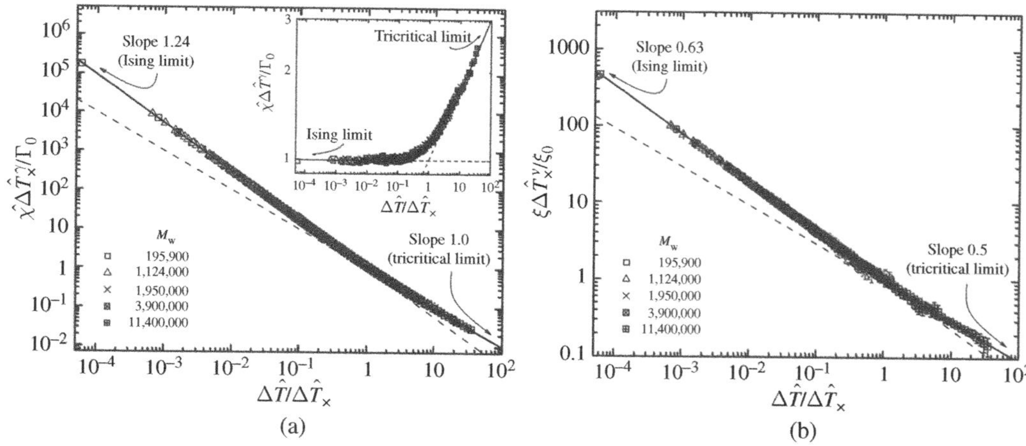

Figure 8.33 Universal scaling behavior of thermodynamic properties obtained by light scattering for the corresponding critical volume fractions (see Figure 8.31). (a) Scaled susceptibility and deviation of the susceptibility from Ising critical behavior (shown in the inset) as a function of the normalized distance from the critical temperature, $\Delta\hat{T}/\Delta\hat{T}_\times$, for polystyrene–cyclohexane solutions with five different molecular weights of polystyrene. The inset shows the deviation of the susceptibility from the asymptotic Ising-like behavior in the vicinity of the Theta-point. (b) The scaled correlation length for the same data. The symbols indicate experimental data, while the solid curve represents the crossover theory. The dashed lines represent the two limiting behaviors: Ising asymptotic behavior and Theta-point (tricritical) behavior. Source: Reproduce with permission from Ref. [80], (2002), © American Physical Society.

8.9 Finite-Size Scaling

For small systems, the size of the system, ℓ, will compete with the correlation length, ξ, in the critical region. The behavior of such a system can be understood through the theory of "finite-size scaling". Consider, for instance, a spatially restricted system, such as a fluid confined to a porous medium with pore size, ℓ. The effects of finite size produce a suppression of the critical singularities, such that the critical point is "rounded" and the thermodynamic properties no longer diverge or vanish, but instead, reach finite values. For example, consider the experimental data for the isobaric heat capacity of a solution of lutidine in water as shown in Figure 8.34. The bulk heat capacity diverges, as predicted by scaling theory (see Figure 8.20), but the finite-size effect causes an increased suppression of the singularity in the heat capacity as the size of the pores decreases.

The prediction from finite-size scaling is that the temperature at which the maximum will occur, $T_m(\ell)$, is shifted relative to the bulk critical temperature by an amount [97]

$$\Delta\hat{T}_m = \frac{T_m(\ell) - T_c}{T_c} \propto \ell^{-1/\nu} \tag{8.134}$$

This prediction is illustrated by the line in Figure 8.35.

The expression for the fluctuation-induced part of the heat capacity near the critical point for a finite-size system is given by

$$\Delta C(\xi, \ell) = \Delta C_\infty(\xi)\Psi(\xi/\ell) \tag{8.135}$$

where $\Delta C_\infty(\xi) = A_0^\pm|\Delta\hat{T}|^{-\alpha} \propto \xi^{\alpha/\nu}$ is the diverging part of the heat capacity in an "infinite-size" system, and $\Psi(\xi/\ell)$ is a universal finite-size scaling function. This scaling function is chosen such

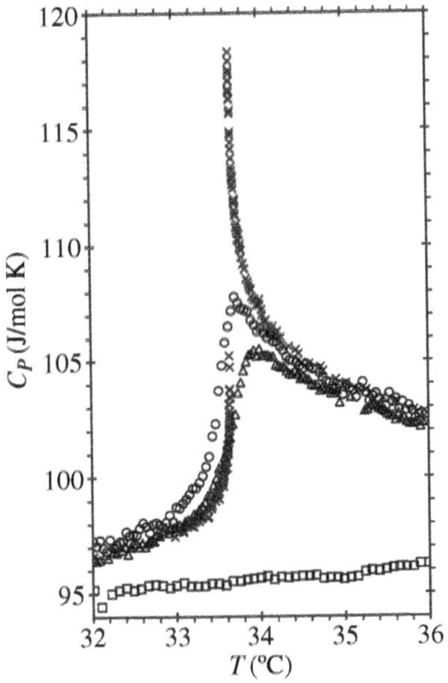

Figure 8.34 Finite-size isobaric heat capacity near the lower critical point of aqueous lutidine solutions. The crosses are the data for the "bulk," a 200 micron-thick calorimeter, the circles are for 250 nm porous nickel, the triangles are for 100 nm porous glass, and the squares are for 10 nm porous glass (see also Figure 8.20). Source: Reproduced with permission from Ref. [97], © 2010, Royal Society of Chemistry.

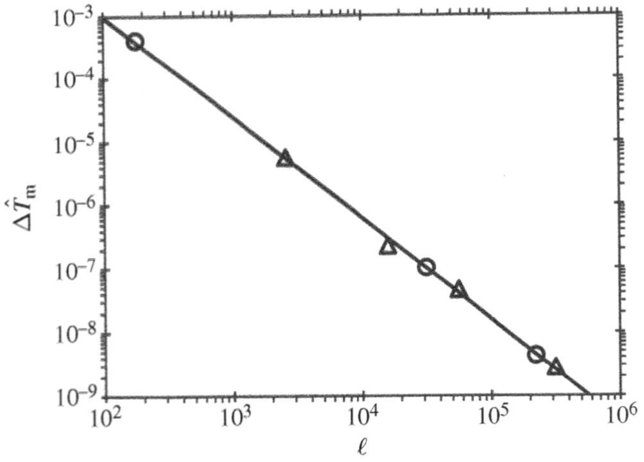

Figure 8.35 Shift in the heat capacity maxima $\Delta\hat{T}_\mathrm{m}$ near the lower critical point of lutidine aqueous solution at the critical point of composition as a function of pore size, ℓ. The triangles are obtained from the experimental data shown in Figure 8.34, while the circles are the results of simulations [98]. The slope of the solid line is $-1/\nu \approx -1.59$. Source: Reproduced with permission from Ref. [97], © 2010, Royal Society of Chemistry.

that it follows the limits

$$\Psi(\xi/\ell) = \begin{cases} \Psi(\xi/\ell) \to 1 & \xi/\ell \to 1 \\ \Psi(\xi/\ell) \to \left(\dfrac{\xi}{a\ell}\right)^{-\alpha/\nu} & \xi/\ell \gg 1 \end{cases} \tag{8.136}$$

where a is a system-dependent constant. Thus, the heat capacity in a finite-size system follows

$$\Delta C(\xi,\ell) = \begin{cases} \Delta C(\xi,\ell) = A_0^{\pm}|\Delta\hat{T}|^{-\alpha} & \xi/\ell \to 1 \\ \Delta C(\xi,\ell) = A_0^{\pm}\left(\dfrac{a\ell}{\xi_0^{\pm}}\right)^{\alpha/\nu} & \xi/\ell \gg 1 \end{cases} \tag{8.137}$$

Based on experimental data, one could consider the scaling function $\Psi(\xi/\ell)$ in the form

$$\Psi(\xi/\ell) = \left[1 + \left(\frac{\xi}{a\ell}\right)^2\right]^{-\alpha/2\nu} \tag{8.138}$$

As shown in Figure 8.36, this crossover function correctly interpolates between the two limits for the behavior of the finite-size heat capacity, given by Eq. (8.136).

Similarly, finite-size scaling predicts the following behavior of the near-critical susceptibility in a finite-size system,

$$\hat{\chi}(\xi,\ell) = \hat{\chi}_{\infty}(\xi)\Upsilon(\xi/\ell) \tag{8.139}$$

where $\hat{\chi}_{\infty}(\xi) = \Gamma_0^{\pm}|\Delta\hat{T}|^{-\gamma} \propto \xi^{\gamma/\nu}$ is the susceptibility of an infinite-size system, and Υ is a scaling function of ξ and ℓ, like $\Psi(\xi/\ell)$, such that it follows the limits

$$\Upsilon(\xi/\ell) = \begin{cases} \Upsilon(\xi/\ell) \to 1 & \xi/\ell \ll 1 \\ \Upsilon(\xi/\ell) \to \left(\dfrac{\xi}{a\ell}\right)^{-\gamma/\nu} & \xi/\ell \gg 1 \end{cases} \tag{8.140}$$

As a result, the finite-size susceptibility is given by

$$\hat{\chi}(\xi,\ell) = \begin{cases} \hat{\chi}_{\infty}(\xi) = \Gamma_0^{\pm}|\Delta\hat{T}|^{-\gamma} & \xi/\ell \ll 1 \\ \hat{\chi}_{\ell} = \Gamma_0^{\pm}\left(\dfrac{a\ell}{\xi_0^{\pm}}\right)^{\gamma/\nu} & \xi/\ell \gg 1 \end{cases} \tag{8.141}$$

In both expressions for the finite-size heat capacity and susceptibility, the first limit is the thermodynamic (infinite-size) limit and the second limit is the mesoscopic (finite-size) limit. The major result of finite-size scaling is that at the mesoscale the thermodynamic properties, which depend on

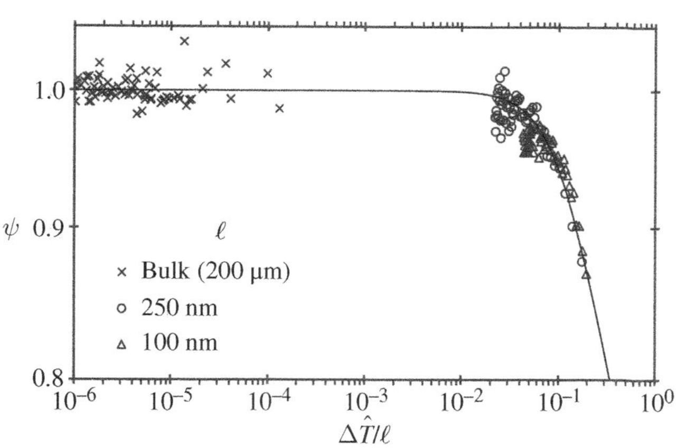

Figure 8.36 Universal crossover function for the heat capacity. The data correspond to those presented in Figure 8.34, while the solid curve is given by Eq. (8.138) where $\xi_0/a \approx 0.1$ nm.

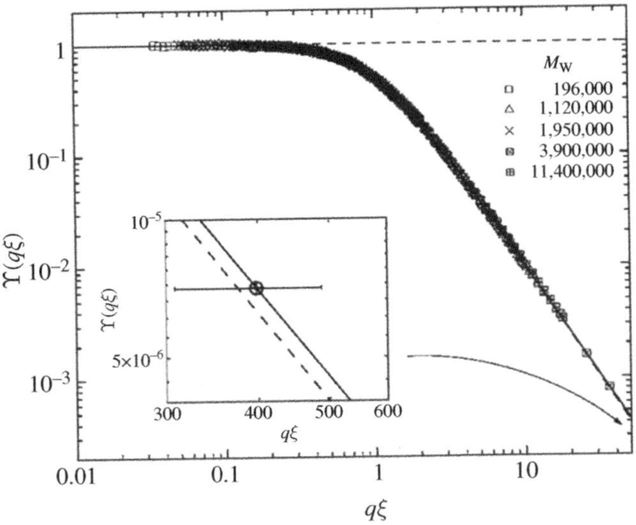

Figure 8.37 Universal finite-size scaling function for the susceptibility, $\Upsilon(q\xi)$, obtained for a system of polystyrene in cyclohexane with different molecular weights at the corresponding critical volume fractions (see Figure 8.29) [81]. The solid curve depicts Eq. (8.143), while the dashed curve corresponds to the Ornstein–Zernike approximation, given by Eq. (4.60). Source: Reproduced with permission from Ref. [81], © 2005, American Institute of Physics.

the correlation length, become explicitly dependent on the size (meaning that, in the limit $\xi/\ell \gg 1$, ℓ replaces ξ).

We have seen how the finite-size effect can be introduced physically through a small system size (like in porous media), but it can also be introduced instrumentally (like in a scattering experiment). Consider, for instance, a scattering experiment conducted with a specific probing wavelength given by Eq. (1.1). In this case, the size dependence is introduced through $\ell = 2\pi/q$ (hence, $a = 1/2\pi$), such that when this size becomes comparable to the correlation length, the finite size effect would be observed. In the meanfield approximation, as well as in the first-order epsilon expansion in RG theory, the scaling function for the q-dependent susceptibility is given through the Ornstein–Zernike formula, see Eq. (4.67), such that

$$\Upsilon(q\xi) \approx \left(1 + q^2\xi^2\right)^{-1} \tag{8.142}$$

More accurately, the finite-size scaling function, $\Upsilon(q\xi)$, is given by the Fisher–Burford approximation, see Eq. (8.36), as

$$\Upsilon(q\xi) = \left(1 + q^2\xi^2\right)^{\eta/2-1} \tag{8.143}$$

such that the finite-size susceptibility has the following limits, as given through Eq. (8.141),

$$\hat{\chi}(q\xi) = \begin{cases} \hat{\chi} \propto \xi^{\gamma/\nu} = \xi^{2-\eta} & q\xi \ll 1 \\ \hat{\chi} \propto q^{-\gamma/\nu} = q^{-2+\eta} & q\xi \gg 1 \end{cases} \tag{8.144}$$

The finite-size scaling of the susceptibility was studied by accurate light scattering experiments as shown in Figure 8.37. When $q\xi \gg 1$, the susceptibility reaches a finite value depending on the wavenumber [94, 99, 100].

One of the important consequences of critical fluctuations in confined systems, not considered in this book, is the emergence of the fluctuation-induced "Casimir forces" [99, 100]. In particular, the attractive Casimir forces are responsible for breaking stability of colloids near critical points of demixing [94].

8.10 Critical Dynamics

As discussed in Sections 8.3 and 8.9, the singularities of the thermodynamic properties in the critical region originate from the divergence of the correlation length of the order-parameter fluctuations. Similarly, the divergence of the correlation length causes anomalies in the dynamic properties in the critical region; thus, it affects the phenomenon of critical slowing down (see Section 8.1), which was described in the meanfield approximation in Section 4.11. In this section, we discuss the key consequences of the effects of critical fluctuations on the transport properties of fluids and fluid mixtures.

8.10.1 Critical Slowing Down and Classes of Universality in Dynamics

The phenomenological theory that describes the dynamics in the critical region, accounting for the critical slowing-down phenomenon, is known as "dynamic scaling" [101–103]. A theoretical method, based on the ideas of dynamic scaling, for calculating the fluctuation-induced enhancements of transport properties in fluids is known as, "mode-coupling theory" [103, 104].

As discussed in Section 4.11, for a system with a nonconserved order parameter (e.g. magnetization or average anisotropy), the critical relaxation time, inversely related to the relaxation rate as $\tau = 1/\Gamma_{\mathrm{R}}$, can be estimated as

$$\tau \sim \tilde{\eta}\tilde{\chi} \tag{8.145}$$

where $\tilde{\eta}$ is an appropriate friction coefficient (like the shear viscosity for liquid crystals), and $\tilde{\chi}$ is the appropriate generalized susceptibility. If the viscosity does not diverge, then with use of Eq. (8.35) for $\tilde{\chi}$, one finds $\tau \sim \xi^{z_\tau}$, where $z_\tau = 2 - \eta \approx 1.97$ is the dynamic critical exponent for the dynamic universality class of an Ising-like nonconserved order parameter in three dimensions. More information about different dynamic universality classes may be found in the review of Hohenberg and Halperin [101]. The key feature of the relaxation of the nonconserved order parameter is the absence of explicit spatial dependence. Contrarily, for systems with a conserved order parameter, the relaxation time explicitly depends on the system's length scale. For the relaxation of the order-parameter fluctuations, the natural length scale is the correlation length ($\xi \sim 1/q$), see Eq. (4.86). The critical relaxation time of the conserved order parameter, inversely related to the relaxation rate $\tau = 1/\Gamma$, is thus estimated as

$$\tau \sim \frac{\xi^2}{\tilde{D}} \tag{8.146}$$

where, in fluids, $\tilde{D}$ is a generalized diffusion coefficient, such as the coefficient of mutual diffusion, the coefficient of temperature diffusivity, or a combination of both.

Specifically, for a single-component fluid, as follows from the discussion in Section 3.11.2, critical slowing down is only associated with the relaxation of the isobaric fluctuations of entropy. The relaxation rate of the entropy fluctuations is determined by the thermal diffusivity, $D_T = \kappa/\rho C_P$, as given by Eq. (3.91), such that the relaxation time is given by

$$\tau \equiv \tau_T = \frac{\xi^2}{D_T} \tag{8.147}$$

Mode-coupling theory [103, 104], developed for describing the dynamics in the critical region of fluids and supported by accurate experimental studies, predicts an anomaly of the thermal conductivity at the liquid–gas critical point, which is induced by the diverging critical fluctuations of entropy. According to the theory, the diverging part of the thermal conductivity, $\Delta\kappa$, reads

$$\Delta\kappa = \frac{k_B T}{6\pi\tilde{\eta}\xi}\rho C_P \tag{8.148}$$

Correspondingly, the thermal diffusivity vanishes at the critical point, following the form of the Stokes–Einstein equation Eq. (2.184), as

$$D_T = \frac{k_B T}{6\pi\tilde{\eta}\xi} \tag{8.149}$$

in which the radius of a particle is replaced by the correlation length. Since, along the critical isochore: $C_p \propto |\Delta\hat{T}|^{-\gamma}$, $\xi \propto |\Delta\hat{T}|^{-\nu}$, and $\tilde{\eta} \propto |\Delta\hat{T}|^{-\psi}$ (see Section 8.2), the diverging part of the thermal conductivity scales as $\Delta\kappa \propto |\Delta\hat{T}|^{-\gamma+\nu+\psi}$. Hence, with $\gamma = 1.24$, $\nu = 0.63$, and $\psi = 0.04$, the thermal conductivity asymptotically diverges along the critical isochore as $\kappa \propto |\Delta\hat{T}|^{-0.57}$.

In accordance with Eqs. (8.147) and (8.149), at the scale of the correlation length, the relaxation time of the critical fluctuations diverges as

$$\tau \sim \tilde{\eta}\xi^3 \propto |\Delta\hat{T}|^{3\nu+\psi} \sim \xi^{z} \tag{8.150}$$

where z_τ is known as the "dynamic critical exponent." For a system, thermodynamically belonging to the Ising-model class of universality and, at the same time, possessing conserved order parameter dynamics, $z_\tau = 3 + \psi/\nu = 3.065$.

Experimental confirmation of the universal value of the dynamic critical exponent was performed near the liquid–liquid critical points of binary-liquid mixtures. The phenomenon of critical slowing down in binary mixtures could be described in the same terms as pure fluids as illustrated in Section 8.10.2.

8.10.2 Isomorphism of Transport Properties in Fluids and Fluid Mixtures

In binary fluid mixtures, the thermal conductivity is expressed through the kinetic coefficients: $\breve{\alpha}$, $\breve{\beta}$, and $\breve{\gamma}$ (not to be confused with the critical exponents α, β, and γ) by $\kappa = \breve{\gamma} - \hat{T}\breve{\beta}^2/\breve{\alpha}$, see Eq. (2.172). Asymptotically close to the critical point, all three kinetic coefficients diverge, as predicted by the mode-coupling theory. The diverging part of the three kinetic coefficients in the hydrodynamic limit ($q = 0$) are obtained by the mode-coupling theory as [67]

$$\Delta\breve{\alpha} = \frac{k_B T\rho}{6\pi\tilde{\eta}\xi}\left.\frac{\partial x}{\partial\mu_{21}}\right|_{T,P} \tag{8.151}$$

$$\Delta\breve{\beta} = \frac{k_B T\rho}{6\pi\tilde{\eta}\xi}\left.\frac{\partial x}{\partial T}\right|_{\mu_{21},P} \tag{8.152}$$

$$\Delta\breve{\gamma} = \frac{k_B T\rho}{6\pi\tilde{\eta}\xi}C_{P,\mu_{21}} \tag{8.153}$$

Keeping in mind that all of the thermodynamic response functions in Eqs. (8.151)–(8.153) diverge strongly as $|\Delta\hat{T}|^{-\gamma}$, see Table 8.6, then all of the diverging parts of the three kinetic coefficients scale as $\Delta\breve{\alpha} \sim \Delta\breve{\beta} \sim \Delta\breve{\gamma} \sim |\Delta\hat{T}|^{-\gamma+\nu+\psi}$, which are the same as the diverging part of the thermal conductivity for a single-component fluid, see Eq. (8.148). Consequently, in binary mixtures, the thermal conductivity, κ, remains finite ($\Delta\kappa = 0$), since all the singularities in the kinetic coefficients, $\Delta\breve{\alpha}$, $\Delta\breve{\beta}$, and $\Delta\breve{\gamma}$, cancel. Note that the Onsager coefficient $\breve{\gamma}$, is isomorphic to the thermal conductivity of a pure fluid, and could be viewed as the thermal conductivity at constant exchange chemical potential.

The dynamic critical behavior was most accurately observed through measurements of the mutual diffusion coefficient, D, by dynamic light scattering in solutions near liquid–liquid critical points. The mode-coupling theory of critical dynamics predicts that, asymptotically close to the critical point, the diffusion coefficient follows a modified Stokes–Einstein–Kawasaki relation of the form [36, 83, 105, 106]

$$D \approx R_{\mathrm{D}} \frac{k_{\mathrm{B}}T}{6\pi\tilde{\eta}\xi} K(q\xi) \tag{8.154}$$

where R_{D} is a universal dynamic amplitude of the order of unity [83, 106]. The wave-number dependent function, $K(q\xi)$, known as the "Kawasaki function," is given by

$$K(x) = \frac{3}{4x^2}(1 + x^2 + [x^3 - x^{-1}] \arctan x) \tag{8.155}$$

where $x \equiv q\xi$. Equation (8.154) satisfies the finite-size scaling limits: $K(q\xi \ll 1) = 1$ and $K(q\xi \gg 1) = q\xi$ [106]. The behavior of the Kawasaki function is illustrated in Figure (8.39a). It follows from Eq. (8.154), that the diffusion coefficient vanishes asymptotically as $D \sim \xi^{-(1+\psi/\nu)}$ at the critical point. Mode-coupling theory also predicts that the diverging mesoscopic (spatial-dependent) viscosity, $\tilde{\eta}$, has the form,

$$\tilde{\eta} \approx \eta_{\mathrm{b}} (Q_0 \xi)^{\psi/\nu} \left[1 + \left(\frac{q\xi}{2} \right)^2 \right]^{-\psi/(2\nu)} \tag{8.156}$$

in which η_{b} is the noncritical background viscosity [107] and Q_0 is an amplitude of the order $1/\xi_0$. The viscosity has two limits: for $(q\xi \ll 1)$ $\tilde{\eta}\,(q\xi \ll 1) \sim \xi^{\psi/\nu}$ (Figure 8.38), while for $(q\xi \gg 1)$ $\tilde{\eta}\,(q\xi \gg 1) \sim q^{-\psi/\nu}$.

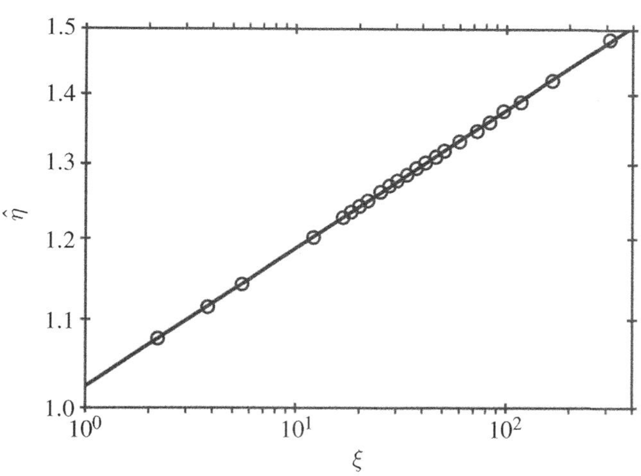

Figure 8.38 The reduced shear viscosity ($\hat{\eta} = \tilde{\eta}/\eta_{\mathrm{b}}$) in a mixture of 3-methylpentane and nitroethane in the vicinity of the liquid–liquid critical point as a function of the correlation length. The symbols represent experimental data obtained from light scattering, while the solid line is the theoretical prediction given by Eq. (8.156). Source: Reproduced with permission from Ref. [105], © 1983, American Physical Society.

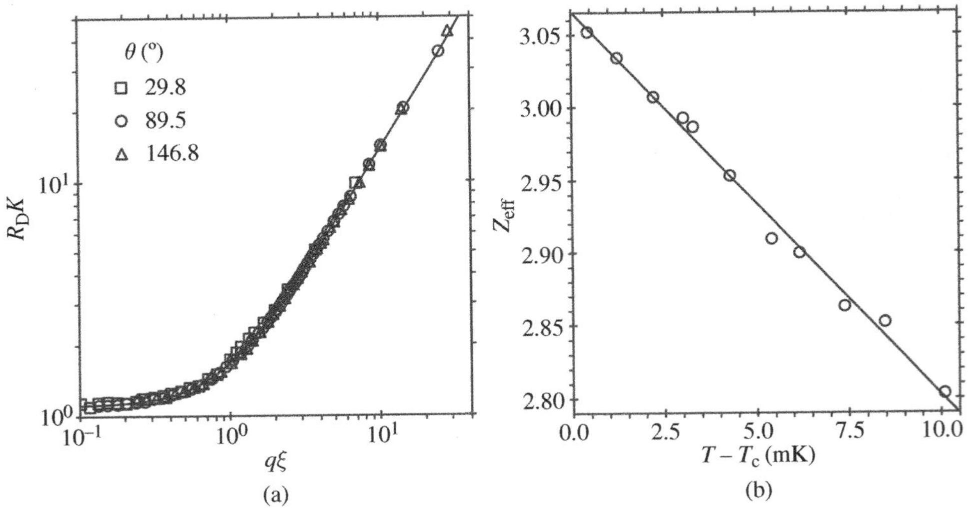

Figure 8.39 (a) The scaled Kawasaki function ($R_D K(q\xi) = 6\pi\tilde{\eta}\xi D/k_B T$) obtained from the asymptotic mutual diffusion coefficient (see Eq. 8.154) as a function of $q\xi$ for a near-critical mixture of 3-methylpentane and nitroethane. (b) The effective dynamic critical exponent, z_{eff}, for the same mixture, obtained from the representation of the Kawasaki function, when $q\xi \to \infty$, as $K(q\xi) \propto \tilde{\eta}\xi D \propto q^{z_{eff}-2}$, which asymptotically reaches the prediction of the mode-coupling theory, $z_\tau = 3.065$ (see Eq. 8.150). Source: Adapted from Ref. [105].

From Eqs. (8.154) and (8.156), the behavior of the diffusion coefficient is determined from the following two limits: In the hydrodynamic regime ($q\xi \ll 1$), the diffusion coefficient becomes independent of q varying with the distance to the critical temperature as $D \propto |\Delta\hat{T}|^{\nu+\psi} = |\Delta\hat{T}|^{z_\tau - 2\nu}$ [103, 108], while in the asymptotic critical limit ($q\xi \gg 1$), D becomes a function of the wavenumber,

$$D \propto q^{-(1+\psi/\nu)} \tag{8.157}$$

such that the relaxation time scales as $\tau = 1/Dq^2 \propto q^{3+\nu/\psi} = q^{z_\tau}$. Figure 8.39b depicts the approach to the asymptotic critical value of the dynamic critical exponent, which can only be observed when $q\xi \gg 1$, such that D depends only on q.

Dynamic light-scattering measurements of Kostko et al. [83] in polymer solutions near their liquid–liquid critical points of demixing, presented in Figure 8.40a, have also confirmed this behavior of the diffusion coefficient. However, the experimental data require an apparent viscosity, $\tilde{\eta}_{app}$, that is different than the viscosity of the solution, $\tilde{\eta}$ used in the application of mode-coupling theory for low-molecular-weight fluids. The apparent viscosity, shown in Figure 8.40b, is in between the viscosity of the solution and the viscosity of a pure solvent (cyclohexane). In polymer solutions, there is an effect of coupling between the viscoelastic and diffusive modes, which, for solutions with a modest molecular weight ($M_w \sim 10^5$), is effectively absorbed through the use of an apparent viscosity. This effect, which can be more pronounced for higher-molecular-weight polymers, will be discussed in Section 8.10.4.

8.10.3 Finite-Time (Frequency) Effects

In experimental practice, measurements are performed during a finite time and, often, in the presence of oscillations with a certain frequency. For example, the speed of sound, the sound attenuation, the thermal conductivity, and the heat capacity are routinely measured with oscillations.

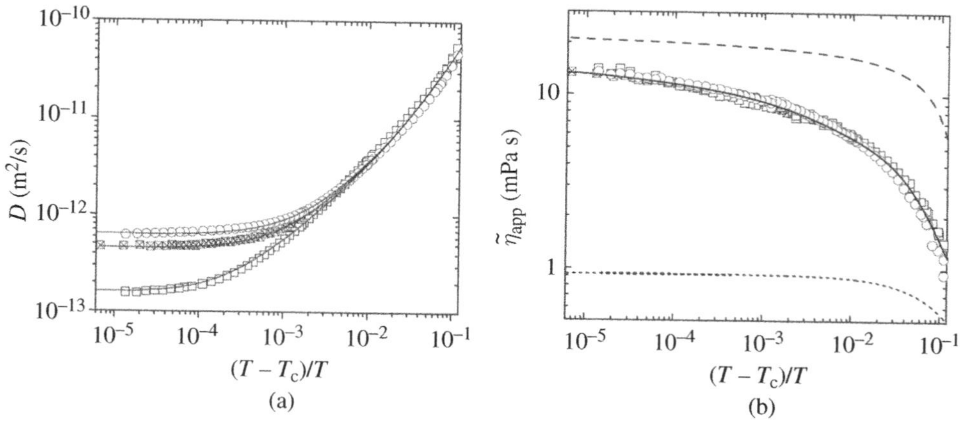

Figure 8.40 (a) Diffusion coefficient of a solution of polystyrene ($M_w = 195,900$ g/mol) in cyclohexane. The data correspond to three light-scattering angles: $\theta = 30°$ (open squares), $\theta = 90°$ (crossed squares), and $\theta = 150°$ (open circles). The curves represent the fit of the mode-coupling theory to the experimental data with use of an apparent viscosity [83]. (b) The apparent viscosity of the same system (symbols and solid curve). The dashed curve represents the viscosity of the solution and the dotted curve represents the viscosity of the solvent. Source: Reproduced with permission from Ref. [83], © 2007, American Physical Society.

Just like how finite-size scaling describes the spatial dependence of physical properties through a competition of the correlation length and the characteristic length scale of the system, finite-time scaling describes the frequency dependence of physical properties through a competition of the relaxation rate of critical fluctuations and the characteristic frequency assigned by the experiment. This effect is observed in the viscosity, presented in Figure 9, in which a larger viscosity was observed on Earth at a higher frequency [109]. The effect of finite time is most readily observed in measurements of the frequency-dependent sound velocity, $u_s(\omega)$, and sound attenuation, $\Delta\alpha_s(\omega)$.

As discussed in Section 3.11.4, the speed of sound has two limits depending on the frequency. In the thermodynamic limit ($\omega\tau \ll 1$), the sound velocity is inversely proportional to the square root of the adiabatic compressibility as $u_s = (M_w \rho \kappa_S)^{-1/2}$, while in the high-frequency limit ($\omega\tau \gg 1$), the sound velocity is usually higher than its thermodynamic value (by about 10–20%). Between these two limits, the speed of sound is qualitatively described by the interpolation given in Eq. (3.97). This effect is known as "sound velocity dispersion." As follows from the Landau–Placzek ratio, Eq. (3.86), the adiabatic compressibility is equal to

$$\kappa_S = \kappa_T \frac{C_V}{C_P} \tag{8.158}$$

In the vicinity of the liquid–gas critical point, since $C_P \sim \kappa_T \sim |\Delta\hat{T}|^{-\gamma}$, then the adiabatic compressibility asymptotically diverges as $\kappa_S \sim C_V \propto |\Delta\hat{T}|^{-\alpha}$. Consequently, the speed of sound in the thermodynamic limit ($\omega\tau \to 0$) vanishes as, $u_s \propto |\Delta\hat{T}|^{\alpha/2} \sim [\tau(\Delta\hat{T})]^{-\alpha\nu/2z_\tau}$ and depends only on τ as given by Eq. (8.150), while in the high-frequency limit ($\omega\tau \to \infty$), u_s remains a constant (see Figure 8.41).

Near critical points in a binary fluid, the adiabatic compressibility in the thermodynamic limit does not diverge; instead, it exhibits a cusp, since $C_{V,x}$ is finite, see Eq. (8.116). However, the frequency-dependent speed of sound, reduced by its value in the thermodynamic limit, exhibits the same crossover between the thermodynamic and high-frequency limits as for single-component

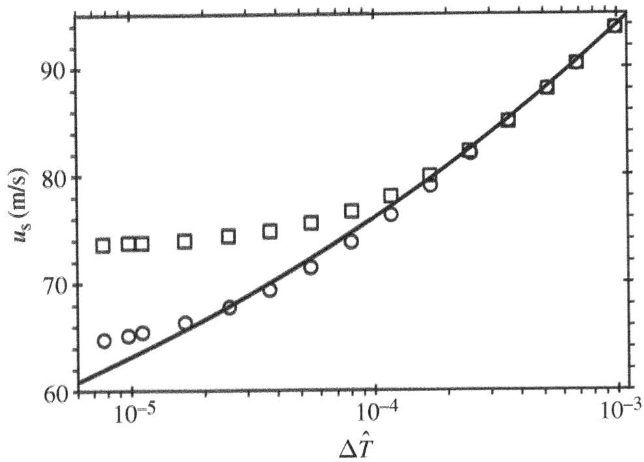

Figure 8.41 The speed of sound in xenon as a function of reduced distance to the critical temperature for two sound frequencies: 5 kHz (squares) and 0.12 kHz (circles). The solid line is the theoretical prediction for the temperature-dependence of the speed of sound in the zero-frequency limit. Source: Reproduced with permission from Ref. [111], © 2005, American Physical Society.

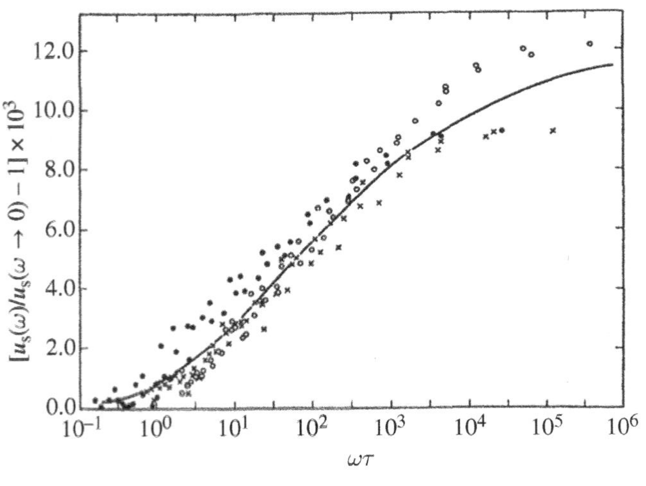

Figure 8.42 Dispersion of sound velocity as a function of relative frequency in a mixture of nitroethane in isooctane. The experimental data (symbols) correspond to frequencies ($\omega/2\pi$): 2 MHz (closed circles), 9.3 MHz (crosses), and 27.7 MHz (open circles), while the solid curve is the interpolation of the dispersion between the two limiting behaviors: $u_s(\omega)/u_s(\omega \to 0) - 1 \propto \omega\tau$ for low frequencies and $u_s(\omega)/u_s(\omega \to 0) - 1 \to$ constant for high frequencies ($\omega \to \infty$). Source: Adapted from Ref. [112]/AIP Publishing.

fluids. An example of the sound velocity dispersion near liquid–liquid critical point of a nitroethane–isooctane mixture as a function of the reduced frequency, $\omega\tau$, is shown in Figure 8.42.

Both the universal behavior of the sound dispersion and the sound attenuation can be understood through the relaxation of the complex bulk viscosity, in which the real part represents the sound attenuation and the imaginary part represents the dispersion of the sound velocity [110]. Correspondingly, as discussed in Section 3.11.4, the sound attenuation primarily depending on the bulk viscosity ζ, as $\Delta\alpha_s\lambda_s \propto\approx \tilde{\zeta}(\omega\tau)/M_w\rho u_s^2$, being a function of one variable ($\omega\tau$), which satisfies

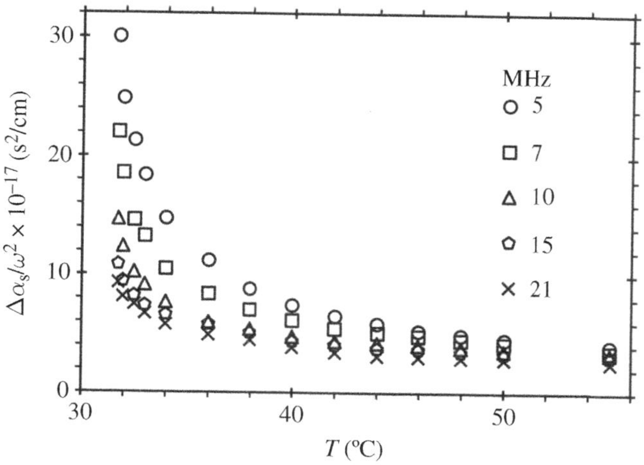

Figure 8.43 Sound attenuation in a near-critical mixture of nitroethane in isooctane as a function of temperature. Source: Adapted from Ref. [113], AIP Publishing/CC BY 4.0.

the following limits

$$\frac{\tilde{\zeta}(\omega\tau)}{M_w \rho u_s^2} \propto \begin{cases} \tau \propto \xi^{z_\tau} & \omega\tau \to 0 \\ 1/\omega & \omega\tau \to \infty \end{cases} \tag{8.159}$$

such that the dimensionless sound attenuation coefficient, associated with the bulk viscosity, has the following limiting behavior,

$$\Delta\alpha_s\lambda_s \propto \begin{cases} \omega\tau & \omega\tau \ll 1 \\ \text{constant} & \omega\tau \gg 1 \end{cases} \tag{8.160}$$

Near a critical point, the bulk viscosity in the zero-frequency limit diverges as $\zeta \propto \xi^{z_\tau - \alpha\nu}$, since $u_s^2 \propto \xi^{-\alpha\nu}$. Consequently, in the same limit, the reduced sound attenuation, $\Delta\alpha_s\lambda_s/\omega \propto \Delta\alpha_s/\omega^2$, diverges as the relaxation time, $\tau \propto \xi^{z_\tau}$, see Eq. (8.150). Since $z_\tau \approx 3$, the near-critical sound attenuation strongly depends on frequency (Figure 8.43). The scaling behavior of sound attenuation is observed for both single-component fluids near the liquid–gas critical point and binary fluids near the liquid–liquid critical point, as demonstrated in Figure 8.44. Note that if the sound attenuation is reduced by its critical value depending on the frequency, then it becomes a universal function of $\omega\tau$, as the corresponding system-dependent amplitude is canceled.

8.10.4 Coupling of Critical Dynamic Modes: Avoided Crossing

In binary mixtures, the order parameter is generally a combination of two conserved quantities: the concentration and the density. Consequently, the rate of relaxation of critical fluctuations of density is determined by the thermal diffusivity, while the rate of relaxation of critical fluctuations of concentration is determined by the mutual diffusion. Therefore, the total effect of critical slowing down will depend on both these two dynamic modes and the coupling between them.

However, if the binary mixture is a solution of polymers in a low-molecular-weight solvent, then there are two dynamic modes: one associated with the conserved order parameter, volume fraction of monomers, and another associated with the polymer-chain dynamics, a nonconserved order parameter. Thus, the critical dynamics between the two physically distinct systems will be affected by both of these modes and the coupling between them.

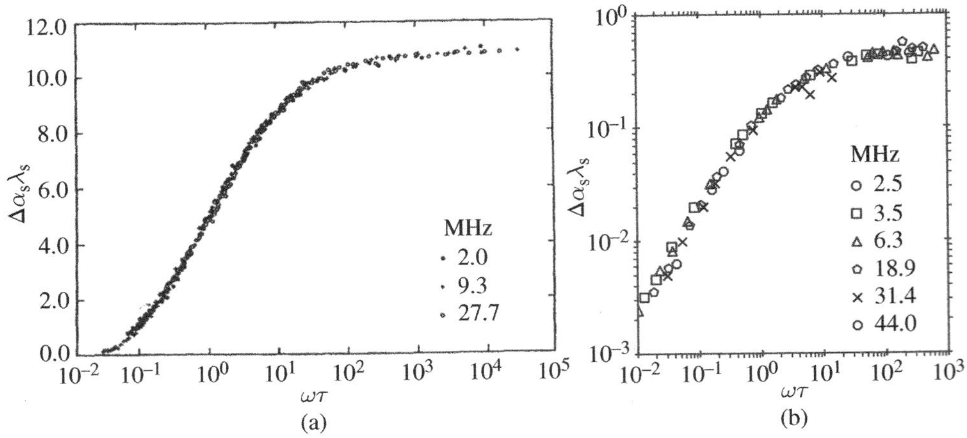

Figure 8.44 (a) Sound absorption in a near-critical mixture of nitroethane and isooctane. Source: Adapted from Ref. [112], AIP Publishing. (b) Sound adsorption in near-critical xenon. Source: Adapted from Ref. [111]. Note the difference between the representation of data in (a) semi-logarithmic scale and (b) double-logarithmic scale.

In this section, we demonstrate how the coupling between two dynamic modes results in the phenomenon of avoided crossing. Meaning that the coupled dynamic modes do not cross each other, and asymptotically close to the critical point, only one of these two modes will dominate in the effect of critical slowing down. The effect of avoided crossing in near-critical binary mixtures could be considered as a classical analog of the Landau–Zenner effect in the transition dynamics of a two-state quantum system [113, 114].

8.10.4.1 Coupling of Diffusive Modes in Binary Fluids

In binary mixtures, the relaxation of the critical fluctuations of density and concentration may be experimentally observed in dynamic light-scattering experiments through the two-exponential decay of the time-dependent correlation function, similar to Eq. (3.94). The rates of the two-exponential decay correspond to the diffusive modes of the thermal diffusivity and mutual diffusion, as defined by Eqs. (3.91) and (3.92), respectively, as well as the coupling between them. The two coupled diffusivities, D_- and D_+, which determine the decay rates of these two modes were derived by Mountain and Deutch [115], as

$$D_{\mp} = \frac{1}{2}\left(D_T + \mathscr{D}\right) \mp \frac{1}{2}\left[\left(D_T + \mathscr{D}\right) - 4D_T D\right] \tag{8.161}$$

where D_T is the thermal diffusivity, given by Eq. (2.180), while $\mathscr{D}$, is an effective mutual diffusion coefficient given in the thermodynamic limit (zero wavenumber) by

$$\mathscr{D} = D\left[1 + \frac{k_T}{TC_{P,x}} \left.\frac{\partial \mu_{21}}{\partial x}\right|_{T,P}\right] \tag{8.162}$$

where k_T is the thermal-diffusion ratio, given by Eq. (2.177). Note that the coupling between D_1 and D_2 vanishes in the infinite-dilute solution limit.

At the critical point of a binary-fluid mixture, the mutual diffusion coefficient, D, vanishes and asymptotically close to the critical point the mutual-diffusion mode is responsible for the critical slowing down effect. The thermal diffusivity, D_T, does not exhibit a significant change in the immediate vicinity of the critical point of a mixture as κ remains finite and $C_{P,x}$ only weakly diverges. Therefore, asymptotically close to the critical point the coupled mode with diffusivity D_- represents

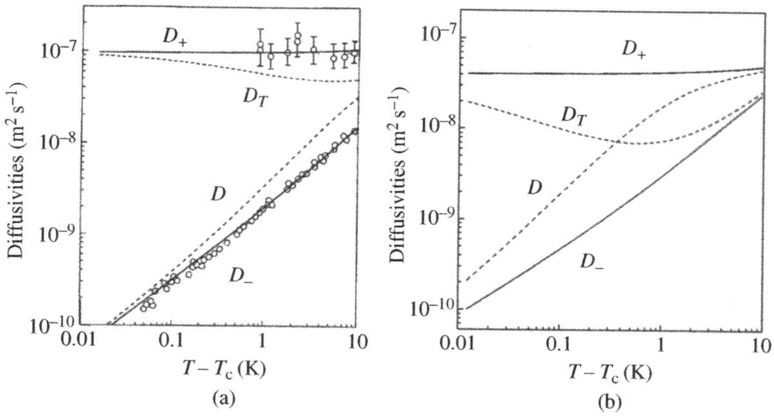

Figure 8.45 Two coupled diffusivities D_- and D_+ for mixtures of methane and ethane near their liquid–gas critical points along the critical isochore at concentrations: (a) 28.93 mol% ethane and (b) 1 mol% ethane. The open circles represent experimental data obtained by dynamic light scattering extrapolated to zero wave number [116]. The solid curves represent the calculated values of D_- and D_+, and the dashed curves represent the predictions for the thermal diffusivity, D_T, and the mutual diffusion coefficient, D. Source: Reproduced with permission from Ref. [117], © 1998, American Physical Society.

a slow (critical) diffusion mode, as D_- tends to the diffusion coefficient D, while the mode with the diffusivity D_+ represents a relatively fast diffusion mode, as D_+ tends to the thermal diffusivity, D_T.

Consequently, experiments performed on binary fluids have commonly interpreted the diffusivity, associated with critical slowing down in mixtures, as mutual diffusion, whereas in pure fluids it was identified with the thermal diffusivity. Indeed, the experimental observation of two diffusive modes in a concentrated mixture of methane and ethane, as shown in Figure 8.45a, confirms this prediction asymptotically close to the critical point. However, in a dilute solution, the theory predicts the phenomenon of avoided crossing of the coupled diffusivities, as the slow and fast diffusive modes exchange their physical meaning away from the critical point, see Figure 8.45b. Thus, further away from the critical point, the thermal diffusivity mainly contributes to the slow-diffusive model, D_-, while the mutual diffusion becomes mainly associated with the fast-diffusive model, D_+. The location where the uncoupled thermal diffusivity and the mutual diffusion cross approximately corresponds to the region where the critical behavior changes from pure-fluid-like to mixture-like (see Section 8.5.3). We emphasize that in experiment only the coupled slow and fast diffusive modes, D_- and D_+, may be observed, while the behavior of the uncoupled thermal diffusivity and mutual diffusion may only be predicted.

8.10.4.2 Dynamic Coupling in Polymer Solutions

A similar phenomenon is observed in polymer solutions in which there is a dynamic coupling between the diffusive relaxation of the critical concentration fluctuations and the relaxation of entanglements/disentanglements of long polymer chains (commonly referred to in the literature as a "viscoelastic relaxation" [117]). Experimental verification of the coupling between these two dynamic modes was demonstrated independently by Tanaka et al. [117] and Kostko et al. [81, 118]. According to the theory of de Gennes and Brochard [119, 120], the normalized time-dependent intensity correlation function, $g_2(t)$, is given in the form

$$g_2(t) - 1 = \left(f_+ e^{-\Gamma_+ t} + f_- e^{-\Gamma_- t}\right)^2 \tag{8.163}$$

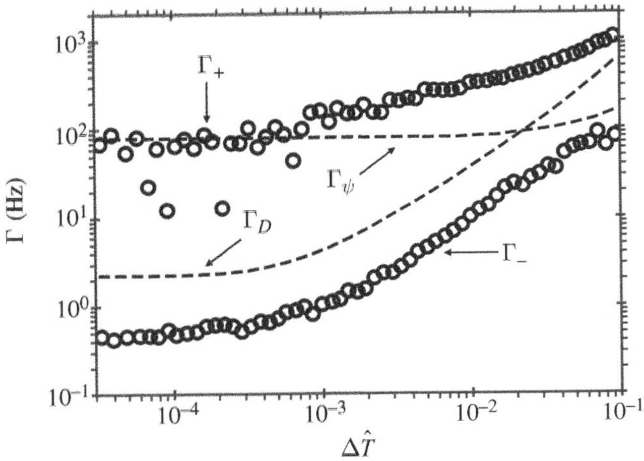

Figure 8.46 Decay rates, $\Gamma_{\pm}$, for a solution of polystyrene in cyclohexane, with $\mathcal{N} = 110,000$, measured at a scattering angle of $\theta = 30^{\circ}$. Source: Adapted from Ref. [83]. The symbols represent the values for the two coupled modes obtained from the fit of the correlation-function data, Eq. (8.163). The points correspond to the faster mode, Γ_{+} (upper), and the slower mode, Γ_{-} (lower). The dashed curves represent the theoretical calculations of the uncoupled viscoelastic mode, Γ_{ψ}, and the uncoupled diffusive mode, Γ_{q}.

with decay frequencies given by

$$\Gamma_{\pm} = \frac{\Gamma_{\psi}}{2}\left[1 + q^2\xi_{\psi}^2 + \frac{\Gamma_q}{\Gamma_{\psi}} \pm \sqrt{\left(1 + q^2\xi_{\psi}^2 + \frac{\Gamma_q}{\Gamma_{\psi}}\right)^2 - 4\frac{\Gamma_q}{\Gamma_{\psi}}}\right] \tag{8.164}$$

and corresponding amplitudes f_{+} and f_{-} given by

$$f_{\pm} = \pm\frac{\Gamma_{\pm} - \Gamma_{\psi}(1 + q^2\xi_{\psi}^2)}{\Gamma_{+} - \Gamma_{-}} \tag{8.165}$$

where $\Gamma_{\pm}$ are the coupled modes, in which the relaxation of the faster mode is designated as, Γ_{+}, and consequently, the slower mode is designated as Γ_{-}.

In the literature, ξ_{ψ} was defined as a mesoscopic viscoelastic length scale, which may be identified with the de Gennes correlation length of the polymer chain, given by Eq. (5.49). In Eqs. (8.164) and (8.165), Γ_q represents the uncoupled q-dependent diffusive relaxation mode, as specified through the diffusion equation, given by Eq. (8.154), and Γ_{ψ} is an uncoupled q-independent viscoelastic relaxation mode associated with the entanglements and disentanglements of the polymer chain. Γ_{ψ} is assumed to be inversely proportional to the viscosity of the solution as $\Gamma_{\psi} \propto \tilde{\eta}(T)^{-1}$ [83], which diverges in the limit $\mathcal{N} \to \infty$. The diffusive relaxation mode, Γ_q is associated with the conserved order parameter, $\phi = \varphi - \varphi_c$, while the viscoelastic mode is associated with the nonconserved polymer-chain order parameter, $|\psi|$ [83, 117, 118]. The dynamic coupling between the two dynamic modes reflects the thermodynamic coupling between these two order parameters, which is responsible for the emergence of Theta-point tricriticality, see Section 5.6.

The effect of coupling between the two dynamic modes is most clearly observed in polymer solutions with a high degree of polymerization (closer vicinity to the Theta point). For instance, Figure 8.46 depicts the dynamic coupling in a solution of polystyrene in cyclohexane with molecular weight of 11.4 million g/mol ($\mathcal{N} = 110,000$) [83]. As illustrated in Figure 8.46, in the vicinity of the critical point, both the uncoupled viscoelastic and diffusive modes saturate. The diffusive relaxation time saturates at a finite wave number, controlled by the scattering angle, while the viscoelastic relaxation time saturates because the degree of polymerization is finite and ξ_ψ reaches a finite value. In this region, the characteristic frequency of the faster-coupled mode scales with the correlation length of the polymer chain as $\Gamma_+ = \Gamma_\psi(1 + q^2\xi_\psi^2)$, while the frequency of the slower coupled mode scales as $\Gamma_- = \Gamma_q(1 + q^2\xi_\psi^2)^{-1}$. In the region far away from the critical temperature, the two modes uncouple as $\Gamma_- \approx \Gamma_\psi$ and $\Gamma_+ \approx \Gamma_q$ [83, 117, 118].

We emphasize that such dynamic light-scattering experiments enable one to obtain the de Gennes correlation length, ξ_ψ, given by Eq. (5.49), which is otherwise not a directly measurable property. In the limit of an infinite degree of polymerization, this correlation length diverges at the Theta point as illustrated in Figure 8.30b and, in this limit, Γ_- vanishes, due to critical slowing down near the tricritical point.

Appendix 8.A Crossover Between Critical Singularities in Pure Fluids and Binary Mixtures

The change from the critical behavior in pure fluids to the behavior in binary fluids is governed by the transformation of the coefficients in the scaling fields. This concept can be illustrated for the liquid–gas critical locus of a binary mixture of symmetric species, which corresponds to the compressible binary lattice model (shown in Figure 2.24). The scaling fields for this case are simplified as the pressure mixing is omitted ($b_2 = 0$, $b_3 = 0$, and $a_3 = 0$), and phase asymmetry disappears in both component limits, such that $b_4 \propto x_c(1 - x_c)$. Along the critical locus, when the critical composition, x_c, changes from 0 to 1, the scaling fields are given by

$$\begin{pmatrix} h_1 \\ h_2 \\ -h_3 \end{pmatrix} = \begin{pmatrix} a_1 & 0 & 0 & a_4 \\ 0 & b_1 & 0 & b_4 \\ 0 & 0 & 1 & 0 \end{pmatrix} \begin{pmatrix} \Delta\hat{\mu}_1 \\ \Delta\hat{T} \\ \Delta\hat{\Pi}_{cr} \\ \Delta\hat{\mu}_{21} \end{pmatrix} \tag{8.A.1}$$

where $h_1 = h_2 = 0$ along the critical locus. Thus,

$$\left.\frac{dh_1}{d\hat{\mu}_{21}}\right|_{CL} = a_1 \left.\frac{d\hat{\mu}_1}{d\hat{\mu}_{21}}\right|_{CL} + a_4 = 0 \quad \text{and} \quad \left.\frac{dh_2}{d\hat{\mu}_{21}}\right|_{CL} = b_1 \left.\frac{d\hat{T}}{d\hat{\mu}_{21}}\right|_{CL} + b_4 = 0 \tag{8.A.2}$$

where the subscript "CL" indicates that the derivatives are to be evaluated along the critical locus and the temperature is reduced by the critical temperature of the solvent, $\hat{T} = T/T_{c1}$. Representing $d\hat{\mu}_1$, with use of the Gibbs–Duhem equation Eq. (2.21), along the critical line as

$$d\hat{\mu}_{1,c} = -\hat{S}_c d\hat{T}_c + \frac{1}{\rho_c}\frac{d\hat{P}_c}{d\hat{\mu}_{21,c}} d\hat{\mu}_{21,c} - x_c d\mu_{21,c} \tag{8.A.3}$$

where the derivative, $d\hat{P}_c/d\hat{\mu}_{21,c}$, may be expressed as

$$\frac{d\hat{P}_c}{d\hat{\mu}_{21,c}} = \frac{d\hat{P}_c}{d\hat{T}_c}\frac{d\hat{T}_c}{d\hat{\mu}_{21,c}} = \frac{d\hat{P}_c}{d\hat{T}_c}\frac{d\hat{T}_c}{dx_c}\frac{dx_c}{d\hat{\mu}_{21,c}} \tag{8.A.4}$$

Then, accounting for the critical-line condition: $\hat{\mu}_{21,c} = \ln[x_c/(1 - x_c)]$ and $dx_c/d\hat{\mu}_{21,c} = x_c(1 - x_c)$, see Section 8.5.3, one obtains, by adopting the critical value of the entropy as $\hat{S}_c = (1/\rho_c)$ $(\partial\hat{P}/\partial\hat{T})|_{h_1=0,\hat{\mu}_{21}}$ (see Section 8.4.1),

$$a_1 \left.\frac{\partial\hat{\mu}_1}{\partial\hat{\mu}_{21}}\right|_{CL} + a_4 = a_1 x_c(1 - x_c)\mathcal{K} - a_1 x_c + a_4 = 0 \tag{8.A.5}$$

where $\mathcal{K}$ is the generalized Krichevskiĭ parameter, given by

$$\mathcal{K} = \frac{1}{\rho_c}\frac{d\hat{T}_c}{dx_c}\left[\frac{d\hat{P}_c}{d\hat{T}_c} - \left.\frac{\partial \hat{P}}{\partial \hat{T}}\right|_{h_1=0,\hat{\mu}_{21}}\right] \tag{8.A.6}$$

Recall that for Type I behavior $b_1 = 1$, then the scaling coefficients a_4 and b_4 can be expressed through two characteristic parameters as

$$a_4 = a_1\left[x_c - x_c(1-x_c)\mathcal{K}\right] = a_1(x_c - \mathcal{K}_1) \tag{8.A.7}$$

$$b_4 = -\frac{d\hat{T}_c}{d\mu_{21,c}} = -x_c(1-x_c)\frac{d\hat{T}_c}{dx_c} = -\mathcal{K}_2 \tag{8.A.8}$$

where $\mathcal{K}_1 \equiv x_c(1-x_c)\mathcal{K}$ and $\mathcal{K}_2 \equiv x_c(1-x_c)(d\hat{T}_c/dx_c)$. In pure-fluid limits, $x_c = 0$ and $1 - x_c = 0$, the solvent (species 1) and the solute (species 2) exchange their roles in the scaling fields. This effect is reflected by changing the coefficients a_1 and a_4 upon moving from CP1 to CP2 along the critical locus. The scaling fields can be also represented in the form

$$h_1 = a_4\Delta\hat{\mu}_2 + a_1\Delta\hat{\mu}_1 - a_4\Delta\hat{\mu}_1 = a_4\Delta\hat{\mu}_2 + (a_1 - a_4)\Delta\hat{\mu}_{12} \tag{8.A.9}$$

Thus, in the limit of pure solvent $a_1 \to 1$ and $a_4 \to 0$, while in the limit of pure solute $a_1 \to 0$ and $a_4 - a_1 \to 1$. Note that the difference $a_4 - a_1x_c = -a_1\mathcal{K}_1 = a_1x_c(1-x_c)\mathcal{K}$ as well as $b_4 = -\mathcal{K}_2 = -x_c(1-x_c)d\hat{T}_c/dx_c$ are symmetric at the two pure-fluid limits.

The relative contribution of the strong susceptibility, $\hat{\chi}_1$, and the weak susceptibility, $\hat{\chi}_2 \ll \hat{\chi}_1$, into the thermodynamic response functions depends on the change of the scaling-field coefficients along the critical locus. In zero ordering field in the one-phase region, the thermodynamic calculus of the relations between the scaling and physical fields (in the case $\rho \approx \rho_c$ and $a_1 = 1$) yields,

$$\kappa_\Pi \propto \left.\frac{\partial x}{\partial \hat{\mu}_{21}}\right|_{\hat{P},\hat{T}} \approx x_c(1-x_c) + \mathcal{K}_1^2\hat{\chi}_1 + \mathcal{K}_2^2\hat{\chi}_2 \tag{8.A.10}$$

$$\kappa_T \propto \left.\frac{\partial \hat{\rho}}{\partial \hat{P}}\right|_{\hat{T},x} \approx \hat{\chi}_1\left[\frac{x_c(1-x_c) + \mathcal{K}_2^2\hat{\chi}_2}{x_c(1-x_c) + \mathcal{K}_1^2\hat{\chi}_1 + \mathcal{K}_2^2\hat{\chi}_2}\right] \tag{8.A.11}$$

see details in Refs. [67, 68]. These functions have four key limits:

- Far away from the critical point or in an extremely dilute solution, both $\mathcal{K}_1^2\hat{\chi}_1$ and $\mathcal{K}_2^2\hat{\chi}_2 \ll x_c(1-x_c)$, then

$$\left.\frac{\partial x}{\partial \hat{\mu}_{21}}\right|_{\hat{P},\hat{T}} \approx x_c(1-x_c) \quad \text{and} \quad \left.\frac{\partial \hat{\rho}}{\partial \hat{P}}\right|_{\hat{T},x} \approx \hat{\chi}_1 \tag{8.A.12}$$

- Closer to the critical point, when $\mathcal{K}_1^2\hat{\chi}_1 \gg x_c(1-x_c)$ but still $\mathcal{K}_2^2\hat{\chi}_2 \ll x_c(1-x_c)$, then

$$\left.\frac{\partial x}{\partial \hat{\mu}_{21}}\right|_{\hat{P},\hat{T}} \approx \mathcal{K}_1^2\chi_1 = x_c^2(1-x_c)^2\mathcal{K}^2\chi_1 \tag{8.A.13}$$

$$\left.\frac{\partial \hat{\rho}}{\partial \hat{P}}\right|_{\hat{T},x} \approx \frac{1}{\mathcal{K}_1^2} = \frac{1}{x_c(1-x_c)\mathcal{K}} \tag{8.A.14}$$

  where the isothermal compressibility does not diverge, being only a function of composition.
- Asymptotically close to the critical point, where $\mathcal{K}_1^2\hat{\chi}_1 \gg \mathcal{K}_2^2\hat{\chi}_2 \gg x_c(1-x_c)$, then

$$\left.\frac{\partial x}{\partial \hat{\mu}_{21}}\right|_{\hat{T},\hat{P}} \approx \mathcal{K}_1^2\hat{\chi}_1 = x_c^2(1-x_c)^2\mathcal{K}^2\hat{\chi}_1 \tag{8.A.15}$$

$$\left.\frac{\partial \hat{\rho}}{\partial \hat{P}}\right|_{x,\hat{T}} \approx \frac{\mathcal{K}_2^2}{\mathcal{K}_1^2}\hat{\chi}_2 = \frac{1}{\mathcal{K}^2}\left(\frac{d\hat{T}_c}{dx_c}\right)^2\hat{\chi}_2 \tag{8.A.16}$$

- At the critical azeotrope, where $\mathcal{K} = 0$ (thus, $\mathcal{K}_1 = 0$) and $\mathcal{K}_2^2 \hat{\chi}_2 \gg x_c(1 - x_c)$, then

$$\left.\frac{\partial x}{\partial \hat{\mu}_{21}}\right|_{\hat{T},\hat{P}} = \left.\frac{\partial x}{\partial \hat{\mu}_{21}}\right|_{\hat{T},\hat{\rho}} \approx \mathcal{K}_2^2 \hat{\chi}_2 = x_c^2(1 - x_c)^2 \left(\frac{d\hat{T}_c}{dx_c} \right)^2 \hat{\chi}_2 \tag{8.A.17}$$

$$\left.\frac{\partial \hat{\rho}}{\partial \hat{P}}\right|_{x,\hat{T}} \approx \hat{\chi}_1 \tag{8.A.18}$$

Note that at the critical azeotrope, the osmotic compressibility weakly diverges as $\kappa_\Pi \sim \hat{\chi}_2$, while the isothermal compressibility strongly diverges as $\kappa_T \sim \hat{\chi}_1$, like in a single-component substance.

Based on the key limits of Eqs. (8.A.10) and (8.A.11), one may consider the conditions that

$$\frac{\mathcal{K}_1^2 \hat{\chi}_1}{x_c(1 - x_c)} = 1 \quad \text{and} \quad \frac{\mathcal{K}_2^2 \hat{\chi}_2}{x_c(1 - x_c)} = 1 \tag{8.A.19}$$

Expressing the strong and weak susceptibilities through the distance to the critical temperature, as follows from Table 8.2, that $\hat{\chi}_1 = \Gamma_0^\pm |\Delta\hat{T}|^{-\gamma}$ and $\hat{\chi}_2 = A_0^\pm |\Delta\hat{T}|^{-\alpha}$, then one could introduce two characteristic temperatures along the path of zero field ($h_1 = 0$) as

$$\hat{\mathcal{T}}_1 = \left[\frac{\Gamma_0^\pm \mathcal{K}_1^2}{x_c(1 - x_c)} \right]^{1/\gamma} = \left[\Gamma_0^\pm x_c(1 - x_c) \mathcal{K}^2 \right]^{1/\gamma} \tag{8.A.20}$$

$$\hat{\mathcal{T}}_2 = \left[\frac{A_0^\pm \mathcal{K}_2^2}{x_c(1 - x_c)} \right]^{1/\alpha} = \left[A_0^\pm x_c(1 - x_c) \left(\frac{d\hat{T}_c}{dx_c} \right)^2 \right]^{1/\alpha} \tag{8.A.21}$$

which determine the transformation of the observable singularities in binary mixtures. The condition, $\Delta\hat{T} \ll \hat{\mathcal{T}}_1$ determines the temperature range of the distortion of the strong singularities of the isothermal compressibility and the isobaric heat capacity as well as the emergence of the divergence of the osmotic susceptibility, while the condition $\Delta\hat{T} \ll \hat{\mathcal{T}}_2 \ll \hat{\mathcal{T}}_1$ determines the range of the emergence of the weak singularities in the response functions, such as the isothermal compressibility and isobaric heat capacity (measured at constant composition). The parameter $\hat{\mathcal{T}}_2$ is also responsible for the renormalization of critical exponents as discussed in Section 8.5.4.

References

1 A. K. Wyczalkowska, J. V. Sengers, and M. A. Anisimov, "Critical fluctuations and the equation of state of van der Waals," *Physica A*, vol. 334, no. 3, pp. 482–512, 2004.

2 M. E. Fisher, "Renormalization group theory: Its basis and formulation in statistical physics," *Rev. Mod. Phys.*, vol. 70, pp. 653–681, 1998.

3 M. A. Anisimov, *Critical Phenomena in Liquids and Liquid Crystals*. Philadelphia, PA: Gordon and Breach, 1991.

4 M. A. Anisimov and J. V. Sengers, "Critical region," in *Equations of State for Fluids and Fluid Mixtures* (J. V. Sengers, R. Kayser, C. Peters, et al., eds.), vol. 5 of Experimental Thermodynamics, pp. 381–434. Amsterdam: Elsevier, 2000.

5 M. R. Moldover, J. V. Sengers, R. W. Gammon, and R. J. Hocken, "Gravity effects in fluids near the gas-liquid critical point," *Rev. Mod. Phys.*, vol. 51, pp. 79–99, 1979.

6 A. Haupt and J. Straub, "Evaluation of the isochoric heat capacity measurements at the critical isochore of SF_6 performed during the German spacelab mission D-2," *Phys. Rev. E*, vol. 59, pp. 1795–1802, 1999.
7 J. A. Lipa, C. Edwards, and M. J. Buckingham, "Precision measurement of the specific heat of CO_2 near the critical point," *Phys. Rev. Lett.*, vol. 25, pp. 1086–1090, 1970.
8 M. A. Anisimov, A. T. Berestov, L. S. Veksler, et al., "Scaling theory and the equation of state of argon in a wide region around the critical point," *Sov. Phys. JETP*, vol. 39, pp. 359–365, 1974.
9 L. D. Landau and E. M. Lifshitz, *Statistical Physics*. New York: Pergamon, 1958.
10 N. A. Shumovskyi, T. J. Longo, S. V. Buldyrev, et al., "Phase amplification in spinodal decomposition of immiscible fluids with interconversion of species," *Phys. Rev. E*, vol. 103, p. L060101, 2021.
11 A. V. Voronel, V. G. Gorbunova, V. A. Smirnov, et al., "Thermodynamic quantities for pure liquids and the applicability of the asymptotic laws near the critical point," *Sov. Phys. JETP*, vol. 36, p. 305, 1973.
12 M. A. Anisimov, "Letter to the editor: Fifty years of breakthrough discoveries in fluid criticality," *Int. J. Thermophys.*, vol. 32, pp. 2001–2009, 2011.
13 M. I. Bagatskiĭ, A. V. Voronel, and V. Gusak, "Measurement of the specific heat c_v of argon in the immediate vicinity of the critical point," *Sov. Phys. JETP*, vol. 16, p. 517, 1963.
14 A. V. Voronel, "Thermal measurements and critical phenomena in liquids," in *Phase Transitions and Critical Phenomena* (C. Domb and M. S. Green, eds.), vol. 5b. New York City: Academic Press, 1976.
15 M. E. Fisher, "Scaling, universality and renormalization group theory," in *Critical Phenomena* (F. J. W. Hahne, ed.), Lecture Notes in Physics, pp. 1–139. Berlin, Heidelberg: Springer-Verlag, 1983.
16 L. D. Landau and E. M. Lifshitz, *Statisticheskaya Fisika (Statistical Physics)*, vol. 5 of Course of Theoretical Physics. Moscow: Nauka, 1964. (in Russian).
17 M. E. Fisher, *The Nature of Critical Points*. Boulder, CO: University of Colorado Press, 1965.
18 L. P. Kadanoff, W. Götze, D. Hamblen, et al., "Static phenomena near critical points: Theory and experiment," *Rev. Mod. Phys.*, vol. 39, pp. 395–431, 1967.
19 M. S. Green and J. V. Sengers, eds., *Critical Phenomena: Proceedings of a Conference Held in Washington, DC, April 1965*. National Bureau of Standards Miscellaneous Publication, 273, 1966.
20 M. S. Green and J. V. Sengers, eds., *The Specific Heat of* He^3 *and* He^4 *in the Neighborhood of Their Critical Points*. National Bureau of Standards Miscellaneous Publication, 273, 1966.
21 J. A. Lipa, J. A. Nissen, D. A. Stricker, et al., "Specific heat of liquid helium in zero gravity very near the lambda point," *Phys. Rev. B*, vol. 68, p. 174518, 2003.
22 D. S. Cannell and J. H. Lunacek, "The long range correlation length and isothermal compressibility of carbon dioxide near the critical point," *J. Phys.*, vol. 33 (C1), pp. C1–91, 1972.
23 I. W. Smith, M. Giglio, and G. B. Benedek, "Correlation range and compressibility of xenon near the critical point," *Phys. Rev. Lett.*, vol. 27, pp. 1556–1560, 1971.
24 J. Zollweg, G. Hawkins, J. W. Smith, et al., "The spectrum and instensity of light scattered from the bulk phases and from the liquid-vapor interface of xenon near its critical point," *J. Phys.*, vol. 33 (C1), pp. C1–135, 1972.
25 J. V. Sengers and A. Michels, *Progress in International Research on Thermodynamic Properties*. New Jersey: ASME, Academic Press, 1962.
26 R. S. Basu and J. V. Sengers, *Thermal Conductivity*. New York: Plenum Press, 1983.
27 J. V. Sengers, "Behavior of Viscosity and Thermal Conductivity of Fluids Near the Critical Point", pages 165–178 in M. S. Green and J. V. Sengers, eds., Critical Phenomena: Proceedings

of a Conference Held in Washington, DC, April 1965. *National Bureau of Standards Miscellaneous Publication*, 273, 1966.

28 R. F. Berg, M. R. Moldover, and G. A. Zimmerli, "Viscoelasticity of xenon near the critical point," *Phys. Rev. Lett.*, vol. 82, p. 920–923, 1999.

29 D. R. Chialvo, "Emergent complexity: What uphill analysis or downhill invention cannot do," *New Ideas Psychol.*, vol. 26, no. 2, pp. 158–173, 2008.

30 X. Zheng, M. A. Anisimov, J. V. Sengers, et al., "Unusual transformation of polymer coils in a mixed solvent close to the critical point," *Phys. Rev. Lett.*, vol. 121, p. 207802, 2018.

31 J. M. J. van Leeuwen and J. V. Sengers, "Gravity effects on the correlation length in gases near the critical point," *Physica A*, vol. 128, pp. 99–131, 1984.

32 K. G. Wilson, "The renormalization group - Introduction," in *Phase Transitions and Critical Phenomena* (C. Domb and M. S. Green, eds.), vol. 6. New York: Academic Press, 1976.

33 K. G. Wilson, "The renormalization group and critical phenomena," *Rev. Mod. Phys.*, vol. 55, pp. 583–600, 1983.

34 L. D. Landau and E. M. Lifshitz, *Statistical Physics: Part 1*, vol. 5 of Course of Theoretical Physics. Oxford: Pergamon, 3rd ed., 1980.

35 M. A. Anisimov, A. T. Berestov, V. P. Voronov, et al., "Critical exponents of fluids," *Sov. Phys. JETP*, vol. 5, p. 49, 1979.

36 H. C. Burstyn, J. V. Sengers, J. K. Bhattacharjee, et al., "Dynamic scaling function for critical fluctuations in classical fluids," *Phys. Rev. A*, vol. 28, pp. 1567–1578, 1983.

37 M. E. Fisher, "Correlation functions and the critical region of simple fluids," *J. Math. Phys.*, vol. 5, p. 944, 1964.

38 M. E. Fisher and R. J. Burford, "Theory of critical-point scattering and correlations. I. The Ising model," *Phys. Rev.*, vol. 156, pp. 583–622, 1967.

39 M. A. Anisimov, A. V. Voronel, and E. E. Gorodetskiĭ, "Isomorphism of critical phenomena," *Sov. Phys. JETP*, vol. 33, pp. 605–612, 1971.

40 M. A. Anisimov, E. E. Gorodetskiĭ, and N. G. Shmakov, "Experimental verification of the isomorphism hypothesis of critical phenomena," *Sov. Phys. JETP*, vol. 36, pp. 1143–1150, 1973.

41 M. A. Anisimov, S. B. Kiselev, J. V. Sengers, et al., "Crossover approach to global critical phenomena in fluids," *Physica A*, vol. 188, pp. 487–525, 1992.

42 H. Behnejad, J. V. Sengers, and M. A. Anisimov, "Thermodynamic behaviour of fluids near critical points," in *Applied Thermodynamics of Fluids* (A. R. Goodwin, C. J. Peters, and J. V. Sengers, eds.), ch. 10, pp. 321–367. Cambridge: Royal Society of Chemistry, 2010.

43 P. Schofield, "Parametric representation of the equation of state near a critical point," *Phys. Rev. Lett.*, vol. 22, pp. 606–608, 1969.

44 P. Schofield, J. D. Litster, and J. T. Ho, "Correlation between critical coefficients and critical exponents," *Phys. Rev. Lett.*, vol. 23, pp. 1098–1102, 1969.

45 M. A. Anisimov, V. A. Agayan, and P. J. Collings, "Nature of the blue-phase-III–isotropic critical point: An analogy with the liquid-gas transition," *Phys. Rev. E*, vol. 57, pp. 582–595, 1998.

46 D. A. Fuentevilla and M. A. Anisimov, "Scaled equation of state for supercooled water near the liquid-liquid critical point," *Phys. Rev. Lett.*, vol. 97, p. 195702, 2006.

47 V. Holten, C. E. Bertrand, M. A. Anisimov, et al., "Thermodynamics of supercooled water," *J. Chem. Phys.*, vol. 136, p. 094507, 2012.

48 J. Luo, L. Xu, E. Lascaris, and S. V. Buldyrev, "Behavior of the Widom line in critical phenomena," *Phys. Rev. Lett.*, vol. 112, p. 135701, 2014.

49 Y. C. Kim, M. E. Fisher, and G. Orkoulas, "Asymmetric fluid criticality. I. Scaling with pressure mixing," *Phys. Rev. E*, vol. 67, p. 061506, 2003.

50 J. Wang and M. A. Anisimov, "Nature of vapor-liquid asymmetry in fluid criticality," *Phys. Rev. E*, vol. 75, p. 051107, 2007.

51 A. K. Wyczalkowska, M. A. Anisimov, J. V. Sengers, et al., "Impurity effects on the two-phase isochoric heat capacity of fluids near the critical point," *J. Chem. Phys.*, vol. 116, pp. 4202–4211, 2022.

52 C. E. Bertrand, J. F. Nicoll, and M. A. Anisimov, "Comparison of complete scaling and a field-theoretic treatment of asymmetric fluid criticality," *Phys. Rev. E*, vol. 85, p. 031131, 2012.

53 C. E. Bertrand and M. A. Anisimov, "Complete scaling for inhomogeneous fluids," *Phys. Rev. Lett.*, vol. 104, p. 205702, 2010.

54 Y. C. Kim, M. A. Anisimov, J. V. Sengers, et al., "Crossover critical behavior in the three-dimensional Ising model," *J. Stat. Phys.*, vol. 110, pp. 591–609, 2003.

55 V. A. Agayan, M. A. Anisimov, and J. V. Sengers, "Crossover parametric equation of state for Ising-like systems," *Phys. Rev. E*, vol. 64, p. 026125, 2001.

56 F. Llovell, L. F. Vega, M. A. Anisimov, et al., "Incorporating critical divergence of isochoric heat capacity into the soft-SAFT equation of state," *AIChE J.*, vol. 61, pp. 3073–3080, 2015.

57 A. A. Povodyrev, M. A. Anisimov, and J. V. Sengers, "Crossover Flory model for phase separation in polymer solutions," *Physica A*, vol. 264, pp. 345–369, 1999.

58 T. J. Bruno and J. F. Ely, *Supercritical Fluid Technology*. Reviews in Modern Theory Application. CRC Press, 1991.

59 Y.-P. Sun, *Supercritical Fluid Technology in Materials Science and Engineering*. Taylor and Francis, 2002.

60 J. Wang, M. A. Anisimov, and J. V. Sengers, "Closed solubility loops in liquid mixtures," *Z. Phys. Chem.*, vol. 219, no. 9, pp. 1273–1297, 2005.

61 L. V. Entov, V. A. Levchenko, and V. P. Voronov, "Specific heat of a liquid mixture near the consolute point in the bulk phase and in a porous medium," *Int. J. Thermophys.*, vol. 14, pp. 221–229, 1993.

62 J. Wang, C. A. Cerdeiriña, M. A. Anisimov, et al., "Principle of isomorphism and complete scaling for binary-fluid criticality," *Phys. Rev. E*, vol. 77, p. 031127, 2008.

63 T. Moriyoshi, S. Kaneshina, K. Aihara, et al., "Mutual solubility of 2-butanol + water under high pressure," *J. Chem. Thermodyn.*, vol. 7, no. 6, pp. 537–545, 1975.

64 S. S. Leung and R. B. Griffiths, "Thermodynamic properties near the liquid-vapor critical line in mixtures of He^3 and He^4," *Phys. Rev. A*, vol. 8, pp. 2670–2683, 1973.

65 M. Smith, H. V. Ness, and M. M. Abbott, *Introduction to Chemical Engineering Thermodynamics*. McGraw-Hill, 9th ed., 2022.

66 K. Abdulkadirova, C. Peters, J. V. Sengers, et al., "An isomorphic Peng–Robinson equation for phase-equilibria properties of hydrocarbon mixtures in the critical region," *J. Supercrit. Fluids*, vol. 55, no. 2, pp. 594–602, 2010. 100th year Anniversary of van der Waals' Nobel Lecture.

67 M. A. Anisimov, E. Gorodetskiĭ, V. Kulikov, et al., "A general isomorphism approach to thermodynamic and transport properties of binary fluid mixtures near critical points," *Physica A*, vol. 220, no. 3, pp. 277–324, 1995.

68 M. A. Anisimov, E. E. Gorodetskiĭ, V. D. Kulikov, et al., "Crossover between vapor-liquid and consolute critical phenomena," *Phys. Rev. E*, vol. 51, pp. 1199–1215, 1995.

69 M. E. Fisher, "Renormalization of critical exponents by hidden variables," *Phys. Rev.*, vol. 176, pp. 257–272, 1968.

70 A. B. Pippard, "XLVIII. Thermodynamic relations applicable near a lambda-transition," *Philos. Mag.*, vol. 1, pp. 473–476, 1956.

71 O. K. Rice, "Statistical thermodynamics of λ transitions, especially of liquid helium," *Phys. Rev.*, vol. 153, pp. 275–279, 1967.

72 J. A. Lipa, J. A. Nissen, D. Avaloff, et al., "Specific heat of helium at constant volume along the lambda line," in *Low Temperature Physics: 24th International Conference on Low Temperature Physics*, vol. 850, pp. 131–132, AIP, 2006.

73 D. L. Connelly, J. S. Loomis, and D. E. Mapother, "Specific heat of nickel near the Curie temperature," *Phys. Rev. B*, vol. 3, pp. 924–934, 1971.

74 M. P. A. Fisher and M. Wortis, "Curvature corrections to the surface tension of fluid drops: Landau theory and a scaling hypothesis," *Phys. Rev. B*, vol. 29, pp. 6252–6260, 1984.

75 J. Kojima, N. Kuwahara, and M. Kaneko, "Light scattering and pseudospinodal curve of the system polystyrene–cyclohexane in the critical region," *J. Chem. Phys.*, vol. 63, p. 333, 1975.

76 M. Nakata, N. Kuwahara, and M. Kaneko, "Coexistence curve for polystyrene–cyclohexane near the critical point," *J. Chem. Phys.*, vol. 62, p. 4278, 1975.

77 M. Nakata, T. Dobashi, N. Kuwahara, et al., "Coexistence curve and diameter of polystyrene in cyclohexane," *Phys. Rev. A*, vol. 18, pp. 2683–2688, 1978.

78 J. Jacob, M. A. Anisimov, J. V. Sengers, et al., "Light scattering and crossover critical phenomena in polymer solutions," *Appl. Opt.*, vol. 40, pp. 4160–4169, 2001.

79 J. S. Hager, M. A. Anisimov, J. V. Sengers, et al., "Scaling of demixing curves and crossover from critical to tricritical behavior in polymer solutions," *J. Chem. Phys.*, vol. 117, p. 5940, 2002.

80 M. A. Anisimov, A. F. Kostko, and J. V. Sengers, "Competition of mesoscales and crossover to tricriticality in polymer solutions," *Phys. Rev. E*, vol. 65, p. 051805, 2002.

81 M. A. Anisimov, A. F. Kostko, J. V. Sengers, et al., "Competition of mesoscales and crossover to theta-point tricriticality in near-critical polymer solutions," *J. Chem. Phys.*, vol. 123, p. 164901, 2005.

82 M. A. Anisimov and J. V. Sengers, "Scaling, tricriticality, and crossover in polymer solutions," *Mol. Phys.*, vol. 103, pp. 3061–3070, 2005.

83 A. F. Kostko, M. A. Anisimov, and J. V. Sengers, "Dynamics of critical fluctuations in polymer solutions," *Phys. Rev. E*, vol. 76, p. 021804, 2007.

84 T. Dobashi, M. Nakata, and M. Kaneko, "Coexistence curve of polystyrene in methylcyclohexane. I. Range of simple scaling and critical exponents," *J. Chem. Phys.*, vol. 72, p. 6685, 1980.

85 K. Xia, "Interfacial tensions of phase-separated polymer solutions," *J. Chem. Phys.*, vol. 97, p. 1446, 1992.

86 K. Xia, X. An, and W. Shen, "Measured coexistence curves of phase-separated polymer solutions," *J. Chem. Phys.*, vol. 105, p. 6018, 1996.

87 P.-G. de Gennes, "Critical opalescence of macromolecular solutions," *Phys. Lett.*, vol. 26A, pp. 313–314, 1968.

88 P.-G. de Gennes, *Scaling Concepts in Polymer Physics*. Ithaca, NY: Cornell University, 1979.

89 I. C. Sanchez, "Critical amplitude scaling laws for polymer solutions," *J. Phys. Chem.*, vol. 93, pp. 6983–6991, 1989.

90 N. Goldenfeld, *Lectures on Phase Transitions and the Renormalization Group*. New York: Addison-Wesley, 1992.

91 A. Pelissetto and E. Vicari, "Critical phenomena and renormalization-group theory," *Phys. Rep.*, vol. 368, pp. 549–727, 2002.

92 Y. B. Melnichenko and G. D. Wignall, "Dimensions of polymer chains in critical semidilute solutions," *Phys. Rev. Lett.*, vol. 78, pp. 686–688, 1997.

93 G. D. Wignall and Y. B. Melnichenko, "Recent applications of small-angle neutron scattering in strongly interacting soft condensed matter," *Rep. Prog. Phys.*, vol. 68, p. 1761, 2005.

94 A. Aharony, O. Entin-Wohlman, D. A. Huse, and L. Radzihovsky, eds., *50 Years of the Renormalization Group: Dedicated to the Memory of Michael E Fisher*. Hackensack: World Scientific, 2024. doi: https://doi.org/10.1142/13571.

95 I. D. Lawrie and S. Sarbach, "Theory of tricritical points," in *Phase Transitions and Critical Phenomena* (C. Domb and J. L. Lebowitz, eds.), vol. 9. New York: Academic Press, 1984.

96 M. A. Anisimov, V. A. Agayan, and E. E. Gorodetskiĭ, "Scaling and crossover to tricriticality in polymer solutions," *JETP Lett.*, vol. 72, pp. 578–582, 2000.

97 M. A. Anisimov and C. E. Bertrand, "Thermodynamics of fluids at meso and nano scales," in *Applied Thermodynamics of Fluids* (A. R. Goodwin, C. J. Peters, and J. V. Sengers, eds.), ch. 7, pp. 172–214. Cambridge: Royal Society of Chemistry, 2010.

98 V. P. Voronov and V. M. Buleiko, "Experimental investigation of the behavior of the specific heat in finite systems in the vicinity of the critical mixing point," *J. Exp. Theor. Phys.*, vol. 86, pp. 586–590, 1998.

99 M. Kardar and R. Golestanian, "The "friction" of vacuum, and other fluctuation-induced forces," *Rev. Mod. Phys.*, vol. 71, pp. 1233–1245, 1999.

100 M. Krech, *The Casimir Effect in Critical Systems*. Singapore: World Scientific, 1994.

101 P. C. Hohenberg and B. I. Halperin, "Theory of dynamic critical phenomena," *Rev. Mod. Phys.*, vol. 49, pp. 435–479, 1977.

102 J. V. Sengers, "Transport properties of fluids near critical points," *Int. J. Thermophys.*, vol. 6, p. 203, 1985.

103 A. Onuki, *Phase Transition Dynamics*. Cambridge: Cambridge University, 2002.

104 T. Ohta and K. Kawasaki, "Mode coupling theory of dynamic critical phenomena for classical liquids. I: Dynamic critical exponents," *Prog. Theor. Phys.*, vol. 55, pp. 1384–1395, 1976.

105 B. I. Halperin, P. C. Hohenberg, and E. D. Siggia, "Renormalization-group calculations of divergent transport coefficients at critical points," *Phys. Rev. Lett.*, vol. 32, pp. 1289–1292, 1974.

106 K. Kawasaki, "Mode coupling and critical dynamics," in *Phase Transitions and Critical Phenomena* (C. Domb and M. S. Green, eds.), vol. 5A. New York: Academic Press, 1976.

107 T. Ohta, "Multiplicative renormalization of the anomalous shear viscosity in classical liquids," *J. Phys. C: Solid State Phys.*, vol. 10, p. 791, 1977.

108 S. K. Das, J. Horbach, K. Binder, et al., "Static and dynamic critical behavior of a symmetrical binary fluid: A computer simulation," *J. Chem. Phys.*, vol. 125, p. 024506, 2006.

109 Zimmerli, "Visoelasticity of Xenon near the Critical Point," *Phys. Rev. Lett.*, vol. 82, pp. 920–923, 1999.

110 K. A. Gillis, I. I. Shinder, and M. R. Moldover, "Bulk viscosity of stirred xenon near the critical point," *Phys. Rev. E*, vol. 72, p. 051201, 2005.

111 M. A. Anisimov, V. P. Voronov, V. M. Malyshev, et al., "Experimental verification of the dynamic scaling theory of the critical point," *JETP Lett.*, vol. 18, p. 133, 1973.

112 R. M. Sadeq, I. Abdelraziq, M. S. Abu-Jafar, et al., "The behavior of sound absorption coefficient for binary mixture nitroethane-isooctane above critical temperature and concentration," *AIP Adv.*, vol. 8, p. 115101, 2018.

113 L. Landau, "On the theory of transfer of energy at collisions II," *Phys. Z. Sowj.*, vol. 2, p. 46, 1932.

114 C. Zener, "Non-adiabatic crossing of energy levels," *Proc. R. Soc. London, Ser. A*, vol. 137, pp. 696–702, 1932.

115 R. D. Mountain and J. Deutch, "Light scattering from binary solutions," *J. Chem. Phys.*, vol. 50, p. 1103, 1969.

116 M. A. Anisimov, V. A. Agayan, A. A. Povodyrev, et al., "Two-exponential decay of dynamic light scattering in near-critical fluid mixtures," *Phys. Rev. E*, vol. 57, pp. 1946–1961, 1998.

117 H. Tanaka, Y. Nakanishi, and N. Takubo, "Nonuniversal nature of dynamic critical anomaly in polymer solutions," *Phys. Rev. E*, vol. 65, p. 021802, 2002.
118 A. F. Kostko, M. A. Anisimov, and J. V. Sengers, "Dynamic crossover to tricriticality and anomalous slowdown of critical fluctuations by entanglements in polymer solutions," *Phys. Rev. E*, vol. 66, p. 020803, 2002.
119 F. Brochard and P.-G. de Gennes, "Dynamical scaling for polymers in theta solvents," *Macromolecules*, vol. 10, p. 1157, 1977.
120 F. Brochard, "Gel-like modes of polymer solutions in ≪θ≫ solvents," *J. Phys. Fr.*, vol. 44, pp. 39–43, 1983.
121 T. Petrova, "Revised release on surface tension of ordinary water substance," *International Association for the Properties of Water and Steam*, 2014.

Exercises

8.1 Based on the Landau–Ginzburg functional, define the correlation length of density fluctuations in the meanfield approximation and in scaling theory. Based on the balance of the thermodynamic term and the gradient term in the Landau–Ginzburg functional, explain the difference between susceptibility in the meanfield approximation and in scaling theory.

8.2 Explain the concept of critical point universality. Compare the Ising model to the lattice-gas model: do they belong to the same universality class in thermodynamics? In dynamics? Explain your answer.

8.3 The isobaric heat capacity is related to the isochoric heat capacity through Eq. (2.65). Based on the relations between critical exponents and that $\kappa_T = -(1/V_t)(\partial V_t/\partial P)_T \sim \hat{\chi}$ will the isobaric heat capacity diverge at the critical point? From this equation, could we predict the behavior of the isochoric heat capacity? If the isochoric heat capacity also diverges, which heat capacity has a stronger asymptotic divergence? Explain the difference between a weak and a strong singularity.
Assuming a weak power-law singularity of the isochoric heat capacity, estimate the dependence of $\gamma = C_P/C_V$ on the distance to the critical temperature (along the critical isochore).

8.4 In the meanfield approximation for the lattice-gas model determine:
a) The critical exponents: α, β, γ, and δ
b) The critical amplitudes: $A_0^{\pm}$, B_0, and $\Gamma_0^{\pm}$
Hint: Use the expressions for the lattice-gas model from Section 2.11.2. Express all relevant quantities through the distance to the critical point. For instance, the free energy in the vicinity of the critical point is represented by the truncated Landau expansion. See Problem 4.2a for details about the construction of the Landau expansion.

8.5 Based on the "fluctuation-cluster" model for a single-component fluid above the liquid–gas critical point along the critical isochore,
a) Derive the universal relation between the critical exponent of the isochoric heat capacity and the critical exponent of the correlation length.
b) Derive the universal relation between the critical amplitudes of the heat capacity and correlation length. Calculate the amplitude A_0^+ for water. Adopt for water the experimental value $\xi_0^+ = 0.13$ nm

8.6 Estimate the surface tension between liquid water and saturated water vapor in the critical region from the "fluctuation-cluster" model. Calculate the amplitude σ_0 from the power law given by Eq. (8.119) for water (where $\vartheta \cong 1.26$). Adopt for water the experimental value $\xi_0^+ = 0.13\,\text{nm}$. What universal relations between critical exponents and between critical amplitudes are obtained in this problem?

8.7 The International Association for the Properties of Water and Steam (IAPWS) has tabulated the surface tension between liquid water and saturated water vapor between the triple point and the critical point (See Ref. [121]). Fit the tabulated data to a two-parameter crossover expression for the surface tension

$$\sigma = \frac{A}{1 + \frac{A}{\sigma_0}\left[(-\Delta\hat{T})^{2\nu} - 1\right]}$$

and compare it with the interpolation performed by IAPWS. Explain the meaning of the surface tension in the two physical limits: $T \to 0$ and $T \to T_c$ in this model.

8.8 Near-critical fluctuations modify not only the temperature dependence of the surface tension, but also, the shape of the density (or concentration) profile. Renormalization Group theory shows that the universal expression for the order-parameter profile near the critical point can be written in terms of a universal scaling function given by Eq. (8.122). On the same diagram, plot the density profile from the van der Waals–Landau meanfield theory, Eq. (6.19), and the fluctuation affected density profile. What effect do fluctuations have on the density profile?

8.9 The correction to the density profile caused by asymmetric fluid phase coexistence is given by Eq. (6.28). For a van der Waals fluid in the vicinity of the critical point, estimate the leading order of the correctional term to the density profile that takes into account both fluctuational effects and fluid asymmetry.

8.10 On the same plot, sketch the coexistence curve (binodal) for an asymmetric fluid in the meanfield approximation (i) and for a real fluid (ii). Using your sketch, identify the features of fluid phase coexistence responsible for the divergence of Tolman's length at the critical temperature.

8.11 Find the temperature of the transition from non-wetting to wetting for the mixture of methanol and n-hexane. The critical temperature of demixing is 33.7 °C. Which phase will wet the liquid–air interface?
Given that at 25 °C the surface tension between the methanol-reach phase and air is 22 mN/m, between n-hexane reach phase and air is 21.5 mN/m, and the interfacial tension between the liquid phases (methanol and n-hexane) is 0.26 mN/m.
Plot the interfacial tension specifically showing the crossing of the interfacial tension between the near-critical liquid phases (methanol and n-hexane) and the difference between the surface tensions with air (methanol-reach and n-hexane reach phases) at the wetting transition temperature.

8.12 Explain the difference between theoretical scaling and experimental thermodynamic variables. For which variables could we observe the universal critical anomalies? What kind of experimental anomalies could be observed in real experiments? In which systems?

8.13 Formulate the relation between theoretical and physical field variables in the theory of complete scaling for a binary fluid. Formulate the experimental conditions which would allow one to observe the universal theoretical anomalies of thermodynamic properties.

8.14 What is the condition of a "pure near-critical fluid" in the presence of an impurity? Explain the "critical dilute solution dilemma."

8.15 According to the complete scaling theory, the scaling fields are analytical functions of physical fields. In particular, the liquid–liquid coexistence of a binary solution, THF + water, exhibits two critical points, an upper and lower critical point. Each branch of the coexistence curve for a symmetric system near the critical point is described by

$$\frac{x - x_c^{\pm}}{x_c^{\pm}} = \pm B_0 |h_2|^{\beta}$$

Develop the relation between the scaling thermal field h_2 and the physical thermal fields, $\Delta \hat{T}_{\pm}$. (*Hint:* consider a product of thermal fields for both critical points.) Plot the "closed loop" connection between the upper and lower critical points, using the parameters: $T_{-} = 345$ K, $T_{+} = 415$ K, $x_c = x_{+} = x_{-} = 0.5$, and $B_0 = 2$.

8.16 Explain why the saturation pressure, at constant composition, splits into two separate branches: "bubble" and "dew". Under which conditions will these branches collapse into a single saturation-pressure curve?

8.17 Formulate the near-critical behavior of the Onsager kinetic coefficient (mobility), $\check{\alpha}$, in the mutual diffusion flux using the dynamic-scaling result for the mutual diffusion coefficient given as

$$D = \frac{k_B T}{6\pi\eta\xi}$$

8.18 (a) Explain the change in the near-critical osmotic susceptibility of a polymer solution upon decreasing the size of the system from macroscopic to microscopic. (b) Explain the crossover in the ℓ-dependent mesoscopic susceptibility in a finite-size fluid. Explain the difference between two physical length scales: the instrumental scale, such as the light-scattering wave number, and physical size of the system. What is the ratio of these two finite-size limits in Figure 8.29? Compare the Fisher–Burford approximation to the Ornstein–Zernike approximation. When does the difference between the two become noticeable?

8.19 Plot the temperature dependence of the normalized light scattering intensity (I/I_0) near the liquid–liquid critical point of a binary fluid for different instrumental light-scattering length scales (different scattering angles, $\theta = 30°, 45°$, and $90°$) in the range of $\Delta\hat{T} = 10^{-1}$ to $\Delta\hat{T} = 10^{-6}$. Given that the wavelength of the incident light $\lambda_0 = 633$ nm, the refraction index $n = 1.5$, the correlation length $\xi = \xi_0 |\Delta\hat{T}|^{-\nu}$ (where $\xi_0 = 0.2$ nm and $\nu = 0.63$), and the critical amplitude of the susceptibility, $\Gamma_0^{+} = 0.1$.

8.20 Explain the concept of "critical slowing down" and the origin of the dynamic critical exponent, z_{τ}. Despite fluids and Ising ferromagnets belonging to the same universality class in thermodynamics, why are their near-critical dynamics described by different dynamic exponents?

Index

Mesoscopic Thermodynamics for Scientists and Engineers, First Edition. Mikhail A. Anisimov and Thomas J. Longo.
© 2024 John Wiley & Sons, Inc. Published 2024 by John Wiley & Sons, Inc.

Printed and bound by CPI Group (UK) Ltd, Croydon, CR0 4YY
17/09/2026
14964315-0002